S0-CAS-492

Transition Metals
in Homogeneous Catalysis

Transition Metals in Homogeneous Catalysis

edited by G. N. Schrauzer

Department of Chemistry
University of California, San Diego
Revelle College
La Jolla, California

1971

MARCEL DEKKER, INC.   New York

CHEMISTRY

COPYRIGHT © 1971 by MARCEL DEKKER, Inc.

ALL RIGHTS RESERVED

No part of this work may be reproduced or utilized in any
form or by any means, electronic or mechanical, including
Xeroxing, photocopying, microfilm, and recording, or by any
information storage and retrieval system, without permission
in writing from the publisher.

MARCEL DEKKER, INC.
95 Madison Avenue, New York, New York, 10016

LIBRARY OF CONGRESS CATALOG CARD NUMBER:
74-162281

ISBN: 0-8247-1608-6

PRINTED IN THE UNITED STATES OF AMERICA

Q D
501
T7571
CHEM

Preface

Catalytic processes will play an ever increasing part in the chemical industry of the future. To satisfy the growing needs for a wide variety of industrial chemicals or energy sources and to reduce the formation of industrial waste products, current industrial research already concentrates heavily on the study and development of new catalytic processes. If these efforts are to be successful, chemical education will have to place greater emphasis on catalysis and related subjects at all levels of university chemistry teaching. Since chemistry curricula at present are still subject to the traditional subdivision into inorganic, organic or physical chemistry it is difficult to accommodate multidisciplinary subjects such as catalysis into any but special topics courses. Many young graduates entering industrial positions for this reason face a period of painful adjustment from academic to real-world industrial research and philosophy.

Whereas heterogeneous catalysis is a well-defined field for which numerous excellent introductory and advanced texts are available, homogeneous catalysis is much more general and cannot be easily identified with any branch of chemistry. In recognizing the need for an introductory book on aspects of homogeneous catalysis I have therefore accepted the invitation of the Publisher to edit the present Volume. In planning the outline of the book I restricted the topics to homogeneous catalysis involving transition metals. With eight invited authors a compilation of reviews on timely subjects was accomplished, covering catalytic hydrogenation and dehydrogenation (Jack Kwiatek), reactions involving -allylic systems (W. Keim), homogeneous catalyzed oxidations of organic compounds (E. W. Stern), carbonylation (D. T. Thompson and R. Whyman), the catalysis of electron transfer reactions (R. G. Linck), and the catalysis of symmetry forbidden reactions (F. D. Mango and J. Schachtschneider).

iii

006

In an introductory chapter I have outlined some thoughts concerning the nature and scope of the catalytic phenomenon and attempted a new definition of catalysis. Other than suggesting the topic I have made no attempt to influence the opinions of the authors in any way. However, I am to be held responsible for any omissions on their behalf, as it was I who suggested that their contributions need not be comprehensive.

I hope this book will stimulate its readers to enter the exciting field of research in catalysis.

G. N. Schrauzer

Contributors to this Volume

WILLI KEIM, *Shell Development Company, Emeryville, California*

JACK KWIATEK,* *The Weizmann Institute of Science, Rehovot, Israel*

ROBERT G. LINCK, *Department of Chemistry, Revelle College, University of California, San Diego, La Jolla, California*

FRANK D. MANGO, *Shell Development Company, Emeryville, California*

JERRY H. SCHACHTSCHNEIDER, *Shell Development Company, Emeryville, California*

GERHARD N. SCHRAUZER, *Department of Chemistry, University of California, San Diego, Revelle College, La Jolla, California*

ERIC W. STERN, *Engelhard Industries, A Division of Engelhard Minerals and Chemicals Corporation, Newark, New Jersey*

DAVID T. THOMPSON, *Imperial Chemical Industries Limited, Petrochemical & Polymer Laboratory, The Heath, Runcorn, Cheshire, England*

ROBIN WHYMAN, *Imperial Chemical Industries Limited, Petrochemical & Polymer Laboratory, The Heath, Runcorn, Cheshire, England*

* Present address: U.S. Industrial Chemicals Co., Division of National Distillers and Chemical Corp., Cincinnati, Ohio.

Contents

7. ELECTRON-TRANSFER CATALYSIS 297

R. G. Linck

Transition Metals
in Homogeneous Catalysis

Catalysis : Fundamental Aspects and Scope

G. N. SCHRAUZER

Department of Chemistry
University of California, San Diego
Revelle College
La Jolla, California

I. Historical Introduction

Even in more specialized texts, little space is usually devoted to the discussion of fundamental aspects of catalysis. What is the nature and scope of the catalytic phenomenon? Which are the basic functions of catalysts in chemical reactions? These questions obviously require more detailed elaboration than is usually found in the literature. The term catalysis (from the Greek *katalysis*, meaning breaking down or loosening) was introduced in its present meaning by J. J. Berzelius in 1836 (*1*). To describe the activating effects in a number of chemical reactions of certain substances

1

called catalysts, Berzelius assumed that these exert a special catalytic force upon the reactants. This force was not meant to be anything mystical, however, as Berzelius cautiously remarked that he considered it nothing but another manifestation of electrochemical affinity. His definition, therefore, is still essentially valid; there can be no question that catalysts exert a force upon the substrates, except that it can now be more clearly defined. With the development of physical chemistry in the latter part of the nineteenth century, particularly with the discovery of the laws of chemical equilibrium and the beginnings of the theory of reaction rates, catalysis became a kinetic phenomenon. In 1902 W. Ostwald (2) defined catalysts as agents that speed up chemical reactions without affecting the chemical equilibrium. The precision of Ostwald's definition and its often tested validity have made it the most frequently cited of all definitions. It is, however, only valid for reversible reactions and does not incorporate any form of autocatalysis. P. Sabatier (3), recognizing the shortcomings of Ostwald's definition, considered catalysis simply as a mechanism causing, or accelerating, certain chemical reactions by substances which themselves are not irreversibly altered. This phenomenological definition is less restrictive than Ostwald's but unduly stresses that catalysts must not undergo irreversible changes during reaction. Most catalysts, however, eventually become inactive due to side reactions or contaminants. Certain reactions can in fact only be carried out stoichiometrically, as the catalyst is chemically changed into an inactive form. The process can nevertheless be made " catalytic " by coupling it with a suitable catalyst regeneration step. If this step is carried out simultaneously with the principal reaction the net effect is indistinguishable from a true catalytic process.

With the development of the theories of reaction rates it became clear that catalysts in general lower the energy of activation of a reaction. This in principle is a useful definition, although it is not really commensurate with the complexity and multitudinous aspects of the phenomenon. It furthermore must be specified that the lowering of the activation energy occurs through the direct interaction of the catalyst with the substrates. In more complex catalytic systems (e.g., enzymes) certain nonstoichiometric additives (e.g. metal ions) may cause reaction rate enhancements indirectly (e.g., by causing conformational changes of the enzyme protein and without actually participating in the chemical conversion of the substrates). Obviously such effectors or activators are not catalysts in the true sense, although a strict distinction may not be always possible.

We will in the following develop the fundamental aspects of catalytic phenomena by first considering the reasons why certain chemical reactions, though thermodynamically possible, exhibit a remarkable lack of kinetic reactivity. The understanding of the reasons for this behavior are intimately related to all catalytic phenomena.

II. Fundamental Aspects of Chemical Reactions

A. PRODUCT FORMATION IN COLLISION REACTIONS

Although elements in their atomic states are usually thermodynamically highly unstable, their lifetimes in the gas phase are often surprisingly long. This is due to the ineffectiveness of binary collisions in producing molecular species in the ground state. The recombination of hydrogen atoms, e.g. by binary collisions, produces molecular hydrogen in the excited $^3\Sigma_u^+$ state whose conversion into the $^1\Sigma_g^+$ ground state by emission of radiation is only possible through a violation of the spin selection rule ($\Delta S = 0$). The emission of radiation, therefore, is a process of very low probability, occurring only in one of about 10^5 binary collisions (4). Successful termination reactions occur almost entirely by three-body collisions. The third body may be any atomic or molecular species present in the reaction phase, including the wall of the reaction vessel. The efficiency of the third body depends on the type of interaction with the substrates. There are two fundamental mechanisms of product-forming energy-transfer processes in three-body collisions (5). The first represented by Eqs. (1) and (2) and is the inverse of the main secondary process in photolysis reactions:

$$A + B + C \longrightarrow AB^* + C \tag{1}$$

$$AB^* \longrightarrow AB + \text{energy} \tag{2}$$

The recombination of nitrogen atoms which is associated with characteristic chemiluminescence (6) proceeds by this mechanism, or the formation of excited CO_2 in carbon monoxide flames. The second mechanism of three-body recombination is the reverse of a photosensitized dissociation:

$$A + B + C \longrightarrow AB + C^* \tag{3}$$

$$C^* \longrightarrow C + \text{energy} \tag{4}$$

Here it is not always certain if the excitation proceeds directly as shown in Eq. (3) or whether the excited third body C^* is formed by collisional energy transfer from an excited molecules AB^*. A variant of this mechanism is the radical-molecule complex mechanism symbolized by Eqs. (5) and (6):

$$A + C \longrightarrow AC \tag{5}$$

$$AC + B \longrightarrow ABC \longrightarrow AB + C^* \tag{6}$$

The efficiency of the third body C in this case will depend on the structure and stability of the activated complex ABC and the reactive cross section of the formation of AC by binary collisions. Although no general quantitative rules can be given it is obvious that there will be greater efficiency of formation of AC if the collisions are " soft " rather than perfectly elastic, if there is a separation of charge in AC due to differences in electronegativity and if C is highly polarizable. In the catalysis of hydrogen recombination by alkali metal ions, e.g., metal hydrides are formed [Eqs. (7) and (8)] (7):

$$H + M \longrightarrow HM \tag{7}$$

$$HM + H \longrightarrow MH_2 \longrightarrow M^* + H_2 \tag{8}$$

The overall reaction is governed by selection rules of spin and of angular momentum. The spin selection rule $\Delta S = 0$ is the most important but holds less strictly in all atoms with strong spin-orbit coupling, during collisions with another molecule, or in the presence of charged species or molecules with asymmetric charge distribution. The selection rule of angular momentum is violated more easily, often already by slightly changing the structure of the intermediate complex from higher to lower symmetry, as will be outlined in the next paragraphs. The efficiency of the third body in Eqs. (3)–(6) finally also depends on the lifetime of the excited species C^*. If its conversion to the ground state is quantum mechanically forbidden and product AB is present in sufficiently high concentration, redissociation of AB will occur in reverse of Eqs. (3)–(6). Depending on the energy of the reactant species some of the collision partners will merely exchange energy without product formation. The theoretical analysis of collisions of atoms under gas kinetic conditions cannot be reviewed here, however, and this analysis is not of direct importance in catalytic processes.

B. Adiabatic and Nonadiabatic Correlations

It is a well-known fact that the potential curves of two electronic states of the same species cannot cross each other. It is sometimes noted, however, particularly in crude calculations, that two states appear to cross, but this effect is entirely due to the approximations used and the crossing of the states disappears in all cases if the approximation is improved (8). If the two states have the same symmetry the off-diagonal matrix element describing the interaction of the two states $\int \psi_1 H \psi_2 \, dv$ is usually large, causing a considerable resonance splitting or " repulsion " of the two states as shown in Fig. 1.

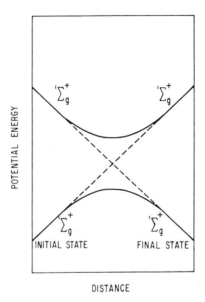

Fig. 1. Schematic potential energy diagram of two states undergoing an adiabatic transition. Adapted from Ref. (5).

In this case the lower electronic states pass smoothly from one to another, and the probability of the reaction system to pass into the upper potential function (i.e., to become electronically excited) is low. Since no change in the S or L values occurs, reactions of this type will have a transmission coefficient close to unity, and are often called *adiabatic*. States with different symmetry properties cannot interact very strongly ($\int \psi_1 H \psi_2 \, dv$ is small), leading to a very small resonance splitting (Fig. 2). This in turn increases the probability of crossing from the lower to the upper state or vice versa, in fact, the system in this case behaves as if the potential functions really cross (8). Reactions of this type are usually called nonadiabatic, although this nomenclature is not entirely unambiguous. The potential energy diagrams of actual chemical reactions are much more complicated and require representation in three rather than two dimensions. States of the same symmetry in these diagrams intersect the cones of potential functions

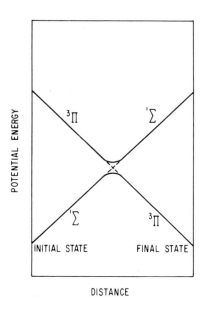

Fig. 2. Schematic potential energy diagram of two states undergoing a nonadiabatic transition. Adapted from Ref. (5).

whereas line intersection occurs for states of different symmetry. Details of the theory of nonadiabatic transitions have recently been outlined by Nikitin (9). In concluding this brief paragraph we return to the definition of catalysis. Catalysts, in nonadiabatic systems, are evidently any species capable of increasing the transition probability of the reaction, by acting as a third body to remove or relax quantum mechanical selection rules or to provide an efficient mechanism for energy transfer. Catalysis in adiabatic reactions is also possible. If the transmission coefficient is small for steric reasons the catalyst may bring the reactants in close contact or in a sterically favorable fashion to facilitate the chemical combination.

C. ADIABATIC CORRELATION RULES

The correlation rules of spin ($\Delta S = 0$) and of angular momentum in elementary chemical reactions have been formulated by Shuler (10) and will be briefly exemplified. A reaction which could be part of a catalytic process, $A + BC \rightarrow AB + C$, may be written configurationally as shown in Eq. (9):

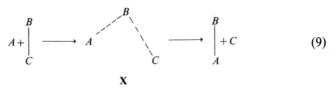

$$(9)$$

X

In this reaction the reactants change from spherical symmetry (atom A) and linear symmetry (molecule BC) to the new atomic species C and the diatomic molecule AB of linear symmetry via the intermediate ABC or complex **X**. The resulting electronic states of ABC are found by forming the direct products of the representations of A and of BC in the assumed symmetry of **X**. A similar correlation must now be made by forming the direct product of AB and C. Given states of reactants and products will correlate adiabatically if and only if the intermediate complex **X** has at least one representation in common with those arising from the reactant and product states. This is most frequently the case if the intermediate complex **X** is the least symmetrical, as will be illustrated by the following concrete examples. Following Ref. (10) we consider the reaction $CH + O_2 \rightarrow CO + OH^*$ which possibly accounts for the presence of excited OH radicals in hydrocarbon flames. Writing the reaction configurationally we assume that the reaction has the choice of either passing through the symmetrical intermediate of symmetry C_{2v}, or the unsymmetrical intermediate of symmetry C_s [Eq. (10)]:

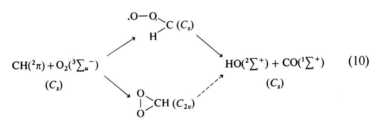

$$(10)$$

Applying correlation Tables 1–3 it is found that the combination of $CH(^2\Pi)$ and $O_2(^3\Sigma_u^-)$ to the intermediate O_2CH of symmetry C_s gives rise to the states $^2A'$, $^2A''$, $^4A'$, $^4A''$, since $^2\Pi$ and $^3\Sigma_u^-$ resolve in C_s symmetry to the representations $^2A'$, $^2A''$, and $^3A''$, and, because $^2A' \times {}^3A'' = {}^2A''$, $^4A''$, and $^2A'' \times {}^3A'' = {}^2A'' \times {}^3A'' = {}^2A'$, $^4A''$, respectively (Table 3). The products $CO(^1\Sigma^+)$ and $HO(^2\Sigma^+)$ transform as $^1A'$ and $^2A'$, however, whose direct product is $^2A'$. It follows that intermediate O_2CH of symmetry C_s correlates adiabatically since it contains the representation $^2A'$. For the intermediate CHO_2 of symmetry C_{2v}, however, the representations

TABLE 1

REPRESENTATIONS OF STATES IN SYMMETRY GROUPS

Initial state	Resolved state		
	C_{2v}	C_s	C_1
S_g	A_1	A'	A
S_u	A_2	A''	A
P_g	$A_2 + B_1 + B_2$	$A' + 2A''$	$3A$
P_u	$A_1 + B_1 + B_2$	$2A' + A''$	$3A$
D_g	$2A_1 + A_2 + B_1 + B_2$	$3A' + 2A''$	$5A$
D_u	$A_1 + 2A_2 + B_1 + B_2$	$2A' + 3A''$	$5A$
Σ_g^+, Σ_u^-	A_1	A'	A
Σ_g^-, Σ_u^-	A_2	A''	A
Π_g, Π_u	$B_1 + B_2$	$A' + A''$	$2A$
Δ_g, Δ_u	$A_1 + A_2$	$A' + A''$	$2A$

TABLE 2

DIRECT PRODUCTS FOR C_{2v}

	A_1	A_2	B_1	B_2
A_1	A_1	A_2	B_1	B_2
A_2	A_2	A_1	B_2	B_1
B_1	B_1	B_2	A_1	A_2
B_2	B_2	B_1	A_2	A_1

TABLE 3

DIRECT PRODUCTS
FOR C_2

	A'	A''
A'	A'	A''
A''	A''	A'

$^2B_1$, $^2B_2$, $^4B_1$, and $^4B_2$ result from the reactants, whereas the products transform as $^2A_1$ and $^4A_1$, respectively. The conversion of the reactants via the symmetrical intermediate (C_{2v}) hence is orbitally forbidden. This fact is of interest since the initially formed products in chemical reactions are expected to be symmetrical, and these will also be the more stable ones. The conversion of symmetrical intermediates to the less stable unsymmetrical species, therefore, consumes a part of the total energy of activation. If we now could promote the unpaired electron of CH from the $^2\Pi$ ground state to the first excited $^2\Delta$ state, the correlation rules reveal that the reaction would become orbitally allowed even for the symmetrical intermediate.

Orbital correlations such as outlined above become less restrictive in reactions of polyatomic molecules, particularly if vibronic interactions are taken into account. Symmetry arguments are very important in concerted electrocyclic reactions, however, as they provide extremely useful means of classification for a large group of organic chemical reactions. The rules deduced by R. B. Woodward and R. Hoffman (11) have recently been applied to fundamental problems in catalysis (12). It has been shown, for example, that the adiabatically forbidden conversion of two molecules of ethylene to cyclobutane becomes allowed if taking place within the coordination sphere of a transition metal atom (12). The application of symmetry arguments to problems of catalysis is described in detail in Chapter 6. The principle of spin conservation is more restrictive than orbital correlation. Many unimolecular decomposition reactions, for example, are spin forbidden. A well-known reaction is the decomposition of nitrous oxide [Eq. (11)]:

$$N_2O(^1\Sigma) \longrightarrow N_2(^1\Sigma) + O(^3\Sigma_u^-) \tag{11}$$

Other spin forbidden decompositions involve substrates such as CH_2N_2, HN_3, COS, H_2O_2, O_3, of organic azo compounds, peroxides, etc. Many endothermic compounds owe their kinetic stability to the fact that the thermodynamically favored formation of decomposition products in the ground state is a process with $\Delta S \neq 0$. Some of the most impressive effects of catalysis are observed in systems of this kind. A detailed analysis of the efficiency of catalysts in relaxing the principle of spin conservation would reveal it to be closely associated with the extent of catalyst interaction with the substrate and the magnitude of the spin-orbit coupling parameters.

III. Principal Functions of Catalysts

We are now in the position to summarize the principal functions of catalysts as follows:

1. Catalysts increase reaction rates by their ability to relax restrictions imposed by quantum mechanical selection rules of spin and angular momentum.
2. Catalysts bring reaction participants together in energetically and sterically favorable fashion (proximity effect).
3. Catalysts introduce efficient alternative reaction pathways by virtue of specific interactions with the substrate(s).

It is obvious that each substance can have catalytic propensies of one form or another, if only to provide a weak perturbation as the third body in a collision reaction. Although the catalytic phenomenon is hierarchial due to the chemical individuality of the catalysts and the occurrence of highly specific interactions between certain catalysts and substrates, all catalytic effects can always be traced back to the three fundamental functions given above. These definitions are also valid for electron-transfer catalysis, even though this topic has not been discussed here.

IV. Selective Catalysis

The formation of the active complex in catalytic reactions makes a negative contribution to the activation entropy which is offset by a larger decrease of ΔF^{+}. In simple reversible systems catalysts increase the rate of attainment of equilibrium and have no effect on the product distribution. By suitable modifications of the catalysts or the addition of selective inhibitors it is possible to develop catalysts "instructed" to perform only one reaction, or to prevent undesirable side reactions. Selective catalysts may be regarded as kinds of Maxwell Demons endowed with information. The information content of a catalyst can at least in principle be calculated by using Brillouin's Negentropy Principle of Information [Eq. (12)] (13):

$$I = S_o - S_s = k \ln P_o - k \ln P_s \qquad (12)$$

In Eq. (12) S_o and S_s are the entropies, P_o and P_s are the number of possibilities (or probabilities) in the general (nonselective) and the selective system, respectively, k is the Boltzmann constant, $1.38 \cdot 10^{-16} \, \text{erg} \cdot \text{deg}^{-1}$. Assume that a catalyst yields a 1:1 mixture of diastereomers of a product.

After selectivization the catalyst produces one diastereomers in 100% optical purity. The catalyst thus has been instructed to recognize the conformation of the starting materials or to perform the chemical reaction stereoselectively. The actual contribution of the information content of the catalyst to the total entropy of activation would nevertheless be vanishingly small. Catalysts or systems of catalysts could, therefore, be packed with vast amounts of information without seriously increasing $\Delta S^{\neq}$. This situation is evidently verified in enzymes, but thus far few if any catalysts have been constructed by chemists from the viewpoint of information theory. One way of endowing catalysts with information is to assure the binding of a certain number of the substrate molecules around the active site. Metal ions or atoms, by virtue of their electronic structure and tendency to form complexes with defined coordination geometries have been shown in some cases to act as templates for the selective synthesis of complicated ring systems in an elegant fashion, both in stoichiometric and catalytic reactions. The development of more sophisticated template systems for the extension of this synthesis principle is only in a beginning stage, but the results achieved so far are quite encouraging.

REFERENCES

1. J. J. Berzelius, *Traité de Chimie I*, 110 (1845).
2. W. Ostwald, *Rev. Sci.*, 1, 640 (1902); *Physik. Z.*, 3, 313 (1902).
3. Cf., P. H. Emmett, P. Sabatier, and E. E. Reid, *Catalysis Then and Now*, Franklin, 1965.
4. See, e. g., G. Herzberg, *Spectra of Diatomic Molecules*, Van Nostrand, 2nd ed., 1961, p. 401.
5. K. J. Laidler, *The Chemical Kinetics of Excited States*, Oxford, Clarendon, 1955, p. 126.
6. K. R. Jennings and J. W. Linnett, *Quart. Revs.* (*London*), 12, 116 (1958), and references cited therein.
7. K. F. Bonhoeffer, *Ergeb. Exakt. Naturwiss.*, 6, 201 (1927).
8. See G. Herzberg, *Spectra of Diatomic Molecules*, Van Nostrand, 2nd ed., p. 295 for more detailed discussion.
9. Nikitin, in *Chemische Elementarprozesse* (H. Hartmann ed.), Springer, Berlin, Heidelberg, N.Y., 1968, p. 43.
10. K. E. Shuler, *J. Chem. Phys.*, 21, 624 (1953).
11. R. B. Woodward and R. Hoffman, *Acc. Chem. Res.*, 1, 17 (1968).
12. F. Mango and J. Schachtschneider, *J. Amer. Chem. Soc.*, 89, 2484 (1967).
13. L. Brillouin, *Scientific Uncertainty and Information*, Academic, New York, 1964, pp. 13–15.

2

Hydrogenation and Dehydrogenation

JACK KWIATEK*

The Weizmann Institute of Science
Rehovot, Israel

* Present address: U.S. Industrial Chemicals Co., Division of National
Distillers and Chemical Corp., Cincinnati, Ohio.

I. Introduction

Hydrogenation and dehydrogenation have been the most basic functions of metals since the early days of catalysis study. Yet, in spite of the extensive effort made toward understanding the nature of these surface-catalyzed hydrogen transfer reactions, mechanistic interpretations have suffered from the inherent difficulty of observing chemisorbed intermediates directly. The recent development of homogeneous hydrogenation systems provides us with models which are helpful in understanding the various factors involved in catalytic hydrogenations. Reactive intermediates are more amenable to identification, and various features in common with heterogeneous systems are revealed.

As in the case of the metals themselves, complexes of the group VIII elements have been found to be especially reactive. Such complexes have a number of properties which influence their ability to function as catalysts. Those properties which undergo gradual change from species to species (electron transferability, bond stability, ligand substitution) contribute to catalyst activity. Those features which change abruptly (number of transferable electrons, coordination sites available, electron configuration) contribute to catalyst selectivity. Both types of properties are further subject to steric effects.

In view of these variables meaningful comparisons of catalytic activity and selectivity can be made only between those complexes with isoelectronic structures and identical coordination numbers and modifying ligands. At the same time it must be recognized that the complexes employed in catalytic reactions are often, strictly speaking, catalyst precursors. They may undergo considerable modification (ligand dissociation and substitution, as well as metal valence changes) before being converted into the "active" species directly involved in the catalytic cycle. Such modification may be signalled by an induction period. Although any of the components of the catalytic cycle may be termed a catalyst, it is convenient to designate that component which does not contain the elements of either of the reactants as the active species.

The structure of the active species, as well as that of all other intermediates directly involved in the catalytic cycle, must be established in order to study the various factors influencing catalytic behavior. Unfortunately, such intermediates, by the very nature of the catalytic process, are often too unstable to isolate, and indirect evidence of their constitution must be relied on. Few systems have been so thoroughly investigated that the mechanisms proposed for them have been generally accepted. Nevertheless, the preliminary classification made here for those systems for which adequate data is available is useful.

Discussion in this chapter will be restricted to catalytic reactions of complexes of the nine group VIII elements. Hydrogenation and dehydrogenation at carbon-carbon unsaturated bonds are emphasized. Ziegler-Natta and Grignard type systems (*1*), which remain rather poorly characterized at the present time, are not discussed, and reactions such as hydroformylation and hydrogenolysis are mentioned only where they relate to hydrogenation. Further information on other hydrogenation systems and substrates may be found in several excellent reviews (*2*) and articles (*3*) from which this author has drawn freely.

II. Hydrogenation Cycles

Though details of the catalytic process, involving expansion and contraction of the coordination sphere and alternate oxidation and reduction of the central metal, differ from system to system, similar elements are present in all hydrogenation cycles. These components are examined in this section.

A. Hydrogen Activation

The essential feature in catalytic hydrogenation is the activation of hydrogen. Yet, the precise manner in which molecular hydrogen is attacked by the active catalyst species is not known. One suggestion (for square d^8 complexes) is that an anti-bonding orbital of the hydrogen molecule accepts an electron from a filled metal orbital; another is that the bonding electrons of hydrogen attack a vacant metal orbital. In any event, a complex containing a hydride ligand is formed. Three types of hydrogen activation have been distinguished [Eqs. (1)–(3)]; Eqs. (1) and (2) are *homolytic* and Eq. (3) is *heterolytic*.

$$M^n + H_2 \rightleftharpoons M^{n+2}H_2 \tag{1}$$

$$2M^n + H_2 \rightleftharpoons 2M^{n+1}H \tag{2}$$

$$M^nX + H_2 \rightleftharpoons M^nH + HX \tag{3}$$

Homolytic cleavages involve the transfer of electrons from the metal to hydrogen atoms. Type 1, in which two electrons are transferred thereby forming a *cis*-dihydride, is one example of a class of reactions known as oxidative-additions. Though an oxidation number two units higher is formally assigned to the metal, there is evidence that such integral oxidation states are not meaningful in labile hydridocomplexes. Type 2, in which a single electron is transferred, results in an increase in the formal oxidation number of the metal by one unit. Heterolytic cleavage is in effect a substitution reaction in which hydride, originating from molecular hydrogen, displaces an anionic ligand from the metal which thereby undergoes no change in oxidation state.

Coordination number and electron configuration play an important role in hydrogen activation. Practically all group VIII metal complexes which are hydrogenation catalysts have a d^6 to d^8 configuration. Coordinatively saturated complexes are unreactive to hydrogen unless the ligands present are labile. Where the complexes are stable in solution, the dissociation of ligands may be promoted by employing elevated temperatures or irradiation. Coordinatively unsaturated complexes have available vacant (active) sites which may react with hydrogen depending on the nature of the other ligands present and the metal.

Of the three types of hydrogen activation, homolytic type 1 appears to be the most commonly encountered. The general order of reactivity toward the oxidative-addition of hydrogen by d^8 complexes increases from nickel to iron, and from iron to osmium. Ligands having both donor and acceptor properties (phosphines, carbon monoxide) stabilize the metal-hydrogen bond. Thus, for a given metal, the weaker the π-acidity of the ligand, the greater will be the electron density at the metal atom and its ability to interact with hydrogen. In this respect, the metal appears to behave as a nucleophile. The energy involved in oxidizing the metal is compensated for by the formation of the metal-hydrogen bond, the strength of which is particularly dependent on the ligand *trans* to it. The greater the *trans* effect of the ligand, the weaker the bond. The hydride ligand itself, in turn, serves to labilize the ligand *trans* to it.

It is generally recognized that a basic requirement for catalytic activity is that the hydrogen cleavage step be reversible. Not only must the hydridocomplex be of sufficient stability that it is readily formed, it must also be labile enough that subsequent transfer of the hydride ligand to a substrate can occur. Nevertheless, many hydridocomplexes which are catalytically active are stable enough to be isolated and characterized.

B. Substrate Activation

It is rather generally accepted that coordination of olefins at a vacant site on the metal is necessary for their hydrogenation to proceed. The formation of a π-olefin complex serves both to lessen the double bond character of the olefin (activation) and to place the substrate in a favorable position (*cis*) for interaction with a hydride ligand.

$$\text{MH} + \text{C=C} \rightleftharpoons M{-}\underset{\displaystyle C}{\overset{\displaystyle \underset{|}{\overset{H}{}}\ C}{\|}} \tag{4}$$

The hydride ligand may be present in the active catalyst species, or may be formed via hydrogen activation. In most cases, the lack of sufficient kinetic data makes it difficult to determine the relative order of hydrogen and olefin activation.

Though several stabilized π-olefin hydridocomplexes, which serve as models for such postulated intermediates, have been isolated, there is little direct evidence for their formation in actively hydrogenating systems. However, observations on the course of hydrogen and hydrogen halide additions to π-olefin complexes support the assumption that transfer of the hydride ligand to coordinated olefin is so rapid that this intermediate cannot be detected.

Observations that excess ligand poisons a catalyst have been noted in support of the hypothesis that a vacant site is required for substrate activation. That coordinately saturated hydridocomplexes, even those which lose molecular hydrogen rapidly and reversibly as required for an active catalyst (see previous section), are unable to hydrogenate olefins has also been cited as evidence that the olefin must be coordinated to the metal in order to enter the catalytic cycle.

One such complex, $Co(CN)_5H^{3-}$, however, has been shown to catalyze the hydrogenation of olefins containing electron-withdrawing (activating) substituents. Thus, the requirement of substrate coordination may depend on the nature of the substrate itself, which in turn suggests that coordinatively saturated complexes with labile hydride ligands should be able to distinguish between activated and nonactivated olefins. In other words, selective hydrogenation may be expected in systems which cannot coordinate olefins (the term "activation selectivity" may be useful in classifying such types). Whether or not hydrides such as $RhClpy(PPh_3)_2H_2$ or

$IrCl(CO)(PPh_3)_2H_2$ which are inactive toward simple olefins can indeed hydrogenate activated olefins, however, remains to be determined. Other factors such as the nature of the hydride ligand may play a dominant role.

C. Hydride Transfer

Group transfers between *cis* sites on metal complexes (insertion reactions) are known to occur readily in carbonylation reactions. A similar mechanism has been proposed for the transfer of hydride to coordinated olefin.

$$
\begin{array}{ccc}
\overset{\text{H}}{\underset{M-\parallel}{|}}\overset{\text{C}}{\underset{\text{C}}{}} & \rightleftharpoons & \overset{\text{H}}{M\cdots}\overset{\text{C}}{\underset{\text{C}}{}} \rightleftharpoons M-C-C-H
\end{array}
\tag{5}
$$

The reversible migration of hydride to the olefin via a four-center transition leads to formation of a metal alkyl complex whose coordination number is reduced by one unit. That such a pathway is energetically preferred over hydrometallation of noncoordinated olefins finds support in the observation that the reverse reaction (decomposition of metal alkyls) occurs most readily with coordinatively unsaturated complexes. All evidence to date indicates that such hydrometallations occur stereospecifically *cis*. Again, direct evidence for the formation of metal alkyls is lacking for most hydrogenation systems. Indirect evidence for the reversible formation of such an intermediate may be found, however, by observing substrate hydrogen-atom exchange and/or isomerization. The insertion of alk-1-enes may proceed either by an anti-Markownikoff or by a Markownikoff addition, only the latter mode being capable of effecting isomerization.

$$
MH + CH_2{=}CHCH_2R
\begin{array}{c}
\nearrow MCH_2CH_2CH_2R \\
\\
\searrow \underset{\underset{M}{|}}{CH_3CHCH_2R}
\end{array}
\tag{6}
$$

$$
\Updownarrow
$$

$$
CH_3CH{=}CHR + MH
$$

Steric effects may be decisive in determining which mode predominates. The proximity of bulky ligands to the site at which the insertion reaction takes place inhibits the formation of *sec*-alkyl groups. With such catalysts the rate of isomerization of alk-1-enes to alk-2-enes, which requires the formation of a *sec*-alkyl species by Markownikoff addition, is found to be much slower than hydrogen atom exchange, which can take place via reversible formation of a primary alkyl by anti-Markownikoff addition. Furthermore, the approach of an alk-2-ene to the coordination site may be more difficult than that of an alk-1-ene as evidenced by alk-2-ene rates of hydrogen-atom exchange which are much slower than alk-1-ene rates of isomerization. Such hydrogenation catalysts are selective for alk-1-enes ("steric" selectivity as distinguished from "activation" selectivity).

The reaction of $Co(CN)_5H^{3-}$ with noncoordinated olefins containing electron-withdrawing substituents, X (see previous section), appears to take place via a hydrogen atom transfer mechanism $[M = Co(CN)_5^{3-}]$:

$$
\begin{array}{c}
\overset{\cdot}{C}=\overset{\cdot}{C}X \\
| \qquad \vdots \\
H\cdots \overset{\cdot}{M}
\end{array}
\underset{\longleftarrow}{\overset{\longrightarrow}{\rightleftharpoons}}
\begin{array}{c}
CH-CXM \\
\\
CH-\overset{\cdot}{C}X + M
\end{array}
\qquad (7)
$$

Either a coordinatively saturated organometal complex, formed by collapse of the radical cage, or an organic radical is generated, depending on substituent X. The hydrogen atom adds to the carbon atom beta to activating group X. Since hydrogen-atom exchange and isomerization may take place via radical intermediates as well as metal alkyls, knowledge of the stereospecificity of hydrogen addition is important in determining mechanism (see Section IID).

D. PRODUCT FORMATION

Four types of hydrogen transfer to alkyl groups leading to formation of hydrogenated product and regeneration of the active catalyst have been distinguished [Eqs. (8)–(11)] (SH = semihydrogenated substrate); Eqs. (8)–(10) are *homolytic* and Eq. (11) is *heterolytic*.

$$M^n H(SH) \longrightarrow M^{n-2} + SH_2 \qquad (8)$$

$$M^n H + M^n(SH) \longrightarrow 2M^{n-1} + SH_2 \qquad (9)$$

$$M^n H + \cdot SH \longrightarrow M^{n-1} + SH_2 \qquad (10)$$

$$HX + M^n(SH) \longrightarrow M^n X + SH_2 \qquad (11)$$

In homolytic transfers, the formal oxidation number of the metal atom is decreased. That hydridoalkyl complexes [as seen in Eq. (8)] have not been detected as intermediates has been ascribed to the rapidity of hydride

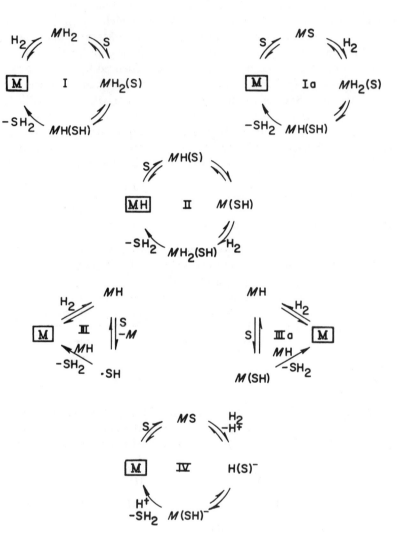

Fig. 1. Classification of homogeneous hydrogenation catalytic cycles. Types I–III involve the homolytic cleavage of molecular hydrogen; Type IV involves heterolytic cleavage. M or MH = active catalyst species; S = substrate.

transfer to an adjacent alkyl group. It has also been suggested that *cis*-dihydridocomplexes may transfer both hydride ligands to an olefin simultaneously, without intermediate formation of a hydridoalkyl complex.

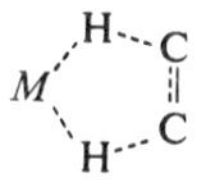

$$
M\begin{array}{c} \overset{H\cdots C}{} \\ \\ \underset{H\cdots C}{} \end{array}\Big\|
$$

Many systems have been found to add hydrogen stereospecifically *cis* to a double bond. Such a result is expected either for simultaneous transfer of two hydride ligands, or for *cis* hydrometallation followed by cleavage of the alkyl group with retention of configuration. Formation of an organic radical intermediate (Eq. 10) sufficiently long-lived to become planar is eliminated in such cases.

Transfers shown in Eqs. (9) and (10) have been proposed only in the case of $Co(CN)_5H^{3-}$ hydrogenations. It is possible that activated olefins are reduced via similar intermediates in other systems as well. Equation (9) has been suggested only in the hydrogenation of conjugated dienes, in which an allylic ligand intermediate presents a site (C—C=C*) for hydrogen attack.

Heterolytic transfers, in which no metal valency change is involved, is limited to those systems which activate hydrogen heterolytically. Electrophilic displacement (by the proton) of the metal-bonded alkyl group (with retention of configuration) has been suggested.

E. SUMMARY

The various steps involved in homogeneous hydrogenations may combine in several ways in completing a catalytic cycle. The main combinations which have been distinguished for group VIII metal complex catalysts are shown in Fig. 1. A discussion of the operation of such hydrogenation cycles in specific catalyst systems is presented in the following section.

III. Group VIII Metal Complex Systems

The group VIII metal complex catalysts discussed are divided into three triads, those of iron, cobalt, and nickel. For each element a further division according to electron configuration is made. Each catalyst is given, followed by the proposed active species (with a question mark for the more speculative ones) in brackets and the conditions under which reactions are usually carried out. Closely related complexes are discussed with that catalyst for which the most experimental data has been obtained.

A. THE IRON TRIAD

1. Iron (d^8)

$$Fe(CO)_5 \ [Fe(CO)_4?] \ 180°, \ 25 \ atm \ pH_2$$

Iron pentacarbonyl catalyzes the hydrogenation of unsaturated fatty esters, both monounsaturated as well as conjugated diene types (4,5). Simple olefins have apparently not been investigated. The elevated temperature required suggests that the active catalyst species is formed by loss of carbon monoxide from $Fe(CO)_5$. The tetracarbonyl formed, thereby, is presumably involved in a type I catalytic cycle. Unfortunately, reactions of the hydridocarbonyl, $FeH_2(CO)_4$, which would be an intermediate in such a process, have not been explored. Also $Fe(CO)_3(1,3\text{-diene})$ may be an intermediate in hydrogenations of fatty ester dienes catalyzed by $Fe(CO)_5$, from which it is isolated. This complex also catalyzes the hydrogenation of fatty ester dienes, at a somewhat lower temperature than that required for $Fe(CO)_5$ (4).

The phosphine complexes, FeH_2L_3 and $FeH_2N_2L_3$ ($L = PRAr_2$), have been prepared (6) but have not been investigated as hydrogenation catalysts.

2. Ruthenium (d^8)

$$Ru(CO)_3(PPh_3)_2$$

Though ruthenium carbonyl is less stable than iron carbonyl toward carbon monoxide dissociation, it has not been studied as a hydrogenation catalyst. The bisphosphine tricarbonyl, however, has been shown to react with hydrogen on irradiation to form a product capable of hydrogenating olefins (7). The nature of the unstable photoproduct has not been determined.

3. Ruthenium (d^7)

$$RuCl_3 \ [RuCl(dma)_x?] \ CH_3CON(CH_3)_2, \ 80°, \ 1 \ atm \ pH_2$$

A ruthenium (I) complex, formed in situ from ruthenium (III) chloride, catalyzes the hydrogenation of activated olefins (maleic acid) (8). The hydrogenation of nonactivated olefins by ruthenium (III) chloride in dimethylformamide (9) may follow a similar course. There is no evidence for a hydridocomplex or substrate complexation, and the mechanism is obscure.

4. Ruthenium (d^6)

$$RuCl_2 \ [RuCl_n^{2-n}] \ aq \ HCl, \ 80°, \ 1 \ atm \ pH_2$$

Aqueous solutions of ruthenium (II) chloride catalyze the hydrogenation of α,β-unsaturated acids, but not simple olefins (10). A detailed study of the reaction kinetics, as well as the use of tracers, indicated a type IV catalytic cycle in which the heterolytic cleavage of hydrogen is the rate-determining step. The postulate that hydrogen is activated heterolytically is based on an analogy with the mechanism of D_2-H_2O exchange by ruthenium (III) chloride (11) which was elucidated by studying the dependence of HD/H_2 on H^+ and Cl^- concentration. Hydrogen addition is stereospecifically *cis*. Though both simple and activated olefins are coordinated by this catalyst, only the latter are hydrogenated in what appears to be a special case of "activation selectivity." Presumably, coordination of a simple olefin is not sufficient to lower the energy barrier for hydride transfer to it. It has been suggested that the fact that such olefins are hydrogenated by ruthenium species in dimethylformamide (9) may be due to the action of the solvent as a base. Thus, removal of the proton, formed in the heterolytic cleavage of molecular hydrogen, drives the hydride transfer step forward (for an alternate view, see Section IIIA3).

$$Ru_2(CO_2Me)_4Cl \ [Ru_2(PPh_3)_4^{5+} \ ?] \ PPh_3-H^+-CH_3OH, \ 25°, \ 1 \ atm \ pH_2$$

The bridged ruthenium (II-III) carboxylate yields a binuclear cationic species upon protonation with HBF_4. In the presence of triphenylphosphine it catalyzes the hydrogenation of alkenes and alkynes either homogeneously or, on cation-exchange materials, heterogeneously (12). No study of mechanism has been made with this system.

$$RuCl_2(PPh_3)_3 \ [RuClH(PPh_3)_2 \ ?] \ C_6H_6-C_2H_5OH, \ 25°, \ 1 \ atm \ pH_2$$

This complex catalyzes the hydrogenation of terminal olefins and acetylenes (13) [but not α,β-unsaturated aldehydes (14)] only in the presence of ethanol which was found to be required for the formation of $RuClH(PPh_3)_3$. Thus, the latter hydridocomplex may be the precursor of the active species (see following paragraph). An arene, $[RuCl_2(C_6H_6)]_n$, is also active ($14a$).

$$RuClH(PPh_3)_3 \ [RuClH(PPh_3)_2] \ C_6H_6, \ 25°, \ 1 \ atm \ pH_2$$

The hydridocomplex, formed readily from $RuCl_2(PPh_3)_3$ and hydrogen in the presence of triethylamine, was found to be the most active homogeneous hydrogenation catalyst known (15). It is extremely selective for the hydrogenation of alk-1-enes of structure $RCH=CH_2$. Hydrogenation is strongly inhibited by excess triphenylphosphine, and pyridine completely poisons the system. Based on similarities between the selectivity, the rate law dependence upon alkene concentration, and the relative rates of isomerization and hydrogen-atom exchange observed for this system and those observed for $RhH(CO)(PPh_3)_3$, a catalyst for which more complete rate data has been obtained, a type II catalytic cycle was proposed. Hydrogenation rates are over ten times faster and selectivity is greater than those obtained with the rhodium catalyst under comparable conditions. [The acetato analog, $RuH(ac)(PPh_3)_3$, is also a highly specific catalyst for the hydrogenation of alk-1-enes (15a)].

The high selectivity observed appears to arise from the interaction of bulky *trans* phosphine ligands with the substrate in the alkyl formation step (steric selectivity). However, the ready hydrogen atom exchange observed with internal olefins, which are not hydrogenated, indicates steric hindrance to the oxidative addition of hydrogen (i.e., after alkyl complex formation). The hydrogenation of conjugated and nonconjugated dienes with terminal unsaturation is slow. It is interesting that buta-1,3-diene yields a mixture of but-1-ene and but-2-ene which is exclusively *cis* (16). Such stereo-selectivity toward the formation of *cis*-isomer has been ascribed to simultaneous 1,4-addition of two hydride ligands to a *cisoid* complexed diene in a chromium system (16a). The only other activated olefin examined as a substrate appears to be acrylonitrile which yields a linear dimer as well as propionitrile (1:4) at elevated temperatures (17). No reason has been given for the lack of hydrogenation of alkynes. The fact that $RuCl_2(PPh_3)_3$ does effect the hydrogenation of such substrates in alcoholic solution indicates a special role for the solvent.

The phosphine complex, $RuH_4(PPh_3)_3$, has recently been prepared (18), but its activity has not been determined. The related anion, $[RuCl_4(bipy)]^{2-}$, catalyzes the hydrogenation of maleic acid, apparently via a type II catalytic cycle (18a).

5. Osmium (d^8)

$$Os(CO)_5 \quad \text{and} \quad Os(CO)_3(PPh_3)_2$$

Both of these carbonyls react with hydrogen to form the stable dihydridocomplexes, $OsH_2(CO)_4$ (19) and $OsH_2(CO)_2(PPh_3)_2$ (20); however, their reactivity as hydrogenatation catalysts has not been determined.

6. Osmium (d^6)

$$OsClH(CO)(PPh_3)_3$$

This complex catalyzes the hydrogenation of ethylene and acetylene (21). Although it was suggested that an eight-coordinated trihydridocomplex, $OsClH_3(CO)(PPh_3)_3$, might be an intermediate, it would appear more likely that the phosphine-dissociated, $OsClH(CO)(PPh_3)_2$, is involved as the active species in a type II catalytic cycle.

B. THE COBALT TRIAD

1. Cobalt (d^8)

$$Co_2(CO)_8 \; [CoH(CO)_3 \quad or \quad Co(CO)_4 ?] \; 110°, \; 200 \; atm \; pH_2 + CO$$

Hydrogenation is usually a minor competing reaction in the hydroformylation of olefins. It may become the major reaction, however, when the olefin is suitably branched ($CH_2{=}CRR'$) (22) or activated (23). Polynuclear aromatics are selectively hydrogenated as well (24). The presence of carbon monoxide is required to prevent decomposition of the catalyst to metallic cobalt. However, it has been shown that the homogeneous hydrogenation of unsaturated fatty esters may be carried out at somewhat lower temperatures (75°–100°) in a pure hydrogen atmosphere (25).

While there still remains disagreement in the mechanistic interpretation of the considerable experimental data which has been obtained (26), there can be little doubt that $HCo(CO)_4$, which is rapidly formed from $Co_2(CO)_8$ at oxo conditions, is involved in the hydroformylation and hydrogenation reactions. In fact, much of our understanding of these reactions is based on studies of the stoichiometric reaction of $HCo(CO)_4$ with olefins at low temperatures and low pressures of carbon monoxide.

As a basis for further discussion, one interpretation of the interrelation of the hydrogenation and hydroformylation reactions is shown in the following scheme in which $CoH(CO)_3$, formed by dissociation of carbon monoxide from $CoH(CO)_4$, is involved as the common active species. This coordinatively unsaturated complex has been invoked in rationalizing the inhibition of alkyl-cobalt carbonyl formation from olefins and $CoH(CO)_4$ by carbon monoxide [$M = Co(CO)_3$].

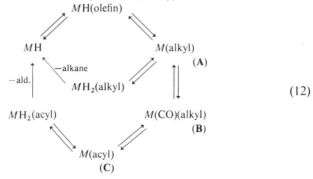

(12)

According to this scheme, in which intermediates of a type II hydrogenation cycle are common to the larger hydroformylation cycle, the ratio of hydrogenation to hydroformylation product depends on the relative concentrations of (A) and (C) as well as the susceptibility of these species to hydrogen addition.

The product distribution observed for various olefins is informative. Thus, hydrogenation of isobutylene to isobutane is favored (22), the accompanying hydroformylation occurring principally at the primary rather than tertiary carbon (97% of the aldehyde product is isovaleraldehyde) (27). Similarly, α-methylstyrene yields mainly isopropylbenzene and the hydroformylation product is exclusively β-phenyl butyraldehyde (28). The observations on isobutylene indicate that two isomeric butylcobalt carbonyls are formed in the reaction; i.e., both Markownikoff and anti-Markownikoff hydrometallation take place. Assuming the above reaction scheme to be valid, the conclusion may be reached that species A predominates for the tert-butyl complex, species C, for the iso-butyl complex (22). The large steric acceleration of carbon monoxide dissociation observed for trimethylacetylcobalt tetracarbonyl (29) may be related to this preferential hydrogenation at tertiary carbons. However, the nature of the relationship, if any, is not apparent. The possibility that such hydrogenations occur by hydrogen transfer from $HCo(CO)_4$ to tertiary alkyl or aralkyl radicals via a type III hydrogenation cycle should be considered. Radical intermediates have also been suggested in the hydrogenolysis of benzylic alcohols (29a).

Hexafluoropropylene yields approximately equal quantities of hydrogenation and hydroformylation products (30). As electronegative substituents are known to stabilize alkylcobalt carbonyls (31), the competitive hydrogenation is in accord with the above reaction scheme (conversion of species A to species C is retarded). A somewhat different line of reasoning has been used to account for the exclusive hydrogenation observed with α,β-unsaturated ketones and aldehydes (32). The formation of a π-oxapropenyl group may stabilize the hydrometallation product toward carbon monoxide insertion (33).

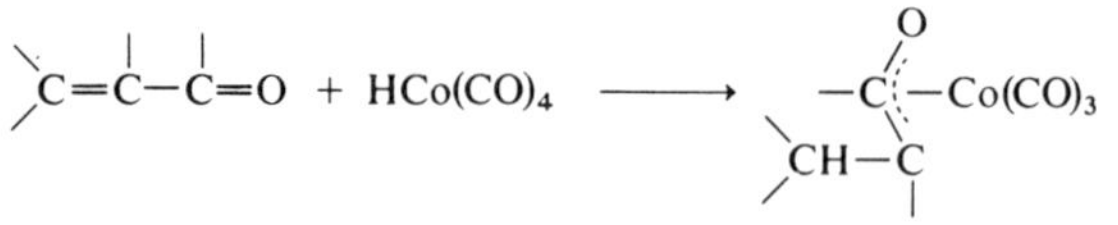

$$\underset{/}{\overset{\backslash}{C}}{=}\underset{|}{\overset{|}{C}}{-}\underset{|}{\overset{|}{C}}{=}O \ + \ HCo(CO)_4 \ \longrightarrow \ \underset{\underset{/}{\overset{\backslash}{CH}}{-}\underset{|}{C}}{-}\overset{\overset{O}{\parallel}}{C}{-}Co(CO)_3 \qquad (13)$$

Whether or not such complexes may be formed has not been established (cf. 34). It is also plausible that a sigma-bonded α-acyl (or formyl) alkylcobalt tetracarbonyl is initially formed. Ready attack by the hydride of

$HCo(CO)_4$ on the acyl (or formyl) oxygen would be expected (cf. *35*), leading to the rapid formation of hydrogenation product via a type IIIa catalytic cycle in which $Co(CO)_4$ is the active species.

$$(CO)_4Co-\underset{\underset{\diagup\diagdown}{\overset{|}{CH}}}{\overset{|}{C}}-\overset{|}{C}=O \quad HCo(CO)_4$$

$$\diagdown\!\!\diagup{-Co_2(CO)_8} \tag{14}$$

$$\underset{\diagup}{\diagdown}CH-\overset{|}{C}=\overset{|}{C}-OH \quad\xrightarrow{\;\;o\;\;}\quad \underset{\diagup}{\diagdown}CH-\overset{|}{CH}-\overset{|}{C}=O$$

According to this view, hydroformylation is prevented from taking place by rapid removal of the initially formed adduct. A similar mechanism, in which an allylic cobalt carbonyl is involved, may be operative in the hydrogenation of conjugated dienes (*23*). The exclusive hydroformylation of α,β-unsaturated esters may thus be due to the lack of susceptibility of α-carboalkoxy groups to such cleavage.

The $Co_2(CO)_8$-catalyzed hydrogenation of aldehydes to alcohols is briefly discussed in Section IVA on dehydrogenation.

$$Co_2(CO)_6(PR_3)_2 \; [Co(CO)_3(PR_3)?] \; 110°, \; 25 \text{ atm } pH_2$$

Cobalt carbonyl phosphines catalyze the competitive hydrogenation and hydroformylation of olefins in the presence of hydrogen and carbon monoxide (*36*). Such complexes ($R = Bu$ or cyclo-C_6H_{11}) however, may also be employed strictly as hydrogenation catalysts (carbon monoxide absent) as they are stabilized toward decomposition by the strong sigma-donor phosphines (*37*). The selective hydrogenation of conjugated dienes to monoolefins observed may take place via a type IIIa catalytic cycle with $Co(CO)_3(PR_3)$ as the active species. The low reactivity of simple olefins (*38*) is probably due to the difficulty of dissociating a carbon monoxide or phosphine ligand, required to provide a coordination site either for olefin activation or hydrogenolysis of an alkyl intermediate. Thus, the catalyst exhibits activation selectivity.

$$Co_3(CO)_6(PBu_3)_3, \; Co(CO)_2(PBu_3)(\pi\text{-}C_4H_7) \; [CoH(CO)_2(PBu_3)?] \; 66°, \; 15 \text{ atm } pH_2$$

This π-allylic complex, as well as the trinuclear cluster compound formed from it on treatment with hydrogen, catalyzes the hydrogenation of simple olefins (internal and terminal) as well as conjugated dienes (*39*). The hydrogenation of simple olefins most probably follows a type II hydrogenation cycle in which the monomeric $CoH(CO)_2(PBu_3)$ is involved as the active species (the cluster compound does not appear to be the active species as it is considerably less active than the π-allylic complex).

The lack of selectivity is nicely explained by the available coordination site [cf. $Co(CO)_3(PBu_3)$ in previous paragraph]. This site may also play a role in determining the course of conjugated diene hydrogenation. Thus, while such hydrogenations may proceed via σ-allylic intermediates in the case of the $Co_2(CO)_6(PR_3)_2$ catalyst, they most probably involve π-allylic intermediates when $Co_3(CO)_6(PR_3)_3$ is employed. Such a difference should be reflected in the stereoselectivities of the respective reactions (40). Unfortunately, monoene product distributions have not been determined, although the decomposition of $Co(CO)_2(PBu_3)(\pi\text{-}C_4H_7)$ itself was studied under oxo conditions (41). The monocarbonyl complex, $CoH(CO)(PPh_3)_3$, is active at 150° and 50 atm pH_2 (41a).

$$CoH_3(PPh_3)_3 \quad \text{and} \quad CoX[P(OR)_3]_{3,4}$$

A number of exceptionally stable hydridophosphine and phosphite complexes which are presumably unreactive have been reported (42–44). However, $CoH_3(PPh_3)_3$ catalyzes the hydrogenation of olefins (41a, 45) and aldehydes (46), while the halophosphite complexes, in the presence of alcohols, are reactive toward vinyl esters and ethers, as well as alkynes (47).

2. Cobalt (d^7)

$$Co(CN)_5^{3-} \quad [Co(CN)_5^{3-}] \, H_2O \quad \text{or} \quad ROH, \, 25° \, 1 \, \text{atm} \, pH_2$$

Alkali metal salts of this complex catalyze the hydrogenation of activated olefins as well as a variety of other substrates, but not simple olefins (48). It is generally agreed that hydrogen activation by this complex yields $CoH(CN)_5^{3-}$ (49), salts of which have been isolated (50–52). The mechanism by which the hydridocomplex is formed has been determined kinetically (53).

Activated olefins, with the exception of conjugated dienes, appear to be hydrogenated via a type III catalytic cycle involving a free radical intermediate. Support for this mechanism is found in kinetic studies (54, 55), radical trapping experiments (40), and the nonstereospecificity of hydrogen addition (56). Although the radical intermediate (R) may be in equilibrium with an organocobalt complex $RCo(CN)_5^{3-}$, the latter species is not directly involved in the catalytic cycle (54). A number of such unstable organocobalt complexes have been isolated (40), and observed spectroscopically (55, 57). The hydridocomplex adds to activated olefins as though it were polarized $Co^+ - H^-$. The fact that such additions to substrates forming stable adducts are stereospecifically cis (56) and that hydrogenation is not observed with such substrates even in the presence of excess $CoH(CN)_5^{3-}$ (40) suggests either a concerted addition via a four-center transition (A), or collapse of a " cage " (B) at a rate substantially greater than that of rotation about the carbon-carbon single bond (X = activating group; $M = Co(CN)_5^{3-}$).

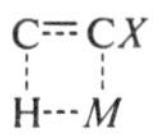

$$
\begin{array}{cc}
\underset{\substack{\vdots\\ H\cdots M}}{C\!=\!CX} & \underset{\substack{\vdots\\ H\cdots M}}{C\!=\!\overset{\centerdot}{C}X} \\[4pt]
\textbf{(A)} & \textbf{(B)}
\end{array}
$$

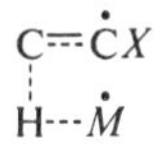

The kinetic trends observed indicate that **(B)** best describes the initial interaction of $CoH(CN)_5^{3-}$ with most activated olefins (*55*). Presumably, such a transition is involved in $Co(CN)_5^{3-}$ abstractions of hydrogen from radicals (*40*) as well. That precoordination of the substrate to the metal is not required is indicated by the fact that cyanide concentration does not affect the rate of hydrometallation (*55,56,58*). Catalytic hydrogenolysis of organic halides by $Co(CN)_5^{3-}$ also takes place via radical intermediates (*40*).

As the catalyst is unreactive with simple monoolefins, the hydrogenation of conjugated dienes is completely selective (activation type). Furthermore, a high degree of stereoselectivity, dependent on cyanide ion concentration (*59*) as well as $CoH(CN)_5^{3-}$ concentration (*58*) and solvent (*60*), is exhibited. Since it is unlikely that these variables could influence the mode of hydrogen transfer to radical species, a type IIIa catalytic cycle, involving organocobalt intermediates, is reasonable for such substrates.

The major product formed from buta-1,3-diene is but-1-ene when hydrogenation is carried out in the presence of excess cyanide, but is *trans*-but-2-ene when little free cyanide is present. A rationale for this dependence of stereoselectivity on cyanide concentration, based on a pmr study of the more stable $Co(CN)_5(CH_2CH\!=\!CH_2)^{3-}$ complex (*34,61*), involves equilibria between σ- and π-butenylcobalt species (*40,59*).

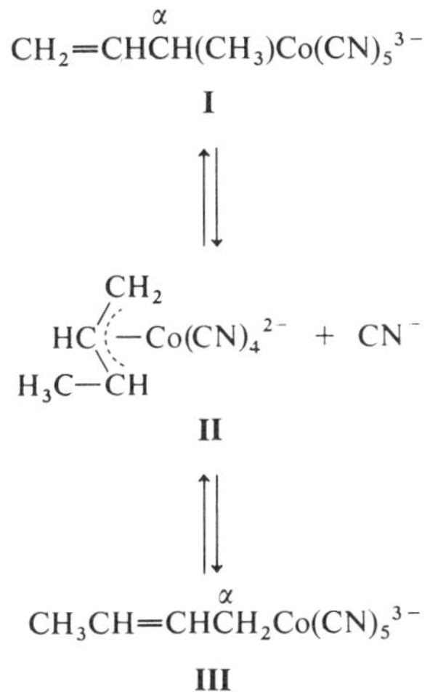

$$
\overset{\alpha}{CH_2}\!=\!CHCH(CH_3)Co(CN)_5{}^{3-}
$$

I

$$\big\updownarrow$$

$$
\begin{array}{l}
\quad\;\; CH_2 \\
HC\!\!\diagup\!\!\diagdown\!\!-Co(CN)_4{}^{2-} \;+\; CN^- \\
H_3C\!-\!CH
\end{array}
$$

(15)

II

$$\big\updownarrow$$

$$
CH_3CH\!=\!\overset{\alpha}{CH}CH_2Co(CN)_5{}^{3-}
$$

III

The intermediate first formed on interaction of $CoH(CN)_5^{3-}$ with buta-1,3-diene is reasonably expected to be I (58). The isolation of III from solutions (34) indicates that if I is indeed initially formed, isomerization occurs readily. That III is isolated even from solutions containing excess cyanide suggests that isomerization may occur, not only via species II, but also via intimate-ion or radical-pair transitions which do not require ligand dissociation. It has been proposed that but-1-ene formation is the result of attack by the $CoH(CN)_5^{3-}$ hydride ligand on the γ-carbon of III (40). A second proposal ascribes the formation of but-1-ene to hydride attack on the α-carbon of I (58). The major source of trans-but-2-ene is probably a preferential attack on the methylene group of the syn-π-(1-methylallyl) species (II). This complex may have been isolated recently (52).

Although this coordination control of stereoselectivity is rather general for conjugated dienes, reduction of the homoconjugated norbornadiene does not exhibit a cyanide dependence. Presumably a π-homoallylic complex (61) cannot be formed from this substrate. Further clarification of the detailed mechanism of this rather complex system, especially determination of the species forming but-1-ene, should be helpful in understanding heterogeneous stereoselective hydrogenations.

$$Co(CN)_n(amine)_m^{2-n}$$

The substitution of amines for cyanide (up to three ligands displaced) results in even more active catalysts for the hydrogenation of activated olefins (62,63,63a). Complexes of ethylenediamine and the like catalyze the hydrogenation of buta-1,3-diene, yielding trans-but-2-ene as a major product (62). However, when the amine is dipyridyl, but-1-ene formation dominates. These results support the mechanism suggested for the $Co(CN)_5^{3-}$ system. Thus, the presence of readily dissociated amine ligands probably allows for the formation of a π-allylic intermediate, while the more strongly held chelating ligand does not. Substitution by orthophenanthroline leads to the formation of coordinatively saturated $Co(CN)_2(phen)_2$ which is inactive (64). Cyanophosphine cobalt (II) complexes have been prepared (65), but they have apparently not been tested as catalysts.

$$Co(dmgH)_2L \quad and \quad vitamin \ B_{12r}$$

Dimethylglyoxime ($dmgH_2$) complexes of cobalt (II) in which the metal is coplanar with nitrogen-containing ligands (Fig. 2) undergo a number of reactions (66) paralleling those exhibited by $Co(CN)_5^{3-}$. Those reactions involving the cleavage of molecular hydrogen differ, however, in that the hydride ligand initially formed with "cobaloximes" is readily lost as a proton, the metal being reduced to the univalent state (67) $[Co=Co(dmgH)_2L]$.

$$Co^{III}H + OH^- \; \rightleftharpoons \; Co^I + H_2O \hspace{2cm} (16)$$

Thus, hydrogen transfers are believed to involve the strongly nucleophilic cobalt (I) species [for a similar reaction of $Co(CN)_5^{4-}$ in alkaline solution, see Ref. (68)].

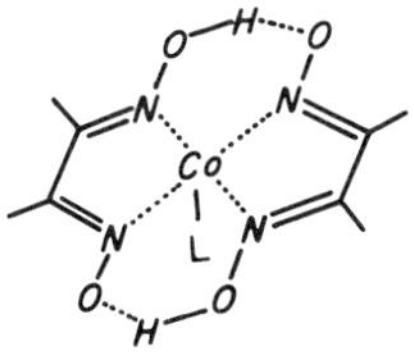

Fig. 2. Cobaloximes(II).

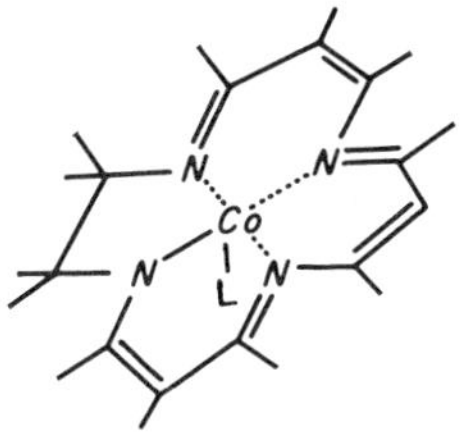

Fig. 3. Vitamin B_{12r} (simplified).

Various activated olefins, as well as propylene, react with cobaloximes (II) in the presence of molecular hydrogen with formation of stable organocobaloximes (III) (69,70) some of which undergo slow reductive cleavage of the organic moiety (66). However, catalytic hydrogenolysis of disulfides (71) and reductive methylation of amines and thiols by formaldehyde (72) have been demonstrated (cf. catalysis by rhodium analog).

Though vitamin B_{12r} (Fig. 3) is not readily reduced by molecular hydrogen in the absence of a catalyst (73), its reactions, including catalysis of the reductive methylations mentioned above, are strikingly similar to those of the bis-dimethylglyoximato model compound. A bacterium extract containing a related complex has been observed to catalyze the formation of methane by hydrogenolysis of methylcobalamine or methylcobaloxime (74). Such reactions indicate possible pathways for hydrogen transfer in biological systems and are being studied as enzyme models (75–77).

3. Rhodium (d^8)

$$Rh_4(CO)_{12} \; [RhH(CO)_3 ?] \; 130°, 300 \text{ atm } pH_2 + CO$$

Although rhodium carbonyls catalyze the hydroformylation of olefins at somewhat milder conditions than does $Co_2(CO)_8$ (78) little competitive hydrogenation takes place, even with α-methylstyrene (79) which is readily hydrogenated by the latter catalyst under oxo conditions.

$$RhH(CO)(PPh_3)_3 \; [RhH(CO)(PPh_3)_2] \; C_6H_6, 25°, 1 \text{ atm } pH_2$$

This hydridocomplex is a highly selective catalyst for the hydrogenation of alk-1-enes (80,81). In solution the catalyst loses one or two of its phosphine ligands depending on its concentration. The bisphosphine complex is the active species responsible for the selective hydrogenations observed; the monophosphine complex is able to hydrogenate alk-2-enes slowly. The presence of excess phosphine greatly depresses hydrogenation rates.

Considerable evidence has been obtained for a type II catalytic cycle which involves the coordination of olefin by the active species, $RhH(CO)$-$(PPh_3)_2$, in the initial step. There is no evidence for prior activation of hydrogen to form $RhH_3(CO)(PPh_3)_2$, which in any case would not be expected to interact with olefin as it is coordinatively saturated. The square planar alkyl complex $RhR(CO)(PPh_3)_2$ formed in the following hydride transfer step has been observed directly only when tetrafluoroethylene was employed as the substrate (82). It is suggested that a *trans* arrangement of the phosphine ligands in this intermediate accounts for the selectivity exhibited by the catalyst. Thus, the formation of a secondary alkyl group, required when the substrate is an alk-2-ene, adjacent to two bulky phosphine ligands is sterically hindered (steric selectivity). Support for this proposal is found in observations of hydrogen-atom exchange and isomerization rates. Alk-1-ene exchange rates are 200 times faster than those of alk-2-enes. On the other hand, rates of isomerization of alk-1-enes, which can only take place via secondary alkyl species, are considerably slower than their rates of exchange. An additional factor which may contribute is the relative instability of secondary alkyls toward olefin elimination. No direct test of the steric factor (i.e., variation of the bulk of the phosphine ligand) appears to have been made. Oxidative addition of hydrogen to the rhodium-alkyl complex is considered to be the rate-determining step. The dihydridocomplex so formed rapidly eliminates alkane to complete the hydrogenation cycle. The formation of RH on treatment of $RhR(CO)(PPh_3)_2$ (82) and $RhR(PPh_3)_3$ (83) with hydrogen presumably follows such a course, illustrating the instability of *cis* hydrido and alkyl ligands.

While terminal olefins, $RCH{=}CH_2$, are rapidly hydrogenated, those of structure $R_2C{=}CH_2$ are not. Conjugated, internal and cyclic olefins are not hydrogenated, nor are acetylenes, though the latter do interact with the catalyst. Functional groups in the olefin substrate, such as hydroxyl in allyl alcohol and nitrile in allyl cyanide, are not affected. The low rate of hydrogenation of styrene, compared to that of alk-1-enes, is ascribed to the directive influence of the phenyl group (Markownikoff addition) coupled with the steric effect preventing such addition.

$$Rh(C_6H_5)(1,5\text{-}C_8H_{12})(PPh_3) \; [RhH(1,5\text{-}C_8H_{12})(PPh_3)\,?]$$

It may be expected that $RhH(PPh_3)_3$, which is formed by the hydrogenolysis of $RhR(PPh_3)_3$ (*83*) or possibly dissociation of $RhH(PPh_3)_4$ (*43*), should be an active catalyst. Though this complex has not been tested, the related $RhH(1,5\text{-}C_8H_{12})(PPh_3)$ may be reasonably assumed to be the active species in hydrogenations of olefins, dienes and acetylenes catalyzed by $Rh(C_6H_5)(1,5\text{-}C_8H_{12})(PPh_3)$ (*84*).

$$RhCl(CO)(PPh_3)_2 \; [RhH(CO)(PPh_3)_2\,?] \; 70°, 60 \text{ atm } pH_2$$

The chloride analog of the active species, $RhH(CO)(PPh_3)_2$, has been shown to catalyze the hydrogenation of olefins at elevated temperatures and pressures of hydrogen (*85–87*). α,β-Unsaturated aldehydes are reduced to saturated alcohols (*14*). The activity of $RhX(CO)L_2$ decreases in the order $X = Cl > Br > I$ and $L = PPh_3 > P(C_6H_{11})_3 > P(OC_6H_5)_3$ (*87*). Reproducible results are obtained only after preheating the catalyst in the presence of substrate (nitrogen atmosphere) at 90° for five hours (*87*); treatment at 70° for eighteen hours does not appear to effect any change in the catalyst (no isomerization of olefin occurs) (*86*). The removal of the inhibition period (observed when the complex was employed as a catalyst for the oxo reaction) by base (*86*) suggests that the active species formed from $RhCl(CO)(PPh_3)_2$ is $RhH(CO)(PPh_3)_2$. However, the effect of alcohols or triethylamine on catalyst activity in hydrogenations, surprisingly enough, has not been examined [cf. $RuCl_2(PPh_3)_3$].

$$RhCl(PPh_3)_3 \; [RhCl(PPh_3)_2] \; C_6H_6 \, , \; 25°, 1 \text{ atm } pH_2$$

Unlike $RhCl(CO)(PPh_3)_2$, the tris-phosphine complex $RhCl(PPh_3)_3$ is a very efficient catalyst for the hydrogenation of alkenes and alkynes (*88,89*) as well as activated olefins (*16,90–92*). Though functional groups such as keto and nitro are not reduced, decarbonylation takes place with allyl alcohol (*91*) and, as a side reaction, with α,β-unsaturated aldehydes (*14, 92a*). The catalyst loses a phosphine ligand readily in solution. There is evidence that this dissociation occurs in a stepwise fashion, the species $[RhCl(PPh_3)_2]PPh_3$, in which the departing ligand is loosely held in a secondary coordination sphere, being initially formed (*93*). The presence of excess ligand inhibits hydrogenations. That catalysts prepared so as to contain two moles of phosphine per rhodium atom are consistently more active than those containing three (*93–95*) supports the formulation of the bis-phosphine complex $RhCl(PPh_3)_2$ as the active species. The activation of $RhCl(PPh_3)_3$ by oxygen has been ascribed to the removal of a phosphine ligand as its oxide (*95a*); this view has been questioned (*95b*).

The evidence obtained for this system indicates that either a type I catalytic cycle, in which oxidative addition of hydrogen occurs in the initial step, or a type Ia cycle, in which olefin is initially coordinated, is operative. While hydrogen adds to $RhCl(PPh_3)_2$ to form a cis-dihydrido-complex $RhClH_2(PPh_3)_2$ and ethylene adds to the active species to form $RhCl(C_2H_4)(PPh_3)_2$, the tendency of other olefins to coordinate is slight. The observation that the ethylene complex does not react with hydrogen, but the dihydridocomplex readily transfers hydrogen to ethylene indicates that hydrogen activation preceeds olefin coordination (type I cycle). Thus, if there is competition by olefin for the active species, it is believed to be external to the catalytic cycle. Though it has been stated that ethylene acts as a catalyst poison due to its high complex formation constant (88), the catalytic hydrogenation of ethylene has been demonstrated (16). The observation that coordinatively saturated complexes such as $RhClH_2py$ $(PPh_3)_2$, $[RhClH_2(PPh_3)_2]_2$ (a halogen-bridged dimer), and $[RhH_2$ $(cis\text{-}Ph_2AsCH{=}CHAsPh_2)_2]Cl$ (95c) do not transfer hydrogen to olefins indicates that a vacant (or solvent-occupied) site on the dihydridocomplex is essential for catalytic activity. Coordination of the olefin at the vacant site available on $RhClH_2(PPh_3)_2$ appears to be the rate-determining step (89).

Two pathways have been considered for hydride transfer to the coordinated olefinic bond, which has been shown to occur stereospecifically cis (96, 97): (1) stepwise addition via alkyl complex $RhClHR(PPh_3)_2$, and (2) simultaneous transfer of both hydrogen ligands to the coordinated olefin. Though formation of a metal alkyl has not been directly observed, other observations indicate the involvement of such a species. An evaluation of isomerization products formed from pent-1-ene in the presence of hydrogen and $RhX(PPh_3)_3$ with various halogen ligands (98), as well as deuteration products formed from various cyclohexenes (99), suggests that metal-alkyls may be intermediates. An intermediate in which rhodium is bonded to carbon alpha to a carbonyl group has been proposed in the hydrogenation and isomerization of α,β-unsaturated lactones (92). The mechanism of isomerization, however, is not obvious, as this process has been shown to be dependent on the presence of oxygen and on the solvent employed (99a). In any case, the manner in which hydride is transferred in hydrogenations remains unresolved.

Terminal olefins, $RCH{=}CH_2$, are all hydrogenated at the same rate. Internal and cyclic olefins as well as terminal alkynes are hydrogenated somewhat less readily; sterically hindered olefins and conjugated dienes are still slower (the latter may be reduced at higher pressures of ca. 60 atm). It is suggested that the lack of substantial differences in the rates of hydrogenation of alk-1-enes and alk-2-enes [cf. the high selectivity of RhH(CO)-$(PPh_3)_3$] is due to a *cis* arrangement of the phosphine ligands in the intermediate π-olefin complex $RhClH_2(PPh_3)_2$(alkene) (see, however, asymmetric hydrogenation by an analogous complex later in this section). Solvent plays an important role in the hydrogenation of alkynes, the rate of which is faster than that of alkenes in the presence of an acidic alcohol cosolvent (*16*) [cf. solvent effect for $RuCl_2(PPh_3)_2$]. It is interesting that *cis*-penta-1,3-diene yields *cis*-pent-2-ene as the initial reduction product (*16*). Whether or not the internal double bond of the diene migrated during reaction is not known, however, as the stereoselectivity of buta-1,3-diene hydrogenation has not been determined [the but-2-ene fraction is exclusively *cis* with the $RuClH(PPh_3)_3$ catalyst]. Since the possibility of forming both a sigma and a pi-allylic intermediate exists [cf. $Co(CN)_5^{3-}$], it would be of interest to examine the effect of phosphine to rhodium ratio on stereoselectivity.

Changes in catalytic activity as a function of the ligands have been observed (*94,95,97,100*). The activity of $RhX(PPh_3)_3$ increases in the order $X = Cl < Br < I$, paralleling the increasing *trans* effect of the halide ligands. However, when X is $SnCl_3$, a ligand with a large *trans* effect, the catalyst is much less active. It is suggested that the optimum thermodynamic stability of an intermediate in the catalytic cycle is overreached in this case. Hydrogenations have been catalyzed by $Rh(NO)(PPh_3)_3$ (*101*) which may formally be considered to be a d^{10} complex, but no comparison of its activity with the halogen analogs has been made. The cyanocomplex has also been prepared (*102*), but its activity has not been reported.

The use of *para*-substituted triphenylphosphines allows the observation of electronic effects in a constant steric environment. Electron-releasing groups $[CH_3, CH_3O, (CH_3)_2N]$ increase, and electron-withdrawing groups (F,Cl) decrease, hydrogenation rates. Presumably the lower rates observed when bulky phosphines (tri-α-naphthyl) are employed is due to steric interference with the incoming olefin substrate. The effect of such interference on the stereochemistry of methylenecyclohexane hydrogenation (*102a*) has been reported (for an interesting interaction of the tri-*o*-tolylphosphine ligand with the rhodium atom itself see Section IVA).

Replacement of aryl groups in the phosphine by alkyl groups illustrates the difficulty of evaluating electronic effects in this system. The observation that $[Rh(Ph_2PCH_2CH_2PPh_2)_2]Cl$ does not react with hydrogen though $[Rh(Me_2PCH_2CH_2PMe_2)_2]Cl$ does (*103*) shows that increasing electron density at the metal increases its ability to add molecular hydrogen. On the other hand, dihydridocomplexes [RhClH$_2$(phosphine)$_3$] formed from fairly basic alkyl phosphines, being less readily dissociated, may not present a vacant site for olefin activation. Indeed, complexes of monoalkyldiaryl phosphines have greatly reduced catalytic activities unless prepared so as to contain only two moles of phosphine per rhodium atom. Complexes of dialkylmonoaryl and trialkyl phosphines are fairly unreactive even under such conditions. Although a site is available for olefin activation, apparently transfer of hydrogen to the olefin is more difficult, and this step becomes rate-determining.

Nevertheless, such complexes have been proven useful in carrying out asymmetric hydrogenations. Thus, the catalyst, formed *in situ* by treating RhCl$_3$(PPhMePr)$_3$ (center of asymmetry at the phosphorus atom) with hydrogen in the presence of triethylamine, hydrogenated α-phenylacrylic acid forming (+)-hydratropic acid of 15 % optical purity (*104*). The same catalyst, formed *in situ* by reacting the phosphine with [Rh(hexa-1,5-diene)Cl]$_2$, hydrogenated α-ethyl and α-methoxy styrenes forming products of 8 % and 4 % optical purity respectively (*105*). Employment of a phosphine in which the asymmetric center was in the alkyl group [PhP(CH$_2$CHMeEt)$_2$] led to a much smaller asymmetric effect (for a much larger effect see rhodium d^6 complexes).

The lack of activity of RhCl[P(OPh)$_3$]$_3$ (cf. active cobalt analogs) has been ascribed to its lack of dissociation in solution. However, mono- and di-N-piperidyl and morpholyl aryl phosphines form systems which are considerably more reactive than RhCl(PPh$_3$)$_3$ (*94,106*). Although both complexes dissociate in solution and absorb hydrogen, RhCl(AsPh$_3$)$_3$ is much less reactive than the triphenylphosphine analog and RhCl(SbPh$_3$)$_3$ is practically unreactive (*106,107*). The lack of reactivity of these complexes appears to be due to the stability of the dihydridospecies formed (hydrogen addition is irreversible at room temperature).

4. Rhodium (d^7)

Rh$_2$(CO$_2$Me)$_4$ [Rh$_2$(PPh$_3$)$_4^{+}$?] PPh$_3$—H$^+$—CH$_3$OH, 25°, 1 atm pH$_2$

The bridged rhodium (II) carboxylate yields Rh$_2^{4+}$ upon protonation. As in the case of the analogous ruthenium complex, Ru$_2$(CO$_2$Me)$_4$Cl, the binuclear cationic species catalyzes the hydrogenation of alkenes and alkynes in the presence of triphenylphosphine (*12*).

Rh(dmgH)$_2$(PPh$_3$) [Rh(dmgH)$_2$(PPh$_3$)] C$_2$H$_5$OH, 20°, 1 atm pH$_2$

The rhodium analog of the better known cobalt dimethylglyoxime complex is reported to catalyze the selective hydrogenation of butadiene rather inefficiently to yield an equimolar mixture of *cis*- and *trans*-but-2-enes

(*108*). In view of the fixed planar structure of bis-dimethylglyoximato ligands (cf. Fig. 2) it may be expected that intermediate butenylrhodium complexes are exclusively sigma-bonded, there being no available coordination site for π-allylic complex formation. The characterization of such species may help to resolve the question posed in the $Co(CN)_5^{3-}$ hydrogenation system, namely, whether allylic groups are preferably cleaved by attack of hydride ligands on α- or on γ-carbon atoms of allylic systems. Evidence for γ-attack in the catalyzed reduction of conjugated dienes by borohydride has been presented (*70*). The cyanorhodium complex $RhH(CN)_4(H_2O)^{3-}$ does not catalyze the hydrogenation of butadiene (*102*).

5. Rhodium (d^6 ?)

Miscellaneous

A number of rhodium (III) systems have been reported to be hydrogenation catalysts. These include $RhCl_3(PPh_3)_3$ and $RhCl_3(AsR_3)_3$ in benzene-ethanol (*109*), $RhCl_3py_3$ in ethanol (*110*), $RhCl_3(SR_2)_3$ (*111*), $RhCl_3$ in dimethylacetamide (*112,113*), and $RhCl_2(BH_4)py_2(dmf)$ in dimethylformamide (*114,115*). A suggested mechanism for the $RhCl_3$-dimethylacetamide system involves oxidative addition of hydrogen to a rhodium (I) olefin complex. It is possible that the alcohol solvent employed with the phosphine, arsine and pyridine complexes is necessary for reduction of the metal or hydridocomplex formation (*115a*) [cf. $RuCl_2(PPh_3)_3$], though this has not been established. The activity of $RhCl_2H(PPh_3)_3$ (*116*) has not been determined.

The borohydride complex is extremely useful as a catalyst for asymmetric hydrogenations (*117*). When the catalyst, prepared in $(+)$- or $(-)$-1-phenylethylformamide, is used to hydrogenate methyl 3-phenylbut-2-enoate, $(+)$- or $(-)$-methyl 3-phenylbutanoate is formed in better than 50% optical yield [cf. $RhCl_3(PPhMePr)_3$ in Rhodium (d^8) section]. Presumably, the semi-hydrogenated substrate is bonded to the rhodium atom *cis* to an amide ligand, such an arrangement resulting in steric selectivity.

6. Iridium (d^8)

$Ir(CO)_x$ [$IrH(CO)_3$?]

Unlike rhodium carbonyls, it is claimed that iridium carbonyls are even more active than cobalt carbonyls in the competitive hydrogenation of olefins under oxo conditions (*79,118*).

IrH(CO)(PPh$_3$)$_3$ [IrH(CO)(PPh$_3$)$_2$?]

The carbonyl phosphine complex catalyzes the hydrogenation of ethylene and acetylene (21). Although seven-coordinate complexes were originally suggested as intermediates in the catalytic cycle, it has been observed that this catalyst may operate in a manner similar to that of the rhodium analog; i.e., by dissociation of a phosphine ligand (type II catalytic cycle) (119). In support of this suggestion, olefinic complexes of the dissociated species, IrH(CO)(PPh$_3$)$_2$, have been isolated (120,121) and spectroscopic evidence of an ethyl intermediate has been obtained (121a). The low activity of this catalyst compared to that of its rhodium analog may be due either to its greater stability toward ligand dissociation or a decreased tendency for hydride transfer to coordinated substrate (81).

IrCl(CO)(PPh$_3$)$_2$ [IrCl(CO)(PPh$_3$)] CH$_3$CON(CH$_3$)$_2$, 80°, 1 atm pH$_2$

The chloride analog of the above-suggested active species, IrH(CO)(PPh$_3$)$_2$, is a rather inefficient catalyst for the hydrogenation of alkenes and alkynes in aromatic solvents (85,122,123). [Complexes containing bidentate phosphine ligands are even less reactive, requiring considerably more vigorous reaction conditions (150°, 80 atm pH$_2$) (124).] The catalyst is much more reactive in dimethylacetamide in which activity depends on the halogen ligand according to $X = \text{Cl} < \text{Br} < \text{I}$ (125). Based on kinetic studies carried out with maleic acid, a type Ia catalytic cycle involving initial coordination of the substrate is favored over the alternate type I cycle in which hydrogen is initially activated. The remarkable increase in hydrogenation rate noted when reactions were carried out with a few percent oxygen in the hydrogen has not been explained. Ir(NO)(PPh$_3$)$_3$, which may formally be considered to be a d^{10} complex [cf. Rh(NO)(PPh$_3$)$_3$], is active in benzene at 85° (101). Unlike its rhodium analog, IrCl(PPh$_3$)$_3$ is not an effective catalyst (125a).

IrH$_3$(PPh$_3$)$_2$ [IrH(PPh$_3$)$_2$] CH$_2$Cl$_2$, 25°, 1 atm pH$_2$

This hydridophosphine complex catalyzes the hydrogenation of alk-1-enes (126). However, the selectivity of the catalyst has not been established. A type II catalytic cycle, involving the partially characterized complex IrH(PPh$_3$)$_2$ as the active species, is proposed.

7. Iridium (d^7)

The cyanocomplex IrH(CN)$_5^{3-}$ has been prepared (127) but there is no information as to its catalytic activity.

8. Iridium (d^6)

Miscellaneous

Unlike $IrH_3(PPh_3)_2$, the coordinatively saturated $IrH_3(PPh_3)_3$ does not hydrogenate alk-1-enes either in dichloromethane or in acetic acid (*126*), though it is effective in the latter solvent for the hydrogenation of activated olefins (*128*). This may be an example of activation selectivity in which the activated olefin need not be coordinated to the metal in order for hydrogenation to be effected [cf. $Co(CN)_5^{3-}$]. It would be most interesting to determine whether conjugated dienes are selectively hydrogenated, and, if so, the stereoselectivity of the reaction.

Other iridium (III) complexes, $IrX_2H(MPh_3)_3$ ($M = P$, As, Sb) and $IrXH_2(CO)(PPh_3)_2$, do catalyze the hydrogenation of alk-1-enes at elevated temperatures and pressures (*129*). Evidence [preactivation in hexane at 100° lowers the temperature at which the catalyst may operate (cf. $RhCl(CO)(PPh_3)_2$); the effect of alcohols or triethylamine on catalytic activity was not examined (cf. $RuCl_2(PPh_3)_3$)] indicates that iridium (I) species may be involved in the catalytic cycle.

C. The Nickel Triad

1. Nickel (d^8)

$$NiX_2(PPh_3)_2 \text{ [Ni}XH(PPh_3)_2\text{?] } C_6H_6, 25°, 1 \text{ atm } pH_2$$

The nickel (II) phosphine complexes ($X = Br$, I) catalyze the hydrogenation of alk-1-enes with no accompanying isomerization (*130*). A type II catalytic cycle involving $NiXH(PPh_3)_2$ as the active species is suggested for this as well as other d^8 complexes in this triad which are employed in the presence of $SnCl_2$. Activities varied in the order Pd > Pt > Ni.

2. Palladium (d^8)

$$PdX_2(PPh_3)_2 \text{ [PdH(SnCl}_2X)(PPh_3)_2\text{?] } SnCl_2-C_6H_6-CH_3OH, 25°, 1 \text{ atm } pH_2$$

The palladium complex catalyzes the hydrogenation of alk-1-enes selectively (*130*). Extensive isomerization to internal olefins is observed with $X = Cl$ or Br, but not with $X = I$ or CN. Activity of the catalyst decreased in the order $X = Cl > Br > I > CN$. A solution of $PdCl_2$ in dimethylformamide may be used to hydrogenate cyclic olefins (*9*).

3. Platinum (d^8)

$$PtX_2(PPh_3)_2 \; [PtH(SnCl_2X)(PPh_3)_2 \, ?] \; SnCl_2 - C_6H_6 - CH_3OH, \; 25°, \; 1 \; atm \; pH_2$$

The same activity and isomerization dependence on the halogen ligand found with the palladium complex is observed with the platinum analog (*130*). Isomerization is more extensive if the phosphine ligand is replaced by AsPh₃ (*131*). Conjugated dienes are quite unreactive even at 90° and 3 atm pH₂, though acrylonitrile and α,β-unsaturated ketones (*132*) were readily reduced. Cyclic olefins and acetylenes are also hydrogenated by this catalyst (*133*). The preformed active species, PtClH(PPh₃)₂, catalyzes the hydrogenation of olefins at the same rate as a buffered solution of PtCl₂(PPh₃)₂ (*130*). Isolation of a hydridoolefin complex, PtH(SnCl₃)-(PPh₃)₂(olefin), analogous to those which would be intermediates in a type II catalytic cycle, is further support for the proposed mechanism (*132*). The dihydridocomplex, PtH₂(PPh₃)₂, has also been characterized (*134*), but its activity has not been determined.

4. Platinum (d^6)

$$H_2PtCl_6 \; [PtCl_2H(SnCl_3)^{2-} \, ?] \; SnCl_2 - alcohols \; (or \; HOAc), \; 25°, \; 1 \; atm \; pH_2$$

Chloroplatinic acid catalyzes the hydrogenation of alkenes and alkynes in the presence of SnCl₂ in alcohol (*135*) or acetic acid (*136*). The presence of halogen acids gives enhanced rates of reaction, the order of olefin activity being terminal > 1,2-disubstituted > trisubstituted (*135*). A type II catalytic cycle, involving a hydridoplatinum anionic complex, is suggested for this system (*137*). Various platinum complexes containing the SnCl₃⁻ ligand have been isolated (*138*).

D. SELECTIVITY

Though the mechanisms discussed for most systems in this section are rather speculative, a correlation between selectivity type and structure of the catalyst has become apparent. Table I lists selected systems for which some confidence attends the formulation of the active species. The first three examples, all of which are four-coordinate complexes with two bulky phosphine ligands, are highly selective catalysts for the hydrogenation of alk-1-enes. Species such as CoH(CO)₂(PR₃) (*39*) and RhClH₂(PPh₃)₂ (*89*) fail to exhibit such selectivity, the latter presumably due to a *cis* arrangement of the bulky ligands. Nevertheless, selectivity toward more complex substrates has been observed with RhClH₂(PPh₃)₂ (*16*).

TABLE 1

EXAMPLES OF SELECTIVE HOMOGENEOUS HYDROGENATION

Active species (hydride form)	Coord. No.	Cycle type	Selectivity type	Substrate	Product	References
RuClH(PPh$_3$)$_2$	4	II	steric	alk-1-enes	alkanes	(15)
RhH(CO)(PPh$_3$)$_2$	4	II	steric	alk-1-enes	alkanes	(81)
PdH(SnX$_3$)(PPh$_3$)$_2$	4	II	steric	alk-1-enes	alkanes	(130)
CoH(CN)$_5^{3-}$	6	IIIa	activation	buta-1,3-diene	butenes	(40)
RhH(dmgH)$_2$(PPh$_3$)	6	IIIa	activation	buta-1,3-diene	butenes	(108)

The last two examples, both of which are six-coordinate complexes, catalyze the hydrogenation of conjugated dienes selectively, apparently due to the lack of an available site for monoolefin coordination. Nevertheless, there is evidence, in the case of CoH(CN)$_5^{3-}$, that the σ-allylic species initially formed with butadiene can convert to a π-allylic species with loss of a CN$^-$ ligand. In this manner ligand concentration determines the stereoselectivity of the reaction (coordination control). Similar observations have been made with mixed cyanoamine cobalt (II) complexes (62). Some chromium carbonyls hydrogenate conjugated dienes stereoselectively (16a, 138a,b).

Factors other than coordination site availability must determine the selectivity for activated olefins exhibited by RuCl$_2$ in aqueous HCl which forms complexes with simple olefins (10). Elucidation of the factors determining such selectivity in the case of IrH$_3$(PPh$_3$)$_3$ (128) awaits a mechanistic study. The selective hydrogenation of acetylenes to olefins has also been noted (133,135,138c).†

IV. Dehydrogenation

The energetics of olefin hydrogenation highly favors the formation of the saturated product. Nevertheless, catalytic dehydrogenation of various substrates, including saturated hydrocarbons, is possible via reactions which are the reverse of those encountered in hydrogenation cycles.

With the exception of the product-forming step, the reactions involved in

† Selective hydrogenation has been reviewed by J. E. Lyons, L. E. Rennick, and J. L. Burmeister, *Ind. Eng. Chem., Prod. Res. Develop.*, **9(1)**, 2(1970).

catalytic hydrogenations have been shown to be reversible. Presumably, those complexes capable of cleaving the carbon-hydrogen bond reversibly (hydrocarbon activation) may best serve as dehydrogenation catalysts since the initial products must be labile enough to undergo further reactions. The following mechanistic types of cleavage may be envisioned [Eqs. (17)–(20)] (SH$_2$ = substrate); Eqs. (17)–(19) are *homolytic* and Eq. (20) is *heterolytic*.

$$M^n + SH_2 \rightleftharpoons M^{n+2}H(SH) \tag{17}$$

$$M^n + SH_2 \rightleftharpoons M^{n+1}H + \cdot SH \tag{18}$$

$$2M^n + SH_2 \rightleftharpoons M^{n+1}H + M^{n+1}(SH) \tag{19}$$

$$M^n X + SH_2 \rightleftharpoons M^n(SH) + HX \tag{20}$$

This section describes all reactions involving cleavage of the carbon-hydrogen bond by a group VIII metal complex, including those in which an oxidizing agent is employed as a hydrogen acceptor. A consideration of the following equilibria involving homolytic cleavage as in Eq. (17) suggests that dehydrogenation catalysts should be structurally and electronically closely related to, if not identical with, hydrogenation catalysts:

$$SH_2 + M \rightleftharpoons \overset{H}{\underset{HS}{\diagup}}M \underset{S}{\overset{-S}{\rightleftharpoons}} \overset{H}{\underset{H}{\diagdown}}M \rightleftharpoons M + H_2 \tag{21}$$

The decomposition of alkyl derivatives of group VIII metal complexes clearly indicates this dual function. Thus, it has been suggested that the formation of alkene as well as alkane in aliphatic aldehyde decarbonylations (*139–141*) proceeds via decomposition of a hydridoalkyl complex.

$$2\ RhClH(CO)(PPh_3)_2(CH_2CH_2R) \longrightarrow CH_3CH_2R + CH_2{=}CHR + H_2 \tag{22}$$

The decomposition of 2-butenylpentacyanocobaltate(III) to equimolar quantities of butenes and butadiene (*34*) is also illustrative:

$$2\ Co(CN)_5(C_4H_7)^{3-} \longrightarrow C_4H_8 + C_4H_6 \tag{23}$$

In most cases dehydrogenation is thermodynamically unfavorable, and equilibria must be shifted by removing hydrogen as such, or from a hydridocomplex intermediate. Thus, dehydrogenation often is carried out in the presence of a reducible substrate (hydrogen acceptor), the overall process being termed a hydrogen transfer. Disproportionation is such a process in which the same substrate serves both as hydrogen donor and hydrogen acceptor.

The stability of metal-carbon bonds is affected by the same factors determining the stability of metal-hydrogen bonds. Thus, the thermal stabilities of σ-bonded organic derivatives of nickel, palladium, and platinum increase with increasing atomic number, and increase in the presence of π-bonding ligands. In addition, the nature of the organic moiety itself has a great influence on the stability of the metal-carbon bond. The following order of stability for alkyl derivatives has been observed: Me > Et > iso-Pr > t-Bu. Increasing stability with increasing π-bonding ability (alkynyl > aryl > alkyl) and electronegativity (C—CN > C—COOR > C—R) has also been noted.

In those cases where an organic radical rather than an organometal complex is formed by carbon-hydrogen bond cleavage, further cleavage of an adjacent carbon-hydrogen bond to form olefin, or dimerization (oxidative coupling or dehydrodimerization) may occur. The autoxidation of hydrocarbons, in which the function of the transition metal complex catalysts is initiation of radical chains from hydroperoxides (142), cannot be considered to be a dehydrogenation process. However, the use of oxygen merely to regenerate a higher valent state of a metal complex is included in the following discussion.

A. SATURATED CH

A comparison of bond dissociation energies for H—H, isopropyl-H, allyl-H, and benzyl-H (104, 94, 77 and 78 kcal/mole, respectively) indicates that carbon-hydrogen cleavage might be effected by transition metal complexes. Indeed, it appears that the saturated carbon-hydrogen bond is readily cleaved when placed in proximity to a coordinatively unsaturated metal center, as exemplified by the following reversible intramolecular cleavage (143); $PP = Me_2PCH_2CH_2PMe_2$.

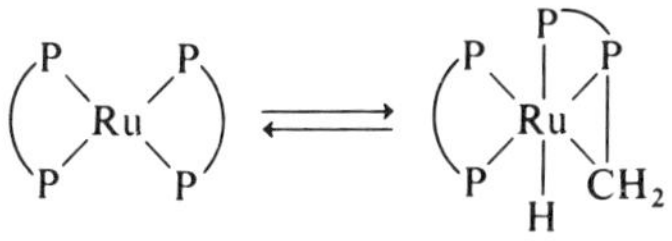

$$\text{(24)}$$

The cleavage of hydrogen from a platinum-coordinated triethylphosphine ligand (144) and from the methyl group of a rhodium-coordinated tri-o-tolylphosphine ligand (145) have also been observed. The strong σ-electron donation from phosphorus atoms to metals apparently withdraws electrons from the phosphorus-substituted alkyl groups, thereby activating them.

The only example reported of carbon-hydrogen bond cleavage in an uncoordinated saturated substrate involves the aromatization of cyclohexane (146).

$$\langle S \rangle \xrightarrow{\text{IrCl(PPh}_3)_3} \langle \text{benzene} \rangle \qquad (25)$$

However, the origin of the benzene detected was not established; it may have derived from a ligand of the complex. The stable organometal hydride isolated from the reaction mixture (83,146) was formed by transfer of an ortho-hydrogen from one of the coordinated phosphine ligands to the metal, as follows:

$$
\begin{array}{c}
\text{Ph}_3\text{P} \quad \text{Cl} \\
\diagdown \text{Ir} \diagup \\
\text{Ph}_3\text{P} \quad \text{PPh}_3
\end{array}
\longrightarrow
\begin{array}{c}
\text{Ph}_3\text{P} \quad \overset{\text{Cl}}{\underset{\text{H}}{|}} \quad \text{PPh}_2 \\
\diagdown \text{Ir} \diagup \\
\text{Ph}_3\text{P}
\end{array}
\qquad (26)
$$

Apparently, the rate of cleavage of the saturated carbon-hydrogen bond by $IrCl(PPh_3)_3$ was too slow to compete effectively with the intramolecular aromatic hydrogen transfer. The reaction suggests, however, that more efficient systems may be found which may serve as models for the biologically catalyzed activation of paraffins (147).

On the other hand, $IrCl(PPh_3)_3$ has been found to catalytically dehydrogenate toluene (148).

$$PhCH_3 \xrightarrow{\text{IrCl(PPh}_3)_3} PhCH_2CH_2Ph + PhCH{=}CHPh + PhC{\equiv}CPh \qquad (27)$$

The rate of cleavage of the benzyl-hydrogen bond is apparently more rapid than intramolecular hydrogen transfer, the benzyl radical or iridium complex so formed producing the products observed. The dihydrido-complex, $IrH_2Cl(PPh_3)_3$, isolated from this reaction is stable, and, therefore, would not be expected to recycle readily. Obviously, an investigation of the nature of the reaction intermediates and the fate of the hydrogen removed is required. A most interesting intramolecular formation of a ligand stilbene derivative from $RhCl_2[P(o\text{-}tol)_3]_2$ has also been observed (145). An unidentified hydridocomplex is formed when $RhCl(PPh_3)_3$ is treated with ethylbenzene at 22° (142). Dehydrogenation of bibenzyl and 9,10-dihydroanthracene is catalyzed at elevated temperatures by $RhCl(PPh_3)_3$, $RhCl_3(AsPh_3)_3$, $IrCl(CO)(PPh_3)_2$, and $RuCl_2(PPh_3)_3$ (149).

The formation of π-allylic complexes by the reaction of palladium (II) chloride with monoolefins having at least one allylic hydrogen proceeds via initial coordination of the carbon-carbon double bond (150–153):

$$
\begin{array}{c}
R \\
\diagup \\
\text{(structure with two Pd centers bridged by Cl, R and R' substituents)}
\end{array}
\underset{\xrightarrow{-HCl}}{\rightleftarrows}
\begin{array}{c}
R \\
\text{(}\pi\text{-allyl Pd dimer structure)} \quad R'
\end{array}
\qquad (28)
$$

The proximity of the allylic hydrogen to the metal may provide a low-energy pathway for the transfer of hydrogen to the metal. The formation of π-allylic metal hydrides as intermediates in the isomerization of olefins is invoked in certain cases (*154*). The observation that the presence of base allows the reaction to take place at room temperature (*155,156*) indicates that direct proton loss from the allylic position may also occur.

Decomposition of π-allylic complexes at elevated temperatures results in the loss of an adjacent hydrogen thereby forming conjugated dienes (*157*).

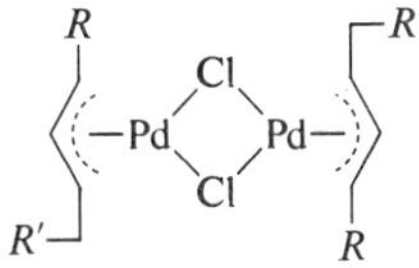

$$\text{(complex)} \xrightarrow{-\text{HCl}} RCH{=}CHCH{=}CHR' + Pd^0 \qquad (29)$$

The overall reaction sequence from monoolefin to diene, constituting a dehydrogenation, appears to have been accomplished in a single step only in the case of an olefin with allylic hydrogens activated by electron-attracting groups (*158*).

$$ROOCCH_2CH{=}CHCH_2COOR + 2CuCl_2 \xrightarrow{PdCl_2}$$
$$ROOCCH{=}CHCH{=}CHCOOR + 2CuCl + 2HCl \qquad (30)$$

This dehydrogenation was made catalytic by employing cupric chloride to recycle the palladium catalyst. The dehydrogenation (*159*) and disproportionation (*150,160*) of cyclohexenes in the presence of palladium (II) salts have also been observed.

Hydrogen transfer reactions are involved in the aromatization of cyclic dienes (*161,162*) and in a curious formylation at a homoallylic position (*163*) catalyzed by cobalt carbonyl. The dehydrogenation of cyclohexa-1,3-diene to benzene in the presence of chloropalladite takes place with concurrent formation of a π-allylic complex, presumably via a palladium hydride intermediate (*164*).

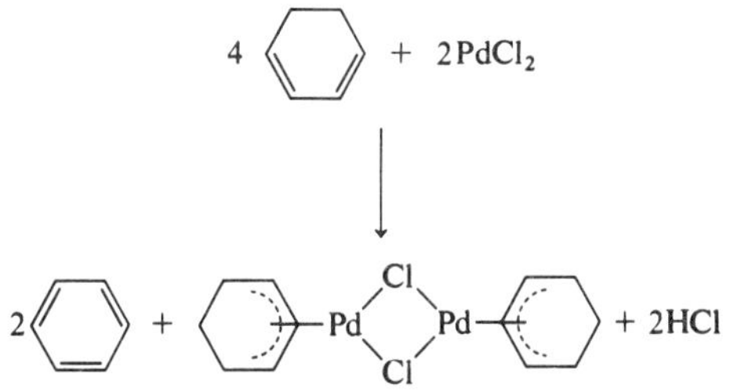

$$4 \,\bigcirc + 2PdCl_2$$
$$\downarrow \qquad\qquad\qquad\qquad (31)$$
$$2\,\bigcirc + \text{(complex)} + 2HCl$$

Cyclohexa-1,3-diene readily undergoes catalytic disproportionation to benzene and cyclohexene in the presence of (cyclohexa-1,3-diene)(pentamethylcyclopentadienyl)rhodium (165). The iridium analog is inactive. On the other hand, cyclohexa-1,4-diene, but not the conjugated 1,3-diene, is catalytically disproportionated to equimolar quantities of benzene and cyclohexene by $IrX(CO)(PPh_3)_2$ ($X = $ Cl or I) (166).

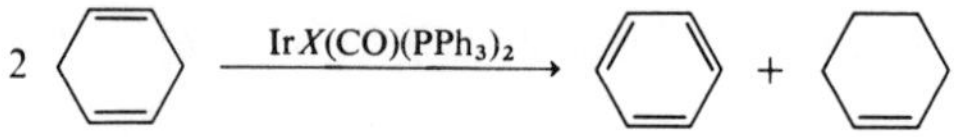

$$2 \; \bigcirc \xrightarrow{\;IrX(CO)(PPh_3)_2\;} \bigcirc + \bigcirc \qquad (32)$$

The following dehydrogenation-hydrogenation sequence was considered as a reaction pathway:

$$\bigcirc + \; IrX(CO)(PPh_3)_2 \longrightarrow \bigcirc + IrH_2X(CO)(PPh_3)_2 \qquad (33)$$

$$\bigcirc + \; IrH_2X(CO)(PPh_3)_2 \longrightarrow \bigcirc + IrX(CO)(PPh_3)_2 \qquad (34)$$

However, the dihydridocomplex proved to be a very poor disproportionation catalyst and it was suggested that hydrogen transfer directly from one molecule of diene to another might occur within a complex containing two π-coordinated diene molecules. The following sequence, in which two molecules of the substrate are σ-bonded to iridium, may also be considered:

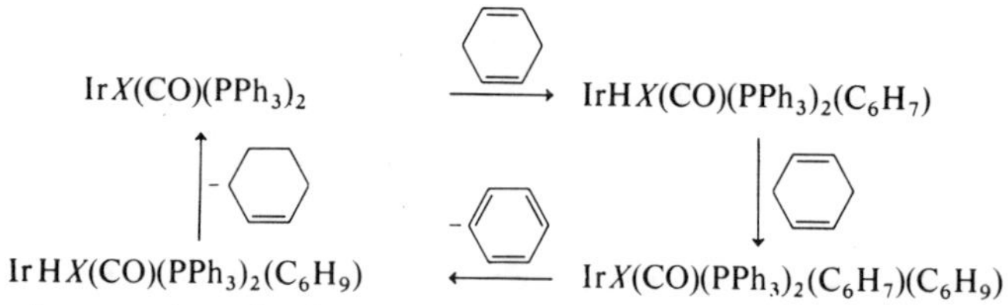

$$
\begin{array}{c}
IrX(CO)(PPh_3)_2 \xrightarrow{\;\;\bigcirc\;\;} IrHX(CO)(PPh_3)_2(C_6H_7) \\[2mm]
\uparrow \;\bigcirc \qquad\qquad \bigcirc \qquad\qquad \downarrow \bigcirc \\[2mm]
IrHX(CO)(PPh_3)_2(C_6H_9) \longleftarrow IrX(CO)(PPh_3)_2(C_6H_7)(C_6H_9)
\end{array}
\qquad (35)
$$

Dehydrogenation of primary alcohols to aldehydes by various group VIII metal complexes is often employed in the preparation of stable hydridocomplexes (167). Here again, proximity of a carbon-hydrogen bond to a metal center may be involved in aiding its cleavage. By employing deuterated ethanol, it has been shown that the metal hydride ligand is formed by abstraction from the carbon bonded to the hydroxyl function (168).

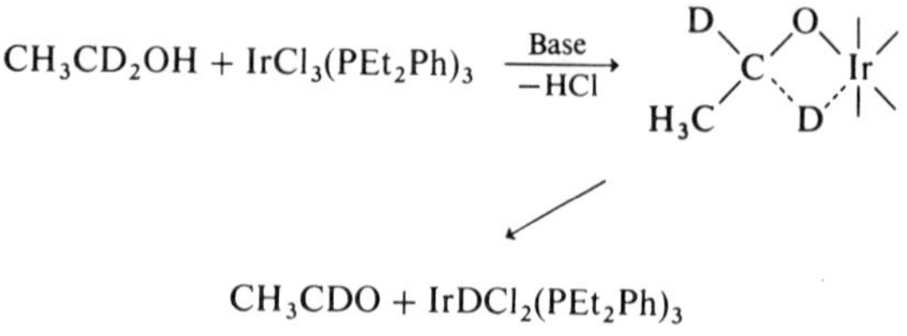

$$CH_3CD_2OH + IrCl_3(PEt_2Ph)_3 \xrightarrow[-HCl]{Base} \begin{array}{c} D \quad O \quad Ir \\ \diagdown C \diagup \diagdown \diagup \\ H_3C \quad D \end{array}$$

$$(36)$$

$$CH_3CDO + IrDCl_2(PEt_2Ph)_3$$

The following reversible reaction sequence has been demonstrated for the catalytic dehydrogenation of isopropanol in the presence of rhodium (III) chloride (*169*), the mechanism of which is presumed to be similar to that shown for the iridium(III) complex:

$$Me_2CHOH + RhCl_6^{3-} \rightleftharpoons Me_2C{=}O + HRhCl_5^{3-} + HCl \qquad (37)$$

$$HCl + HRhCl_5^{3-} \rightleftharpoons H_2 + RhCl_6^{3-} \qquad (38)$$

The addition of stannous chloride, known to form the $SnCl_3$ ligand with chlorides, stabilized the rhodium(III) catalyst toward reduction to the metal.

Alcohol to ketone hydrogen transfers have been catalyzed by iridium(III) complexes (*170,171*). Thus, cyclohexanones are reduced to cyclohexanols by isopropanol with a high degree of stereoselectivity by employing dimethylsulfoxide or trialkylphosphites as ligands.

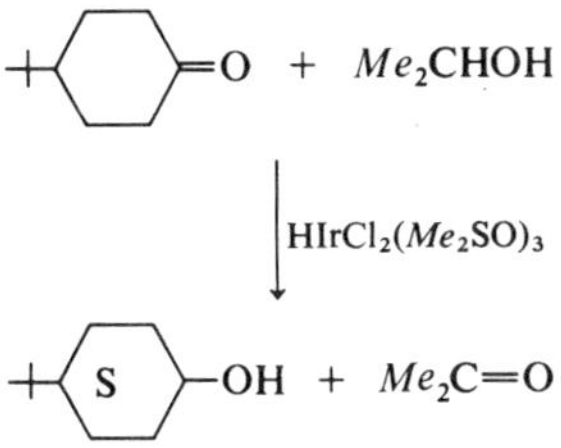

$$(39)$$

The hydrogen attached to the hydroxyl-bonded carbon of isopropanol was shown to be transferred to the carbonyl carbon of the ketone being reduced. Although the mechanism of these reductions has not been elucidated, a hydrogenation-dehydrogenation sequence involving an iridium(III) hydride appears reasonable. Another route, namely, the direct transfer of hydrogen within a complex containing both donor and acceptor molecules, is also possible. For this reason it would be of particular interest to attempt an asymmetric reduction of ketones employing an optically active alcohol as the hydrogen source.

The α,β-unsaturated ketones, such as chalcone in the following example, are reduced to the corresponding saturated ketones (*171*).

$$PhCOCH{=}CHPh + Me_2CHOH \xrightarrow{\;HIrCl_2(Me_2SO)_3\;} PhCOCH_2CH_2Ph + Me_2C{=}O$$

$$(40)$$

It might be expected that the saturated ketone is formed by hydride attack at the *beta*-carbon as in transition state (**A**):

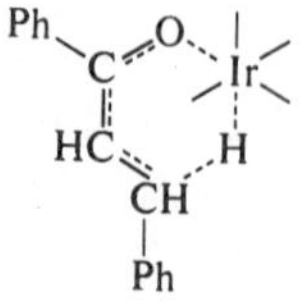

$$\text{(A)} \qquad\qquad \text{(B)}$$

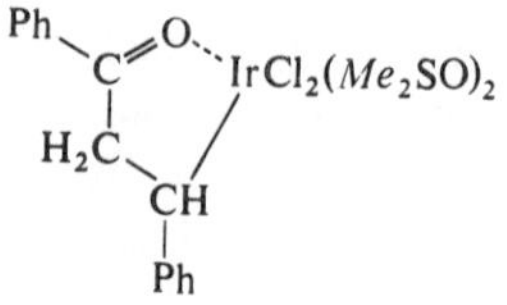

However, the isolation of complex (**B**) from the reaction indicates that an organometal complex may be formed in the catalytic cycle. Further work is required to clarify the reaction mechanism. It should be noted that $IrH_3(PPh_3)_3$ in acetic acid, though inactive toward alkenes, catalyzes the hydrogenation of aldehydes and activated olefins (*128*). The active catalyst, which may be a hydridoiridium acetate similar to the hydrogen transfer catalysts discussed above, also catalyzes the transfer of hydrogen from formic acid to aldehydes.

Platinum complexes have been reported to transfer hydrogen from methanol to olefins (*172*), and cobalt carbonyl is able to utilize isopropanol as a hydrogen source in the hydroformylation reaction (*173*). The latter alcohol dehydrogenation is the reverse process of aldehyde hydrogenation by $Co_2(CO)_8$ which has been suggested as occurring either via an oxygen- (*174,175*) or a carbon- (*176*) bonded cobalt intermediate. An interesting example of an intramolecular hydrogen transfer is provided by the cobalt hydrocarbonyl catalyzed isomerization of allyl alcohol to propionaldehyde (*177*). Exclusive formation of methyl-deuterated propionaldehyde by $DCo(CO)_4$ supports the following reaction path:

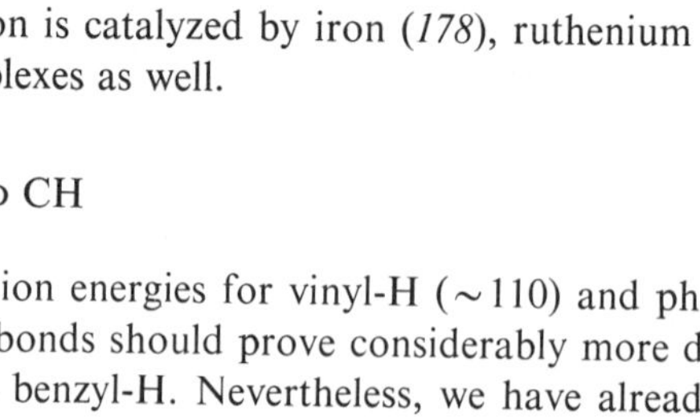

$$\text{(41)}$$

$$CH_2DCH{=}CHOH \xrightarrow{\quad\quad} CH_2DCH_2CHO$$

This isomerization is catalyzed by iron (*178*), ruthenium (*179*) and palladium (*180*) complexes as well.

B. Unsaturated CH

Bond dissociation energies for vinyl-H ($\sim$110) and phenyl-H (102) indicate that such bonds should prove considerably more difficult to cleave than allyl-H and benzyl-H. Nevertheless, we have already noted that an aryl-hydrogen may readily transfer from an iridium-coordinated triphenyl-phosphine ligand (*145*). Similar reversible intramolecular cleavages have also been observed with iron (*6,181*), ruthenium (*15,18*), cobalt (*182*) and rhodium (*183*) phosphine complexes. Mechanistically similar electrophilic substitutions in which the *ortho* hydrogens are eliminated as protons (*183a*) are involved in the formation of ortho-substituted arylmetal complexes from azobenzene (*184,185*) and aromatic Schiff bases (*186,187*).

Aryl-hydrogen bond cleavage of an arene ligand containing no other functional group has also been observed (*143,188*); $PP = Me_2PCH_2CH_2-PMe_2$.

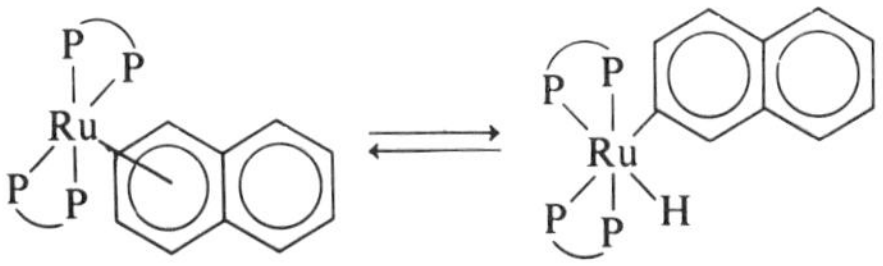

$$(42)$$

A similar intramolecular cleavage has been considered to explain the chloroplatinite-catalyzed aromatic hydrogen exchange (*189,190*).

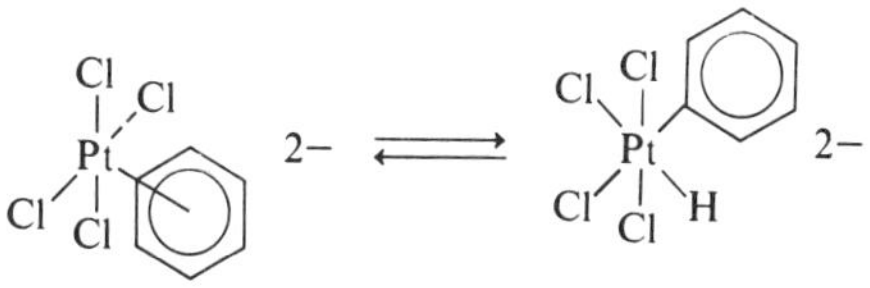

$$(43)$$

Monosubstituted benzenes exhibit ortho deactivation in such exchanges, the rates of which show little dependence on the electronic nature of the substituents. It is of interest that biphenyl and chlorobenzene were obtained as by-products of the exchange under conditions of low acidity.

Analogous substitution effects have been observed in the conversion of benzenes to biaryls in the presence of platinum and palladium acetates (*191,192*). The palladium acetate coupling was made catalytic by carrying the reaction out under fifty atmospheres pressure of oxygen (*193*). The reaction scheme presented (*194*) involves decomposition of an arylpalladium by a concerted mechanism to yield the biaryl and a univalent palladium complex:

$$ArH + Pd^{2+} \longrightarrow ArPd^+ + H^+ \tag{44}$$

$$2ArPd^+ \longrightarrow ArAr + 2Pd^+ \tag{45}$$

It was suggested that the arylpalladium is formed via an electrophilic substitution mechanism. Though this mechanism, in which proton is lost directly from the ring, differs considerably from the homolytic cleavage,

involving formation of a hydridocomplex, proposed for the platinite-cata-lyzed hydrogen exchange (cf. the possible modes of π-allylpalladium com-plex formation), the net result in each case is loss of a proton:
Electrophilic substitution:

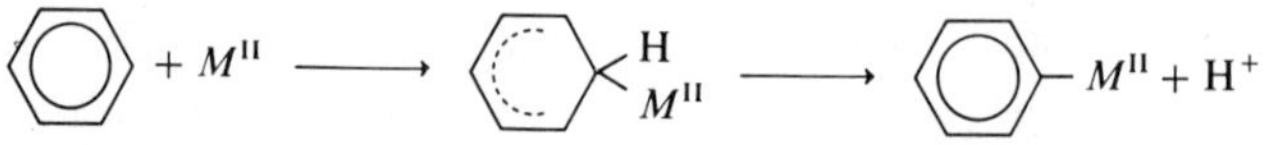

$$\bigcirc + M^{II} \longrightarrow \bigcirc\!\!<^{H}_{M^{II}} \longrightarrow \bigcirc - M^{II} + H^+ \qquad (46)$$

Homolytic cleavage:

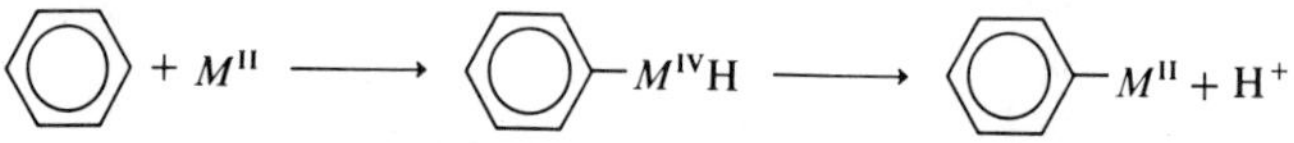

$$\bigcirc + M^{II} \longrightarrow \bigcirc - M^{IV}H \longrightarrow \bigcirc - M^{II} + H^+ \qquad (47)$$

Oxidative coupling reactions have also been observed for α-olefins (150,195), vinyl esters (196) and acrylonitrile (197). It has been suggested that coupling by palladium acetate, involved in the first two cases, proceeds via an acetate-assisted abstraction of hydride by palladium from the coordinated olefin. This is followed by the oxidation of hydride and release of the pro-ton so formed (195).

$$2CH_2{=}CR_2 + Pd(OAc)_2 \longrightarrow R_2C{=}CHCH{=}CR_2 + 2AcOH + Pd^0 \quad (48)$$

Although no mechanism has been presented for the coupling of acrylo-nitrile to mucononitrile by ruthenium (III) salts, a reversible homolytic cleavage of vinylic-hydrogen has been proposed in the ruthenium-catalyzed dimerization of acrylonitrile (198).

$$L_xRu + CH_2{=}CHCN \rightleftharpoons L_xRuH(CH{=}CHCN) \qquad (49)$$

A similar cleavage of propiolate ester by $IrCl(CO)(PPh_3)_2$ has been ob-served (199). It is worth noting that coupling reactions result on decom-position of various σ-bonded aryl and vinyl metal complexes (200).

Palladium acetate has also been employed as a catalyst in the mixed coupling of benzene with olefins (201).

$$ArCH{=}CH_2 + ArH + 2M^n \xrightarrow{\text{Pd(OAc)}_2} ArCH{=}CHAr + 2H^+ + 2M^{n-1} \quad (50)$$

This reaction shows the close relationship of aromatic and olefinic coupling. The intermediate formation of a σ-bonded arylpalladium complex is indi-cated by the reaction of such a species, generated *in situ* from arylmercury salts, with olefins to form styrene derivatives (202).

$$[ArPdX] + CH_2{=}CH_2 \longrightarrow ArCH{=}CH_2 + Pd^0 + HX \qquad (51)$$

V. Concluding Remarks

Many of the components of the various hydrogenation cycles discussed in this chapter have counterparts in accepted mechanisms for the hydrogenation of olefins on metal surfaces (*203,204*). However, extreme care must be taken in drawing analogies between homogeneous and heterogeneous systems (*205*). A single catalytic site (metal atom) is involved in homogeneous hydrogenation cycles other than type IIIa. Such reaction sequences often lead to high selectivities. On the other hand, a variety of reaction sites is available on a solid surface (*206*).

An example of the increased complexity of reactions on metal surfaces is found in attempts to determine the nature of adsorbed olefins. Though the correlation between relative stabilities of such species on various metals and metal-olefin complexes suggests that pi-bonding is involved in both cases (*203*, p. 207), evidence from the isotopic exchange pattern of specially designed hydrocarbons indicates that, at least in some cases, sigma-diadsorbed species on adjacent metal atoms may be involved (*207*). The counterpart of such species in homogeneous hydrogenation systems, necessarily composed of binuclear or metal cluster complexes, has not been observed. Another example is found in the two-site adsorbed species proposed in the heterogeneous hydrogenation of buta-1,3-diene. Not only is there no analogy in homogeneous hydrogenations, there is evidence that precoordination of the diene is not required, at least in some homogeneous systems. Interesting stereoselectivities have been observed in hydrogenations of buta-1,3-diene on metal surfaces (*208*), and parallels may be drawn between conformations of the sigma and pi-allylic chemisorbed species proposed in these reactions and those proposed in homogeneous systems (*40,58*). Hydrogen isotope distributions (*48,209*) and kinetic correlations (*115*) provide further insight into common pathways utilized in the two types of hydrogenation system. Undoubtedly such comparisons will become more useful as our understanding of the relationship between stereochemical considerations and mechanism increases.

REFERENCES

1. M. F. Sloan, A. S. Matlack, and D. S. Breslow, *J. Amer. Chem. Soc.*, **85**, 4014 (1963); S. J. Lapport, *Ann. N. Y. Acad. Sci.*, **158**, 510 (1969); F. Ungvary, B. Babos, and L. Marko, *J. Organometal. Chem.*, **8**, 329 (1967); Y. Tajima and E. Kunioka, *J. Catal.*, **11**, 83 (1968); R. Stern and L. Sajus, *Tetrahedron Lett.*, **1968**, 6313; Y. Takegami, T. Ueno, and T. Fujii, *Bull. Chem. Soc. Jap.*, **42**, 1663 (1969).

2. J. Halpern, *Ann. Rev. Phys. Chem.*, **16**, 103 (1965); C. W. Bird, *Transition Metal Intermediates in Organic Synthesis*, Logos-Academic, London, 1967, Chap. 10; M. E. Volpin and I. S. Kolomnikov, *Usp. Khim.*, **38**, 561 (1969); *Russian Chem. Rev.* **4**, 273 (1969).

3. W. Strohmeier, *Structure and Bonding*, **5**, 96 (1968); J. P. Collman and W. P. Roper, *Advan. Organometal. Chem.*, **7**, 53 (1968).

4. E. N. Frankel, T. L. Mounts, R. O. Butterfield, and H. J. Dutton, *Advan. Chem. Ser.*, **70**, 177 (1968).

5. I. Ogata and A. Misono, *Yukagaku*, **13**, 644 (1964); *CA*, **63**, 17828 (1965).

6. A. Sacco and M. Aresta, *Chem. Commun.*, **1968**, 1223.

7. J. P. Collman, *Accounts Chem. Res.*, **1**, 136 (1968).

8. B. Hui and B. R. James, *Chem. Commun.*, **1969**, 198.

9. P. N. Rylander, N. Himelstein, D. R. Steele, and J. Kreidl, *Englehard Ind. Tech. Bull.*, **3**, 61 (1962).

10. J. Halpern, J. F. Harrod, and B. R. James, *J. Amer. Chem. Soc.*, **88**, 5150 (1966).

11. J. Halpern and B. R. James, *Can. J. Chem.*, **44**, 671 (1966).

12. P. Legzdins, G. L. Rempel, and G. Wilkinson, *Chem. Commun.*, **1969**, 825.

13. D. Evans, J. A. Osborn, F. H. Jardine, and G. Wilkinson, *Nature*, **208**, 1203 (1965).

14. F. H. Jardine and G. Wilkinson, *J. Chem. Soc.*, C, **1967**, 270.

14a. I. Ogata, R. Iwata, and Y. Ikeda, *Tetrahedron Lett.*, **1970**, 3011.

15. P. S. Hallman, B. R. McGarvey, and G. Wilkinson, *J. Chem. Soc.*, A, **1968**, 3143.

15a. A. C. Skapski and F. A. Stephens, *Chem. Commun.*, **1969**, 1008.

16. J. P. Candlin and A. R. Oldham, *Discussions Faraday Soc.*, **46**, 60, 92 (1968).

16a. E. N. Frankel, E. Selke, and C. A. Glass, *J. Amer. Chem. Soc.*, **90**, 2446 (1968).

17. J. D. McClure, R. Owyang, and L. H. Slaugh, *J. Organometal. Chem.*, **12**, P8 (1968).

18. W. H. Knoth, *J. Amer. Chem. Soc.*, **90**, 7172 (1968).

18a. B. C. Hui and B. R. James, *Inorg. Nucl. Chem. Lett.*, **6**, 367 (1970).

19. F. L'Eplattenier and F. Calderazzo, *Inorg. Chem.*, **6**, 2092 (1967).

20. F. L'Eplattenier and F. Calderazzo, *Inorg. Chem.*, **7**, 1290 (1968).

21. L. Vaska, *Inorg. Nucl. Chem. Lett.*, **1**, 89 (1965).

22. L. Marko, *Chem. Ind. (London)*, **1962**, 260.

23. M. Orchin, *Advan. Catal.*, **5**, 385 (1953).

24. S. Friedman, S. Metlin, A. Svedi, and I. Wender, *J. Org. Chem.*, **24**, 1287 (1959).

25. E. N. Frankel, E. P. Jones, V. L. Davison, E. Emken, and H. J. Dutton, *J. Amer. Oil Chem. Soc.*, **42**, 130 (1965).

26. A. J. Chalk and J. F. Harrod, *Advan. Organometal. Chem.*, **6**, 147 (1968).

27. I. Wender, J. Feldman, S. Metlin, B. H. Gwynn, and M. Orchin, *J. Amer. Chem. Soc.*, **77**, 5760 (1955).

28. D. M. Rudkovskii and N. S. Imyanitov, *Zh. Prikl. Khim.*, **35**, 2719 (1962); *CA*, **59**, 2689 (1963).

29. R. F. Heck, *J. Amer. Chem. Soc.*, **85**, 651 (1963).

29a. I. Wender, H. Greenfield, S. Metlin, and M. Orchin, *J. Amer. Chem. Soc.*, **74**, 4079 (1952).

30. D. M. Rudkovskii, N. S. Imyanitov, and V. Y. Gankin, *Inst. Neftekhim. Protsessov.*, **1960**, 121; *CA*, **57**, 10989 (1962).

31. R. F. Heck and D. S. Breslow, *J. Amer. Chem. Soc.*, **84**, 2499 (1962).
32. H. Adkins and G. Kresk, *J. Amer. Chem. Soc.*, **71**, 3051 (1949).
33. R. W. Goetz and M. Orchin, *J. Amer. Chem. Soc.*, **85**, 2782 (1963).
34. J. Kwiatek and J. K. Seyler, *J. Organometal. Chem.*, **3**, 421 (1965).
35. J. Kwiatek and J. K. Seyler, *J. Organometal. Chem.*, **3**, 433 (1965).
36. L. H. Slaugh and R. D. Mullineaux, *J. Organometal. Chem.*, **13**, 469 (1968).
37. I. Ogata and A. Misono, *Discussions Faraday Soc.*, **46**, 72 (1968).
38. F. Piacenti, M. Bianchi, and E. Benedetti, *Chim. Ind. (Milan)*, **49**, 245 (1967).
39. G. Pregaglia, A. Andreetta, and G. Ferrari, *Chem. Commun.*, **1969**, 590.
40. J. Kwiatek and J. K. Seyler, *Advan. Chem. Ser.*, **70**, 207 (1968).
41. W. W. Spooncer, A. C. Jones, and L. H. Slaugh, *J. Organometal. Chem.*, **18**, 327 (1969).
41a. M. Hidai, T. Kuse, T. Hikita, Y. Uchida, and A. Misono, *Tetrahedron Lett.*, **1970**, 1715.
42. A. Sacco, M. Rossi, and C. F. Nobile, *Chem. Commun.*, **1966**, 589.
43. J. J. Levison and S. D. Robinson, *Chem. Commun.*, **1968**, 1405.
44. J. Ellermann and W. H. Gruber, *Angew. Chem. Int. Ed.*, **7**, 129 (1968).
45. A. Misono, Y. Uchida, T. Saito, and K. M. Song, *Chem. Commun.*, **1967**, 419.
46. A. Misono, Y. Uchida, M. Hidai, and T. Kuse, *Chem. Commun.*, **1968**, 981.
47. M. E. Volpin and I. S. Kolomnikov, *Proc. Conf. Hom. Catalysis, Kiev, 1966*; see ref. 2, *Russian Chem. Rev.*
48. J. Kwiatek, *Catal. Rev.*, **1**, 37 (1967).
49. M. G. Burnett, P. J. Connolly, and C. Kemball, *J. Chem. Soc.*, A, **1967**, 800.
50. R. G. Banks and J. M. Pratt, *Chem. Commun.*, **1967**, 776.
51. R. G. Banks and J. M. Pratt, *J. Chem. Soc.*, A, **1968**, 854.
52. G. Pregaglia, D. Morelli, F. Conti, G. Gregorio, and R. Ugo, *Discussions Faraday Soc.*, **46**, 110 (1968).
53. J. Halpern and M. Pribanic, *Inorg. Chem.*, **9**, 2616 (1970).
54. L. Simandi and F. Nagy, *Acta Chim. Acad. Sci. Hung.*, **46**, 137 (1965).
55. J. Halpern and L. Wong, *J. Amer. Chem. Soc.*, **90**, 6665 (1968).
56. L. M. Jackman, J. A. Hamilton, and J. M. Lawlor, *J. Amer. Chem. Soc.*, **90**, 1914 (1968).
57. M. D. Johnson, M. L. Tobe, and L. Wong, *J. Chem. Soc.*, A, **1968**, 929.
58. M. G. Burnett, P. J. Connolly, and C. Kemball, *J. Chem. Soc.*, A, **1968**, 991.
59. J. Kwiatek, I. L. Mador, and J. K. Seyler, *Advan. Chem. Ser.*, **37**, 201 (1963).
60. K. Tarama and T. Funabiki, *Bull. Chem. Soc. Jap.*, **41**, 1744 (1968).
61. J. Kwiatek and J. K. Seyler, *Proc. Eighth Int. Conf. Coord. Chem., Vienna, 1964*, 308.
62. A. Farcas, U. Luca, and O. Piringer, *Proc. Eleventh Int. Conf. Coord. Chem., Haifa, 1968*, 29.
63. G. Schwab and G. Mandre, *J. Catal.*, **12**, 103 (1968).
63a. C. E. Wymore, *155th Meeting, Amer. Chem. Soc., San Francisco, 1968*, M53.
64. G. Mandre, *Proc. Eleventh Int. Conf. Coord. Chem., Haifa, 1968*, 35.
65. P. Rigo, M. Bressan, and A. Turco, *Inorg. Chem.*, **7**, 1460 (1968).
66. G. N. Schrauzer, *Ann. N.Y. Acad. Sci.*, **158**, 526 (1969).
67. G. N. Schrauzer and R. J. Windgassen, *Chem. Ber.*, **99**, 602 (1966).
68. J. Hanzlik and A. A. Vlcek, *Chem. Commun.*, **1969**, 47.
69. G. N. Schrauzer and R. J. Windgassen, *J. Amer. Chem. Soc.*, **88**, 3738 (1966).

70. T. Mizuta and T. Kwan, *Nippon Kagaku Zasshi*, **88**, 471 (1967); *CA*, **67**, 99537y (1967).

71. G. N. Schrauzer and R. J. Windgassen, *J. Amer. Chem. Soc.*, **89**, 3607 (1967).

72. G. N. Schrauzer and R. J. Windgassen, *Nature*, **214**, 492 (1967).

73. G. N. Schrauzer, E. Deutch, and R. J. Windgassen, *J. Amer. Chem. Soc.*, **90**, 2441 (1968).

74. B. C. McBride, J. M. Wood, J. W. Sibert, and G. N. Schrauzer, *J. Amer. Chem. Soc.*, **90**, 5276 (1968).

75. G. N. Schrauzer, *Accounts Chem. Res.*, **1**, 97 (1968).

76. R. J. P. Williams, *R.I.C. Rev.*, **1**, 13 (1968).

77. H. A. O. Hill, J. M. Pratt, and R. J. P. Williams, *Chem. Brit.*, **5**, 156 (1969).

78. B. Heil and L. Marko, *Chem. Ber.*, **101**, 2209 (1968).

79. N. S. Imyanitov and D. M. Rudkovskii, *Inst. Neftekhim. Protsessov*, **1963**, 30; *CA*, **60**, 9072e (1964).

80. D. Evans, G. Yagupsky, and G. Wilkinson, *J. Chem. Soc.*, A, **1968**, 2660.

81. C. O'Connor and G. Wilkinson, *J. Chem. Soc.*, A, **1968**, 2665.

82. G. Yagupsky, C. K. Brown, and G. Wilkinson, *Chem. Commun.*, **1969**, 1244.

83. W. Keim, *J. Organometal. Chem.*, **14**, 179 (1968).

84. M. Takesada, H. Tamazaki, and N. Hagihara, *Nippon Kagaku Zasshi.*, **89**, 1126 (1968).

85. L. Vaska and R. E. Rhodes, *J. Amer. Chem. Soc.*, **87**, 4970 (1965).

86. D. Evans, J. A. Osborn, and G. Wilkinson, *J. Chem. Soc.*, A, **1968**, 3133.

87. W. Strohmeier and W. Rehder-Stirnweiss, *J. Organometal. Chem.*, **18**, P28 (1969).

88. J. A. Osborn, F. H. Jardine, J. F. Young, and G. Wilkinson, *J. Chem. Soc.*, A, **1966**, 1711.

89. F. H. Jardine, J. A. Osborn, and G. Wilkinson, *J. Chem. Soc.*, A, **1967**, 1574.

90. J. F. Biellmann and H. Liesenfelt, *C.R. Acad. Sci.*, Paris, Ser. C, **263**, 251 (1966).

91. A. J. Birch and K. A. M. Walker, *J. Chem. Soc.*, C, **1966**, 1894.

92. J. F. Biellmann and M. J. Jung, *J. Amer. Chem. Soc.*, **90**, 1673 (1968).

92a. R. E. Harmon, J. L. Parsons, D. W. Cooke, S. K. Gupta, and J. Schoolenberg, *J. Org. Chem.*, **34**, 3684 (1969).

93. D. R. Eaton and S. R. Suart, *J. Amer. Chem. Soc.*, **90**, 4170 (1968).

94. Y. Chevallier, R. Stern, and L. Sajus, *Tetrahedron Lett.*, **1969**, 1197.

95. M. Montelatici, A. van der Ent, J. A. Osborn, and G. Wilkinson, *J. Chem. Soc.*, A, **1968**, 1054.

95a. H. van Bekkum, F. van Rantwijk, and T. van de Putte, *Tetrahedron Lett.*, **1969**, 1.

95b. R. L. Augustine and J. F. Van Peppen, *Chem. Commun.*, **1970**, 497, 571.

95c. J. T. Mague and J. P. Mitchener, *Chem. Commun.*, **1968**, 911.

96. A. J. Birch and K. A. M. Walker, *Tetrahedron Lett.*, **1966**, 4939.

97. L. Horner, H. Buthe, and H. Siegel, *Tetrahedron Lett.*, **1968**, 4023.

98. G. C. Bond and R. A. Hillyard, *Discussions Faraday Soc.*, **46**, 20 (1968).

99. A. S. Hussey and Y. Takeuchi, *J. Amer. Chem. Soc.*, **91**, 672 (1969).

99a. R. L. Augustine and J. F. Van Peppen, *Chem. Commun.*, **1970**, 495.

100. M. Montelatici, A. van der Ent, J. A. Osborn, and G. Wilkinson, *Proc. Eleventh Int. Conf. Coord. Chem.*, *Haifa, 1968*, 51.

101. J. P. Collman, N. W. Hoffman, and D. E. Morris, *J. Amer. Chem. Soc.*, **91**, 5659 (1969).

102. D. N. Lawson, M. J. Mays, and G. Wilkinson, *J. Chem. Soc.*, A, **1966**, 52.

102a. T. R. B. Mitchell, *J. Chem. Soc.*, **B**, **1970**, 823.

103. J. Chatt and S. A. Butter, *Chem. Commun.*, **1967**, 501.

104. W. S. Knowles and M. J. Sabacky, *Chem. Commun.*, **1968**, 1445.

105. L. Horner, H. Siegel, and H. Buthe, *Angew. Chem.*, **80**, 1034 (1968).

106. R. Stern, Y. Chevallier, and L. Sajus, *C.R. Acad. Sci., Paris*, **Ser. C**, **264**, 1740, (1967).

107. J. T. Mague and G. Wilkinson, *J. Chem. Soc.*, **A**, **1966**, 1736.

108. B. G. Rogachev and M. L. Khidekel, *Izv. Akad. Nauk SSSR, Ser. Khim.*, **1969**, 141.

109. J. A. Osborn, G. Wilkinson, and J. F. Young, *Chem. Commun.*, **1965**, 17.

110. R. D. Gillard, J. A. Osborn, P. B. Stockwell, and G. Wilkinson, *Proc. Chem. Soc.*, **1964**, 284.

111. B. R. James, F. T. T. Ng, and G. L. Rempel, *Inorg. Nucl. Chem. Lett.*, **4**, 197 (1968).

112. B. R. James and G. L. Rempel, *Can. J. Chem.*, **44**, 233 (1966).

113. B. R. James and G. L. Rempel, *Discussions Faraday Soc.*, **46**, 48 (1968).

114. I. Jardine and F. J. McQuillin, *Chem. Commun.*, **1969**, 477.

115. I. Jardine and F. J. McQuillin, *Chem. Commun.*, **1969**, 502, 503.

115a. B. Hudson, P. C. Taylor, D. E. Webster, and P. B. Wells, *Discussions Faraday Soc.*, **46**, 37 (1968).

116. A. Sacco, R. Ugo, and A. Moles, *J. Chem. Soc.*, **A**, **1966**, 1670.

117. P. Abley and F. J. McQuillin, *Chem. Commun.*, **1969**, 477.

118. N. S. Imyanitov and D. M. Rudkovskii, *Neftekhim.*, **3**, 198 (1963); *CA*, **59**, 7396c (1963).

119. G. Yagupsky and G. Wilkinson, *J. Chem. Soc.*, **A**, **1969**, 725.

120. K. W. Muir and J. A. Ibers, *J. Organometal. Chem.*, **18**, 175 (1969).

121. W. H. Baddley and M. S. Fraser, *J. Amer. Chem. Soc.*, **91**, 3661 (1969).

121a. G. Yagupsky, C. K. Brown, and G. Wilkinson, *J. Chem. Soc.*, **A**, **1970**, 1392.

122. G. G. Eberhardt and L. Vaska, *J. Catal.*, **8**, 183 (1967).

123. W. Strohmeier and T. Onada, *Z. Naturforsch.*, **B**, **24**, 461 (1969).

124. K. A. Taylor, *Advan. Chem. Ser.*, **70**, 195 (1968).

125. B. R. James and N. A. Memon, *Can. J. Chem.*, **46**, 217 (1968).

125a. J. P. Collman, M. Kubota, F. D. Vastine, J. Y. Sun, and J. W. Kang, *J. Amer. Chem. Soc.*, **90**, 5430 (1968).

126. M. Giustiniani, G. Dolcetti, M. Nicolini, and U. Belluco, *J. Chem. Soc.*, **A**, **1969**, 1961.

127. K. Krogmann and W. Binder, *Angew. Chem. Int. Ed.*, **6**, 881 (1967).

128. R. S. Coffey, *Chem. Commun.*, **1967**, 923.

129. M. Yamaguchi, *Kogyo Kagaku Zasshi*, **70**, 675 (1967); *CA*, **67**, 99542w (1967).

130. P. Abley and F. J. McQuillin, *Discussions Faraday Soc.*, **46**, 31 (1968).

131. R. W. Adams, G. E. Batley, and J. C. Bailar, *Inorg. Nucl. Chem. Lett.*, **4**, 455 (1968).

132. H. A. Tayim and J. C. Bailar, *J. Amer. Chem. Soc.*, **89**, 3420, 4330 (1967).

133. I. Jardine and F. J. McQuillin, *Tetrahedron Lett.*, **1966**, 4871.

134. L. Malatesta and R. Ugo, *J. Chem. Soc.*, **1963**, 2080.

135. H. van Bekkum, J. van Gogh, and G. van Minnen-Pathuis, *J. Catal.*, **7**, 292 (1967).

136. L. P. vant Hof and B. G. Linsen, *J. Catal.*, **7**, 295 (1967).

137. G. C. Bond and M. Hellier, *J. Catal.*, **7**, 217 (1967).

138. R. D. Cramer, R. V. Lindsey, C. T. Prewitt, and U. G. Stolberg, *J. Amer. Chem. Soc.*, **87**, 658 (1965).

138a. A. Miyake and H. Kondo, *Angew. Chem. Int. Ed.*, **7**, 631 (1968).

138b. A. Rejoan and M. Cais, *Proc. Eleventh Int. Conf. Coord. Chem.*, *Haifa, 1968*, 32.

138c. K. Sonogashira and N. Hagihara, *Bull. Chem. Soc. Jap.*, **39**, 1178 (1966).

139. R. H. Prince and K. A. Raspin, *Chem. Commun.*, **1966**, 156.

140. K. Ohno and J. Tsuji, *J. Amer. Chem. Soc.*, **90**, 99 (1968).

141. M. C. Baird, C. J. Nyman, and G. Wilkinson, *J. Chem. Soc.*, A, **1968**, 348.

142. J. Blum, J. Y. Becker, H. Rosenman, and E. O. Bergmann, *J. Chem. Soc.*, B, **1969**, 1000.

143. J. Chatt and J. M. Davidson, *J. Chem. Soc.*, **1965**, 843.

144. S. Bresadola, P. Rigo, and A. Turco, *Chem. Commun.*, **1968**, 1205.

145. M. A. Bennett and P. A. Longstaff, *J. Amer. Chem. Soc.*, **91**, 6266 (1969).

146. M. A. Bennett and D. L. Milner, *Chem. Commun.*, **1967**, 581.

147. J. B. Davis, *Petroleum Microbiology*, Elsevier, 1967.

148. D. J. Cardin, M. F. Lappert, and N. F. Travers, *Proc. Eleventh Int. Conf. Coord. Chem.*, *Haifa, 1968*, 821.

149. J. Blum and S. Biger, *Tetrahedron Lett.*, **1970**, 1825.

150. R. Huttel, J. Kratzer, and M. Bechter, *Chem. Ber.*, **94**, 766 (1961).

151. R. Huttel and H. Christ, *Chem. Ber.*, **96**, 3101 (1963).

152. M. Donati and F. Conti, *Inorg. Nucl. Chem. Lett.*, **2**, 343 (1966).

153. H. C. Volger, *Preprints Div. Petrol. Chem.*, *Amer. Chem. Soc.*, **14(4)**, F12 (1969).

154. M. Orchin, *Advan. Catal.*, **16**, 1 (1966).

155. D. Morelli, R. Ugo, F. Conti, and M. Donati, *Chem. Commun.*, **1967**, 801.

156. A. D. Ketley and J. Braatz, *Chem. Commun.*, **1968**, 169.

157. M. Donati and F. Conti, *Tetrahedron Lett.*, **1966**, 4953.

158. J. Tsuji, *Accounts Chem. Res.*, **2**, 148 (1969).

159. J. M. Davidson, R. G. Brown, and C. Triggs, *Preprints Div. Petrol. Chem.*, *Amer. Chem. Soc.*, **14(4)**, F6 (1969).

160. Y. Odaira, T. Oishi, T. Yukawa, and S. Tsutsumi, *J. Amer. Chem. Soc.*, **88**, 4105 (1966).

161. T. Rull, *Bull. Soc. Chim. Fr.*, **1964**, 2680.

162. W. H. Clement and M. Orchin, *Ind. Eng. Chem.*, *Prod. Res. Develop.*, **4**, 283 (1965).

163. F. Piacenti, S. Pucci, M. Bianchi, R. Lazzaroni, and P. Pino, *J. Amer. Chem. Soc.*, **90**, 6847 (1968).

164. S. D. Robinson and B. L. Shaw, *J. Chem. Soc.*, **1964**, 5002.

165. K. Moseley and P. M. Maitlis, *Chem. Commun.*, **1969**, 1156.

166. J. E. Lyons, *Chem. Commun.*, **1969**, 564.

167. J. Chatt, R. S. Coffey, and B. L. Shaw, *J. Chem. Soc.*, **1965**, 7391.

168. L. Vaska and J. W. DiLuzio, *J. Amer. Chem. Soc.*, **84**, 4989 (1962).

169. H. B. Charman, *J. Chem. Soc.*, B, **1967**, 629.

170. Y. M. Y. Haddad, H. B. Henbest, J. Husbands, and T. R. B. Mitchell, *Proc. Chem. Soc.*, **1964**, 361.

171. J. Trocha-Grimshaw and H. B. Henbest, *Chem. Commun.*, **1967**, 544.

172. J. C. Bailar and H. Itatani, *J. Amer. Chem. Soc.*, **89**, 1592 (1967).

173. G. Natta, P. Pino, and R. Ercoli, *J. Amer. Chem. Soc.*, **74**, 4496 (1952).

174. L. Marko, *Proc. Chem. Soc.*, **1962**, 67.

175. R. W. Goetz and M. Orchin, *J. Org. Chem.*, **27**, 3698 (1962).

176. C. L. Aldridge and H. B. Jonassen, *J. Amer. Chem. Soc.*, **85**, 886 (1963).

177. R. W. Goetz and M. Orchin, *J. Amer. Chem. Soc.*, **85**, 1549 (1963).

178. G. F. Emerson and R. Pettit, *J. Amer. Chem. Soc.*, **84**, 4591 (1962).

179. J. K. Nicholson and B. L. Shaw, *Proc. Chem. Soc.*, **1963**, 282.

180. A. Marbach and Y. L. Pascal, *C. R. Acad. Sci., Paris*, Ser. C, **268**, 1074 (1969).

181. G. Hata, H. Kondo, and A. Miyake, *J. Amer. Chem. Soc.*, **90**, 2278 (1968).

182. G. W. Parshall, *J. Amer. Chem. Soc.*, **90**, 1669 (1968).

183. W. Keim, *J. Organometal. Chem.*, **19**, 161 (1969).

183a. G. W. Parshall, *Accounts Chem. Res.*, **3**, 139 (1970).

184. J. P. Kleiman and M. Dubeck, *J. Amer. Chem. Soc.*, **85**, 1544 (1963).

185. A. C. Cope and R. W. Siekman, *J. Amer. Chem. Soc.*, **87**, 3272 (1965).

186. M. M. Bagga, P. L. Pauson, F. J. Preston, and R. I. Reed, *Chem. Commun.*, **1965**, 543.

187. P. E. Baikie and O. S. Mills, *Chem. Commun.*, **1966**, 707.

188. S. D. Ibekwe, B. T. Kilbourn, U. A. Raeburn, and D. R. Russell, *Chem. Commun.*, **1969**, 433.

189. R. J. Hodges and J. L. Garnett, *J. Phys. Chem.*, **72**, 1673 (1968).

190. R. J. Hodges and J. L. Garnett, *J. Catal.*, **13**, 83 (1969).

191. R. van Helden and G. Verberg, *Rec. Trav. Chim. Pays-Bas*, **84**, 1263 (1965).

192. J. M. Davidson and C. Triggs, *Chem. Ind.* (*London*), **1966**, 457.

193. J. M. Davidson and C. Triggs, *Chem. Ind.* (*London*), **1967**, 1361.

194. J. M. Davidson and C. Triggs, *J. Chem. Soc.*, A, **1968**, 1324.

195. H. C. Volger, *Rec. Trav. Chim. Pays-Bas*, **86**, 677 (1967).

196. C. F. Kohll and R. van Helden, *Rec. Trav. Chim. Pays-Bas*, **86**, 193 (1967).

197. Neth. patent 6,803,864 (1968) (to Halcon Intnl.).

198. E. Billig, C. B. Strow, and R. L. Pruett, *Chem. Commun.*, **1968**, 1307.

199. J. P. Collman and J. W. Kang, *J. Amer. Chem. Soc.*, **89**, 844 (1967).

200. M. Tsutsui, J. Aryoshi, T. Koyano, and M. N. Levy, *Advan. Chem. Ser.*, **70**, 266 (1968).

201. Y. Fujiwara, I. Moritani, M. Matsuda, and S. Teranishi, *Tetrahedron Lett.*, **1968**, 3863.

202. R. F. Heck, *J. Amer. Chem. Soc.*, **90**, 5518 (1968).

203. G. C. Bond and P. B. Wells, *Advan. Catal.*, **15**, 92 (1964).

204. S. Siegel, *Advan. Catal.*, **16**, 124 (1966).

205. R. L. Augustine and J. Van Peppen, *Ann. N.Y. Acad. Sci.*, **158**, 482 (1969).

206. S. Carra and R. Ugo, *Inorg. Chimica Rev.*, **1**, 49 (1967).

207. R. L. Burwell and K. Schrage, *J. Amer. Chem. Soc.*, **87**, 5253 (1965).

208. J. J. Phillipson, P. B. Wells, and G. R. Wilson, *J. Chem. Soc.*, A, **1969**, 1351.

209. H. Simon, O. Berngruber, and S. K. Erickson, *Tetrahedron Lett.*, **1968**, 707.

3

π-Allyl System in Catalysis

WILLI KEIM

Shell Development Company
Emeryville, California

59

I. Introduction

Over the last decade homogeneous and heterogeneous reactions based on π-allyl complexes as catalysts or catalytic intermediates have had a prodigious growth. Although often not well understood they represent an exciting development in coordination chemistry and are finding increasing interest in both the academic and industrial worlds. On the academic side this interest is explained by the unusual bond of the allyl ligand with a transition metal giving rise to a variety of unique complexes. In addition a number of new stoichiometric organic reactions, exemplified by allylic coupling (*1,2*) or allylic insertion (*3,4*), have demonstrated the usefulness of allyl complexes for the syntheses of organic compounds.

Industrial interest predominantly centers around the effectiveness of π-allyl complexes or intermediates for stereospecific oligomerization and polymerization reactions which are of immense commercial importance.

The subject of this chapter is to elaborate a general understanding of the diversified role that π-allyl systems can play in homogeneous catalysis involving alkenes and alkynes. Special emphasis has been given to present these reactions within their currently discussed mechanism. Many patent references have been included, and it is hoped that this chapter bridges academic and industrial research in this area.

Because of the limited space, many works of importance to the field of π-allyl complexes in catalysis are not mentioned. For instance, relevant topics such as allylic oxidation (*5–8*), allylic carbonylation (see Chapter 5 in this book), and π-allyl complexes in hydroformylation reactions (*9–10*) have not been discussed. More detailed accounts are listed in the references.

II. Oligomerization and Cooligomerization of Monoalkenes

A. π-ALLYLNICKEL HALIDE/LEWIS ACID CATALYSTS

There has been considerable interest in recent years in the oligomerization and cooligomerization of monoalkenes. Among the many systems studied reactions involving π-allyl complexes or π-allyl intermediates have attracted great industrial attention (*11–18*). Extensive pioneering research in applying π-allyl complexes to the dimerization of monoolefins has been carried out by Wilke, Bogdanovic, and co-workers (*19–23*). They obtained highly active catalysts of the composition (π-allyl Ni)AlX$_4$ (X = halide) by reacting π-allylnickel halides with Lewis acids such as aluminum halides. These systems converted about 30,000 moles of propylene per mole of nickel. The direction of this dimerization toward linear or branched products could be controlled by the addition of phosphine ligands (20). Table 1 shows the effect of various phosphines.

TABLE 1

Ligand	n-hexenes	2-methylpentenes	2,3 dimethylbutenes
none	21.3	73.2	5.5
ϕ_3P	21.6	73.9	4.5
(i-pr)(t-but)$_2$P	1.0	81.5	17.3
(i-pr)$_3$P	1.8	30.3	67.9

Catalysts with strongly basic phosphines yielded predominantly a mixture of 2,3-dimethylbutenes and 2-methylpentenes. The structure of the dimers formed is not significantly affected by modification of the allyl group or of the Lewis acids. The addition of strongly coordinating ligands such as CO, cyclooctadiene-1,5, or excess phosphine led to inactive complexes of the form (1).

$$\left| CH \begin{array}{c} CH_2 \\ CH_2 \end{array} Ni \begin{array}{c} L_1 \\ L_2 \end{array} \right|^+ \quad AlX_4^-$$

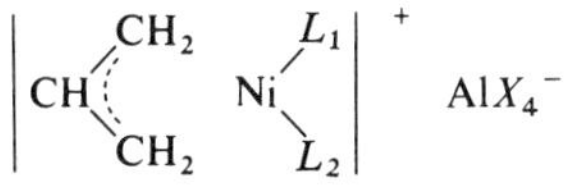

(a) $L_1 = L_2 = $ CO, PR_3, cyclooctadiene-1,5;

(b) $L_1 = $ CO, $L_2 = PR_3$

(1)

It has been concluded (24) that these ligands block available coordination sides, thus preventing coordination of olefins.

Besides catalysts based on π-allyl complexes, many systems employing Ziegler-Natta type catalysts have been disclosed for the oligomerization of alkenes. (25–31). The products obtained by adding various phosphines to these Ziegler-Natta type catalysts parallel those listed in Table 1. It is worthwhile to point out that π-allyl complexes without a cocatalyst (Lewis acid) lack catalytic activity (32).

Both the Ziegler-Natta type catalysts and the π-allyl based catalysts form similar catalytically effective intermediates. Wilke and Bogdanović successfully demonstrated that Ziegler-Natta type catalysts yielded π-allyl complexes which were also active for the propylene dimerization (*20*). From the Ziegler-Natta type catalyst nickelacetylaceteonate/ethylaluminum sesquichloride which dimerized cyclooctene to (**2**) and (**3**), complex (**4**) was isolated in high yields. Interestingly, upon addition of aluminum halides to (**4**) a very active dimerization catalyst was obtained, which also dimerized cyclooctene to (**2**) and (**3**).

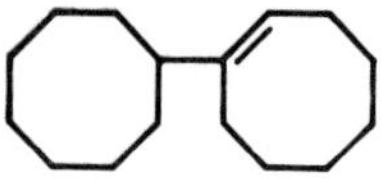

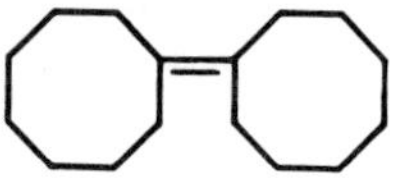

B. MECHANISM OF ALKENE DIMERIZATION

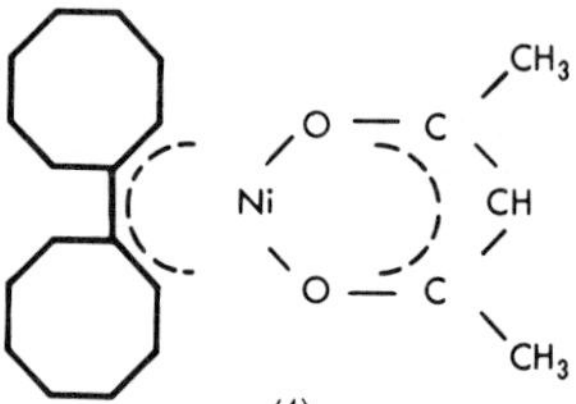

The mechanism of the dimerization of monoolefins is not well understood and various routes have been postulated. Ewers (*33*) speculates that π-complexes of zerovalent nickel represent the active species. For the dimerization of ethylene by nickelocene, Tsutsui and co-workers (*34*) evoke a mechanism in which two ethylene molecules coordinate to nickelocene prior to their dimerization. Ketley et al. (*35*) also favor a mechanism of π-complexing for the dimerization of propylene. He could demonstrate that π-allyl complexes formed during the reaction are catalytically inactive. A frequently discussed mechanism rests on a metal-hydride intermediate (*36–39*). A hydride mechanism is also postulated by Wilke and Bogdanović (*20,40*). Detailed information on the reaction steps is lacking. An equilibrium of the type shown in Eq. (1) may account for the formation of a hydride-species which reacts with an additional propylene molecule.

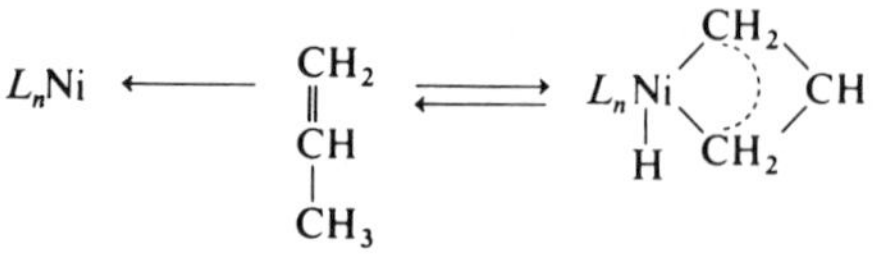

The direction of the hydride addition is determined by the nature of ligand L. In an elegant experiment Bogdanović has demonstrated the influence of the ligand (40). Various nickel-hydride complexes were added to propylene and the resulting alkyl complexes decomposed by reaction with iodine according to Eq. (2).

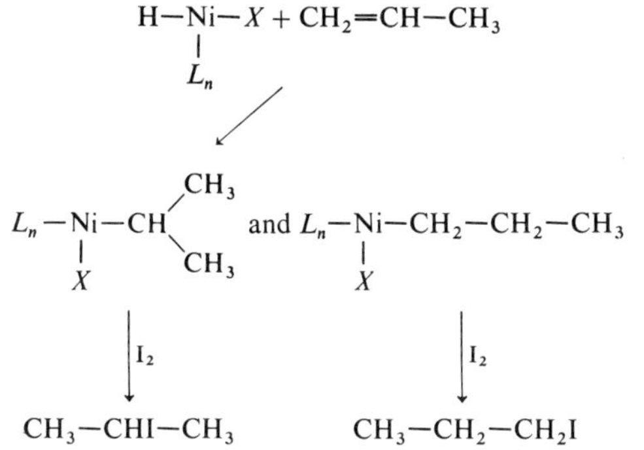

$$H-Ni-X + CH_2=CH-CH_3$$
$$L_n$$

$$L_n-Ni-CH\underset{X}{\overset{CH_3}{\diagdown_{CH_3}}} \quad \text{and} \quad L_n-Ni-CH_2-CH_2-CH_3 \qquad (2)$$
$$X$$

$$\downarrow I_2 \qquad\qquad \downarrow I_2$$

$$CH_3-CHI-CH_3 \qquad CH_3-CH_2-CH_2I$$

Paralleling the results of Table 1, the more basic phosphine gave predominantly 2-iodopropane. Based on these results a general mechanism for the dimerization of propylene via a π-allyl complex/Lewis acid is shown in Fig. 1.

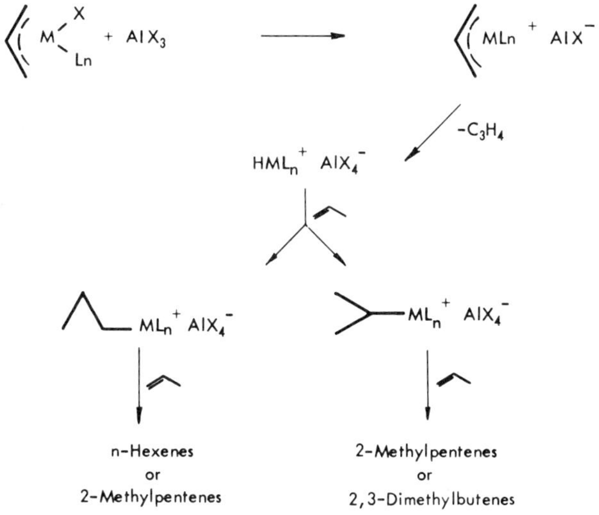

Fig. 1

Nothing is known about the fate of the π-allyl group, but it is conceivable that it has an important function for stabilizing the system.

III. Oligomerization and Cooligomerization of Alkynes

Although the oligomerization of alkynes by a wide variety of catalysts is a well-known reaction, very few papers have been published dealing with the catalysis of alkynes via π-allyl complexes.

Oberkirch (19,41) describes attempts to synthesize hexamethylbenzene chromium by reacting tris-π-allyl chromium with butyne-2. This reaction resulted in the formation of substantial amounts of hexamethylbenzene, but he was unable to isolate an intermediate complex.

An interesting reaction path for the trimerization of alkynes involving a π-allyl intermediate has been indicated by Hübel (42,43). Cobalt carbonyl reacts with one alkyne forming the well-characterized acetylene-dicobalt-hexacarbonyl complex (5). Complex (5) reacts with an additional acetylene molecule giving (6), which has not been isolated. Complexes similar to (6), however, have been described (43–45). Complex (6) yields (7) on reaction with a third molecule of acetylene; the structure of (7), determined by X-ray analysis (46), contains a six-carbon chain which is bonded via a π-allylic system to different cobalt atoms. Upon heating (7) produces a cyclic trimer.

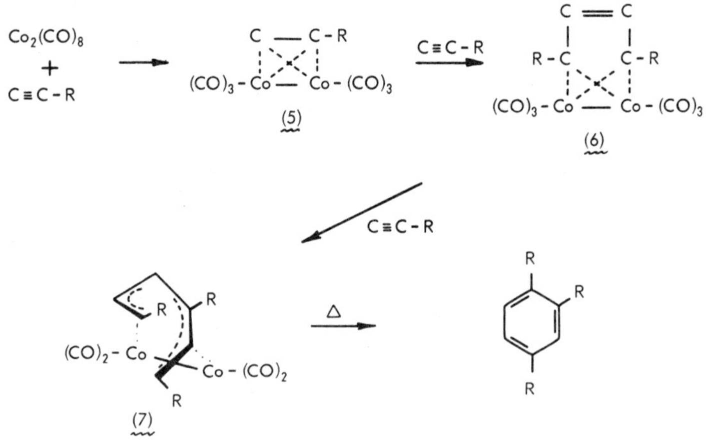

Recently Shier (47,48) reported the cooligomerization of allene with methylacetylene giving substituted vinylacetylenes. The reaction mechanism discussed involves the π-allyl intermediate (8).

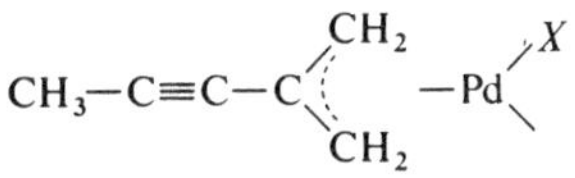

$$CH_3-C\equiv C-C\begin{matrix} CH_2 \\ \\ CH_2 \end{matrix}-Pd\begin{matrix} X \\ \\ \end{matrix}$$

(8)

IV. Oligomerization and Cooliogmerization of 1,3-Dienes

This section describes the use of various π-allyl complexes or π-allyl intermediates for the oligomerization of 1,3-dienes, such as butadiene, isoprene, and piperylene. These catalysts provide a facile route to a vareity of novel cyclic and open chain oligomers.

A. CYCLIC AND LINEAR OLIGOMERIZATION OF BUTADIENE

1. Cyclic Oligomerization of Butadiene

Butadiene, the simplest 1,3-diene, is a large scale industrial chemical of low cost. New synthesis routes will offer even lower cost butadiene and certainly promote wider chemical attention. Perhaps the biggest single incentive to its broader chemical application besides polymers is the development of catalytic routes to C_4, C_5, C_6, C_8, C_{10}, and C_{12} membered ring compounds (Fig. 2).

These materials offer a new series of fundamental starting materials for organic syntheses. They may be converted to dicarboxylic acids or lactams

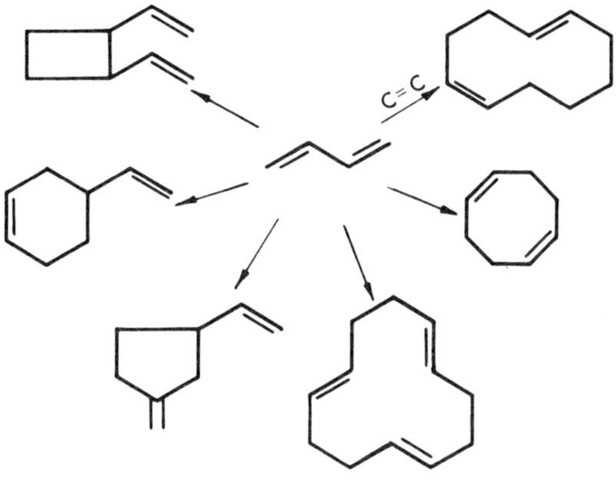

Fig. 2

known to be valuable initial materials for plastics, polyamides, and poly-esters. Today, cyclooctadiene-1,5* and cyclododecatriene-1,5,9† are commercial products (49) and large scale production at moderate cost awaits only suitable market demand. There are indications of a rapidly growing market exemplified in nylon-12 which is based on cyclododecatriene-1,5,9 (50–52).

The elegant syntheses of COD and CDT pioneered and thoroughly investigated by Wilke and co-workers have been reviewed extensively (19,53–60) and Fig. 3 outlines briefly the essential steps leading to COD and CDT.

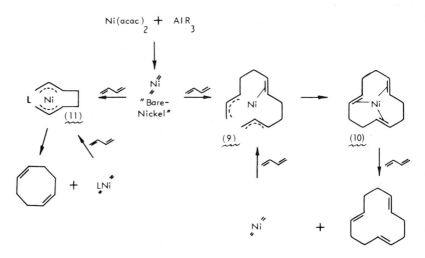

Fig. 3

The key to elucidating the mechanism of cyclic oligomerization is the use of " bare " nickel complexes as catalysts. The " bare " nickel formed by reducing $Ni(acac)_2$ with alkylaluminum compounds trimerizes butadiene forming the isolable intermediate (9), which upon ring closure gives the well characterized complex (10). Incoming butadiene displaces CDT in (10) thus completing the catalytic cycle. In the presence of an electron donor the reaction path changes and only two butadiene molecules co-ordinate and form complex (11), which also has been isolated. The formation of COD from (11) arises by a cyclization and displacement step as

* cyclooctadiene—1,5≡COD.
† cyclododecatriene≡CDT.

depicted. Suitable choice of the donor can ensure a yield of $>95\%$ COD. Various transition metals such as cobalt (*57*), iron (*61,99*), chromium (*19*), titanium (*63*), and palladium (*62*) also catalyze the cyclization of butadiene to COD and CDT, other numerous examples of catalytic systems are listed in the patent literature.

The catalytic dimerization of butadiene to *cis*-1,2-divinylcyclobutane by zerovalent nickel complexes has been described by Heimbach and Brenner (*64*). The yield of the four membered ring is a function of the butadiene conversion. At high butadiene conversion mainly cyclooctadiene-1,5 is formed. Conversely, by destroying the catalyst before all the butadiene has reacted, good yields of divinylcyclobutane can be isolated. Heimbach explains these findings by a Cope-rearrangement catalyzed via zerovalent nickel complexes. Indeed, divinylcyclobutane can easily be converted to cyclooctadiene-1,5 by employing zerovalent nickel complexes. Surprisingly, with the proper choice of reaction conditions this reaction also reconverts divinylcyclobutane to butadiene ($\sim 75\%$ yield). A series of equilibria (*3*) has been proposed (*64*) to account for the complexity of this butadiene dimerization by zerovalent nickel complexes.

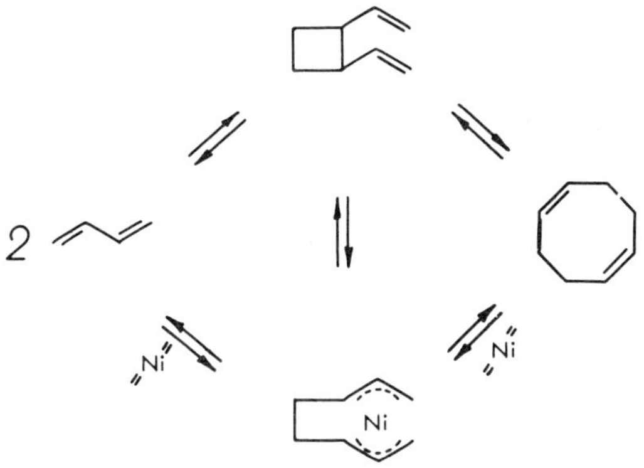

A catalyst exhibiting remarkable specificity to form only one cyclic product, namely vinylcyclohexene, has been described by Candlin (*65*). The complexes $Fe(CO)_2(NO)_2$ or π-allyl-$Fe(CO)_2NO$ convert butadiene in $>95\%$ conversion to 99.5% 4-vinylcyclohexene. The percentage conversion was generally higher with π-allyl-$Fe(CO)_2NO$ compared with $Fe(CO)_2(NO)_2$, indicating that the reaction may proceed through a π-allyl intermediate. A mechanism applying a dipyridyl iron catalyst for the cyclic dimerization of butadiene has been discussed by Yamamoto (*66*). Deuterated butadiene-1,1,4,4-d_4 was used and exclusively cyclooctadiene-

3,3,4,4,7,7,8,8-d$_8$ and 4-vinylβ, β-d$_2$-cyclohexene-3,3,5,5,6,6-d$_6$ were forced
These results are in complete agreement with the mechanism proposed by
Wilke and co-workers as shown in Eq. (4).

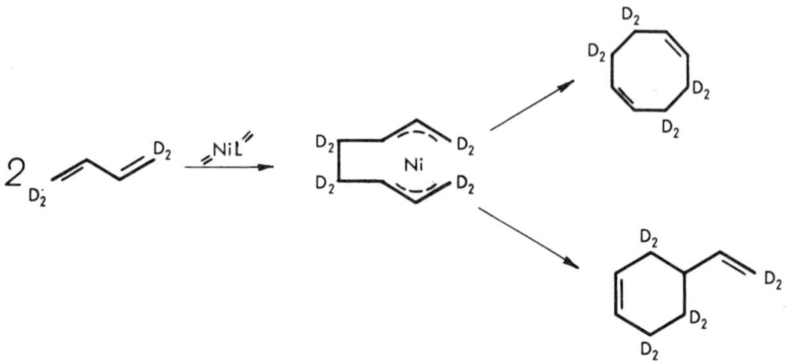

$$\tag{4}$$

The synthesis of a five-membered ring butadiene dimer has been reported
by Müller and co-workers (57) who isolated 1-vinyl-3-methylenecyclopen-
tane in preparative interesting amounts. No further information on the
reaction or the route of formation of this product has appeared.

2. Linear Oligomerization of Butadiene

Interestingly, various catalyst systems have been described which form
both cyclic and linear oligomers from butadiene. The synthesis of linear
oligomers involves a transfer of a hydrogen atom. A catalyst derived from
Fe(acac)$_3$ and triethylaluminum converts butadiene into n-dodecatetraene-
1,3,6,10, 3-methylheptatriene-1,4,6, cyclooctadiene-1,5, cyclododecatriene 1,5,9, and 4-vinylcyclohexene (67). The iron-based catalysts are very
complex. Variation of ligands, mole ratio of aluminum to iron, and
traces of impurity can markedly change the product distribution (68–72).
Catalysts producing most selectively trans-3-methylheptatriene-1,4,6 are
based on cobalt (19,57,71,73–76). On the basis of the dimerization [Eq. (5)]
of butadiene-1,1,4,4-d$_4$, the transfer of the hydrogen from the one to the
four position of butadiene has been proposed (72,92).

$$D_2C=CH-CH=CD_2 \longrightarrow \underset{\underset{CD_3}{|}}{D_2C=CH-CH-CD}=CH-CH=CD_2 \tag{5}$$

Natta and co-workers (77) have been able to isolate a complex Co C$_{12}$H$_{19}$
(**12**) which catalyzes the dimerization of butadiene to 3-methylheptatriene-
1,4,6.

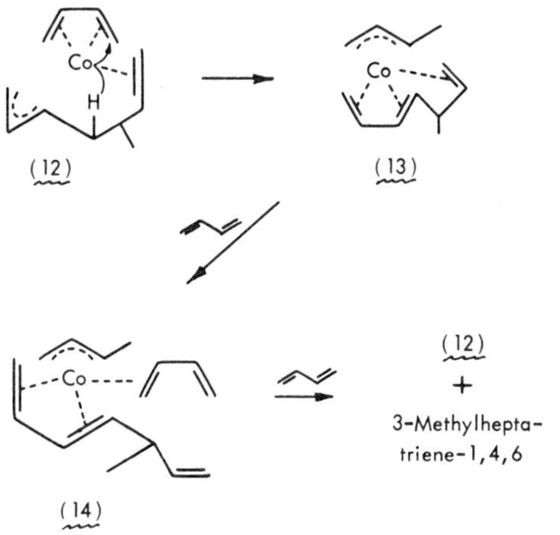

An X-ray analysis (78) confirmed the structure **(12)** proposed by Natta. From the knowledge of the molecular structure and from the chemical evidence accumulated, it has been suggested (78) that a hydrogen shift in **(12)** (indicated by arrow) yields the intermediate **(13)**. Incoming butadiene displaces the vinyl group of the C_8-chain. The driving force resides in the greater tendency for the coordination of butadiene than an isolated vinyl group. The coordinated butadiene in **(14)** adds to the π-crotyl group. By an additional coordination of butadiene **(14)** returns to **(12)** with concomitant displacement of 3-methylheptatriene, thus completing the catalytic cycle. Steric reasons are envoked to explain the indicated addition leading to the branched product. The 3-methylheptatriene-1,4,6 itself is a conjugated 1,3-diene and accordingly higher oligomers can also be formed (57,79).

The linear oligomerization of butadiene has also been described for rhodium (80) and palladium (62,81). In the case of rhodium the linear dimer n-octatriene-2,4,6 is the principal product. Bis-π-allyl palladium converts butadiene to a mixture of trimers which upon hydrogenation yield predominantly n-dodecane. A conceivable explanation rests on the postulation that a palladium complex analogous to the nickel complex **(9)** in Fig. 3 is formed. However, the size of the palladium atom precludes the ring closure of the C_{12}-chain to give CDT (19).

The complex π-allylpalladium acetate, also reacts with butadiene yielding linear dimer octatriene-1,3,7, and trimer, n-dodecatetraene-1,3,6,10 (82). With addition of triphenylphosphine it proved possible to promote the formation of the dimer. Intermediate complexes, in these reactions have been isolated and characterized. They shed some light on the oligo-merization steps involved. π-Allylpalladium acetate reacts with butadiene to form **(15)** via addition of the allyl fragment to the diene molecule (*81–84*).

By allowing the reaction to continue displacement of the C_7-moiety occurs and complex **(16)** is formed. However, if the reaction of **(15)** with butadiene is carried out in the presence of acetic acid (see Section V) the complex **(17)** can be isolated.

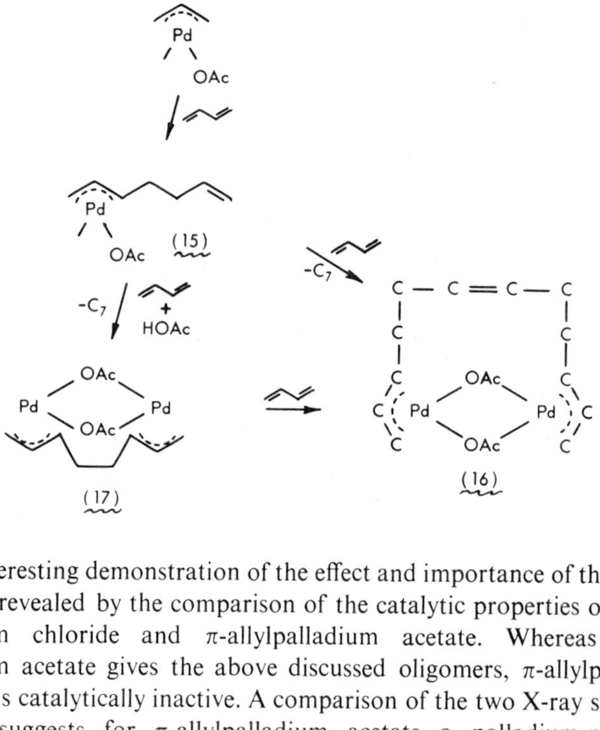

An interesting demonstration of the effect and importance of the anionic group is revealed by the comparison of the catalytic properties of π-allyl-palladium chloride and π-allylpalladium acetate. Whereas π-allyl-palladium acetate gives the above discussed oligomers, π-allylpalladium chloride is catalytically inactive. A comparison of the two X-ray structures (*85–86*) suggests for π-allylpalladium acetate a palladium-palladium interaction. One is tempted to reduce the difference in their catalytic properties to this metal-metal interaction.

Finally, an interesting oligomerization of butadiene, to a tetramer, which also may be explained by considering a π-allyl mechanism, has been reported (*87*). *Trans, trans, trans-n-*hexadecatetraene-1,6,10,14 is produced by electrolytic reduction of a solution containing nickel complexes and butadiene.

B. CYCLIC AND LINEAR OLIGOMERIZATION OF 1,3-DIENES OTHER THAN
   BUTADIENE

The catalytic oligomerization of the simplest 1,3-diene, butadiene, has
been extended to other various 1,3-dienes such as isoprene, piperylene,
2,3-dimethylbutadiene, chloroprene, cyclohexadiene-1,3 and octatriene-
1,3,7. Many examples have been described mainly in the patent literature
(58,88–94). In general it can be stated that catalysis directed at open chain
or cyclic oligomers is complicated by the concomitant formation of the
various, possible isomers. The products and their distribution can be
altered by modifying the catalysts with ligands such as phosphines or
phosphites (58,95). Certainly, this type of oligomerization has not as yet
been thoroughly investigated. Higher specificity and selectivity may be obtainable by variation of the transition metal and the ligand field. Candlin
(65) reports a highly specific dimerization of isoprene to (18) and (19) by
π-allyl-Fe(CO)$_2$NO.

(18)   (19)

The dimerization of octatriene-1,3,7 by bis-π-allylpalladium (97) gives
high yields of three products [73% hexadecapentaene-1,6,9,11,15 and two
products of undetermined structure (27%) which upon hydrogenation
yield predominantly n-dodecane (98% selectivity)]. This reaction can be
generally applied for linear dimerization reactions of conjugated dienes
according to Eq. (6).

$$\text{octatriene-1,3,7} + \text{/\!\!\diagup\!\!\diagdown\!\!\diagup\!\!\diagdown} R \longrightarrow C_{12}-R. \tag{6}$$

Table 2 lists some of the hydrocarbons which have been obtained upon
hydrogenation of the dimerization products.

TABLE 2

HYDROCARBONS OBTAINED UPON HYDROGENATION OF DIMERIZATION PRODUCTS

Octatriene-1,3,7 + butadiene ⟶	n-dodecane
Octatriene-1,3,7 + heptatriene-1,3,6 ⟶	n-pentadecane
Octatriene-1,3,7 + dodecatetraene-1,3,7,11 ⟶	n-eicosane
Octatriene-1,3,7 + hexadecapentaene-1,6,9,11,15 ⟶	n-tetracosane
2 Dodecatetraene-1,3,7,11 ⟶	n-tetracosane

The addition of triphenylphosphine leads to a complex mixture of linear and branched products. In the case of the dimerization of octatriene-1,3,7 with a catalyst consisting of bis-π-allyl palladium and triphenylphosphine (ratio, 1:1) the predominant product formed has the structure 4-butenyl-3-dodecatetraene-1,6,8,11.

A remarkable selectivity to mainly 2,7-dimethyldecatriene-1,4,7 was also observed in dimerizing isoprene with bis-π-allyl palladium containing traces of alcohol (*98*).

Summarizing it may be stated that the cyclic and linear oligomerization of 1,3-dienes other than butadiene still remains an area of challenging, research.

C. COOLIGOMERIZATION OF 1,3-DIENES WITH ALKENES, ALLENES, AND ALKYNES

Much work has been done on the cooligomerization of 1,3-dienes with alkenes, alkynes, allylidenimine, and acrylic esters using π-allyl complexes. In particular, substantial industrial research efforts have been devoted to an understanding and commercial application of these cooligomerizations. They can be divided into two classes:

1. Reactions resulting in cyclic cooligomers.
2. Reactions yielding open-chain cooligomers.

The first class includes Diels-Alder type reactions. Recent examples have been supplied by Wilke, Heimbach, and co-workers who found that two molecules of a 1,3-diene can react with alkenes (*58*), alkynes (*3,96,100–102*), allylidenimine (*3*), and acrylates (*3*) to form ten and twelve-membered ring compounds. A reaction mechanism involving π-allyl intermediates has been proposed (*94*) and is depicted in Fig. 4. (Only the cooligomerization of butadiene and alkenes is shown.)

Two butadiene molecules initially dimerize yielding complexes (**20**) and (**21**), the formation of which depends strongly on the reaction conditions such as temperature, solvent, and added ligands. The complexes (**20**) and (**21**) react with a monoalkene yielding a C_{10}-chain bonded to nickel by a carbon-σ-bond and a π-allylic or σ-allylic group (**22**) and (**23**), respectively. The synthesis of cyclodecadiene can easily be derived from the π-allylic form (**22**), whereas the σ-allylic form (**23**) yields the open-chain product. Indeed, cyclic- and open-chain products are formed during the reaction, but by selecting the right reaction conditions the course of the reaction can be directed to predominantly one product (*96,101–103*).

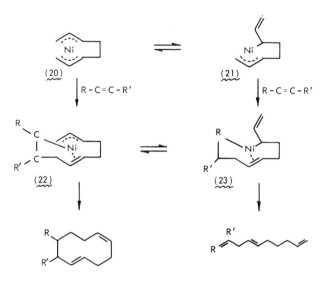

Fig. 4

The codimerization of 1,3-dienes with monoolefins yielding cooligomers of 1,4-diene structure is of great commercial importance because of its utility as the third component in ethylene-propylene-diene terpolymers (EPDM) (*104–109*).

The most frequently studied reaction is the codimerization of butadiene and ethylene. Substitution of the double-bond of a monoolefin appears to lower its reactivity (propylene less reactive than ethylene). The reactivity of the 1,3-dienes depends to a large extent on their structure. By codimerizing ethylene with various 1,3-dienes it appears that ethylene generally adds to the more highly substituted double bond of the 1,3-dienes.

A mechanism for this reaction involving π-allyl complexes as intermediates has been postulated in many papers (*61,80,113–119*). The addition of a metal hydride species to the 1,3-diene has been proposed to account for the π-allylic intermediate which may exist in two isomeric forms, anti-syn- (**24**) and (**25**). The selective reaction of either the syn- or anti-form with ethylene explains the stereochemistry of the 1,4-dienes produced. Palladium (*116*) catalyzed reactions appear to give predominantly the *trans* isomer, while iron (*61,113,117–119*), cobalt (*114*), nickel (*110*), or rhodium (*80,119*) based catalyst yield mainly *cis* isomers or mixtures of *cis* and *trans*.

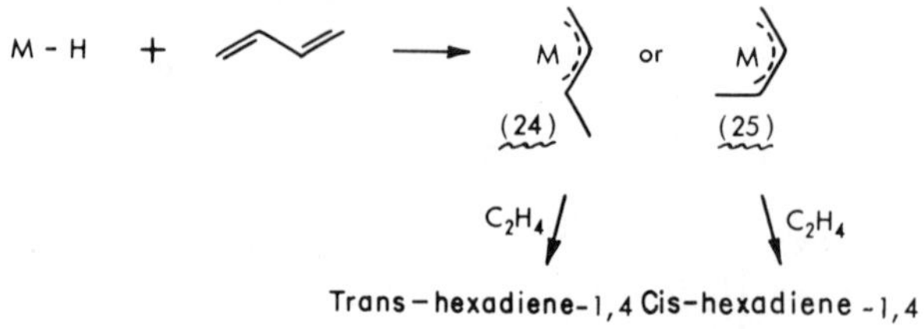

$$M-H \quad + \quad \diagup\!\!\diagdown\!\!\diagup \quad \longrightarrow \quad M \underset{(24)}{\diagdown} \quad or \quad M \underset{(25)}{\diagdown}$$

$$C_2H_4 \swarrow \qquad\qquad \searrow C_2H_4$$

Trans–hexadiene-1,4   Cis–hexadiene -1,4

The coupling of a 1,3-diene with an olefinic compound containing a functional group such as acrylates (*57,120*) or acrylonitrile (*121*) or acrolein has also been reported. Müller and co-workers (*57*) report the codimerization of vinyl aromatic compounds with butadiene to aryl-n-decatrienes. [Eq. (7)].

$$\phi-CH=CH_2 + 2\,CH_2=CH-CH=CH_2 \quad\longrightarrow$$

$$\phi-CH=CH-CH_2-CH=CH-CH_2-CH_2-CH \qquad (7)$$
$$\overset{\|}{\underset{\underset{CH_3}{|}}{CH}}$$

These compounds could be of technical interest for the detergent industry. One mole of the nickel complex used gave 500–1000 kg phenyl decatriene.

V. Telomerization Reactions of 1,3,-Dienes

As discussed previously, nickel containing catalysts give usually cyclic oligomers with 1,3-dienes. However, in alcoholic media the same catalysts catalyze a linear dimerization to *n*-octatriene (*57,122–124*). Recently Heimbach (*94,125*) describes that the hydrogen transfer reaction giving *n*-octatrienes are also effected by dialkylamines. In alcohol solutions different isomers of *n*-octatriene were formed, but with morpholine more than 95% octatriene-1,3,6 has been synthesized. Depending on the reaction condition an alkylation of the amino group was also observed yielding (*26*)

$$\diagup\!\!\!\diagdown N-(CH_2-CH=CH-CH_2)_2H \quad (26)$$

A. TELOMERIZATION OF BUTADIENE

A novel linear 1,3-diene dimerization reaction accompanied by terminal addition of nucleophilic reagents containing OH- and NH-groups has been discovered and extensively explored by Smutny and co-workers at Shell (*82,126–135*). For instance, butadiene and a nucleophile react at high conversion and selectivity according to Eq. (8).

$$2CH_2=CH-CH=CH_2 + \Psi'-H \xrightarrow{\text{catalyst}}$$

$$\Psi'-CH_2-CH=CH-(CH_2)_3-CH=CH_2 \ (>95\%) \qquad (8)$$

$$+$$

$$CH_2=CH-CH-(\Psi')-(CH_2)_3-CH=CH_2 \ (<5\%)$$

$\Psi-H$ = alcohols, amines, and acids.

The reaction is remarkable, because the chain cleanly terminates after oligomerizing only two butadiene units giving nearly quantitative yields of one product with no significant by-products. Most of Group VIII transition metals are active catalysts, but the majority of data were obtained using palladium based catalysts.

The reaction is very rapid and more than 1000 moles of the nuclophile can be converted per mole of catalyst (*127*). A number of nucleophiles have been studied.

Alcohols: methanol, ethanol, isopropanol, butanol, allylalcohol, benzyl alcohol, phenol, *p*-methoxyphenol, *p*-chlorophenol, 2,6-dimethyl-phenol, α,naphthol

Amines: *n*-butylamine, diethylamine, aniline, morpholine, piperidine, diisopropylamine

Acids: acetic, butryic, methacrylic, pivalic.

B. Telomerization of Isoprene and Piperylene

Variation of the diene has also proved successful. When phenol and piperylene were reacted a mixture of phenoxydecadienes were obtained. The predominant product was 2-phenoxy-(5-methyl)-nonadiene-3,8 (*97*). Isoprene also participated in this reaction, however, the reaction was complex and no predominant product was formed.

C. Cotelomerization

An interesting extension of this reaction is the cotelomerization of butadiene, octatriene-1,3,7 and a nucleophile (*127,128*) as indicated in Eq. (9) ("cross" oligomerization).

$$CH_2{=}CH{-}CH{=}CH_2$$
$$+\Psi{-}H$$
$$+CH_2{=}CH{-}CH{=}CH{-}(CH_2)_2{-}CH{=}CH_2$$

$$\xrightarrow{\text{Pd-catalysts}} \quad \begin{array}{c} \Psi{-}CH_2{-}CH{=}CH{-}(CH_2)_3 \\ | \\ CH_2{=}CH{-}(CH_2)_2{-}CH{=}CH \end{array} \quad (9)$$

The "cross" oligomerization of butadiene, n-dodecatetraene-1,3,7,11 and nucleophile also has been described ($127,128$).

An interesting outgrowth of this reaction was the discovery, that the same catalyst which brings about the telomerization of 1,3-diene also catalyzes a highly selective degradation reaction (127) according to Eqs. (10) and (11).

$$\Psi{-}CH_2{-}CH{=}CH{-}(CH_2)_3{-}CH{=}CH_2$$
$$\longrightarrow \quad \Psi{-}H + CH_2{=}CH{-}CH{=}CH_2{-}(CH_2)_2{-}CH{=}CH_2 \quad (10)$$

or

$$\Psi{-}CH_2{-}CH{=}CH{-}(CH_2)_3{-}CH{=}CH{-}(CH_2)_2{-}CH{=}CH_2$$
$$\longrightarrow \quad \Psi{-}H + CH_2{=}CH{-}CH{=}CH{-}(CH_2)_3{-}CH{=}CH{-}(CH_2){-}CH{=}CH_2$$
$$(11)$$

For instance, if the reaction products obtained from converting phenol, butadiene, and bis-π-allyl palladium are aged in air before distillation, pure 1-phenoxyoctadiene-2,7 could be removed by distillation. Conversely, when the distillation was carried out before aging had occured, octatriene-1,3,7 was obtained in nearly quantitative yields. In addition to n-octatriene-1,3,7,n-dodecatetraene-1,3,7,11 and n-hexadecapentaene-1,3,7,11,15 have been synthesized in high conversions and selectivities. This novel reaction leads to a variety of interesting organic compounds based on 1,3-dienes and a nucleophile. It provides a convenient path to straight chain polyolefins with high stereoselectivity.

D. Mechanism of Telomerization of 1,3-Dienes

Principally two mechanisms have been postulated to account for the formation of telomers and dimers by the same catalyst. A reaction mechanism based on the "Wilke-type" intermediate (11) has been entertained by Feldman and co-workers (123), Heimbach (125), Takahashi, Hagihara, and co-workers ($133,134$) and Medema and co-workers (82). In the fashion indicated by arrows the formation of the telomers and dimers has been explained from (27), (28), and (29).

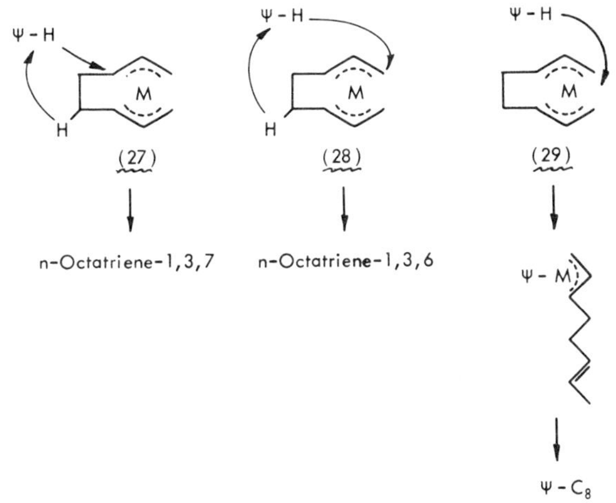

Ψ - H
M
H
(27)
(28)
(29)
n-Octatriene-1,3,7
n-Octatriene-1,3,6
Ψ - M
Ψ - C8

All attempts, however, to isolate catalytic intermediates of type (**11**) and to shed some light on the individual steps occurring have been unsuccessful so far.

A reaction mechanism based on a synergistic effect of two metals has also been discussed as a possible reaction route (*127,97*). In this scheme, a well characterized complex (**30**) which has been isolated from the reaction mixtures of butadiene/phenol/bis-π-allyl palladium or butadiene/acetic acid/bis-π-allyl palladium, is thought to be the key intermediate.

(30)
Pd
Ψ
+ ΨH
(31)
(30) + ΨC8
H - Ψ
Ψ = OØ
= Ac

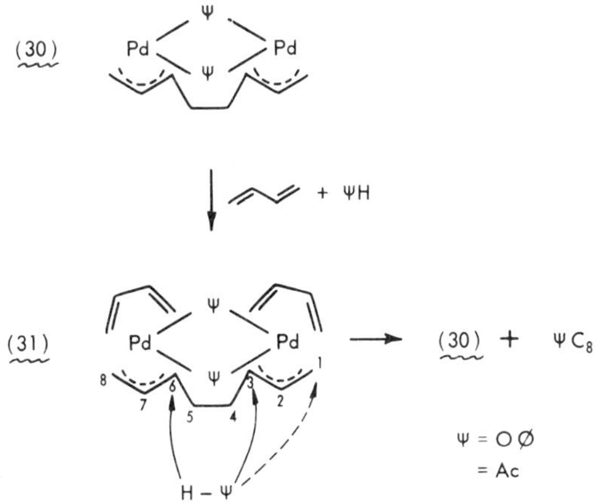

Incoming butadiene coordinates to (**30**) thus forming (**31**). The strong donor butadiene then transforms the pure π-allyl form into either π-σ-allyl form or σ-σ-allyl form. Simultaneously the coordination of the butadiene loosens the C_8-chain thus enabling an attack of the nucleophile. The proton goes on carbon atom 6 as evidenced from labeled experiments (*123,127, 132*). The acetato group has the choice between carbon atom 1 and 3, thus forming the two telomers observed [see also Eq. 8)]. Because of the nucleophilic attack occurs generally at the terminus of a π-allyl system, the 1-derivative is the major product. After the nucleophile has been added to the C_8-chain the two coordinated butadiene molecules dimerize returning to complex (**30**), thus closing the catalytic circle. In the absence of an excess of butadiene the same intermediate can also account for the degradation of the telomer to the diene-dimer and the nucleophile as shown in (**32**).

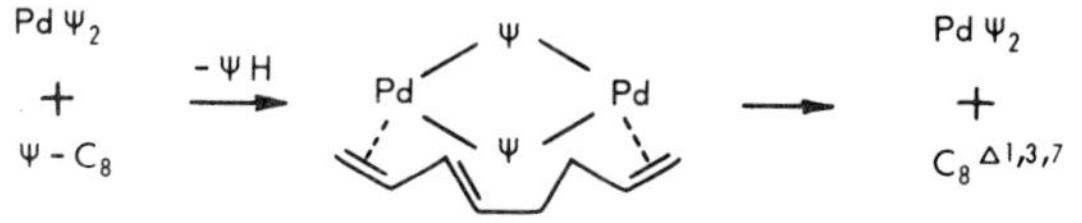

(32)

$$\Psi = O\emptyset$$
$$= OAc$$

Interestingly, when butadiene is reacted with π-allyl palladium acetate in the absence of a nucleophile (*82*) the reaction path changes to the formation of *n*-dodecatetraene-1,3,6,10 only.

Allusion should also be made to the importance of the nature of the bridging ligand. The complex π-allyl palladium chloride is inactive for the telomerization reaction. Exchange of the chloro-bridge for an acetato group leads to a very active catalyst. Another example which very nicely demonstrates the influence of the bridging ligand is based on complex (*33*). The latter does not catalyze the telomerization reaction. The only reaction occurring is the insertion of butadiene yielding (**34**).

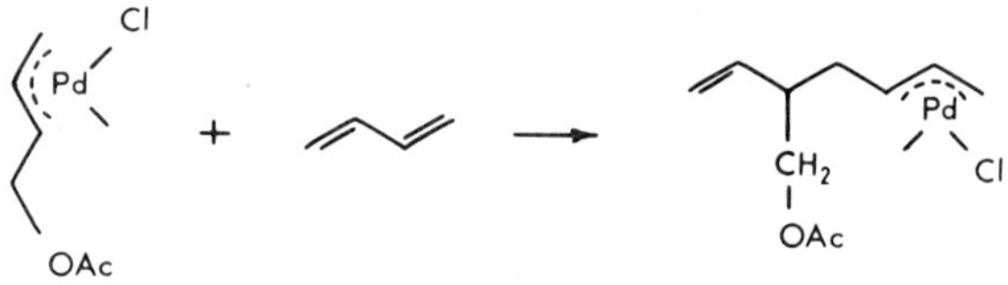

(33)   (34)

But the addition of a co-catalyst such as sodium phenoxide or a direct exchange of the chloride-group with an acetato-group converts (**33**) into an active catalyst.

An interesting extension of this telomerization which may open the route to a variety or organometallic compounds has recently been described by N. Hagihara (*135*). Low valent palladium-phosphine complexes catalysed a novel hydrosilation of butadiene with silicon hydride according to Eq. (12).

$$2\,CH_2{=}CH{-}CH{=}CH_2 + R_3SiH$$

$$\longrightarrow\ R_3Si{-}CH_2{-}CH{=}CH{-}CH_2{-}CH_2{-}CH{=}CH{-}CH_3 \qquad (12)$$

VI. Oligomerization of 1,2-Dienes

Very little is known about the oligomerization of 1,2-dienes. Still less is known about allene oligomerizations catalyzed via π-allyl complexes. The ease with which π-allyl complexes can be generated from 1,2-dienes (136–139) may support the speculation that π-allyl complexes play an important role, but supporting evidence is lacking.

Mainly three groups (47,140,142–145) have published on the oligomerization of allene yielding predominantely cyclic dimers, trimers, tetramers, and pentamers, but a detailed description of the steps involved is lacking. Many of these cylcic oligomers may be derived by postulating π-allylic intermediates of π-allyl or σ-allyl forms.

Interestingly, if the oligomerization of allene by π-allyl-palladium acetate is carried out in glacial acetic acid (47,145–147) 2,3-dimethylbutadiene, 3-methyl-2-hydroxy methyl butadiene acetate, and 2,3 dihydroxymethyl butadiene diacetate are formed. The complex (35) has been

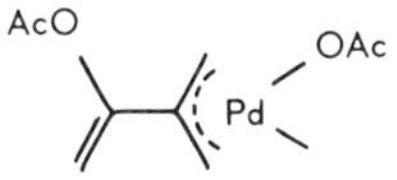

(35)

isolated in this reaction and has been used to deduce the formation of the telomers. Coordination of additional allene with (35) followed by reaction with acetic acid has been proposed as a possible reaction route.

VII. Hydrogenation

Homogeneous catalytic hydrogenation of organic substrates can be achieved by π-allyl complexes of transition metals. Catalytic amounts of bis-π-methallyl nickel, bis-π-allyl platinum or tris-π-allyl rhodium have frequently been used to hydrogenate olefins (148). Pregaglia (9) reports the

use of π-$C_4H_7Co(CO)_2$ (PBu_3) as a hydrogenation catalyst. Also co-
workers at Imperial Chemical Industry disclosed the hydrogenation properties of various π-allyl complexes (149). So far mechanistic studies have
not been disclosed and the function of the π-allyl complex is unknown. It
can be speculated that a reaction according to Eqs. (13) and (14) occurs
and that the catalysis is caused by the fine divided metal or the hydride
species, respectively.

$$M + H_2 \longrightarrow M^\circ + \text{propane} \tag{13}$$

$$ML_m + H_2 \longrightarrow H_nML_m + \text{propane} \tag{14}$$

For the hydrogenation of polyenes, π-allylic intermediates have been proposed and also isolated. An allylic bond can easily be formed when a
hydride adds to a conjugated double bond (151). Misono (150) describes the
hydrogenation of cyclododecatriene to cyclododecene by $[Co(CO)_3PR_3]_2$.
The mechanism is postulated in view of the effect of σ-donor ligands
which modify the equilibrium condition between σ-and π-allyl type reaction intermediates produced by the addition of metal hydride to diene.
Kwiateck and co-workers describe the successful isolation of π-allylcobalt
complexes during the hydrogenation of conjugated dienes by potassium
pentacyanocobaltate (152). On addition of a metal hydride to butadiene
syn- and anti-forms of the corresponding π-crotyl complexes can be
formed, which, in accordance with the experimental findings, give *trans*-
and *cis*- butene-2. Interestingly, the formation of predominantly butene-1
or butene-2 under proper choice of conditions can also be derived by
considering a π- or σ-allyl complex as intermediate. Whereas the σ-crotyl
complex yields butene-1 the π-crotyl complex gives butene-2. Tayim and
Bailar also described the homogeneous hydrogenation of polyolefins
catalyzed by platinum- and palladium-$SnCl_3$ complexes on the basis of
allylic intermediates (153).

   A mechanism based on allylic intermediates has also been considered by
Frankel and co-workers for the hydrogenation of unsaturated fatty esters
by iron pentacarbonyl (154). It should be emphasized, however, that the
hydrogenation of a conjugated diene or polyene not necessarily is limited
to a π-allylic type mechanism. Reduction of methyl sorbate with pure
deuterium catalyzed by tricarbonyl (methyl benzoate) chromium indicates
that 1,4-addition occurs (155) and it is not necessary to invoke a π-allyl
complex as an intermediate.

VIII. Isomerization Reactions

In this part only isomerization reactions of double bonds will be discussed. For skeletal rearrangements via π-allyl intermediates the reader is referred to references (*184,185*).

ISOMERIZATION OF DOUBLE BONDS

Double-bond isomerization is frequently observed during oligomerization reactions of olefins catalyzed by π-allyl complexes. Also, π-allyl intermediates have often been proposed to explain the isomerization of olefins. However, very little is known about the double-bond isomerization directly catalyzed by π-allyl complexes and systematic studies are lacking.

Double-bond migration of olefins necessitates a hydrogen shift (*156–159,179*) and it is tempting to assume a hydrogen transfer via a π-allyl intermediate as depicted in (**36**) to (**38**).

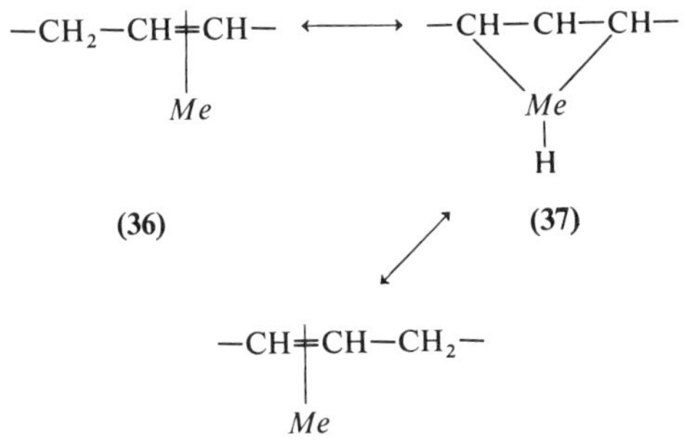

$$-CH_2-CH{=}CH- \quad \longleftrightarrow \quad -CH-CH-CH-$$

$$\overset{|}{Me} \qquad\qquad\qquad \overset{\diagdown\;\;\diagup}{Me}$$

$$\overset{|}{H}$$

$$\textbf{(36)} \qquad\qquad \diagup \qquad \textbf{(37)}$$

$$-CH{=}CH-CH_2-$$

$$\overset{|}{Me}$$

$$\textbf{(38)}$$

Coordination of a double-bond to a metal (**36**) is followed by abstraction of an allylic hydrogen thus forming a π-allyl intermediate (**37**). Hydrogen transfer in (**37**) yields the isomerized species (**38**). The intermediate (**37**) can exist in a syn- and anti-form and evidence (*158*) is in agreement with a postulate that the syn-form gives a *trans* and the anti-form the *cis* olefin. The isomerization of non conjugated diolefins has also been proposed to occur via a π-allylic intermediate. Rinehart and Lasky (*160*) suggest for the isomerization of cycooctadiene-1,5 to cyclooctadiene-1,3 by rhodium complexes the mechanism as shown in Eq. (15).

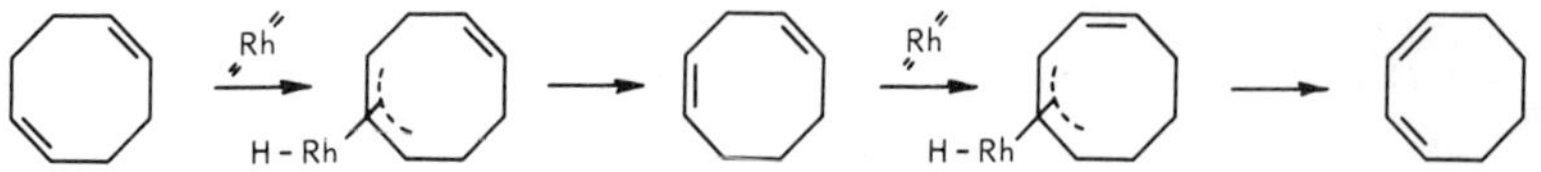

$$\tag{15}$$

The complexity of this isomerization is illustrated by the finding of the reverse reaction (*161,186*) namely the conversion of cyclooctadiene-1,3 to a rhodium complex containing cyclooctadiene-1,5. Pettit and Emerson (*157*) also propose that the isomerization of pentadiene-1,4 to pentadiene-1,3 occurs in an analogous fashion as indicated for the cyclooctadiene isomerization,

Many theoretical and experimental attempts have been made to confirm the existence of π-allylic complexes as intermediates in isomerization reactions. For heterogeneous catalysts kinetic studies and MO-calculations have been carried out (*162*). In homogeneous systems it has been possible to isolate π-allyl complexes. For instance, palladium chloride and butene-1 yield the corresponding π-allyl palladium complex (*8,163,164*). Based on the following observations, however, various authors have challenged the validity of a π-allylic mechanism.

1. π-allyl complexes formed during the isomerization are very often inactive for further isomerization.
2. Branched olefins are most easily converted to π-allyl complexes (*5*). Accordingly branched olefins isomerize less easily than linear ones, a fact explained by the formation of catalytically inactive stable π-allyl complexes (*8*). A comparison of the isomerization of various hexenes under identical reaction conditions revealed that hexenes containing the most branching at the olefinic double bond cause the most rapid catalyst deactivation (*165*).

3. Pentadiene-1,4 does not isomerize under conditions appropriate for pentene-1, because traces of conjugated dienes deactivate the hydridro-palladium species (*166*) according to Eq. (*16*).

$$\tag{16}$$

4. The isomerization of $C_5H_{11}-CD_2-CH{=}CH_2$ showed that no deuterium became attached to the terminal carbon atom, the predicted product of the π-allyl mechanism (*167–169, 180*).

To explain and account for all observations made during isomerization reactions various alternative mechanisms have been proposed, which are briefly summarized:

1. Addition of a metal hydride to a double bond generates a σ-carbon-metal-bond, followed by elimination of a metal-hydride (169–$173,178$) according to Eq. (17).

$$-CH_2-CH=CH_2 \xrightarrow{+MeH} -CH_2-\underset{\underset{Me}{|}}{CH}-CH_3 \xrightarrow{-MeH} CH=CH-CH_3 \qquad (17)$$

2. Chatt suggested a mechanism based on a carbene-complex intermediate (167) as depicted in Eq. (18).

$$R-CH_2-\underset{\underset{Me}{|}}{CH}=CHR \longleftrightarrow RCH_2-\underset{\overset{\|}{Me}}{C}-CH_2R$$

$$RCH=CH-\underset{\underset{Me}{|}}{CH_2R} \qquad (18)$$

3. An attractive mechanism which rationalizes many existing data has been proposed by Coffey (178). A metal hydride coordinates to a double-bond followed by the formation of an dihydrido type π-allyl transition state and an allylic proton shift according to Eq. (19).

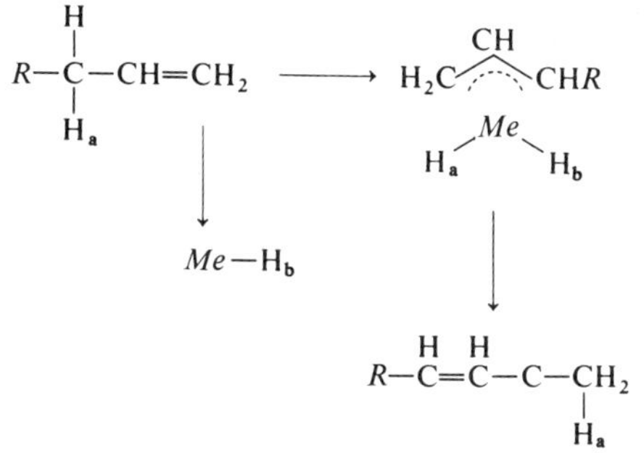

$$R-\underset{\underset{H_a}{|}}{\overset{\overset{H}{|}}{C}}-CH=CH_2 \longrightarrow H_2C\underset{}{\overset{CH}{\diagdown}}CHR$$

$$\underset{Me-H_b}{\Big\downarrow} \qquad \underset{H_a}{\overset{Me}{\diagup}}\overset{}{\diagdown}_{H_b}$$

$$\Big\downarrow \qquad (19)$$

$$R-\overset{H}{C}=\overset{H}{C}-\underset{\underset{H_a}{|}}{C}-CH_2$$

4. The findings that the isomerization of allyl benzene to β-methyl styrene by $DCo(CO)_4$ occurs without incorporation of deuterium has led to the suggestion of a 1,3 sigmatropic hydrogen shift mechanism (174–177).

In this connection, a related reaction namely the isomerization of allyl alcohols to aldehydes should be mentioned. Both a π-allyl- and 1,3 sigma-tropic hydrogen shift mechanism have been proposed ($181,182$).

There seems to be some controversy, and recent results indicate that the isomerization of unsaturated alcohols and olefins is compatible with a π-allyl type mechanism.

Many possibilities may exist outside the given pattern and it can be speculated that different catalyst systems may utilize different mechanistic routes effecting the same isomerization of olefins. Different substrates and reaction media may change the mechanism for the same metal catalyst. For instance, a proton-donor solvent may promote a quite different mechanism than an aromatic solvent. Insufficient attention has been given to the influence of solvents and reactant concentrations. Isotropic tracer techniques have often been applied prematurely. Certainly, many more data are necessary to gain a better understanding of isomerization reactions of double-bonds.

IX. Polymerization Reactions

There is considerable interest, especially in industry, in the utilization of π-allyl complexes as possible polymerization catalysts. Table 3 lists a variety of transition metal π-allyl complexes which have been claimed in

TABLE 3

π-ALLYL COMPLEXES CLAIMED IN POLYMERIZATION REACTIONS

Alkenes-alkynes	Transition metals	References
Alkenes[a]	Cr, Mo, W, Ti, Zr, Hf	(41, 187)
	Pd	(188)
	Ni	(189)
Vinyl monomers[b]	Cr, Mo, W, Ti, Zr, Hf	(190–192)
	Ni, Pd	(193–194)
Allenes	Pd, Ni	(145, 195, 196)
1,3-Dienes[c]	Ni, Pd, Pt	(194, 197–206, 207–208)
	Co, Rh, Ir	(19, 209–210)
	Cr, Mo, W	(19, 209, 211)
	Ti, Zr, Hf	(19, 204, 209)

[a] Ethylene, propylene, cyclobutene, etc.
[b] Styrene, acrylates, vinyl chloride, etc.
[c] Butadiene, isoprene, etc.

polymerization reactions of alkenes and alkynes. It would surpass the frame of this chapter to give a review of all patents and papers in this field.

Investigations of the behavior of π-allyl-type complexes is of particular interest in the stereospecific polymerization of dienes. It is conceivable to assume that π-allyl complexes serve as the active centers. In a general form the active species of butadiene polymerization may be written as shown in (39).

Incoming butadiene can insert into the CH_2-M bond (1,4-*cis* or 1,4-*trans* configuration) or into the $CH-M$ bond (1,2 or 3,4 units are formed). The direction of this insertion is determined by factors such as nature of ligand-field X, kind of transition metal used, oxidation state, etc. A plausible mechanism leading to 1,4-*cis* units can be visualized as follows: Butadiene coordinates to the catalytic intermediate (39) resulting in either syn- or anti-configuration. The formation of the syn- or anti-form may be determined by coordination of one or both double-bonds of the butadiene molecule. The syn- form leads to 1,4-*cis* and the anti-form to 1,4-*trans* poly butadiene.

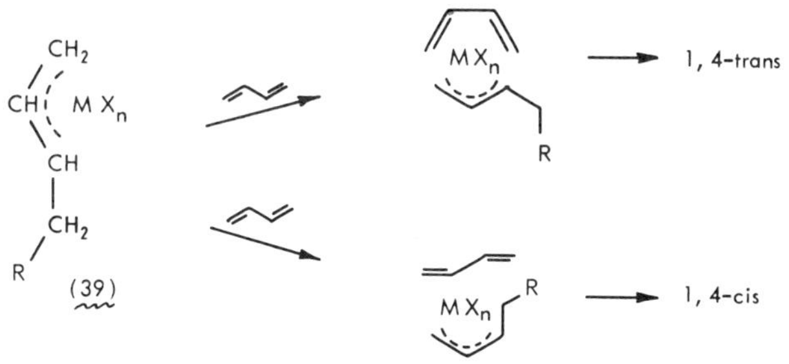

REFERENCES

1. E. J. Corey, M. F. Semmelhack, L. S. Hegedus, *J. Am. Chem. Soc.*, **90**, 2416 (1968).

2. E. J. Corey, M. F. Semmelhack, *J. Am. Chem. Soc.*, **89**, 2755 (1967).

3. G. Wilke, *Pure and Applied Chem.*, **17**, 179 (1968).

4. J. Tsuji, *Accounts of Chem. Res.*, **2**, 144 (1969).

5. R. Huettel and H. Christ, *Chem. Ber.*, **97**, 1439 (1964).

6. W. Kitching, Z, Rappoport, S. Winstein, W. G. Young, *J. Am. Chem. Soc.*, **88**, 2054 (1966).

7. P. M. Henry, *J. Organic Chem.*, **32**, 2575 (1967).

8. E. W. Stern, *Catalysis Rev.*, **1**, 114 (1967).

9. G. Pregaglia, A. Andreetta, and G. Ferarrani, *Chem. Communication*, 590 (1969).

10. W. W. Spooncer, A. C. Jones, and L. H. Slaugh, *J. Orgonometal. Chem.*, **18**, 327 (1969).

11. O. T. Onsager, H. Wang, and U. Blindheim, *Helv. Chim. Acta.*, **52**, 196 (1969).

12. Sentralinstitut for Industriell Forskuing, Oslo, *Neth. Appl.* 6.601.770; 6.612.735 *U.S.* 3.442.971.

13. Sun Oil Company, *Neth. Appl.* 6.707.094.

14. British Petroleum Company, *Br.*, 1245/1967.

15. Farbwerke Hoechst, *Neth. Appl.*, 6.612.735.

16. Phillips Petroleum Company, *Belg.*, 707.477 *U.S.*, 3.457.320.

17. G. Wilke, Max-Planck-Institut f. Kohlen-forschung, *U.S.*, 3.379.706 3.468.921.

18. Shell Oil Company, *U.S.*, 3.424.816.

19. G. Wilke, B. Bogdanović, P. Hardt, P. Heimbach, W. Keim, M. Kroener, W. Oberkirch, K. Tanaka, E. Steinrücke, D. Walter, and H. Zimmermann, *Angew. Chem.*, **78**, 157 (1966) [int. edit. **5**, 151 (1966)].

20. B. Bodganović, G. Wilke, *Brennstoff-Chemie*, 323 (1968).

21. G. Wilke, *Proceeding of R. A. Welch Foundation IX*, Houston (1966).

22. D. Walter, *Dissertation Technische Hochschule Aachen* (1965).

23. D. Walter and G. Wilke, *Angew. Chem.*, **78**, 941 (1966).

24. U. Birkenstock, H. Boennemann, B. Bogdanović, D. Walter, and G. Wilke, *Advances in Chemistry Series*, **70**, 250. (1968).

25. J. Ewers, *Erdoel und Kohle*, **21**, 763 (1968).

26. Shell Internationale, *G.B.*, 640.535, U.S. 3.431.318.

27. British Petroleum Company, *Neth. Appl.* 6.608.574.

28. Imperial Chemical Ind., *Neth. Appl.* 6.707.095.

29. Sun Oil Company, *Neth. Appl.* 6.816.085.

30. Esso Research and Engineering Company, *Br. Pat.* 1.151.015.

31. Feldblyum, W. Sh., Obeshchalcva, N. W., *Dokl. Akad. Nauk USSR*, **172**, 368 (1967).

32. N. W. Obsehchalova, V. Sh. Feldblyum, and N. M. Paschenko, *J. of Org. Chem. (USSR)*, **4**, 982 (1968).

33. J. Ewers, *Angew. Chem.*, **78**, 593 (1966).

34. M. Tsutsui, J. Aryoshi, T. Koyano, and M. N. Levy, *Advances in Chemistry Series* **70**, 266 (1968).

35. A. D. Ketley, L. P. Fisher, A. J. Berlin, C. R. Morgan, E. H., Gorman, and T. R. Steadman, *Inorganic Chem.*, **6**, 657 (1967).

36. K. Noak and F. Caldacrazo, *J. Organometal. Chem.*, **10**, 101 (1967).

37. R. G. Schultz, J. M. Schuck, and B. S. Wildi, *Journal of Catalysis*, **6**, 385 (1966).

38. R. F. Heck, *Accounts of Chemical Research* Vol. 2, 12 (1969).

39. R. Cramer, *Accounts of Chemical Research*, Vol. 1, 186 (1968).

40. B. Bogdanović, H. Henc, H. G. Karmann, H. G. Nüssel, G. Wilke, *Am. Chem. Soc. Meeting, New York, September, 1969.*

41. W. Oberkirch, *Dissertation Technische Hochschule Aachen* (1964).

42. U. Kruerke, C. Hoogzand, and W. Hübel, *Chem. Ber.*, **94**, 2817 (1961).

43. G. M. Whitesides and W. J. Ehmann, *J. Am. Chem. Soc.*, **91**, 3800 (1969).

44. J. P. Collman, *Accounts of Chemical Research*, **1**, 136 (1968).

45. W. Keim, *J. Organometal. Chem.*, **16**, 191 (1969).

46. O. S. Mills and G. Robinson, *Proc. Chem. Soc.*, 187 (1964).

47. G. D. Shier, *Am. Chem. Soc. Meeting, Minneapolis, April 13–18, 1969.*

48. *Chemical and Eng. News,* 21 April, 49 (1969).
49. *Hydrocarbon Processing,* **48**, November, 171 (1969).
50. *European Chemical News,* 10 November, 10 (1969).
51. *European Chemical News,* 31 October, 14 (1969).
52. DuPont, Inc., *Belg. Pat.* 708755, 708756.
53. Isao Ono and Keiichi Kihara, *Hydrocarbon Processing Vol. 46,* August 147 (1967).
54. G. N. Schrauzer, *Adv. in Organometallic Chem.,* Vol. 2, 2–45 (1964).
55. C. W. Bird, *Transition Metal Intermediates in Organic Synthesis,* Logos, London, 1967.
56. J. P. Candlin, K. A. Taylor, D. T. Thompson, "Reactions of Transition Metal Complexes", Elsevier, 1968.
57. H. Müller, D. Wittenberg, H. Seibt, and E. Scharf, *Angew. Chem. Interm.,* **4**, 327 (1965).
58. G. Wilke, B. Bogdanović, P. Borner, H. Breil, P. Hardt, P. Heimbach, G. Hermann, H. J. Kaminsky, W. Keim, M. Kroener, H. Müller, E. Meuller, W. Oberkirch, J. Schneider, K. Tanaka, K. Weyer, *Angew. Chemie Internat. Edit.,* **2**, 105 (1963).
59. M. Sittig, *Catalysis and Catalytic Processes, Noyes Development Corporation* 1967.
60. M. Sittig, *Noyes Development Corporation Diolefins Manufacture and Derivatives,* 1968.
61. A. Curbonaro, A. Greco, G. Dall'Asta, *Tet. Lett.,* **22**, 2037 (1967).
62. W. Keim, *Dissertation Technische Hochschule Aachen,* 1964.
63. B. Bogdanović, *Dissertation Technische Hochschule Aachen,* 1962.
64. P. Heimbach and W. Brenner, *Angew. Chem.,* **79**, 813 (1967), *Belg. Pat.* 713.167.
65. J. P. Candlin and W. H. Janes, *J. Chem. Soc.,* (C), 1856 (1968). *Br. Pat.* 1,085,875.
66. A. Yamamoto, K. Morifuji, S. Ikeda, T. Saito, Y. Uchida, and A. Misono, *J. Am. Chem. Soc.,* **90**, 1878 (1968).
67. M. Hidai, Y. Uchida, and A. Misono, *Bull. Chem. Soc. Japan,* **38**, 1243 (1965).
68. G. Wilke., French Patent 1,320.729.
69. H. Takahasi, S. Tai, and M. Yamaguchi, *J. Org. Chem.* **30**, 1661 (1965).
70. M. Hidai, K, Tamai, Y. Uchida, and A. Misono, *Bull. Chem. Soc., Japan,* **39**, 1357 (1966).
71. T. Saito, Y. Uchida, and A. Misono, *Bull. Chem. Soc. Japan,* **37**, 105 (1964).
72. T. Saito, Y. Uchida, A. Misono, A. Yamamoto, K. Morifugi, S. Ikeda, *J. Organomet. Chem.,* **6**, 572 (1966).
73. S. Otsuka, T. Kikuchi, and T. Taketomi, *J. Am. Chem. Soc.,* **85**, 3709 (1963).
74. Union Carbide, *U.S. Patent,* 3,392,207.
75. Copolymer Rubber and Chemical Corporation, *U.S. Patent,* 3,407,243.
76. S. Tanaka, K. Mubuchi, and N. Shimazaki, *J. Org. Chem.,* **29**, 1626 (1964).
77. G. Natta, U. Giannini, P. Pino, and A. Cassata, *Chimica e Industria,* **47**, 524 (1965).
78. G. Allegra, F. LoGiudice, and G. Natta, *Chem. Comm.,* 1263 (1967).
79. E. W. Duck, D. K. Jenkins, J. M. Locke, and S. R. Wallis, *J. Chem. Soc.,* C, 227 (1969).
80. T. Alderson, E. L. Jenner, and R. V. Lindsay, *J. Am. Chem. Soc.,* **87**, 5638 (1965).
81. R. Van Helden, C. F. Kohl, D. Medema, G. Verberg, and T. Jonkhoff, *Rec. Trav. Chim.* **87**, 961 (1968).
82. D. Medema, R. van Helden, and E. F. Kohl, *Inorganica Chimica Acta,* **225**, (1969).

83. Y. Takahashi, S. Sakai, and Y. Ishi, *Chem. Comm.*, 1093 (1967).

84. D. Medema, R. Van Helden, and C. F. Kohl, *Symposium New Aspects of the Chemistry of Metal Carbonyls and Derivatives*, Venice, Sept. 2–4, (1968).

85. J. M. Rowe, *Proc. Chem. Soc.*, **66**, (1962).

86. M. R. Churchill and R. Mason, *Nature*, **204**, 777 (1947).

87. N. Yamazaki and S. Murai, *Chem. Comm.*, 147 (1968).

88. J. Furukawa, T. Kakuzen, H. Morikawa, R. Yamamoto, and O. Okuno, *Bull. Chem. Soc. Japan*, **41**, 155 (1968).

89. G. Wilke, *Japan Patent* 263197, *British Patent*, 860377, *U.S.* 3,450.732.

90. *Imperial Chemical Industry, British Patent*, 1.148.777.

91. L. Porri, M. C. Gallazi, A. Colombo, G. Allegra, *Tet. Letters*, 4189, (1965).

92. S. Otzuka, K. Taketomi, *Europ. Polym. J.*, **2**, 289 (1966).

93. K. C. Dewhirst, *J. Org. Chem.*, **32**, 1297 (1967).

94. P. Heimbach, *XXI International Congress of Pure and Applied Chemistry*, (London), 223 (1968).

95. P. Heimbach and W. Brenner, *Angew Chem. Internat. Edn.*, **6**, 800 (1967).

96. P. Heimbach, *Angew. Chem.*, **78**, 983 (1966).

97. H. C. Chung, W. Keim and E. J. Smutny, reported *Gordon Res. Conf. on Hydrocarbon*, 1969.

98. H. C. Chung, W. Keim and E. J. Smutny, to be published.

99. A. Carbonaro, A. L. Segre, A. Greco, C. Tosi, G. Dall'Asta, *J. Am. Chem. Soc.*, **90**, 4453 (1968).

100. P. Heimbach, W. Brenner, *Angew. Chem.*, **79**, 814 (1967).

101. P. Heimbach, R. Schimpf, *Angew. Chem.*, **80**, 704 (1968).

102. P. Heimbach, W. Brenner, K. J. Ploner, and F. Thömel, *Angew. Chem.*, **81**, 744 (1969).

103. W. Schneider, B. F. Goodrich, *Comp. Neth. Appl.* 6.613.603, 6.613.476.

104. G. Crespi and G. D. Drusco, *Hydrocarbon Processing*, **48**, February, 103, (1969).

105. DuPont *U.S. Patent*, 3.407.245, 3.407.244, 3.152.195.

106. Columbian Carbon Company, *U.S. Patent*, 3.457.319.

107. Toyo Rayon, *U.S.Patent*, 3.405.194, 3.405.193, 3.408.418, *Japan Publ.*, 2.443/69, 29.725/68.

108. Montecatini Edison, *Neth. Appl.* 67-15290, 67-17024.

109. B. F. Goodrich Company, *U.S. Patent*, 3.398.309, 3.376.358, 3.392.208.

110. R. G. Miller, T. J. Kealy, and A. L. Barney, *J. Am. Chem. Soc.*, **89**, 3756 (1967).

111. R. Cramer, *Accounts Chem. Res.*, **1**, 186 (1968).

112. R. Cramer, *J. Am. Chem. Soc.*, **89**, 1633 (1967).

113. G. Hata and D. Aoki, *J. Org. Chem.*, **32**, 3754 (1967).

114. G. Hata and A. Miyake, *Bulletin of Chem. Soc. of Japan*, **41**, 2443 (1968).

115. W. Schneider, B. F. Goodrich Company, *U.S. Patent*, 3.398,209, 3.441.627.

116. W. Schneider, *Chemical and Engineering News*, 21 April, 50 (1969).

117. M. Jwamoto and S. Yuguchi, *J. Org. Chem.* **31**, 4920 (1966).

118. G. Hata and A. Miyake, *Chem. Soc. of Japan* (*Bulletin*) **41**, 2762 (1968).

119. Y. Tajima, E. Kunioka, *Chem. Comm.* 603 (1968).

120. A Misono Y. Uchida, T. Saito, and K. Uchida, *Bulletin of Chem. Soc. of Japan*, **40**, 1889 (1967).

121. G. Schrauzer, *Adv. Catalysis*, **18**, 377 (1968).

122. *Badische Anilin und Soda Fabrik, U.S. Patent*, 3,277,099.

123. J. Feldman, O. Frampton, B. Saffer, and M. Thomas, presented Division Petroleum Chemical, *Am. Chem. Soc. Chicago*, August 30-September 4 (1964).
124. Phillips Petroleum Company, *U.S.Patent*, 3.435.088.
125. P. Heimbach, *Angew. Chem.*, **80**, 967 (1968) *Neth. Appl.* 6613675, *Belg. Pat.* 687,398.
126. E. J. Smutny, *J. Am. Chem. Soc.*, **89**, 6794 (1967).
127. E. J. Smutny, H. Chung, K. C. Dewhirst, W. Keim, T. M. Shryne, and H. E. Thyret, *Am. Chem. Soc. Meeting, Minneapolis, April 13–18, 1969, Chemical and Engineering News*, 21 April, (1969), *Chemical and Engineering News*, 11 Dec. 21 (1967).
128. Shell Oil Company U.S. Patents 3.267.169, 3.350.451, 3.407.224, 3.364.176, 3.432.465, 3.444.202, 3.499.042, 3.518.318, 3.518.315, 3.489.813, 3.502.725, 3.493.617, 3.398.168, 3.444.258.
129. Mitsubishi Chemical Ind., *Neth. Appl.*, 68.16008.
130. E. J. Smutny, H. Chung, K. C. Dewhirst, W. Keim, and T. M. Shryne, *11th Conf. on Coord. Chem.*, *Sept. 8–18, Haifa, 1968*.
131. S. Takahashi, T. Shibano, and N, Hagihara, *Tetrahedron Letters*, 2451 (1967).
132. S. Takahashi, H. Yamazaki, and N. Hagihara, *Bull. Chem. Soc. Japan*, **41**, 254, (1968).
133. S. Takahashi, T. Shibano, and N. Hagihara, *Bull. Chem. Soc. of Japan*, **41**, 454 (1968).
134. S. Takahashi, H. Yamazaki, and N. Hagihara, *Mem. Inst. Sci. and Ind. Res.*, Osaka Univ., **25**, 125 (1968).
135. S. Takahashi and N. Hagihara, *Chem. Comm.*, **161**, (1969).
136. R. G. Schultz, *Tetrahedron*, **20**, 2809 (1964).
137. H. S. Lupin, J. Powell, and B. L. Shaw, *J. Chem. Soc.*, **A**, 1687 (1966).
138. W. Keim, *Angew. Chem.*, **80**, 279 (1968).
139. C. U. Pittman, *J. Chem. Comm.*, **127**, (1969).
140. F. W. Hoover and R. V. Lindsey, Jr., *Org. Chem.*, 3051 (1969).
141. F. N. Jones and R. V. Lindsey, Jr., *Org. Chem.*, 3838 (1968).
142. R. E. Benson and R. V. Lindsey, Jr., *J. Am. Chem. Soc.*, **81**, 4239 (1959).
143. S. Otsuka, A. Nakamura, and H. Ninamida, *Chem. Comm.*, **191** (1969).
144. S. Otsuka, A. Nakamura, K. Tani, and S. Ueda, *Tet. Lett.*, **297** (1969).
145. G. D. Shier, *J. Organom. Chem.*, **10**, 15 (1967).
146. G. D. Shier, *Chemical and Engineering News*, April 21, **49** (1969).
147. Dow Chemical, *Belgian Patent*, 712,545.
148. W. Keim, unpublished results.
149. *Imperial Chemical Industry, Netherlands Application* 68-05418.
150. I. Ogata and A. Misono, presented *Meeting of Faraday Soc. London 1968*.
151. B. R. James, *Coordin. Chem. Rev.*, **1**, 505, (1966).
152. J. Kwiatek, *Adv. in Chemistry Series*, **70**, 207 (1968).
153. H. A. Tayim and J. C. Bailor, *J. Am. Chem. Soc.*, **89**, 4330 (1967).
154. E. N. Frankel, T. L. Mounts, R. O. Butterfield, and H. J. Dutton, *Adv. in Chem. Series*, **70**, 177.
155. E. N. Frankel, E. Selke, C. N. Glass, *J. Am. Chem. Soc.*, **90**, 2446 (1968).
156. T. A. Manuel, *J. Org. Chem.*, **27**, 3941 (1962).
157. R. Pettit and G. F. Emerson, *Adv. Organometal. Chem.*, **1**, 17 (1964).
158. J. J. Rooney and G. Webb, *J. of Catalysis*, **3**, 488 (1964).

159. M. Orchin, *Adv. in Catalysis*, **16**, 1, (1966).

160. R. E. Rinehart and J. S. Lasky, *J. Am. Chem. Soc.*, **86**, 2516 (1964).

161. J. K. Nicholson and B. L. Shaw, *Tet. Let.*, 2533 (1965).

162. S. Carra and R. Ugo, *Inorganica Chimica Acta Reviews*, **1**, 56–57, (1967).

163. D. Morelli, R. Ugo, F. Conti, and M. Donati, *Chem. Comm.*, 801 (1967).

164. A. D. Ketley and J. Braatz, *Chem. Comm.*, 169 (1968).

165. M. B. Sparkle, L. Turner, and A. J. M. Wenham, *J. of Catalysis*, **4**, 332 (1965).

166. G. C. Bond and M. Hellier, *J. of Catalysis*, **4**, 1, (1965).

167. N. R. Davies, *Nature*, **201**, 490 (1964).

168. N. R. Davies, *Australian J. of Chem.*, **17**, 212 (1964).

169. R. Cramer and R. V. Lindsey, Jr., *J. Am. Chem. Soc.*, **88**, 3534 (1966).

170. R. F. Heck and D. S. Breslow, *J. Am. Chem. Soc.*, **83**, 4023 (1961).

171. G. C. Bond and M. Hellier, *Chem. Ind.* (London), **35**, (1965).

172. J. F. Harrod and A. J. Chalk, *J. Am. Chem. Soc.*, **88**, 3491 (1966).

173. J. F. Harrod and A. J. Chalk, *J. Am. Chem. Soc.*, **86**, 1776 (1964).

174. R. B. Woodward and R. Hoffmann, *J. Am. Chem. Soc.*, **87**, 2511 (1965).

175. J. A. Berson, *Accounts of Chem. Res.*, **1**, 152 (1968).

176. G. V. Smith and J. R. Swoop, *J. Organic Chem.*, 3904 (1966).

177. L. Roos and M. Orchin, *J. Am. Chem. Soc.*, **87**, 5502 (1965).

178. R. S. Coffey, *Tet. Lett.*, 3809 (1965).

179. I. I. Moiseev, A. A. Grigorev, and S. V. Pestrikov, *J. of Organic Chemistry* (USSR), 346 (1968).

180, J. F. Harrod, A. J. Chalk, *Nature*, **205**, 280 (1965).

181. G. F. Emmerson and R. Pettit, *J. Am. Chem. Soc.*, **84**, 4591 (1961).

182. W. T. Hendrix, F. G. Gowherd, and J. L. von Rosenberg, *Chem. Comm.*, 97 (1968).

183. F. G. Gowherd, J. L. von Rosenberg, *J. Am. Chem. Soc.*, **91**, 2157 (1969).

184. H. Frye, E. Kuljian, and I. Viebrock, *Inorg. Chem. Nucl. Lctt.*, 119 (1966).

185. M. Tsutsui, M. Hancock, J. Ariyoshi, and M. N. Levy, *Angew. Chem.*, **81**, 453 (1969).

186. J. C. Trebellas, J. R. Olechowski, and H. B. Jonassen, *J. Organometal. Chem.*, **6**, 412 (1966).

187. *Imperial Chemical Industry*, *Netherlands Applications*, 68-03022, 68-08004, 69-1819, 60-2740, 69-4684, 69-8042.

188. A. D. Ketley, J. A. Bratz, *Polym. Lett.*, **6**, 341 (1968).

189. L. Porri, G. Natta, and M. C. Galazzi, *Chimica e Industria*, **46**, 428 (1964).

190. D. G. H. Ballard, W. H. Janes, and T. Medinger, *J. Chem. Soc.*, **B**, 1168 (1968).

191. D. G. H. Ballard and T. Medinger, *J. Chem. Soc.*, **B**, 1176 (1968).

192. *Imperial Chemical Company*, *U.S. Patent*, 3,436,383.

193. R. Huettel, C. Koenig, *157th Am. Chem. Soc. Nat'l. Meeting, Minneapolis, April 13–18, 1969.*

194. *Badische Anilin und Soda Fabrik*, *Netherlands Application*, 6502247.

195. Japanese Synthetic Rubber, *British Patent*, 1,134,332.

196. S. Otzuka, K. Mori, F. Imaizumi, *J. Am. Chem. Soc.*, **87**, 3017 (1965).

197. L. Porri, G. Natta, and M. C. Gallazzi, *Chim. Ind.* (Milan) **46**, 428 (1964).

198. J. P. Durand, F. Dawans, and Ph. Teyssie, *Polym. Lit.*, **5**, 785 (1967).
    Polym. Lit., **6**, 757 (1968), Belg. Pat. 713,415.
    Polym. Lit., **7**, 111 (1969).

199. SNAM Progett. S.p.A., *Netherlands Application*, 6709706.
200. G. Lugil, W. Marconi, A. Mazzei, and N. Palladino, *Inorg. Chimicu Acta*, 151 (1969).
201. Y. Ostrovskaya, K. L. Mar Kovetskic, *Dokl. Chemistry*, **181**, 761 (1968).
202. A. Vasilevich, U.S. Patent, 3,468,866.
203. A. V. Volkov, B. A. Dolgoplosk, *Dokl. Akad. Nauk SSSR*, **183**, 1083 (1968).
204. N. P. Simanova, *Vysokomol. Soedin. Ser. B* **10**, 8, 590 (1968).
205. B. A. Dolgoplosk, *J. of Polym. Sci.*, (C), **16**, 3685 (1968).
206. E. I. Tinyakova, *J. of Polym. Sci.*, (C), **16**, 2625 (1967).
207. W. R. Grace and Co., *O.L.S.*, 1,813,035.
208. G. M. Klvostik, V. N. Sokolov, *Dokl. Akad. Nauk SSSR*, **186**, 894 (1968).
209. G. Wilke, U.S. Patents, 3,424,777; 3,422,128
 3,432,530; 3.379,706
210. R. E. Rinehart, H. P. Smidt, H. S. Witt, and H. Romeyn, *J. Am. Chem. Soc.*, **83**, 4864 (1961).
211. I. A. Orershin, *Vysokomol. Soedin*, **11**, 1840 (1969).

4

Homogeneous Metal Catalyzed Oxidation of Organic Compounds

ERIC W. STERN

Engelhard Industries
A Division of Engelhard Minerals
and Chemicals Corporation
Newark, New Jersey

I. Introduction

The oxidation of organic compounds is a large and important area of chemical research encompassing numerous degradative and synthetic reactions. While metals are widely employed in these reactions (*1*), their involvement is frequently as stoichiometric reagents rather than as catalysts. Only when the reduced metal returns readily to its original active oxidation state can such reactions be termed catalytic. It is these cases which are of greatest interest in commercial applications and in providing suitable models for biological oxidation processes.

Preferably, the participation of metal catalysts in organic oxidations should be viewed against the general framework of oxidation mechanisms (*2,3*). Such a comprehensive approach to the subject is well beyond the scope of this brief discussion which is limited to some examples of homogeneous catalysis. However, before proceeding to these, it would be well to review some of the ways in which oxidation reactions have been treated and to outline the functions of and requirements for oxidation catalysts.

In general, organic chemists have tended to discuss oxidations in terms of the fate of particular functional groups or molecular types. Broadly speaking, an organic molecule is oxidized when hydrogen is removed and/or oxygen or halogen added. Removal of hydrogen with one electron is designated as a homolytic oxidation. The remaining organic species, having an unpaired electron, can initiate a chain reaction by removing a radical from an undissociated molecule or by adding to an undissociated electrophile. It can also react with another radical in the system leading to coupling, or can lose another hydrogen atom resulting in disproportionation or dehydrogenation.

When hydrogen is removed with two electrons, the process is a heterolytic oxidation. The resulting positively charged (acidic) species may abstract hydride from a neutral molecule or add to a nucleophilic molecule in a chain reaction, react with a negatively charged (basic) molecule, or lose a proton. Frequently, the initially formed positively charged species will decrease its energy content by internal rearrangement (isomerization) prior to subsequent reaction.

Distinction between homolytic and heterolytic processes merely on the basis of the types of products formed is not always a simple matter. Frequently it is necessary to resort to other criteria. For example, homolytic reactions can be initiated by addition of radicals and heterolytic reactions by acids or bases. The rates of heterolytic reactions are often increased in solvents of high dielectric constant which decrease the electrostatic work necessary for heterolytic fission, Reaction of intermediates with radicals or inhibition by materials known to react with radicals is typical of homolytic reactions, while reactions of intermediates with bases indicates heterolytic reactions.

The approach of inorganic chemists to the subject of metal participation in organic oxidations quite naturally has been from the point of view of electron transfer to metal ions or complexes; i.e., reduction of the metal. Current concepts of electron transfer are based largely on studies involving electron transfer between metal ions (4–7). These are termed "inner sphere" if electron transfer is preceded by substitution into the first coordination shell of the reacting ion and "outer sphere," if transfer is accomplished without prior substitution. Here too, difficulties sometimes arise in distinguishing between these categories, since electrons can be transferred via ligand bridges (inner sphere) without permanent transfer of the bridging group. The metal ion can undergo a one or two electron change, and, if the change is permanent, it is not difficult to decide which of these has occurred. Reactions are complementary, if the oxidant and reductant undergo equivalent electron changes and noncomplementary, if they do not.

Clearly, an adequate description of metal catalysis in organic oxidation reactions must include information concerning the fate of both the organic and inorganic moieties involved. Formally, as stated earlier, oxidation of an organic compound consists either of hydrogen abstraction or oxidant addition. In the former case, the prime role of the catalyst would appear to be to facilitate such removal by substrate "activation." In the latter instance, either oxidant or substrate "activation" may occur. The nature of "activation," in these cases, will likely depend on bonding in intermediates or the activated complex. This will be a function of the metal, its oxidation state, and the type and number of ligands in its coordination shell.

Catalysis will also depend on the ease with which the metal, if its ability to participate in the product forming reaction has been altered in this process, can be returned to its original active oxidation state. This will depend on whether the metal can be reoxidated at a reasonable rate, either by reducing a component of the reaction mixture, or by an externally applied oxidizing agent. Metal reoxidation must, of course, be possible under conditions which will not lead to degradation of the organic substrate or to undesirable side reactions. Also, the ability of the metal to interact with substrate and/or oxidant must not be impaired irreversibly by formation of stable complexes. Most important of all, for homogeneous catalysis, the metal must remain soluble. Ideally, these conditions would be met, if the catalyst were to function merely as a "bridge" for transfer of electrons from organic substrate to oxidant.

Admittedly, these are oversimplifications. To date, the intimate mechanisms of most homogeneous metal-catalyzed reactions remain largely speculative. It is, therefore, necessary to examine, in some detail, the evidence on which such speculation is based. Hence, the following discussion of homogeneous catalysis of oxidation of organic compounds is limited to relatively few examples. These, however, should serve to illustrate the status of research in this area.

II. Oxidation of Olefins by Reaction with Nucleophiles

Much of the recent interest in homogeneous metal catalysis can be traced to announcement of the now well known Wacker process for production of acetaldehyde from ethylene (8). This combines the stoichiometric reduction of Pd(II) [Eq. (1)] with an *in situ* reoxidation of the metal [Eq. (2)] into an overall process in which Pd(II) functions as a catalyst [Eqs. (1)–(4)].

$$C_2H_4 + H_2O + PdCl_2 \longrightarrow CH_3CHO + Pd + 2HCl] \tag{1}$$

$$Pd + 2CuCl_2 \longrightarrow PdCl_2 + 2CuCl \tag{2}$$

$$2CuCl + 2HCl + \tfrac{1}{2}O_2 \longrightarrow 2CuCl_2 + H_2O \tag{3}$$

$$C_2H_4 + \tfrac{1}{2}O_2 \longrightarrow CH_3CHO \tag{4}$$

Subsequent research has shown that the product forming step is a general reaction of olefins with nucleophiles which yields carbonyl compounds in water and unsaturated esters, ethers, amines, amides, nitriles, and carbon compounds in nonaqueous media. Various aspects of the reaction have been studied and summarized in the review literature (9–11). Essential features are as follows:

A. PRODUCT DISTRIBUTIONS

1. Aqueous Systems

The reaction in which ethylene is converted to acetaldehyde yields mixtures of aldehydes and ketones from propylene and higher terminal olefins and ketones from internal olefins. Formation of carbonyl compounds having the same carbon skeleton as the original olefin is, however, essentially limited to olefins having at least one hydrogen on each of the olefinic carbon atoms. Substitution of both hydrogens on one of the carbons by alkyl or aryl groups generally results in hydration, oxidative degradation, and rearrangements (*8,12*). The effect on aldehyde/ketone ratios in products derived from terminal olefins with side chains of increasing length and complexity is unclear, and in any event, appears to be minor (*8,12–14*). Aromatic substituents on the double bond appear to favor aldehyde formation (*13*). This effect is enhanced by electron withdrawing substituents on the ring. Electron donating substituents facilitate ketone formation (*15*).

Increased acid and chloride concentrations increase the formation of aldehyde relative to ketone from propylene (*13*) and styrene (*15*). The effect is not due to increased ionic strength (*15*).

Increasing temperatures appear to favor aldehyde formation from propylene (*13*). However, with propylene, as well as with higher olefins, increasing temperatures lead to formation of π-allyl complexes which yield unsaturated carbonyl products on reacting with water (*12,16*).

Aldehyde/ketone ratios have also been found to depend on the particular Pd(II) salt employed and on the particular acid used, if the medium is acidified (*13*). In dilute nitric acid, $PdCl_2$ and $Pd(NO_3)_2$ catalyze the conversion of ethylene to glyoxal in a reaction which consumes HNO_3 (*17*).

The presence of reagents which facilitate reoxidation of Pd(O) to Pd(II), such as $CuCl_2$ and *p*-benzoquinone, can also affect product distributions. While *p*-benzoquinone apparently has no effect on either aldehyde/ketone ratios (*15*) or by-product formation (*18*), presence of $CuCl_2$ leads to substantial chlorohydrin formation from ethylene at high copper and chloride concentrations (*19*) and to chlorinated aldehydes and ketones (*20,21*).

2. Nonaqueous Systems

With respect to product distributions, the reaction of olefins with nucleophiles in nonaqueous media is far more complex than the corresponding aqueous reaction. The type of product formed depends not only on the particular nucleophile employed, but for any given nucleophile, may consist of monosubstituted vinyl or allyl compounds or disubstituted saturated products. Since several stereoisomers of unsaturated products are possible, and such side-reactions as double bond isomerization and oligomerization are frequently encountered, the number of products obtained from any olefin can be very large indeed. In fact, in many cases, complete product identification has not been possible. Because of the great commercial interest in the production of vinyl acetate from ethylene (*22,23*), ester formation has been studied more extensively than other nonaqueous reactions. It will, therefore, be the basis for subsequent discussion.

The effect of olefin structure on product distributions is frequently masked by other variables and is, therefore, difficult to assess. However, in the case of terminal olefins, secondary vinyl ester formation seems to be favored over primary ester formation (*24–27*). The tendency to primary vinyl ester formation apparently is increased by aromatic substitution (*25,28*). Allyl ester formation increases with increasing chain length (*25–27,29,30*) and with shift of the double bond from the terminal position (*27–29,31*). Terminal olefins form primary allyl esters while 2-olefins form secondary allyl esters (*27*). Cyclic monoolefins yield allylic and other ester products (*32–34*).

Products are strongly affected by solvents. For example, primary vinyl ester formation is markedly increased when propylene reacts with acetic acid in 1,2-dimethoxyethane rather than isooctane (*10*). Similar effects have been noted on addition of benzonitrile to an acetic acid medium (*25*). Dimethyl sulfoxide promotes allyl ester formation (*27*). With respect to saturated products, 1,1-diacetates are generally observed when reactions are carried out in acetic acid (*22,35,36*). Their formation has been shown to be a function of acetic acid concentration in such solvents as N,N-dimethylformamide, N,N-dimethylacetamide, sulfonamides, sulfoxides, sulfones, nitriles, ketones, and esters (*36*). No diacetate formation was observed at low concentrations of acetic acid in hydrocarbon (*23,24*) and ether (*10*) solvents.

Effects on product distribution as a function of the nature and concentration of anions in reaction media have also been noted. Increasing acetate or chloride concentrations relative to palladium in acetic acid increased terminal substitution in 1-hexene (*31*). The addition of nitrates in the reaction of ethylene with $PdCl_2$ in glacial acetic acid led to formation of glycol di- and monoacetates (*37*). This reaction has also been observed when $Pd(NO_3)_2$ was used (*39*). In chloride free acetic acid containing perchloric acid, $Pd(OAc)_2$ converts cyclohexane to cyclohexanone, cyclohexyl acetate, 1-cyclohexene-1-yl acetate, 2-cyclohexen-1-yl acetate, benzene, cyclohexanol, and phenol. In the absence of perchloric acid, the major product is benzene with only traces of other products formed (*39*). Reaction of styrene with $Pd(OAc)_2$ in acetic acid leads to α and β- acetoxystyrenes and 1,4-diphenylbutadiene. With $Pd(NO_3)_2$, acetophenone is formed in addition to these while with $PdSO_4$, no unsaturated acetates are formed. Products were acetophenone, α-phenethyl acetate, and 1,3-diphenyl-1-butene (*28*).

Regenerating agents, again, affect product distributions. The presence of
p-benzoquinone is reported to increase allyl products (26). The position of
substitution and relative amounts of vinyl and allyl esters and 1,1-diesters
vary as a function of chloride and acetate concentrations in regeneration
by CuCl₂ (31) and Cu(OAc)₂ (40). At high chloride concentrations in
copper containing systems, formation of unsaturated esters is essentially
suppressed. Products consist of 1,2-di- and monoacetates and chloro-
acetates (34,38,40–42). The relative amounts of these materials is a function of chloride, acetate, and water concentrations (41). In addition, 1,3
and 1,4-disubstituted products have been found (41).

B. Rates

In both aqueous and nonaqueous systems, variables which influence
product distributions also influence rates. The reactions are complex,
consisting of a series of steps each of which, when they are separable, may
be affected to varying degrees.

1. Aqueous Systems

When rates have been measured by decreases in olefin volume or pressure, it has been found that this occurs in two distinct steps: a rapid
initial olefin absorption greater than necessary to saturate the solution
followed by a slower absorption. The first of these steps is presumed to be
due to complex formation. The second is reaction (13,43). Overall rates
are generally first order in both olefin and palladium salt (11,18,43–45).

Olefin structure has been found to affect both the fast complex forming
as well as the slower complex decomposition reactions. Qualitatively,
terminal olefins react faster than internal cis olefins which react faster than
internal trans olefins (8,13,46). The complex forming equilibrium is affected
by olefin structure to a considerably greater extent than is complex
decomposition (43,44).

Studies of the effect of temperature on rate, over a limited range,
indicate that both complex formation and decomposition are affected.
However, in this instance, decomposition is affected more than formation
(43,47).

Rate inhibition by acid has been found for ethylene, propylene, the
butenes (21,43,44), and styrene (15), but not for cyclohexene (18). In
general, the inhibition is first order in proton and affects complex decomposition but not complex formation. At very low acid concentrations, the
rate increases linearly with acid concentration (15,48).

Halides affect the reaction rate strongly in the order $I^- > Br^- > Cl^-$ (21). At low chloride concentrations, rate increases as a function of chloride concentration have been found (15,48). At intermediate chloride levels, rates are inhibited, showing inverse second order dependence on chloride concentration (18,43,44,49). At high chloride and Pd(II) concentrations, the rate dependence on all components of the reaction mixture, olefin, Pd(II), acid, and chloride, becomes more complicated (11). In the absence of chloride, the rate dependence on Pd(II) becomes second order (50). Other salts, such as phosphates, fluorides, nitrates, and sulfates, have less effect but do exert some inhibiting effect with increasing concentration (21). Some of this effect may be the result of producing insoluble Pd(II) species. The effect of ionic strength, measured by addition of perchlorates, is to increase rate at low ionic strength and to decrease rate at higher ionic strength (15,43,50). The entire effect, however, is small.

The reaction of regenerating agents, such as $CuCl_2$ (21) or p-benzoquinone (50), with Pd(0) is fast relative to the reaction of olefins with water and Pd(II). Therefore, presence of such materials as a rule, does not cause difficulties in rate studies, if care is taken to allow for changing acid and halide concentrations (21). However, as mentioned earlier, at high chloride and copper concentrations, the presence of $CuCl_2$ leads to formation of saturated 1,2-disubstituted products. This may be responsible for the rate increases observed in the presence of Cu(II) (51,52).

In general, rate expressions derived by various workers for reactions at intermediate proton and chloride concentrations are in agreement (11,43,45) [Eq. (5)]. The second order

$$-d(\text{olefin})/dt = k[PdX_2][\text{olefin}]/[H^+][X^-]^2 \tag{5}$$

rate constant is the product of the equilibrium constant for complex formation and the rate constant for decomposition of the complex.

At low acid and chloride concentrations, results are accommodated by the following rate expression [Eq. (6)] at constant olefin partial pressure and Pd(II) concentration (regeneration by p-benzoquinone):

$$\text{rate} = a[H^+][Cl^-]/b + [H^+]^2[Cl^-]^3 \tag{6}$$

At high chloride and Pd(II) concentrations, the following rate law applies (11) [Eq. (7)]:

$$\text{rate} = k_1[PdCl_4^{-2}][C_2H_4]/[H^+][Cl^-]^2 +$$

$$k_2[PdCl_4^{-2}]^2[C_2H_4]/[H^+][Cl^-]^3 \tag{7}$$

2. Nonaqueous Systems

As in the case of aqueous systems, olefin absorption in nonaqueous systems also occur in two steps (36), indicating complex formation followed by complex decomposition. Qualitative rate comparisons indicate that terminal olefins react more rapidly than internal olefins (31). Rates have been found to be generally first order with respect to both olefins and Pd(II) (53,54). However, in a system containing Pd(OAc)$_2$, NaOAc, and p-benzoquinone in acetic acid, this dependence is reported to hold only in the presence of a three to fourfold excess of the other reagent. Lower values were obtained when the concentrations of olefin and Pd(II) are comparable (55,56).

The rate dependence on acetate is complex. The reaction can be carried out in the absence of added acetate in acetic acid at elevated temperatures and pressures (57) or in N,N-dimethylacetamide (36), but the rate is slow and is accelerated markedly by addition of acetate (22,36,53,56). Rates increase linearly with increasing acetate concentration to a maximum value and then decrease sharply (53,56). In the region of rate decrease, the rate is inversely proportional to the acetate concentration squared (56).

The precise effect of chloride concentration on the rate of ester formation in acetate containing systems is difficult to evaluate because of changes in product distribution. The addition of small amounts of chloride to acetate systems apparently increases the rate of vinyl acetate formation (38). At high chloride concentrations, vinyl ester formation is suppressed (42).

A second order dependence of rate on acetic acid concentration has been reported (53).

Rates appear to be independent of p-benzoquinone concentrations (11,53). They are also independent of Cu(II) concentrations at low chloride levels, but become first order in Cu(II) at high chloride concentrations. Under these conditions, glycol mono and di-esters are formed (40). Use of Fe(OAc)$_3$ for regeneration of Pd(II) is reported to increase overall rates without glycol ester formation (38).

C. ISOTOPE EFFECTS

Ethylene reacts more slowly in D$_2$O than in H$_2$O (21,49), with the ratio, $k_{H_2O}/k_{D_2O} = 4.05 \pm 0.15$ (49). The acetaldehyde produced in D$_2$O contains no deuterium (21). In a comparison of reaction rates of C$_2$H$_4$ and C$_2$D$_4$ in aqueous medium, no primary isotope effect was found; $k_{C_2H_4}/k_{C_2D_4} = 1.07$ (43).

No comparable quantitative results are available for non-aqueous systems. However, no deuterium was incorporated into ethylidene diacetate formed in CH_3COOD (62a). Reaction of propene-2-d with $PdCl_2$ in isoctane containing acetic acid and Na_2HPO_4 resulted in formation of isopropenyl and n-propenyl acetates, 75% of which retained one deuterium (24). A comparison of relative reaction velocities between C_3H_6 and C_3H_5D indicated the possibility of an isotope effect.

D. Mechanism

It is difficult to draw generally valid mechanistic conclusions, since both product distributions and rates are affected strongly by a number of variables. Therefore, while most workers agree on the interpretation of the major features of the reaction, no such accord is found concerning mechanistic details. On the basis of available data, it has been concluded that both olefin and nucleophile are complexed by Pd(II) prior to reacting and that a hydride shift in the olefin moiety occurs at some point during the decomposition of this complex.

The olefin-Pd(II) complexes formed initially can be observed or isolated under conditions which inhibit further reaction. In all cases (8,12,21,36), these complexes have been shown to be identical to the dimeric π-complexes, $(C_nH_{2n}PdCl_2)_2$, which can be prepared by reaction of olefin with $(\phi CN)_2PdCl_2$ (58). The complexes react rapidly with water (8,59) or nonaqueous nucleophiles (22,23,36) to yield the same products obtained from the olefin directly. On the other hand, π-allyl complexes, $(C_nH_{2n-1}PdCl)_2$, are less reactive and give rise to other products (12,27,33). Moreover, π-allyl complex formation would not be consistent with the formation of different products from 1- and 2-olefins (27) as the same complex should be formed from both isomers.

While it appears reasonably certain that the reacting complexes are, indeed, π-olefin complexes, their eaxct nature, monomer, dimer, di-olefin PdX_2 or olefin-dipalladium, must depend on reaction conditions and medium compositions. In aqueous systems, at sufficiently high chloride concentrations to assure conversion of most palladium species to $PdCl_4^{-2}$, π-complex formation and nucleophile incorporation can be represented as follows for reaction of ethylene (11,43,44,60) [Eqs. ((8)–(10)]:

$$C_2H_4 + PdCl_4^{-2} \rightleftharpoons [C_2H_4PdCl_3]^- + Cl^- \tag{8}$$

$$[C_2H_4PdCl_3]^- + H_2O \rightleftharpoons [C_2H_4PdCl_2(H_2O)] + Cl^- \tag{9}$$

$$[C_2H_4PdCl_2(H_2O)] + H_2O \rightleftharpoons [C_2H_4PdCl_2(OH)]^- + H_3O^+ \tag{10}$$

(1)

Equilibrium (*8*) which, as discussed, corresponds to initial olefin absorption, was found to be inhibited by Cl^- (*43*). Equilibrium (*9*) accounts for the additional rate inhibition by Cl^-, and (*10*) is in accord with rate inhibition by acid as well as the isotope effect observed in D_2O.

The rate accelerations by H^+ and Cl^- at low acid and chloride concentrations have been explained by postulating formation of less active or totally inactive species under these conditions (*15*). For example [Eq. (11)]:

$$[C_2H_4PdCl_2(H_2O)] + 2H_2O \rightleftharpoons [C_2H_4Pd(OH)_2(H_2O)] + 2H^+ + 2Cl^- \quad (11)$$

Another explanation of this effect involves rearrangement of the presumably *trans*-hydroxo-olefin complex (**1**) to a more reactive *cis*-hydroxo-olefin species (**2**) (*15,48*) [Eqs. (12) and (13)]:

$$trans\ [C_2H_4PdCl_2(OH)]^- + H_2O \rightleftharpoons [C_2H_4PdCl(OH)_2]^- + H^+ + Cl^- \quad (12)$$

$$[C_2H_4PdCl(OH)_2]^- + H^+ + Cl^- \rightleftharpoons cis\ [C_2H_4 PdCl_2(OH)]^- + H_2O \quad (13)$$

$$(2)$$

While the question of complex stereochemistry has not been considered widely (*10*), it would appear that most authors at least tacitly assume a *cis*-hydroxo-olefin complex as vital for subsequent steps (*11,44,61*).

The additional rate term at high palladium and chloride concentrations (*11*) has been accounted for by postulating formation of a di-palladium-olefin species (**3**) [Eq. (14)].

$$[C_2H_4PdCl_2(OH)]^- + PdCl_4^{-2} \rightleftharpoons \begin{bmatrix} Cl \diagdown Cl \diagdown \overset{\displaystyle Cl}{|} OH \\ PdPd \\ Cl \diagup Cl \diagup C_2H_4 \end{bmatrix}^{-2} + Cl^- \quad (14)$$

$$(3)$$

Reaction of olefins and acetate in acetic acid and other nonaqueous media is thought to involve similar equilibria. Formation of species such as $[C_nH_{2n}PdCl_2(OAc)]^-$ (*38,40*), $[C_nH_{2n}PdCl(OAc)_2]^-$ (*38*), or $[C_nH_{2n} Pd(OAc)_2(HOAc)]$ (*11*) has been proposed, depending on the relative chloride and acetate concentrations.

The dependence of rate on Cu(II) at high copper and chloride concentrations has been accomodated by postulating a chloride bridged, binuclear complex $[C_nH_{2n}PdCl_3CuCl_2]^-$ (*40,42*).

Rearrangement of the initially formed π-olefin complex to a σ-complex, in which the olefin has inserted into the nucleophile–palladium bond, is the generally accepted next step in the reaction in both aqueous and non-aqueous systems (*27,33,40,43,48,60,62*) [Eqs. (15),(16)].

$$[C_2H_4(PdCl_2(OH)]^- \longrightarrow [(HO-CH_2CH_2)-PdCl_2]^- \qquad (15)$$
$$\textbf{(4)}$$

$$[C_2H_4PdCl_2(OAc)]^- \longrightarrow [(AcOCH_2CH_2)-PdCl_2]^- \qquad (16)$$
$$\textbf{(5)}$$

Oxymetallation products are known to form in similar reactions involving Tl(III), Hg(II), and Pb(IV) (*63*). However, stable Pd(II) products have been obtained only from certain cyclic dienes by reaction with Pd(II) salts and alcohols in the presence of weak bases (*64,65*). Attempts to prepare (**4**) and (**5**) by reaction of $PdCl_2$ with the corresponding Hg(II) compounds led to formation of Pd(0) and acetaldehyde and vinyl acetate, respectively. This has been taken as evidence that the desired oxypalladation products formed and decomposed and that the σ complexes are, therefore, reasonable intermediates (*62*).

The small magnitude of the effect of olefin structure on the rate of π complex decomposition has been interpreted as indication that the transition state for the π to σ rearrangement has little carbonium ion character. In contrast, olefin structure exerts a large effect in the formation of oxymetallation products of Tl(III) and Hg(II) (*66*). In the case of palladium, the rearrangement is pictured as proceeding as a concerted, four center addition (*44*) [Eq. (17)].

$$(17)$$

This *cis* insertion of ethylene contrasts with the *trans* stereochemistry found in stable oxypalladation products (*65*). The latter has been explained as resulting from attack of noncomplexed nucleophile from solution. However, kinetics of the reaction under discussion indicate that the nucleophile is complexed prior to insertion, and the transition state appears to have little carbonium ion character. Thus, *cis* addition is preferred (*11,44*). Such stereochemistry requires a planar transition state and, therefore, a *cis* disposition of olefin and nucleophile in the complex prior to insertion. Planarity is achievable by rotation of the olefin around the coordinate bond. Evidence for such rotation has been obtained for Rh(I) (*67*) and Pt(II) (*68*) ethylene complexes but not, as yet, for Pd(II).

Questions concerning complex rearrangements to achieve *cis* configuration and olefin rotation can be avoided by assuming a pentacoordinate intermediate (6) in which the attacking nucleophile occupies an axial coordination site prior to insertion (*10,15*). Feasibility of this concept has been demonstrated in the case of 1,5-cyclooctadiene insertion into an Ir-H bond (*69*). Alternatively,

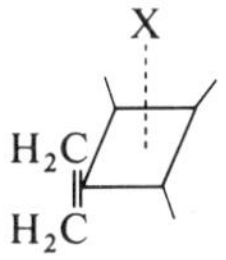

(6) 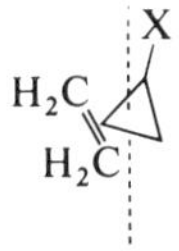   (7)

a trigonal bipyramidal configuration (7) in which the positions originally *trans* and axial become coplanar with and equidistant from the double bond also makes olefin rotation or rearrangements unnecessary (*10*).

A number of suggestions have been made as to the manner in which the σ oxypalladation products (4) and (5) are converted to products. All of these take into account the necessity for postulating a hydride shift in the course of this process. The simplest suggestion involves heterolytic cleavage of the Pd-C bond in (4) or (5) (*60,62*) [Eqs. (18) and (19)].

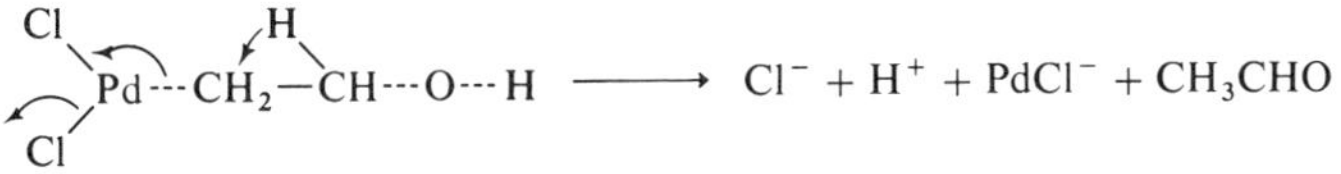

$$\text{Cl} \backslash \text{Pd---CH}_2\text{--CH---O---H} \longrightarrow \text{Cl}^- + \text{H}^+ + \text{PdCl}^- + \text{CH}_3\text{CHO} \quad (18)$$

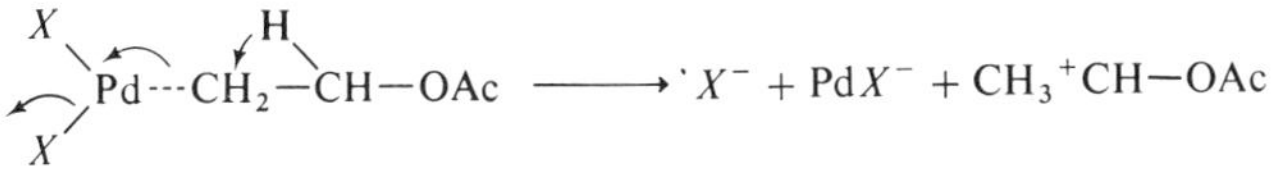

$$X \backslash \text{Pd---CH}_2\text{--CH--OAc} \longrightarrow {}^\cdot X^- + \text{Pd}X^- + \text{CH}_3{}^+\text{CH--OAc} \quad (19)$$

In the nonaqueous case [Eq. (19)], this leads to formation of a carbonium ion which can lose a proton to yield vinyl acetate or react with acetate (or acetic acid) to yield ethylidene diacetate.

It would appear reasonable that, if oxypalladation occurs and such intermediates were prone to heterolysis, at least some solvolysis of the Pd-C bond should occur with formation of 1,2-disubstituted products. This is indeed what occurs in the case of Tl(III) and Hg(II) reactions (*63*). As stated earlier however, 1,2-disubstituted products are only found in the presence of high Cu(II) and Cl$^-$ concentrations or in the presence of nitrates. Therefore, except in these cases, heterolysis would not appear to be a reasonable suggestion. Moreover, formation of a "free" carbonium ion species has the further serious drawback that no products of skeletal rearrangement have been found (*24,27,31*).

Since there is no compelling reason (other than to account for products) to expect a concerted heterolytic cleavage, it has been suggested that the initially formed 1,2-oxypalladation product rearranges to a 1,1-oxypalladation product (48,70) in a series of equilibria involving palladium hydride elimination and readdition (48) [Eqs. (20) and (21)] followed by solvolytic Pd-C cleavage [Eqs. (22) and (23)].

$$\sigma[(HO-CH_2CH_2)-PdCl_2]^- \rightleftharpoons \pi[(HO-CH=CH_2)PdHCl_2]^- \tag{20}$$

$$\textbf{(8)}$$

$$\pi[(HO-CH=CH_2)PdHCl_2]^- \rightleftharpoons \sigma[CH_3-CH-OH)PdCl_2]^- \tag{21}$$

$$\sigma[CH_3-CH-OH)PdCl_2]^- + H_2O \longrightarrow CH_3-HC\begin{smallmatrix}\overset{+}{O}H\cdot2\\\\OH\end{smallmatrix} + (PdCl_2)^{-2} \tag{22}$$

$$CH_3-HC\begin{smallmatrix}\overset{+}{O}H_2\\\\OH\end{smallmatrix} \longrightarrow CH_3CHO + H_3O^+ \tag{23}$$

Alternatively, a 1,1-insertion product can be formed directly from the π-olefin complex in a single step involving a hydride shift in either a 4-(31) [Eq. (24)] or 5-(10) [Eq. (25)] coordinate intermediate.

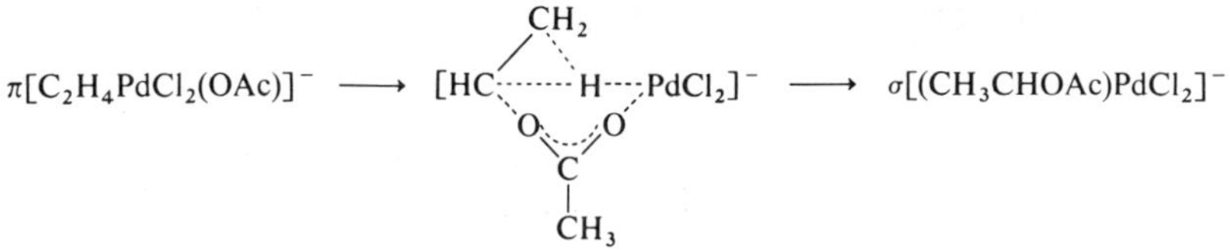

$$\pi[C_2H_4PdCl_2(OAc)]^- \longrightarrow \left[HC\begin{smallmatrix}CH_2\\\\ \end{smallmatrix}\cdots H\cdots PdCl_2\right]^- \longrightarrow \sigma[(CH_3CHOAc)PdCl_2]^- \tag{24}$$

$$\longrightarrow CH_3-\underset{\underset{OAc}{|}}{\overset{\overset{H}{|}}{C}}-Pd- \tag{25}$$

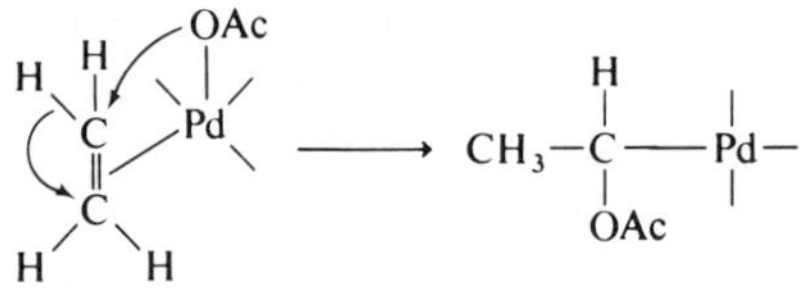

The elimination-addition of Pd-H is consistent with the known ability of Pt(II) hydrides to add reversibly to olefins (71) and the apparent β-hydrogen elimination of palladium alkyls (62,72). This proposal, however, would require that an intermediate, such as (8) must have sufficient stability to undergo further reaction. Since the only palladium hydride prepared to date is unstable to oxidizing and acidic conditions (73), which prevail in the present case, questions as to the feasibility of this route arise. Direct elimination of Pd-H from a 1,2-oxypalladation product to yield vinyl or allyl products (27) is inconsistent with lack of deuterium incorporation or exchange when the reactions are carried out in deuterated solvents. On the other hand, direct 1,2-hydride shifts in organic rearrangements are known to proceed without isotope exchange (*74*). This had led to suggestions of hydride shifts in decomposition of 1,2-insertion products with (*43*) [Eq. (26)] and without (*61*) [Eq. (27)] palladium participation.

$$
\begin{array}{c}
\underset{\underset{\underset{\textstyle Cl}{\diagup}}{\overset{\textstyle H}{\underset{\textstyle Pd\cdots H}{|}}}{\overset{\textstyle H}{\underset{\textstyle |}{C}}} {-} \overset{\textstyle H}{\underset{\textstyle |}{C}} {=} O \cdots H \longrightarrow CH_3CHO + HCl + Pd(O)
\end{array}
\qquad (26)
$$

$$
\begin{array}{c}
\underset{Cl}{\overset{Cl}{\diagdown}} Pd \underset{}{\overset{}{\diagup}} \overset{H}{\underset{}{C}} \overset{H}{\diagdown} H \\
\;\;\;\;\;\;\;\; C \\
\;\;\;\; H{-}O \;\; H
\end{array}
\xrightarrow{\;Cl^-\;} [HPdCl_3]^{-2} + CH_3CHO
\qquad (27)
$$

These suggestions focus attention on another unanswered question concerning the decomposition reaction. Namely, is Pd(II) reduced by electron or hydride transfer? Also unknown is whether the rate determining step in the overall reaction does (*11,24*) or does not (*43*) involve the hydride shift. Finally, the mechanisms cited largely seek to explain only vinylic products. Other steps have been proposed to account for allylic (*33,40*) and other (*27,31*) products.

Thus, no unique mechanism can be written, at this time, to account for all products obtained in the reaction of olefins with nucleophiles and Pd(II). This situation, however, may merely indicate that π-olefin complexes can be classed with such reactive intermediates as carbonium ions and radicals which can undergo a variety of reactions depending on their environment.

III. Oxidation of Alcohols to Carbonyl Compounds

A number of metal salts, notably those of group VIII noble metals, have been found capable of oxidizing alcohols to the corresponding aldehydes or ketones. The reaction has received relatively little study, either with respect to its preparative utility or its mechanism. However, it has been employed in the preparation of stable platinum, iridium, ruthenium, and osmium hydrides (*75–79*) [Eq. (*28*)].

$[PtCl_2(PEt_3)_2] + CH_3CH_2OH + KOH$

$$\longrightarrow [PtHCl(PEt_3)_2] + CH_3CHO + KCl + H_2O \qquad (28)$$

The reaction was studied briefly on the basis of an early suggestion that the production of acetaldehyde in the hydrolysis of Zeise's salt, $K[C_2H_4PtCl_3]$, might arise from hydration of ethylene to ethanol followed by oxidation to acetaldehyde (78). It was shown that this mechanistic route could be discounted. No ethanol was detected among hydrolysis products of Zeise's salt. Moreover, the rate of ethanol oxidation by K_2PtCl_4 was much slower than hydrolysis of the olefin complex under comparable conditions (79). Similar results were obtained in comparisons of the corresponding reactions in palladium systems (13,80).

More recent investigations indicate that the reaction of alcohols with salts of Pd(II), which leads to stoichiometric reduction of the metal to Pd(O) [Eq. (29)] can be made catalytic by an *in situ* reoxidation of Pd(II).

$$RCH_2OH + PdX_2 \longrightarrow RCHO + Pd(O) + 2HX \tag{29}$$

For systems involving $PdSO_4$ in 20% H_2SO_4, the regenerating agent is oxygen (39), while $PdCl_2$ in alcohol can be regenerated by $CuCl_2$, $Cu(NO_3)_2$, and 1,4-naphthoquinone (81). All of these, in turn are regenerable by oxygen.

The aldehydes formed from primary alcohols can be oxidized further to carboxyllic acids. These are either isolated as such (39) or as esters of the starting alcohols (39,81,82). Product aldehydes also react with alcohols to form acetals. The ratio of products seems to depend primarily on the particular alcohol reacted and the regenerating system used. When carried out in neat alcohols, the reaction appears to be inhibited at water concentrations greater than 20 mole % (81).

In order for reaction to occur, the presence of at least one α hydrogen is required. Thus 2-methyl-2-propanol fails to react while 2,2-dimethyl-1-propanol is converted readily to aldehyde (81).

Since no rate data are available, no detailed mechanism can be written at this time. However, on the basis of analogy to systems in which stable metal hydrides are formed, it seems reasonable to suppose that palladium hydride is formed by α hydrogen abstraction from the alcohol in either a concerted or stepwise manner [Eq. (30)]. Palladium hydride then decomposes to Pd(O) [Eq. (31)].

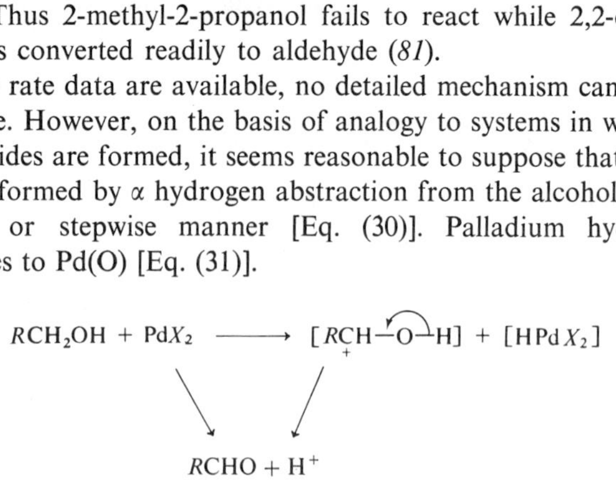

$$RCH_2OH + PdX_2 \longrightarrow [RCH-O-H] + [HPdX_2] \atop +$$

$$RCHO + H^+ \tag{30}$$

$$[HPdX_2]^- \longrightarrow Pd(O) + HX + X^- \tag{31}$$

IV. Oxidative Coupling Reactions

Broadly speaking, oxidative coupling reactions occur via the loss of a molecule of hydrogen from two molecules of substrate. When a metal salt catalyst is employed, it is reduced in this process, but can be returned to its active oxidation state by reaction with oxygen [Eqs. (32) and (33)]. The overall reaction, then, is one in which two substrates fragments have combined, and the hydrogen lost from them has been converted to water [Eq. (34)].

$$2RH + M^{+n} \longrightarrow R - R + M^{+(n-2)} + 2H^+ \quad (32)$$

$$M^{+(n-2)} + 2H^+ + \tfrac{1}{2}O_2 \longrightarrow M^{+n} + H_2O \quad (33)$$

$$2RH + \tfrac{1}{2}O_2 \longrightarrow R - R + H_2O \quad (34)$$

From the organic chemical point of view, such reactions have been rationalized as either radical or ionic in nature, depending primarily on the products formed and whether the "catalyst" has undergone an overall one or two electron change prior to reoxidation. However, in most cases which have been investigated thoroughly, the precise nature (radical or ionic) of organic intermediates is far from clear.

A comprehensive treatment of this reaction type is not possible due to its wide scope. Discussion is, therefore, limited to some recent investigations of homogeneous metal catalysis in this area.

A. COUPLING OF AROMATIC COMPOUNDS

Aromatic compounds react with Pd(II) salts to yield, primarily, coupled products and Pd(O). Thus, biphenyl is formed from benzene in acetic acid solutions containing $PdCl_2$ and sodium acetate (83) or with $Pd(OAc)_2$ only (84) [Eq. (35)].

$$2C_6H_6 + PdX_2 \longrightarrow C_6H_5-C_6H_5 + Pd(O) + 2HX \quad (35)$$

The reaction becomes catalytic with respect to palladium when carried out under oxygen in acetic acid containing perchloric acid (85).

1. Products and Rates

Products and rates depend on the aromatic compounds involved and the compositions of reaction media. Monosubstituted benzenes form mixtures of substituted biphenyls in which para and meta isomers predominate. Only small amounts of ortho isomers are found, presumably due to steric hindrance (83,84). However, some influence on isomer ratios by inductive effects of substituents also appears to be indicated (83).

In chloride-free media, phenyl acetate is formed from benzene in addition to biphenyl (84). Toluene yields benzyl acetate, Addition of lithium acetate increases acetoxylation relative to coupling. Acetate formation can be suppressed by addition of perchloric acid (84,85), and no phenyl acetate is found in the presence of oxygen (85). No coupling is obtained with $PdBr_2$ and PdI_2 (83). Use of $PdSO_4$ in 20% H_2SO_4 leads to benzene and toluene coupling, but also side chain oxidation, in the latter case (84).

The rate at which benzene is converted to biphenyl is influenced by a number of factors. No reaction was observed in acetic acid solutions containing $PdCl_2$ in the absence of sodium acetate (83). In reactions involving $Pd(OAc)_2$ in acetic acid, the rate was accelerated by $HClO_4$ (84,85). With monosubstituted benzenes, coupling rates were somewhat increased when the substituents were electron donating and decreased when they were electron withdrawing (83). In acetic acid, the rate of biphenyl formation was found to be first order in benzene and $PdCl_2$ but independent of sodium acetate concentration (83) [Eq. (36)].

$$d(\text{biphenyl})/dt = k_2[\text{PdCl}_2][\text{C}_6\text{H}_6] \tag{36}$$

Similar kinetics apply in chloride free systems (85).

No ring deuteration was found when reactions were carried out with $Pd(OAc)_2$ in DOAc. A considerable isotope effect was noted in the reaction of C_6D_6, $k_{C_6H_6}/k_{C_6D_6} = 5.0$ (85).

2. Mechanism

The above results indicate that benzene and Pd(II) combine in a rate determining step having the characteristics of an electrophilic aromatic substitution reaction. That is, products and rates are determined in the initial complex formation (83) [Eq. (37)].

$$\text{C}_6\text{H}_6 + \text{Pd}X_2 \xrightarrow{k} \langle\cdots\rangle \tag{37}$$

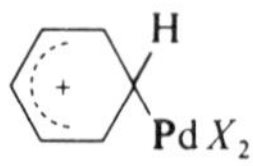

Irreversibility of this step is shown by lack of exchange in DOAc.

On the basis of analogy to reactions of Tl $(OAc)_3$ and $Hg(OAc)_2$ with benzene, it was suggested originally (84) that, in subsequent fast steps, the initial benzene-palladium (II) adduct was converted to a σ-phenyl palladium (II) compound [Eq. (38)] which then decomposed to biphenyl and palladium.

$$\text{(ring structure with H, PdX}_2\text{)} + NaOAc \longrightarrow \phi PdX + NaX + HOAc \qquad (38)$$

Decomposition of ϕPdX by a radical mechanism was eliminated by the observation that no phenol was formed when the reaction was carried out under oxygen (83). However, it does appear that a one electron transfer is involved. When benzene reacts with $Pd(OAc)_2$ in acetic/perchloric acid, under anaerobic conditions, at less than $60°C$, no metallic palladium precipitates. Instead, a magenta material forms which has been isolated and identified as a Pd(I) species, $[Pd(C_6H_6)(H_2O)(ClO_4)]_n$ (85). This material disproportionates to Pd(O) and Pd(II) in the presence of halides, and is reoxidized readily to Pd(II) by oxygen or other oxidizing agents.

Since decomposition of ϕPdX apparently involves a one electron transfer to palladium, but there is no evidence for phenyl radical formation, a rapid bimolecular decomposition reaction [Eq. (39)] has been proposed (85).

$$2\phi PdX \longrightarrow \phi-\phi + 2PdX \qquad (39)$$

While the precise mode of decomposition of σ-aryl palladium complexes remains somewhat obscure, their intermediacy in aromatic coupling reactions appears reasonably certain. Thus, stable p-tolyl mercury (II) acetate and bis-p-tolyl mercury react with palladium (II) acetate to produce bitolyl and Pd(O) (86). These reactions are envisioned as entailing displacement of mercury and formation of the more reactive aryl palladium species. In the case of the very rapid reaction of bis-p-tolyl mercury with palladium (II) acetate, the observed stoichiometry suggests that the intermediate, p-tolyl palladium (II) acetate attacks a second mole of bis-p-tolyl mercury [Eqs. (40)–(42)].

$$ArHgAr + Pd(OAc)_2 \longrightarrow ArPd(OAc) + ArHg(OAc) \qquad (40)$$

$$ArPd(OAc) + ArHgAr \longrightarrow Ar-Ar + Pd + ArHg(OAc) \qquad (41)$$

$$2ArHgAr + Pd(OAc)_2 \longrightarrow Ar-Ar + Pd + 2ArHg(OAc) \qquad (42)$$

B. Coupling of Olefins with Aromatic Compounds

A reaction which appears to be closely related to the coupling of aromatic compounds is the coupling of aromatic compounds with olefins. In a typical example of this reaction, ethylene and benzene couple to yield styrene in refluxing acetic acid containing $Pd(OAc)_2$ or $PdCl_2$ and sodium acetate. Similarly, styrene reacts with benzene to form *trans*-stilbene (*87,88*) [Eq. (43)].

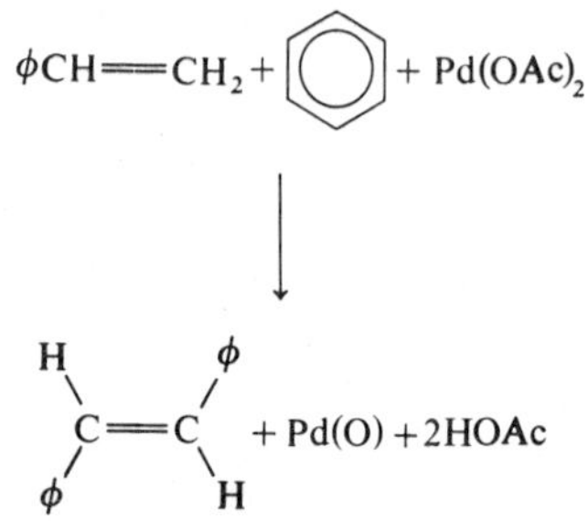

$$\phi CH = CH_2 + \bigcirc + Pd(OAc)_2$$

$$\downarrow \qquad\qquad (43)$$

$$\begin{array}{c} H \\ \diagdown \\ C \end{array} = \begin{array}{c} \phi \\ \diagup \\ C \end{array} + Pd(0) + 2HOAc$$

In the reaction of styrene, yields of stilbene increase with increasing acetate. In fact, the presence of acetate appears to be essential, as the reaction will not occur in the presence of $PdCl_2$, unless carried out in acetic acid (*89*). Yields of coupling products are somewhat better when $Pd(OAc)_2$ is used rather than a combination of $PdCl_2$ and sodium acetate.

As indicated, Pd(II) is reduced to Pd(0). The latter, however, can be regenerated with silver(I) acetate or copper(II) acetate (*90*). The reaction can, therefore, be considered as catalyzed by Pd(II).

An increase of the number of aromatic substituents on ethylene affects further coupling only in terms of rate. Thus the reactivity of phenyl substituted ethylenes toward benzenes is styrene > 1,1-diphenylethylene > *trans*-stilbene > triphenylethylene (*91*). Substituents on benzene exert a directing effect on the coupling position but do not exert much influence on overall rates (*92*). Benzene couples to the terminal carbons of propylene and 1-butene. With olefinic compounds, such as acrylonitrile or vinyl acetate, the β-carbon, i.e. the terminal olefinic carbon, is again involved (*92*).

While the mechanism of the reaction has not been investigated specifically, rationalization of observations reported to date is possible by assuming formation of a σ-arylpalladium complex from the aromatic compound and Pd(II). Insertion of olefin into the Pd-C bond in such a complex, followed by palladium hydride elimination, would lead to the observed products [Eq. (44)].

$$\text{ArPd}X + \overset{H}{\underset{H}{>}}C=C\overset{H}{\underset{H}{<}}$$

$$\underset{\underset{H\ \ H}{|\ \ \ |}}{\overset{H\ \ H}{\underset{|\ \ \ |}{Ar-C-C-PdX}}} \longrightarrow \text{ArCH}=\text{CH}_2 + [\text{HPd}X] \tag{44}$$

Indeed, such a mechanism has been proposed for the reaction of alkyl or aryl mercury (II) salts with group VIII metal salts and olefins (93). This interesting reaction has been investigated extensively (93–99). Examples include alkylation, arylation, and carboxylation of olefinic compounds by reaction with a mercury, tin, or lead alkyl or aryl salt or ester, and a salt of Pd(II), Ru(III), Rh(III), Fe(III), or Ni(II) in polar solvents, such as methanol, acetonitrile, or acetic acid. Best results are obtainable with Pd(II) salts. If alkylation is to be carried out, it is required that the alkyl group have no β-hydrogens. The only other restrictions seem to be the stability of the organometallic and the solubility of the group VIII metal salt.

Most examples of the reaction involve ϕHgCl and Li$_2$PdCl$_4$. The reactions proceed readily at room temperature and are unaffected by moisture or air. The Pd(II), which is reduced to Pd(O), can be regenerated, as usual, with CuCl$_2$ and air. However, presence of CuCl$_2$ leads to some product chlorination.

Other observations, which are particularly pertinent to the Pd(II) catalyzed coupling of olefins and aromatic compounds, have been made. Thus, in reactions of unsymmetrically substituted olefins, the alkyl or aryl group couples exclusively to the less substituted olefinic carbon. Olefin reactivity decreases with increasing substitution on the olefin.

Mechanistic details remain unknown, at this time. It has been suggested that the first step [Eq. (45)] is σ-aryl palladium complex formation by exchange between the Pd(II) salt and aryl mercury. This step is followed by olefin insertion and palladium hydride formation [Eqs. (46) and (47)] in the same manner as suggested for the palladium-only system.

$$\phi\text{HgCl} + \text{PdCl}_2 \rightleftharpoons [\phi\text{PdCl}] + \text{HgCl}_2 \tag{45}$$

$$[\phi\text{PdCl}] + \text{C}_2\text{H}_4 \longrightarrow \phi\text{CH}_2\text{CH}_2\text{PdCl} \tag{46}$$

$$\phi\text{CH}_2\text{CH}_2\text{PdCl} \longrightarrow \phi\text{CH}=\text{CH}_2 + [\text{HPdCl}] \tag{47}$$

Reversibility of σ-aryl palladium formation in exchange reactions appears reasonable in view of the formation of mercury allyls from π-allyl palladium salts and mercury (*100*). This step, if viewed as an insertion, may involve a metal-metal bonded intermediate [Eq. (48)].

$$\phi HgCl + PdCl_2 \; \rightleftharpoons \; \underset{Cl \quad Cl}{\overset{\phi}{Hg \quad Pd-Cl}} \; \rightleftharpoons \; HgCl_2 + [\phi PdCl] \qquad (48)$$

Neither the palladium-only or the palladium-mercury reaction appears to involve either radical or ionic intermediates. Decomposition of the olefin insertion product by palladium hydride elimination is compatible with the known instability of such σ-alkyls with β-hydrogens relative to olefins. However, this may be system dependent. Thus the reaction of ethylene, $ClHgCOOCH_3$ and $PdCl_2$ yields vinyl acetate in acetonitrile (*93*) and methyl propionate in tetrahydrofuran at 90°C (*101*).

C. COUPLING OF β-SUBSTITUTED α-OLEFINS

The coupling of β-alkyl or aryl-α-olefins to 1,1,4,4-tetra-substituted butadienes (*28,102*) and of vinyl acetate to 1,4-diacetoxy-butadiene (*103*) are further examples of oxidative coupling of unsaturated materials promoted by Pd(II). Both reactions are reported to require $Pd(OAc)_2$ as the palladium species. In the case of vinyl acetate, substitution of $PdCl_2$ and lithium acetate or $Cu(OAc)_2$ for $Pd(OAc)_2$ leads exclusively to the Pd(II) catalyzed decomposition to acetaldehyde and acetic anhydride (*104*) which is also observed as a minor side reaction when coupling predominates. This indicates, once again, that in homogeneous metal catalyzed reactions, the composition of the coordination sphere is a prime factor in determining products.

As before, Pd(II) is reduced to Pd(0) stoichiometrically in these reactions, but can be regenerated with $Cu(OAc)_2$ (*103*).

In the coupling of vinyl acetate, increasing acetic acid concentrations in reaction media lead to rate decreases and formation of acetic acid adducts, such as 1,1,4-triacetoxy-2-butene and 1,1,4,4-tetraacetoxybutane, at the expense of primary coupling product (*103*). Similar reactions are probably responsible for complications in product analyses in reactions of β-acetoxy- or β-chloro-α-olefins (*102*). Formation of 1-acetoxybutadiene in the coupling of vinyl acetate has also been observed (*103*).

If β-alkyl- or aryl-α-olefins are coupled in the presence of excess $Pd(OAc)_2$, further oxidation to di or tetrasubstituted benzenes takes place

(*102*). For example, isobutene in glacial acetic acid containing Pd(OAc)$_2$ and sodium acetate couples at 85°C to 2,5-dimethyl-2,4-hexadiene and 2,5-dimethyl-1,5-hexadiene. These, in turn, are converted to p-xylene [Eq. (49)].

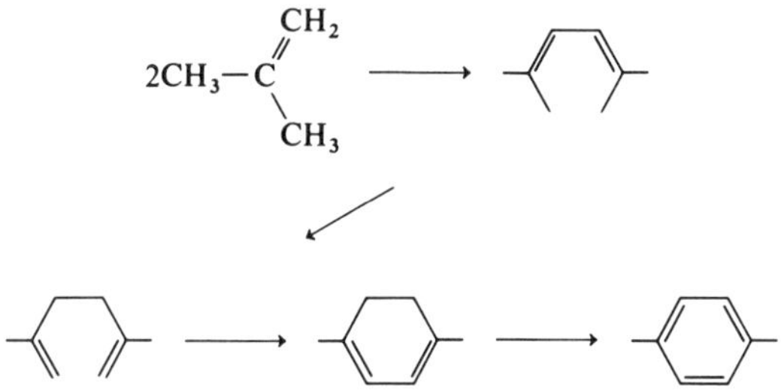

$$
\begin{array}{c}
\qquad\quad CH_2 \\
\qquad\quad \parallel \\
2CH_3-C \\
\qquad\quad \backslash \\
\qquad\quad CH_3
\end{array}
\longrightarrow
\quad \text{(p-xylene diene)}
$$

(49)

$$
\text{(structure)} \longrightarrow \text{(structure)} \longrightarrow \text{(structure)}
$$

Since it could be shown that 2,5-dimethyl-1,5-hexadiene was converted to xylene more rapidly than the 2,4-isomer, it was concluded that isomerization preceded cyclization. The final step was presumed to consist of dehydrogenation of the cyclohexadiene which should be rapid in the presence of freshly formed Pd(O).

The reaction is not notably affected by solvents and can be carried out in such media as DMF, 1,2-dimethoxyethane or acetic acid. While omission of sodium acetate from acetic acid media resulted in nonoxidative dimerization, e.g., formation of 4-methyl-2,4-diphenyl-2-pentene rather than 2,5-diphenyl-2,4-hexadiene from 2-phenylpropene, no such effect was noted in aprotic solvents.

The loss of hydrogens from vinylic carbons only was confirmed by formation of 2,5-diphenyl-2,4-hexadiene-1,6-d$_3$ from 2-phenyl-1-propene-3-d$_3$ without loss or scrambling of deuterium [Eq. (50)].

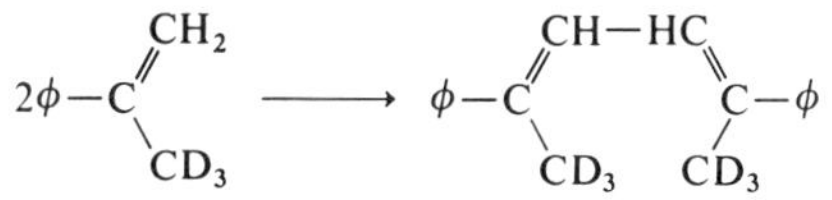

$$
\begin{array}{c}
\qquad CH_2 \\
\qquad \parallel \\
2\phi-C \\
\qquad \backslash \\
\qquad CD_3
\end{array}
\longrightarrow
\begin{array}{c}
CH-HC \\
\phi-C \qquad\quad C-\phi \\
\quad\backslash \qquad\quad / \\
\quad CD_3 \quad CD_3
\end{array}
$$

(50)

No incorporation of deuterium into coupling products was observed when the reaction was carried out in DOAc. Therefore, it appears that π-allyl intermediates are not involved and that mechanisms involving stable organic radical or ionic intermediates are also unlikely.

The observed inability of PdCl$_2$ to participate in these reactions, has led to suggestions that the active palladium species responsible for coupling of both vinyl acetate and of β-alkyl and aryl-α-olefins is an acetate bridged dimer in which the palladium nuclei are in close proximity. Complexing of a substrate molecule to each of the palladium nuclei would bring the vinylic CH_2 groups into position for coupling. Suggested details (*102*) are that approach of acetate to one of the π-complexed olefin moieties facilitates hydride abstraction from the methylene group by palladium. Synchronous oxidation of hydride to proton with the electron pair being transferred back to the olefin results in formation of a vinyl carbanion. The latter attacks the neighboring olefin with hydride displacement to Pd, leading to coupling and formation of one molecule of Pd(O) from the dimer [Eq. (51)].

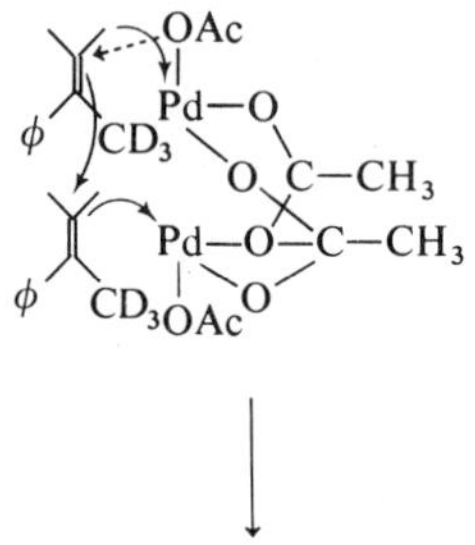

$$
\phi-C=CH-CH=C-\phi \ + \ Pd(OAc)_2
$$
$$
\underset{CD_3}{|} \qquad \underset{CD_3}{|}
$$
$$
+ \ HOAc \ + \ [HPd \ OAc]
$$

(51)

There are several problems with this suggestion. First, the process would have to be entirely synchronous in order to explain the non-incorporation of deuterium from DOAc. Second, acetate, present in 2/1 excess over palladium in these cases, should tend to convert palladium acetate dimers to monomeric species. This may explain the decrease in coupling products relative to enol acetates observed in reactions of styrene and methyl styrene (*28*).

An alternate reaction scheme in closer accord with mechanisms suggested for the previously discussed coupling reactions promoted by Pd(OAc)$_2$ might involve the formation of vinyl acetate species from olefin, as a first step [Eq. (52)].

$$
\phi-\underset{CD_3}{\underset{|}{C}}=CH_2 + Pd(OAc)_2 - \phi-\underset{CD_3}{\underset{|}{C}}=CHOAc + Pd(O) + HOAc \qquad (52)
$$

Indeed, enol acetates are observed as by-products in coupling reactions (*102*). Moreover, it is known that no deuterium is incorporated into vinyl acetate formed in DOAc (*62a*).

The next step would be oxidative addition of enol acetate to Pd(O)

[Eq. (53)]. Reactions of this type have been observed to occur with certain vinyl halides (*105*).

$$\phi-\underset{\underset{\text{CD}_3}{|}}{\text{C}}=\text{CHOAc} + \text{Pd(O)} \longrightarrow \phi\underset{\underset{\text{CD}_3}{|}}{\text{C}}=\text{CH}-\text{Pd}-\text{OAc} \qquad (53)$$

The resulting σ-vinyl palladium complex may then undergo olefin insertion followed by palladium hydride elimination [Eqs. (54) and (55)].

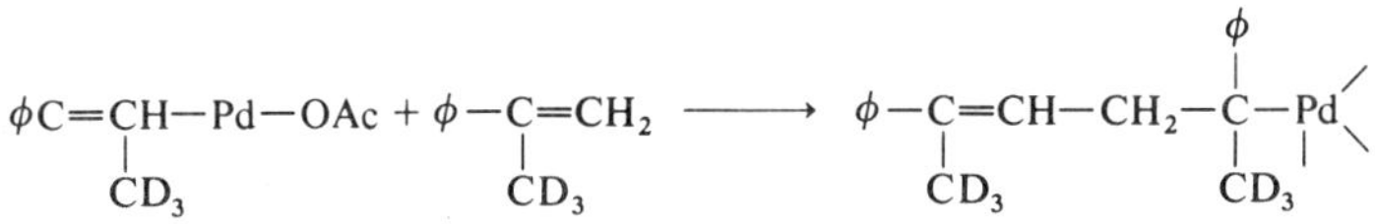

$$\phi\text{C}=\text{CH}-\text{Pd}-\text{OAc} + \phi-\text{C}=\text{CH}_2 \longrightarrow \phi-\text{C}=\text{CH}-\text{CH}_2-\text{C}-\text{Pd} \qquad (54)$$

$$\phi-\text{C}=\text{CH}-\text{C}-\text{C} \longrightarrow \phi\text{C}=\text{CH}-\text{CH}=\text{C}\phi + [\text{H}-\text{Pd(OAc)}] \qquad (55)$$

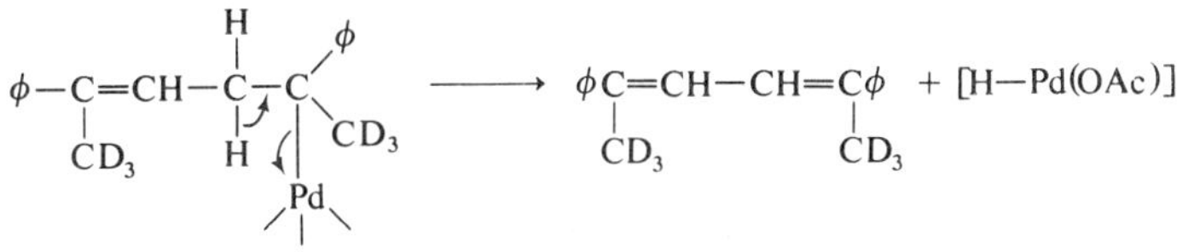

This scheme, while accounting for products formed from β-alkyl- or aryl-α-olefins, accounts only for the 1-acetoxybutadiene by-product observed in the coupling of vinyl acetate. On the other hand, the scheme which involves vinyl carbanion formation would also not seem particularly appropriate for vinyl acetate. Therefore, it may be that vinyl acetate coupling proceeds by yet another mechanism.

D. Coupling of Thiols

The oxidative coupling of alkyl and aryl thiols to disulfides has been the subject of several detailed investigations (*106–108*). The reactions most frequently are carried out in aqueous media and will not proceed in the absence of a base, such as NaOH (*106*). They can be catalyzed by a variety of metal salts, those of copper, cobalt, and nickel generally being most effective (*107*). Overall, oxygen consumption is stoichiometric for the reaction [Eq. (56)] and the high selectivity to disulfides implied by this has been confirmed by product analyses (*107*).

$$4R\text{SH} + \text{O}_2 \longrightarrow 2R\text{SS}R + 2\text{H}_2\text{O} \qquad (56)$$

Qualitative rate comparisons indicate that the nature of the organic portion of thiols has some effect on reaction rates (*106,107*). However, the extent to which this may be a steric or electronic phenomenon is uncertain.

In all cases examined, the rate of reaction has been found to depend on oxygen partial pressure but not on base concentration (*107,108*). The dependence of rate on thiol concentration appears to be a function of the catalyst and the particular thiol employed, but in the overwhelming number of cases studied, rates do not depend on thiol concentrations (*107,108*). Since many of the metal salts or complexes used as catalysts are converted to insoluble species, few measurements of rate dependence on catalyst concentration have been possible. In general, a rate increase with increasing catalyst/substrate ratio has been noted (*106*). In the case of ferricyanide catalysis, first order rate dependence on catalyst concentration was found (*108*).

The following reaction scheme has been proposed to account for the gross features of the reaction (*106*) [Eqs. (56')–(60)].

$$RSH + OH^- \longrightarrow RS^- + H_2O \tag{56'}$$

$$RS^- + M^{+n} \longrightarrow RS\cdot + M^{(n-1)} \tag{57}$$

$$2RS\cdot \longrightarrow RSSR \tag{58}$$

$$2M^{+(n-1)} + O_2 \longrightarrow 2M^{+n} + O_2^{-2} \tag{59}$$

$$O_2^{-2} + H_2O \longrightarrow \tfrac{1}{2}O_2 + 2OH^- \tag{60}$$

The requirement of basic conditions clearly indicates that thiol is converted to thyil anion prior to reaction, and there is general agreement on this step. However, considerable discussion has been devoted as to which step in the sequence is rate determining, the details of electron transfer from thiyl anion to metal, and whether or not thiyl free radicals in the ordinarily accepted sense are, indeed, involved.

In cases in which the overall rate was found to be independent of thiol, the conclusion that the reoxidation of metal is rate controlling (*107*) appears justified. Nonetheless, this does not come to grips with the unanswered questions concerning the details of catalyst action. It has been possible to demonstrate, in a number of instances (*108*), that metal-thiol complexes are formed in solution. When strong complexing agents, such as cyanide or ethylenediamine are added, no thiol complexes can be detected. Under these conditions, reaction rates are severely retarded and only small amounts of disulfides are found among products. This observation suggests that there may be two catalytic mechanisms for thiol coupling: an "inner sphere" mechanism in which substitution of thiyl anion into the metal coordination sphere precedes electron transfer, and an "outer sphere" mechanism in which such substitution is blocked by stronger ligands.

Under conditions in which the "inner sphere" mechanisms is thought to apply, no thiyl radicals could be detected by addition of trapping agents. Therefore, the possibility of involvement of two metal centers has been suggested. The proposed scheme, in which disulfide formation is a consequence of radical coupling, may thus be valid only for the "outer sphere" mechanism. However, since the coupling of thiyl radicals is known to be extremely rapid (109), it is not known whether their intermediacy can be eliminated from consideration in "inner sphere" mechanisms.

Very likely the presence of such ligands as cyanide or ethylenediamine affects the rate of metal reoxidation. Nonetheless, in the case of catalysis by ferricyanide, it has been shown that the overall rate of reaction, as measured by oxygen absorption, is still several orders of magnitude slower than the rate of Fe(III) reduction. This observation, together with the finding that "outer sphere" reaction rates are largely independent of thiol concentration, again indicates that metal reoxidation is overall rate controlling.

While the suggestion of an "outer sphere" electron transfer accounts for coupling under conditions in which thiol complexing appears to be prevented, a considerable influence of thiol structures on measured rates has been noted in these cases as well. Since electron transfer does not appear to be rate controlling, this effect has been explained as due to approach of thiyl anions to the metal catalyst complexes in some preferred orientation.

As can be seen, mechanistic details remain obscure even in relatively simple reactions involving only one product. However, since most of the difficulties associated with the interpretation of results, in this instance, have arisen from the superposition of metal reoxidation on the coupling reaction, it seems that a study of the reaction under anaerobic conditions (i.e. stoichiometric in metal) might prove fruitful.

E. COUPLING OF PHENOLS

Homogeneous metal catalyzed oxidative coupling of phenols has received considerable attention due to the reaction's synthetic applications and its possible involvement in some natural product biosyntheses (110). Products generally are either carbon-carbon or carbon-oxygen coupled materials. Selectivity depends on reaction conditions, the catalyst, and the particular phenol employed.

A catalyst system of particular interest, since it is also effective in acetylene coupling (111–113), consists of copper salts and amines. When used with 2,6-disubstituted phenols, the type of product (C—C or C—O coupled) depends primarily on the ratio of amine to copper in the catalyst Thus, for a pyridine-CuCl catalyst in polar or nonpolar solvents, high pyridine/Cu ratios favor coupling of 2,6-dimethylphenol to poly-2,6-dimethyl-1,4-phenylene ether. Low pyridine/Cu ratios favor formation of 3,3',5,5'-tetramethyldiphenoquinone (114). At constant pyridine/Cu, carbon-carbon coupling is favored by bulky ortho substituents on the phenol (115). In all cases, oxygen is absorbed until the phenolic substrate has been exhausted.

Recent studies of the reaction have centered around attempts to characterize the catalyst. In view of the known ability of oxygen to convert Cu(I) rapidly, in the presence of basic ligands, to Cu(II), the possibility that simple Cu(II) complexes could effect phenol coupling was examined (116,117). It was found that such materials as $CuCl_2$, Py_2CuCl_2, $Cu(OH)_2$, and $Py_2Cu(OR)_2$ were inactive. On the other hand, materials formed in situ by oxidation of pyridine-CuCl catalysts were active under noncatalytic (anaerobic) conditions, producing the same products and yields from substrates, depending on the pyridine/Cu ratio (117). Some deactivation was noted when catalysts were stored in air prior to use (116). Addition of chelating amines, such as ethylenediamine or diethylenetriamine, had no effect, while triethylenetramine at amine/Cu ratios ≥ 1 was completely deactivating.

For the coupling of 2,6-di-t-butyl phenol to 3,3',5,5'-tetra-(t-butyl) diphenoquinone in pyridine, the overall reaction rate was found to have the following dependence on phenol, copper, and oxygen (116) [Eq. (61)]:

$$-d[O_2]/dt = k[Cu]^2[ArOH][P_{O_2}] \tag{61}$$

Absorption spectra of the active catalyst resembled those of $[Py_2CuCl_2(OH)]^-$. Electron spin resonance spectra clearly showed the presence of Cu(II) in the active systems. Addition of phenol decreased the ESR signal, indicating reduction of Cu(II) to Cu(I). However, some residual resonance remained in the presence of phenol, raising the possibility of radical species formation from the substrate.

These results have been interpreted as indicating that initially present pyridine-Cu(I) complexes are oxidized to species containing O_2^- or OH^- [Eq. (62)], which, by virtue of their basicity, are able to abstract a proton

$$2Cu(I) + 2Py + O_2 \longrightarrow PyCu—O_2—CuPy \longrightarrow 2PyCu(II)—O^- \tag{62}$$

from complexed phenol. The phenolate anion produced then reduces Cu(II) to Cu(I), and the resulting radical couples.

Spectroscopic investigations of pyridine-CuCl catalysts in methanol suggested the formation of two compounds at low pyridine/CuCl. One of these was shown to increase at the expense of the other as the pyridine/Cu ratio was increased (*117*) in line with previous predictions concerning the necessity of two catalyst species to explain C—C or C—O coupling (*114*).

Oxidation of CuCl in pyridine/methanol resulted in formation of a material with the empirical composition, PyCuCl (OCH$_3$), which was active in phenol coupling. While it was not possible to determine the molecular weight of this material, a chloride-bridged, dimeric structure (**9**) was proposed.

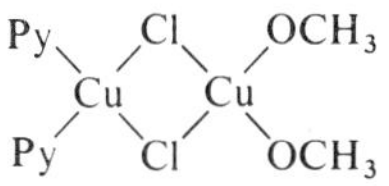

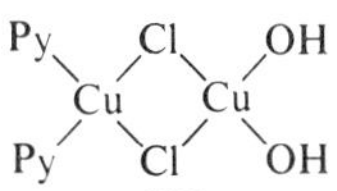

This suggestion is consistent with the previously cited study in which the active species might be assumed to be [PyCuCl(OH)]$_2$ (**10**) or [PyCuCl(O^{-2})]$_2$.

Attempts to exchange the methoxy groups in (9) with phenoxy groups, led either to no reaction or coupling.

On the basis of these results, the following mechanism for phenol coupling at low pyridine/Cu ratios was suggested [Eqs. (63)–(66)]:

$$4Cu^{I}Cl + 4Py + 4CH_3OH + O_2 \tag{63}$$

$$\downarrow$$

$$[PyCu^{II}Cl(OCH_3)]_2 + 2H_2O$$

$$[PyCuCl(OCH_3)]_2 + 2ArOH \longrightarrow [PyCuCl(OAr)]_2 + 2CH_3OH \tag{64}$$

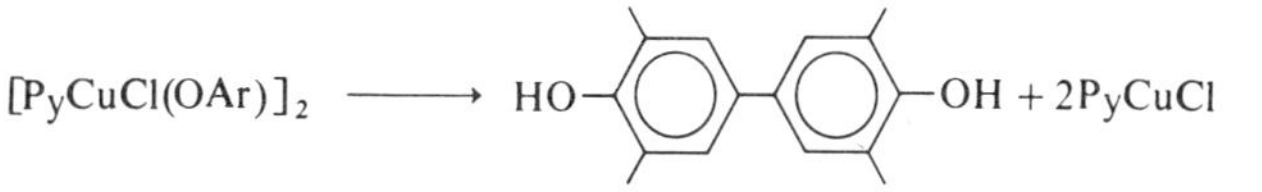

$$(65)$$

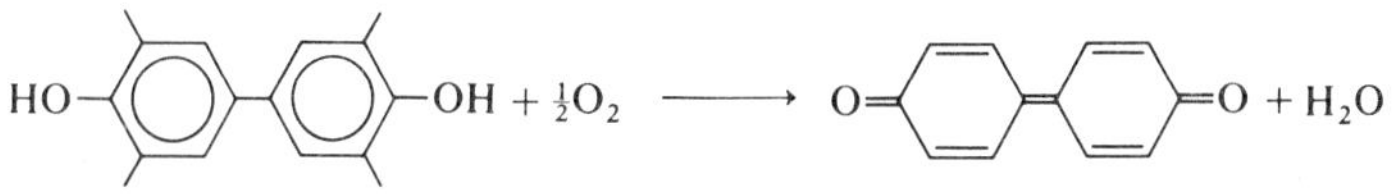

$$(66)$$

At high pyridine/Cu ratios, it was assumed that the bridged, dimeric Cu(II) species (**9**) was cleaved by pyridine [Eq. (67)] to produce a mononuclear species, Py$_2$CuCl(OCH$_3$) (11), which, again, exchanges with

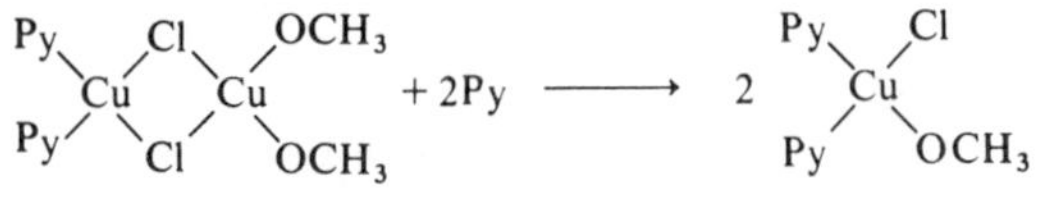

$$Py\diagdown_{Cu}\diagup^{Cl}\diagdown_{Cu}\diagup^{OCH_3} + 2Py \longrightarrow 2 \quad Py\diagdown_{Cu}\diagup^{Cl}_{OCH_3} \qquad (67)$$

(11)

phenol. Formation of the phenol exchanged material is then followed by C—O coupling and reduction of Cu(II). The details of coupling are not discussed. However, it seems unlikely that such high specificity as a function of catalyst composition would result if phenoxy radicals formed and coupled in an independent step. Therefore, the reactions should probably be envisioned as insertion processes in which free radicals, as such, do not separate from the complexes.

It is of interest to note here that while much remains unknown about mechanistic details of homogeneous metal catalyzed coupling reactions, most investigations have resulted in similar mechanistic suggestions. Thus, a number of reactions of this type are thought to involve binuclear complexes and one electron transfers to metal, and coupling of organic moieties within the complexes rather than formation of independent free radical species.

V. Acetoxylation of Alkylbenzenes

Acetoxylation was mentioned briefly as a side reaction observed in the oxidative coupling of aromatic compounds. Under certain conditions, this reaction becomes predominant. For example, in acetic acid containing $PdCl_2$ and sodium acetate, the ratio of bitolyl to benzyl acetate and benzylidene diacetate obtained from toluene was found to change from 38 to 0.0074 as the $NaOAc/PdCl_2$ ratio was varied from 5 to 20 (*118*). Overall yields decreased sharply at intermediate acetate/$PdCl_2$ ratios. The reaction of toluene with $Pd(OAc)_2$ in acetic acid yields benzyl acetate primarily (*84*). Addition of potassium or sodium acetate to this system increases the rate of benzylic oxidation product formation and almost completely suppresses the small amount of coupling product formed at low acetate/Pd(II) [Eq. (68)].

$$\phi CH_3 + KOAc + Pd(OAc)_2 \xrightarrow[HOAc]{100^\circ} \phi CH_2OAc + \phi CH(OAc)_2 + Pd(0) \qquad (68)$$

As mentioned in the discussion of coupling reactions, acetoxylation is essentially eliminated by the presence of oxygen at low acetate/Pd(II) ratios (*84,85,119,120*). However, this effect is not noted at high acetate/Pd

(II) ratios, and, under these conditions, acetoxylation can be made catalytic by reoxidizing Pd(O) to Pd(II) with air (*121*). The reoxidation tends
to be rather sluggish in these systems. It can, however, be accelerated
greatly by the addition of $Sn(OAc)_2$ and high surface area solids such as
charcoal or alumina. The latter presumably keep precipitated metallic
palladium well dispersed.

The acetoxylation of tolune is considerably faster than that of benzyl
acetate (*121*).

Xylenes yield mixtures of xylyl acetate (12), xylylene diacetate (13) and
methylbenzylidene diacetate (14) (*122*) [Eq. (69)]

(69)

In this case, considerably more α,α-diacetate (**14**) is formed than α,α'-
diacetate (**13**). During initial reaction periods, the ratio of xylyl acetate to
xylylene diacetate was found to remain fairly constant after which the reaction appeared to become sequential. Ortho and meta-xylenes reacted
more slowly than para-xylene but yielded similar products. Toluene reacted more rapidly than the xylenes which reacted more rapidly than
mesitylene, durene, and hexamethylbenzene. Ethylbenzene yielded α and
β-acetoxyethylbenzenes, styrene, *cis* and *trans*-β-acetoxystryene and small
amounts of acetophenone and benzaldehyde (*121*). Benzaldehyde was also
observed as a by-product in reactions of toluene (*121,123*). It was shown
to result from hydrolytic and thermal decomposition of benzylidene
diacetate.

The reaction has been found to be enhanced by electron donating ring
substituents, such as alkoxy groups (*123*). Such substituents also give rise
to some nuclear acetoxylation (*121*). Electron withdrawing ring substituents, such as NO_2, Cl, or OAc, greatly decrease reaction rates.

Acid inhibition in perchloric-acetic acid mixtures has been noted (*84,85*). However, conversion of hexamethylbenzene to pentamethylbenzyl acetate was not affected by acid. Under these conditions, and in the absence of air, formation of Pd(I) rather than Pd(O) was found (*39*). Under oxygen, the reaction became catalytic. Residual acetoxylation of toluene by Pd(OAc)$_2$ in HClO$_4$-HOAc was also unaffected by oxygen.

Under regenerating conditions (high OAc/Pd, Sn(OAc)$_2$, and air), acetoxylation rates were found to be a function of Pd(II) but not toluene at low palladium concentrations and became diffusion limited at high Pd(II) concentrations (*121*). This was interpreted as indicating that, overall, Pd(II) regeneration was rate controlling.

No complete reaction mechanism has been suggested. However, it appears that three reaction types are involved (*39*). These are, reaction at low acetate/Pd(II), which is acid and O$_2$ inhibited, reaction at high acetate/Pd (II) which is unaffected by oxygen, and acetoxylation under strongly acidic conditions.

Inhibition of acetoxylation at low acetate/Pd(II) ratios suggests a radical mechanism under these conditions. This had led to comparisons with acetoxylations by Pb(OAc)$_4$ which are also oxygen inhibited (*120*). In the latter case, ring methylation [Eqs. (70)–(72)] accompanies acetoxylation. Such products are not found in reactions involving Pd(OAc)$_2$. Therefore, it has been suggested (*120*) that

$$Pb(OAc)_4 \longrightarrow Pb(OAc)_3 + OAc\cdot \qquad (70)$$

$$OAc\cdot \longrightarrow CH_3\cdot + CO_2 \qquad (71)$$

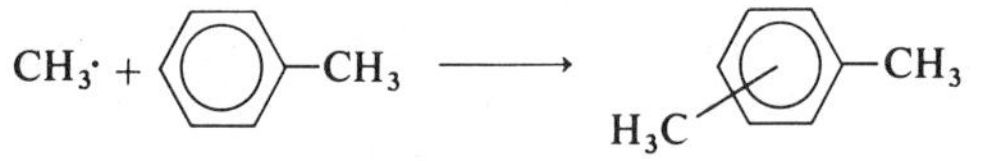

$$\qquad (72)$$

it has been suggested (*120*) that Pd(I), which is formed as a product of coupling, abstracts an electron from hydrocarbon in a chain process leading to acetoxylation [Eqs. (73)–(76)].

$$RH + Pd(OAc) \longrightarrow RH^+ + AOc^- + Pd(O) \qquad (73)$$

$$RH^+ \rightleftharpoons R\cdot + H^+ \qquad (74)$$

$$R\cdot + Pd(OAc)_2 \longrightarrow R^+ + AOc^- + Pd(OAc) \qquad (75)$$

$$R^+ + OAc^- \longrightarrow ROAc \qquad (76)$$

Acid inhibition would be explained by Eq. (74), if reversible, and by (75) and (76) as acetate concentrations would be greatly reduced under these circumstances. Oxygen inhibition may be the result of rapid oxidation of Pd(I) to Pd(II). Since Pd(I) is expected to be a weaker oxidizing agent than Pd(II), Eq. (73) has been explained as due to a kinetic effect on electron transfer, possibly because Pd(I) is paramagnetic.

A drawback of this scheme is that Pd(I) has been shown to disproportionate rapidly in the presence of Cl⁻ which, therefore might be expected to have the same effect on Eq. (73) as oxygen. This effect has not been noted. Secondly, a carbonium ion, if formed [Eq. (75)] (in this case, a benzyl carbonium ion) can be expected to result in products formed by electrophilic attack on a second aromatic molecule, as has been shown for reactions of this type (124). This difficulty, however, can be avoided by postulating ligand transfer to the radical species in Eq. (75).

Since explanations involving either radical or ionic organic species, in the generally accepted sense, do not appear to be entirely satisfactory, an alternative mechanism is suggested. This involves rearrangement of a σ-phenyl palladium (II) species (which was the suggested intermediate in aromatic coupling) to a σ-benzylpalladium (II) species, the latter undergoing reductive elimination [Eqs. (77)–(80)].

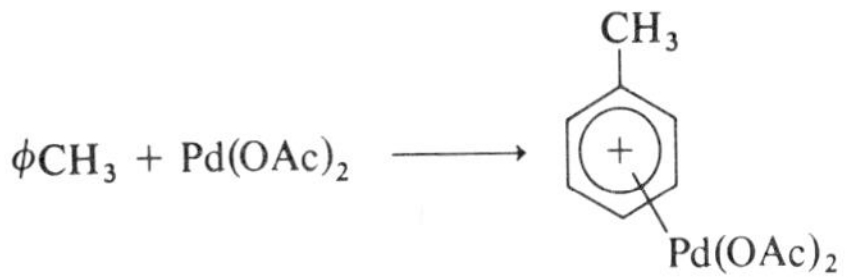

(77)

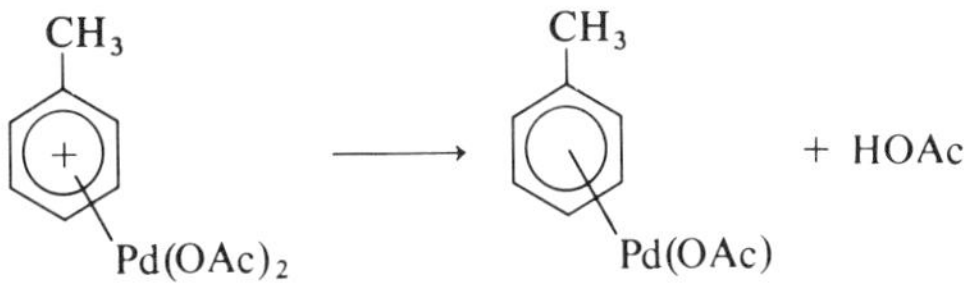

(78)

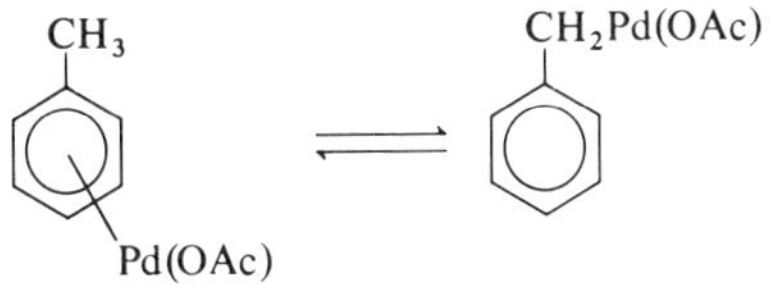

(79)

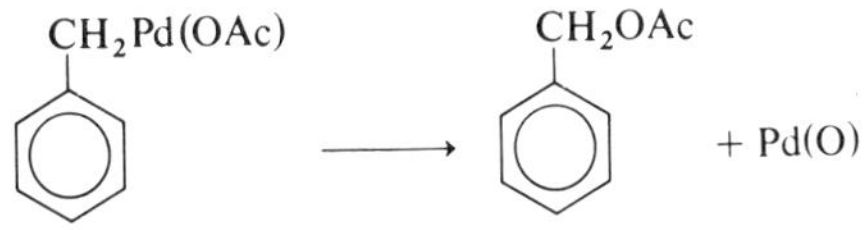

(80)

Support for this scheme comes from the observation that a benzylpalladium (II) species formed from tetrakistriphenylphosphine-palladium (O)

and benzyl chloride reacts with OAc⁻ to yield benzyl acetate and benzylidene diacetate (*125*). It is, however unclear why oxygen would interfere
with this route, unless the formation of benzyl palladium originates from
Pd(I) or Pd(O) rather than Pd(II). Acid inhibition can be accounted for
either by reversibility in Eq. (78) or by cleavage of benzyl palladium acetate
to toluene.

In order to account for results at high acetate/palladium ratios, it has
been suggested that suppression of coupling at increasing acetate concentration may indicate that coupling requires a bridged dimeric palladium
species which is converted by acetate to a mononuclear complex active for
acetoxylation (*118*). This suggestion is in accord with previously cited suggestions concerning the bimolecular nature of coupling catalysts. However,
it does not shed light on the mechanism of acetoxylation under these conditions. It is possible, of course, that some of the previously cited schemes
may apply in this instance since oxygen is not an inhibitor.

VI. Autoxidation

Reactions of organic compounds with oxygen have been studied ex-
haustively and discussed extensively (*2,3,126–128*). Various aspects of
metal catalysis in these reactions have also received considerable attention
(*129–132*). It would, therefore, be neither possible nor desirable to attempt
anything but a very brief summary of this area in the context of this
review.

Autoxidations of organic compounds generally have the characteristics
of radical chain processes. For uncatalyzed reactions, these have been discussed in terms of the following steps [Eqs. (81)–(87)]:

$$RH \longrightarrow R\cdot \qquad \text{Initiation} \qquad (81)$$

$$R\cdot + O_2 \longrightarrow RO_2\cdot \quad \left.\right\} \quad \text{Progagation} \qquad (82)$$

$$RO_2\cdot + RH \longrightarrow ROOH + R\cdot \qquad\qquad\qquad (83)$$

$$ROOH \longrightarrow RO\cdot + OH\cdot \qquad \text{Branching} \qquad (84)$$

$$R\cdot + R\cdot \longrightarrow R{-}R \qquad\qquad\qquad\qquad (85)$$

$$RO_2\cdot + R\cdot \longrightarrow ROOR \quad \left.\right\} \text{Termination} \qquad (86)$$

$$RO_2\cdot + RO_2\cdot \longrightarrow ROH + RCOR + O_2 \qquad (87)$$

The overall process can be accelerated greatly by a variety of metal salts
and complexes. Detailed investigations of catalyzed reactions have shown
that metals can participate in any of the steps outlined above. However,

there are frequently difficulties in establishing the precise role of a metal catalyst in a particular reaction. These stem not only from the ability of metals to participate in a number of oxidation states and to promote a number of steps in the same reaction but also from the ability of metals to react with radicals to form nonchain carrying products, thereby diverting or inhibiting the reaction. Moreover, since primary oxidation products frequently can also form radicals and trace metals are difficult to remove from reaction media, the interpretation of experimental results is far from simple. Therefore, mechanistic conclusions should be viewed with skepticism, particularly as to their general applicability. This is especially necessary when detailed mechanisms are proposed on the basis of product analyses only or are interpretations of overall kinetics.

The following discussion merely seeks to illustrate the manner in which homogeneous metal catalysts may be involved in various stages of the chain process. It will deal primarily with the autoxidation of hydrocarbons, with illustrative examples taken from the recent literature.

A. INITIATION

One of the characteristic features of uncatalyzed autoxidation processes is an induction period. That is, in the absence of a sufficient concentration of radicals to initiate the chain reaction, no oxygen absorption is noted. In many instances, the addition of metal salts notably shortens or eliminates the induction period. This indicates that the metal salts promote the formation of radicals which are then capable of reacting with oxygen and carrying on the chain.

The most frequently encountered manner in which metals induce initiation is by catalyzing the decomposition of hydroperoxides which have formed in the system but which are too stable to decompose spontaneously at the particular temperatures involved. This mode of initiation is the prime manner in which metals function as autoxidation catalysts and will be discussed under chain branching.

In some instances, induction periods can be eliminated by metal catalysts even when hydroperoxides have been excluded carefully. In these cases, radical formation can be envisioned to occur either by direct electron transfer from substrate to metal [Eqs. (88) and (89)], by hydrogen abstraction from substrate, possibly by species formed

$$RH + M^{+n} \longrightarrow RH\cdot + M^{+(n-1)} \tag{88}$$

$$RH^+ \longrightarrow R\cdot + H^+ \tag{89}$$

by interaction of the catalyst with oxygen [Eqs. (90) and (91)] or by decomposition of the catalyst in a manner which generates radicals

$$M^{+n} + O_2 \longrightarrow M^{+(n+1)}O_2^- \cdot \qquad (90)$$

$$M^{+(n+1)}O_2^- \cdot + RH \longrightarrow M-O-O-H + R \cdot \qquad (91)$$

by electron transfer from a ligand to the metal [Eqs. (92) and (93)].

$$M^{+n}X \longrightarrow M^{+(n-1)} + X \cdot \qquad (92)$$

$$X \cdot + RH \longrightarrow HX + R \cdot \qquad (93)$$

The direct abstraction of electrons from substrates, as pictured in Eq. (88), would not appear to have wide applicability in hydrocarbon autoxidation. Nevertheless, this mode of radical generation is cited frequently, particularly when the catalyst is a metal in a high oxidation state or is known to be a very strong oxidizing agent. When substrates consist of saturated hydrocarbons, this route would seem to be especially unfavorable. Radical generation by the scheme in Eqs. (92) and (93), particularly when the catalyst is a metal carboxylate, is then more likely. The scheme may have some applicability with unsaturated hydrocarbons capable of complex formation, such as olefins or aromatics. Thus, the autoxidation of toluene is catalyzed by Co(III) perchlorate in 50% aqueous acetonitrile 1 M in HClO$_4$ (133). The reaction was found to be first order with respect to both Co(III) and toluene and, in addition to oxidation products, yielded some bibenzyl, evidence of the likelihood of radical formation. The rate of reaction was, however, quite slow. Addition of carboxylic acids led to considerable rate enhancement indicating the relative inefficiency of radical generation by direct electron abstraction in comparison with generation by catalyst decomposition.

The generation of radicals by species resulting from the interaction of oxygen with catalysts [Eqs. (90) and (91)], was originally proposed to explain initiation by metals in low oxidation states in the absence of initial hydroperoxides (134). Detailed investigations of this process are hampered by peroxide decomposition which is frequently catalyzed by the same metal in the same oxidation state, and, being more efficient, rapidly masks other forms of initiation (135,136). If the catalyst is not active in promoting hydroperoxide decomposition, i.e., if the final reaction product is hydroperoxide, the situation is less ambiguous. Thus, catalysis of cumene autoxidation to cumene hydroperoxide by copper phthalocyanine has been explained in this fashion (*137*). Similarly, the oxidation of styrene to polymeric peroxides, catalyzed by Co(II) stearate or acetylacetonate, was shown to be first order in styrene, oxygen, and Co(II) [Eq. (94)].

$$\text{rate} = k[\text{Styrene}][\text{P}_{\text{O}_2}][\text{Co(II)}] \tag{94}$$

This finding was used to support a scheme in which an initially formed Co(II)—O_2 complex attacked styrene, giving rise to a radical species (138) [Eqs. (95) and (96)].

$$\text{Co(II)} + O_2 \rightleftharpoons [\text{Co} \cdots O_2] \tag{95}$$

$$[\text{Co} \cdots O_2] + \phi\text{CH}=\text{CH}_2 \longrightarrow \text{Co}-\text{O}-\text{O}-\text{CH}_2-\overset{\bullet}{\text{CH}}\phi \tag{96}$$

In this case, distinguishing between this path and one in which peroxy radical attacks styrene is difficult.

Since a number of stable transition metal-oxygen complexes are known, their reactions, which certainly bear on this type of initiation, will be discussed separately.

Results of an increasing number of careful investigations of metal catalyzed autoxidations indicate that radical formation by electron transfer from a catalyst ligand to metal [Eq. (92)] is a very likely initiation path. This, as stated earlier, is a particularly favorable route for metal carboxylates or when reactions are carried out in carboxylic acids. Thus, the previously mentioned acceleration of toluene autoxidation by addition of carboxylic acids (*133*) very likely is due to formation of Co(III) carboxylates which decompose to radical species capable of hydrogen abstraction from toluene [Eqs. (97)–(99)].

$$\begin{array}{c} R-\text{C}-\text{O}-\text{Co(III)} \\ \parallel \\ \text{O} \end{array} \longrightarrow R\text{COO} \cdot + \text{Co(II)} \tag{97}$$

$$R\text{COO} \cdot \longrightarrow R \cdot + \text{CO}_2 \tag{98}$$

$$R \cdot + \phi\text{CH}_3 \longrightarrow R\text{H} + \phi\text{CH}_2 \cdot \tag{99}$$

Formation of products which incorporate radicals formed in this manner is further evidence for this type of catalyst function. For example, when toluene autoxidation in acetic acid is catalyzed by $Pb(OAc)_4$, products, in addition to benzyl acetate, consist of methylbenzylacetates, xylenes, and methylphenylacetic acid (*139*) [Eq. (100)]. These products can be accounted for by decomposition

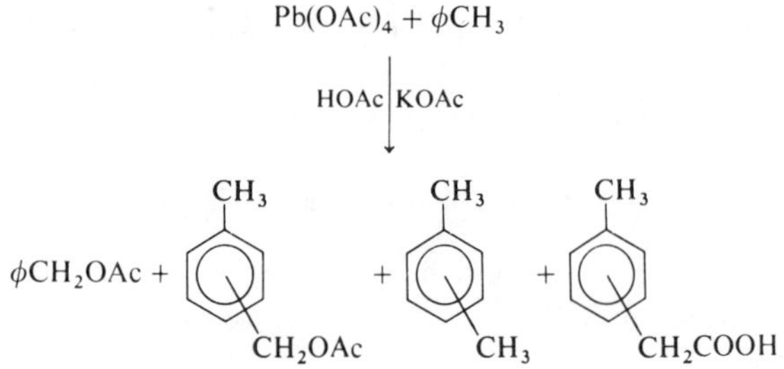

$$Pb(OAc)_4 + \phi CH_3$$

$$\downarrow \text{HOAc} | \text{KOAc}$$

$$(100)$$

of Pb(OAc)$_4$ to methyl radicals [Eqs. (101) and (102)] which abstract hydrogen from either toluene or acetic acid [Eqs. (103) and (104)]. The resulting radicals then add to toluene.

$$Pb(OAc)_4 \longrightarrow Pb(OAc)_3 + CH_3 \cdot + CO_2 \qquad (101)$$

$$Pb(OAc)_3 \longrightarrow Pb(OAc)_2 + CH_3 \cdot + CO_2 \qquad (102)$$

$$CH_3 \cdot + \phi CH_3 \longrightarrow CH_4 + \phi CH_2 \cdot \qquad (103)$$

$$CH_3 \cdot + CH_3COOH \rightleftharpoons CH_4 + \cdot CH_2COOH \qquad (104)$$

Catalysis of toluene autoxidation by Mn(III) acetate (*140,141*) does not result in xylene formation. Results have, therefore been rationalized by reaction schemes [Eqs. (105) and (106)] involving a carboxymethyl radical formed directly from Mn(OAc)$_3$ (*141*).

Added evidence for this scheme is found in reactions of Mn(OAc)$_3$ with olefins which yield the corresponding butyrolactone, presumably by addition of carboxymethyl radical to olefin.

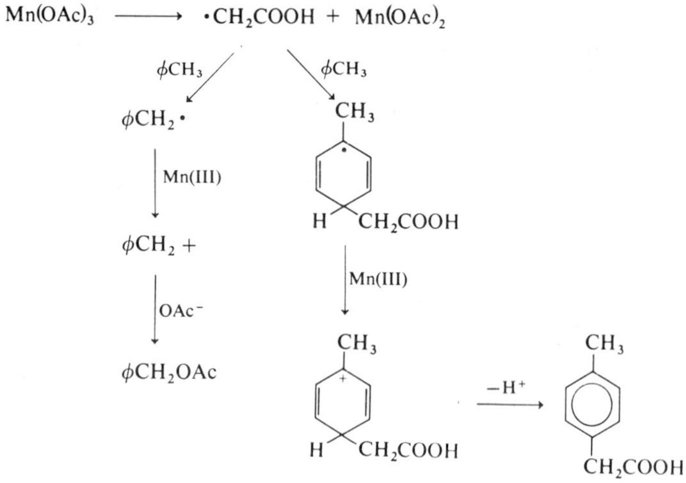

$$(105)$$

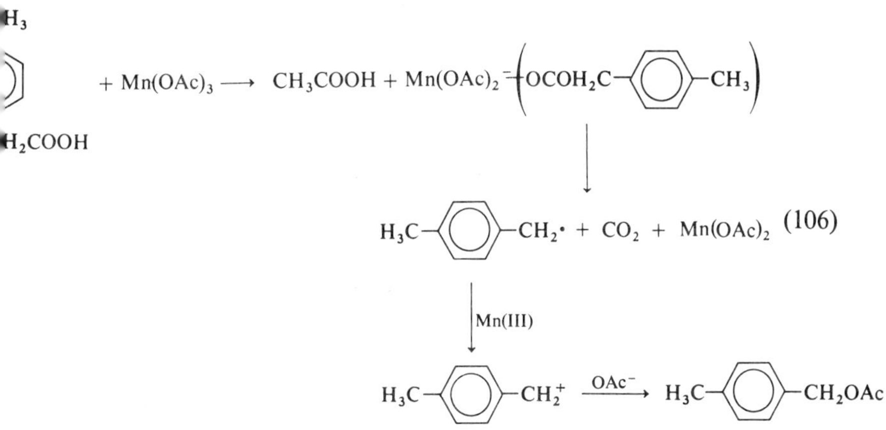

H_3

$+ Mn(OAc)_3 \longrightarrow CH_3COOH + Mn(OAc)_2 \left[OCOH_2C - \bigcirc - CH_3 \right]$

H_2COOH

$$H_3C - \bigcirc - CH_2\cdot + CO_2 + Mn(OAc)_2 \quad (106)$$

$$\downarrow Mn(III)$$

$$H_3C - \bigcirc - CH_2^+ \xrightarrow{OAc^-} H_3C - \bigcirc - CH_2OAc$$

B. BRANCHING

As mentioned, the most commonly encountered form of metal catalyst participation in hydrocarbon autoxidation is in the promotion of hydroperoxide decomposition to radical species. There is general agreement that this proceeds via the so-called Haber-Weiss mechanism [Eqs. (107) and 108)] first proposed to explain Fe(II) catalysis of the decomposition of hydrogen peroxide (142).

$$ROOH + M^{+n} \longrightarrow RO\cdot + OH^- + M^{+(n+1)} \quad (107)$$

$$ROOH + M^{+(n+1)} \longrightarrow RO_2\cdot + H^+ + M^{+n} \quad (108)$$

This reaction is extremely efficient and very often controls the overall reaction rate once a sufficient concentration of hydroperoxide has been reached (143). Since the process is cyclic as far as either of two oxidation states of the metal is concerned, regeneration of the catalyst in a separate step is not required. Catalysts usually are long lived, each mole of metal taking part in numerous radical generating peroxide decompositions (144, 145). Deactivation occurs by conversion to species unable to complex or decompose peroxides or by formation of insoluble species (144, 146). Generation of nonradical species from peroxides can also be brought about by the same catalyst and leads to inhibition (145).

Catalysis of the decomposition of hydroperoxides to radicals is thought to involve complexing of the hydroperoxide by metal prior to electron transfer, i.e., an "inner sphere" process (144, 147). This type of catalysis can, therefore, be expected to depend on the particular metal employed, its oxidation state, and the number and nature of other ligands in its co-ordination sphere ($145,146$).

In its detailed aspects, the action of the catalyst closely resembles initiation by catalyst decomposition (148) [Eqs. (109)–(111)].

$$ROOH + M^{+n} X \longrightarrow ROOM^{+n} + HX \qquad (109)$$

$$ROOM^{+n} \longrightarrow ROO\cdot + M^{+(n-1)} \qquad (110)$$

$$ROOM^{+n} \longrightarrow RO\cdot + {}^-OM^{+(n+1)} \qquad (111)$$

Other possibilities include oxidative addition of the hydroperoxide to metal followed by decompositions [Eqs. (112) and (113)].

$$ROOH + M^{+n} \longrightarrow RO-M^{+(n+2)} - OH \qquad (112)$$

$$RO-M^{+(n+2)} - OH \longrightarrow RO\cdot + HO-M^{+(n+1)} \qquad (113)$$

C. Termination

Radical chains are terminated when radicals are converted into non-radical species. When this is promoted by a metal catalyst, the catalyst is more properly termed an inhibitor. However, in many instances, the type of products obtained in a particular reaction is a function of how initially formed radicals interact with the catalyst and/or substrate. Understanding of this process is, therefore, required for an adequate description of radical oxidations.

The ability of metal catalysts to promote coupling has been discussed already. Another way in which radical chains may be terminated or diverted is by oxidation of radicals to ionic species [Eq. (114)] which has been mentioned briefly in connection with initiation by catalyst decomposition [Eqs. (105) and (106)]. This step can

$$R\cdot + M^{+n} \longrightarrow R^+ + M^{(+n-1)} \qquad (114)$$

occur with or without ligand transfer. For instance, in the conversion of n-butyl radicals, formed by peroxide decomposition, 1-butene formation, in the presence of $Cu(OAc)_2$, has been explained as due to radical oxidation without ligand transfer [Eq. (115). The formation of butyl chloride when butyl radicals are oxidized by $CuCl_2$ is thought to occur with ligand transfer (149) [Eq. (116)].

$$n\text{Bu} \cdot\ + \text{Cu(OAc)}_2 \longrightarrow [\ >\!\!\text{Cu(II)} \cdot\cdot \text{Bu}\ ]$$

$$\diagup\!\!\!\diagup \tag{115}$$

$$[\ >\!\!\text{Cu(I)Bu}^+] \xrightarrow{\ -\text{H}^+\ } 1 - \text{C}_4\text{H}_8 + \text{Cu(I)}$$

$$n\text{Bu} \cdot\ + \text{CuCl}_2 \longrightarrow [\text{Cl}-\text{Cu}-\text{Cl} \cdot\cdot \text{Bu}] \longrightarrow \text{BuCl} + \text{Cu(I)} \tag{116}$$

Catalysis by Pb(IV) or Cu(II) of the conversion of phenylisopropyl radicals to isopropenyl benzene and 2-phenyl-2-acetoxypropane has been rationalized along similar lines (150). The ease of radical oxidation by metal salts or complexes correlates reasonably well with the ionization potentials of the particular radicals involved.

From the foregoing, it can ben see that such steps as initiation by catalyst decomposition, hydroperoxide decomposition, and radical oxidation are relatable in a continuum of ligand-metal electron transfer processes [Eqs. (117) and (118)].

$$R^- + M^{+n} \longrightarrow R-M \longrightarrow R\cdot + M^{+(n-1)} \tag{117}$$

$$R\cdot + M^{+n} \longrightarrow R-M \longrightarrow R^+ + M^{+(n-1)} \tag{118}$$

VII. Oxygen Complexes and Their Reactions

Oxygen complexes have received discussion both with respect to their utility as oxygen "carriers" (151) and from the standpoint of oxygen "activation" (152,153). In the former category are materials such as bis(salicylaldehydeimine) Co(II) complexes (151) and bis(triphenylphosphine) carbonly Ir(I) chloride (154) which complex reversibly with oxygen. Oxygen activation deals with alterations in O—O bonding which render the complexed oxygen more reactive toward oxidizable substrates. While the variables affecting the activation of oxygen are only incompletely understood, the fact that complexed oxygen is reactive toward a variety of substrates enhances the plausibility that such activation may be involved in a variety of catalyzed autoxidations.

The ability of Co(II) salts to initiate radical chains in autoxidation reactions has been mentioned. Explanation of such observations have involved complexing of oxygen with paramagnetic Co(II) to produce a radical species capable of hydrogen atom abstraction from substrate [Eqs. (119) and (120)]. It is, therefore, of

$$\text{Co(II)} + \text{O}_2 \longrightarrow \text{Co(III)O}_2^- \cdot \tag{119}$$

$$\text{Co(III)O}_2^- \cdot\ + R\text{H} \longrightarrow \text{Co(III)OOH} + R\cdot \tag{120}$$

interest that new paramagnetic species are formed by the interaction of oxygen with such Co(II) complexes as $[(CN)_5CO(II)]^{-3}$ (155) and vitamin B_{12r} (156). Changes in ESR spectra observed during formation of these species were interpreted as indication that the unpaired electron in the oxygen complexes no longer resided principally on the metal and that the complexes could thus be represented most conveniently as $Co(III)O_2^-$ · moieties. Formation of this superoxo species was rapid and reversible, in the case of vitamin B_{12r}. In the reaction of $[(CN)_5Co(II)]^{-3}$, the stable, binuclear, oxygen bridged complex, $[(CN)_5Co-O-O-Co(CN)_5]^{-6}$ was formed rapidly by reaction of the mononuclear complex with a second mole of pentacyanocobaltate(II). The binuclear peroxo complex is diamagnetic. Paramagnetic oxygen bridged complexes of the type,

$$[(NH_3)_4Co \underset{\underset{O_2}{\diagdown}\diagup}{\overset{\overset{H_2}{\overset{N}{\diagup}\diagdown}}{}} Co(NH_4)]^{+4}$$

are also known. In these, the bridging oxygen resembles a superoxide ion. They are generated from the corresponding diamagnetic species by reaction with such strong oxidizing agents as Cr(IV) or $KMnO_4^-$ (157).

Of further interest with respect to involvement of Co(II) in autoxidations is the isolation of a mononuclear hydroperoxide complex, $[(CN)_5 Co-OOH]^{-3}$. This material is formed either by acid hydrolysis of $[(CN)_5 Co-OO-Co(CN)_5]^{-6}$ (158) or by oxygen insertion into the Co—H bond in $[(CN)_5 Co-H]^{-3}$ (159). The methods of preparation indicate that the compound is an intermediate in both hydrolysis of the bridged binuclear complex to H_2O_2 and in catalysis of hydrogen oxidation by Co(II) (158). Thermal decomposition in solution was found to be first order with respect to the hydroperoxide in the range 35–60°C. The activation energy for this process is about 23 Kcal/mole. Products are oxygen and $[(CN)_5 CoOH]^{-3}$. A homolytic bond cleavage has been suggested (159,160) [Eq. (121)]:

$$Co(III)OOH \longrightarrow Co(II) + HO_2 \qquad (121)$$

The hydroperoxide, $[(CN)_5 CoOOH]^{-3}$, reacts with alcohols to form aldehydes; this reaction has not been investigated. Products derived from reaction with 1-octene in aqueous systems indicate hydrolysis to H_2O_2 which then reacts with the olefin. Similarly, in acetic acid, it was shown that high yields of epoxides could be obtained in stoichiometric reactions of olefins with either the mononuclear hydroperoxide or the binuclear peroxide by solvolysis with formation of per-acetic acid (*159*) [Eq. (122)] followed by reaction of the latter with olefin.

$$[(CN)_5CoOOH]^{-3} + 2HOAc \longrightarrow [(CN)_5CoOAc]^{-3} + CH_3COOOH + H_2O$$

(122)

The overall reaction could not be made catalytic, as regeneration of hydride from Co(III) acetate requires strong reducing agents, such as $NaBH_4$.

The intermediacy of Co(III) superoxo species in autoxidations catalyzed by Co(II) has not yet been demonstrated. However, the existence of such species lends plausibility to the thought that materials of this type may, indeed, be involved.

Direct participation of metal-oxygen complexes in both stoichiometric and catalystic autoxidations has been found in the case of oxygen complexes of zerovalent nickel, palladium, and platinum and monovalent ruthenium, rhodium, and iridium. The complexes are prepared readily from oxygen and triaryl or tricycloalkylphosphine (*161–163*) or isocyanide (*163*) complexes in the case of the nickel triad (*161,164*), from the bis(triarylphosphine) carbonyl halide or tris(triarylphosphine) halide complexes of iridium (*154,165,166*) and with slightly more difficulty from similar rhodium complexes (*163,166,167*). A stable oxygen complex is formed from bis(triphenylphosphine) carbonyl (nitrosyl) ruthenium(I) chloride by carbonyl displacement (*168*).

Stability of the complexes is a function of both metal and ligands. Thus, in $(\phi_3P)_2MO_2$, the stability is Ni < Pd ≪ Pt. In $(R_3P)_2NiO_2$, the stability is phenyl < piperidyl < cyclohexyl (*162*). The stability as a function of halides in $(\phi_3P)_2 IrCOX(O_2)$ is in the order I ≫ Br > Cl with oxygen complexing being completely reversible for the chloride and irreversible for the iodide (*169*).

Rates of formation of the complexes in solution depend on complex concentration and oxygen partial pressure. For $(\phi_3P)_2IrCOX(O_2)$ the order is I > Br > Cl, with the iodide reacting ca. ten times as fast as the chloride. The $(\phi_3P)_2 PtO_2$ complex is formed almost 100 times as rapidly as $(\phi_3P)_2 IrCOCl(O_2)$ (*105*). Activation energies are, however, similar (ca. 10 Kcal/mole) (*105,165,170*).

The compounds are mononuclear and diamagnetic. Because they are formed from diamagnetic d^8 and d^{10} metal complexes and yield hydrogen peroxide on acid hydrolysis (*154,162*), it was assumed in the early stages of their investigation that they were mononuclear peroxides, i.e., that reaction between oxygen and the metal involved a two electron transfer.

This suggestion appeared reasonable in view of the characteristic strong infrared absorption between 800 and 900 cm^{-1} shown by these materials (*154,161,162,164,167,168,171*). This absorption, assigned to O—O stretching is found in mononuclear peroxides in which the peroxy group acts as a bidentate ligand (*172*). X-ray structural determinations would seem to substantiate the surmised structures. The oxygen molecule is coplanar with the metal atom with both oxygen atoms equidistant from it, i.e., the two oxygen atoms and the metal are the corners of an isoceles triangle (*169, 173–175*). However, it is not possible to decide whether oxygen should be considered as a mono or bidentate ligand on the basis of this structure (*169,173*). Moreover, O—O bond lengths, in a number of cases, differ substantially from those normally encountered in peroxides (Table I).

TABLE 1

O—O Bond Lengths in Oxygen Complexes and Related Compounds

Compound	O—O (A°)	References
$(\phi_3P)_2IrCOCl \cdot O_2$	1.30	(*173*)
$(\phi_3P)_2IrCOI \cdot O_2$	1.47	(*169*)
$(\phi_2PCH_2P\phi_2)IrO_2$	1.66	(*166*)
$(\phi_3P)_2PtO_2$	1.45(1.26)	(*174* and *175*)
$[(NH_3)_5Co—O—O Co(NH_3)_5]^{+4}$	1.30	(*176*)
O_2	1.21	(*173*)
O_2^-	1.28	(*173*)
O_2^{-2}	1.49	(*173*)

There is an apparent rough correlation between O—O bond length and the reversibility of complex formation. For example, in $(\phi_3P)_2IrCOCl \cdot O_2$, which on the basis of complete reversibility of oxygen complexing is classed as a carrier, the O—O distance, while somewhat greater than in molecular oxygen, is substantially below that in typical peroxides. It is, however, similar to that found in O_2 bridged binuclear peroxides. In the corresponding iodide and in phosphine chelate derivatives, in which the oxygen is bound irreversibly, the O—O bond length is in the same order or longer than in typical peroxides. On the other hand, there is, at the moment, uncertainty concerning the O—O distance in the stable $(\phi_3P)_2PtO_2$ complex and, therefore, some question concerning the validity of correlating this measurement with stability. Similar questions have been raised concerning correlations of O—O stretching frequencies with bond lengths and stabilities (169). Also while, as will be seen, the reactivity of oxygen in these complexes may resemble that of superoxides, peroxides, or excited oxygen, correlation of this behavior with either stability or O—O bond distance remains poor.

On the basis of the limited data available, then, it is clear only that the O—O bond distances in oxygen complexes is greater than in uncomplexed oxygen. The current tendency is to consider this effect in terms of a π-complexed O_2 molecule (173,176).

While there are many unanswered questions concerning the nature of complexed oxygen, there is no doubt that it can be considered activated. For example, platinum, palladium, and iridium oxygen complexes have been found to participate in stoichiometric autoxidations in which the O—O bond is cleaved and oxygen is transferred to substrate under extremely mild conditions. Both $(\phi_3P)_2PtO_2$ and $(\phi_3P)_2PdO_2$ are converted to the corresponding carbonates by CO (177) and to sulfates by SO_2(166, 179,180) [Eq. (123)].

$$(\phi_3P)_2PtO_2 + SO_2 \longrightarrow (\phi_3P)_2PtSO_4 \qquad (123)$$

The platinum complex reacts with NO and NO_2 to yield nitrite (178) and nitrate (178,179), respectively. Similarly, $(\phi_3P)_2IrCOI—O_2$ is converted to nitrate by NO_2 and to sulfate by SO_2(166).

The intermediacy of oxygen complexes also is implicated strongly in the autoxidation of phosphines to phosphine oxides and of isocyanides to isocyanates which is catalyzed by complexes of Ni, Pd, and Pt(O) (161–163, 181). Thus, $(\phi_3P)_4$ Pd and (ϕ_3P_4) Pt react, as discussed, with oxygen to yield the corresponding oxygen complexes and triphenylphosphine oxide (161,162,181) [Eq. (124)].

$$(\phi_3P)_4Pd + 2O_2 \longrightarrow (\phi_3P)_2PdO_2 + 2\phi_3PO \qquad (124)$$

The oxygen complexes, in turn, react with excess phosphine to yield the parent zerovalent complex and phosphine oxide (161,162) [Eq. (125)].

$$(\phi_3P)_2PdO_2 + 4\phi_3P \longrightarrow (\phi_3P)_4Pd + 2\phi_3PO \qquad (125)$$

Similar observations have been made in the conversion of t-butylisocyanide to isocyanate catalyzed by Ni(O) (163).

Phosphine autoxidation catalysis was found to be inhibited by ϕ_3Sb, $(EtO)_3P$, and CO (161), possibly due to formation of stable complexes of oxidation products. No inhibition by ϕ_2S (161,181) or $(CH_3)_2S$ (181) was found. However, neither of these materials was oxidized catalytically. Pyridine inhibition results from formation of an insoluble pyridine oxide complex (181).

Reaction rates have been shown to depend on both substrate and metal. For catalysis by Pt(O), relative rates as a function of substrate were ϕ_3P > cyclohexyl isocyanide > Bu_3P, while for Pd(O), Bu_3P > cyclohexylisocyanide > ϕ_3P (161). In the case of Pd(O) catalysis, rates appear to be influenced by the electron donating or withdrawing character of ligands. Thus, tris(tri-*p*-tolyl-phosphine) palladium(O) is considerably more active than tetrakis (triphenylphosphine) palladium(O). Tetrakis(tri-*p*-chlorophenylphosphine) palladium(O) reacts with oxygen to form the corresponding phosphine oxide but rapidly loses its ability to function as a catalyst for phosphine autoxidation (181a).

The rate of reduction of $(\phi_3P)_2PtO_2$ by ϕ_3P has been studied and was found to be first order both in complex and phosphine (182). Absorption spectra of products of ethylene displacement from $(\phi_3P)_2Pt(C_2H_4)$ by ϕ_3P in solution were invariant when the ratio of phosphine to complex equalled or was greater than one. It was, therefore, assumed that the Pt(O) species in solution was $(\phi_3P)_3Pt$, The reduction was depicted as a "dissociative oxygen insertion" [Eq. (126)].

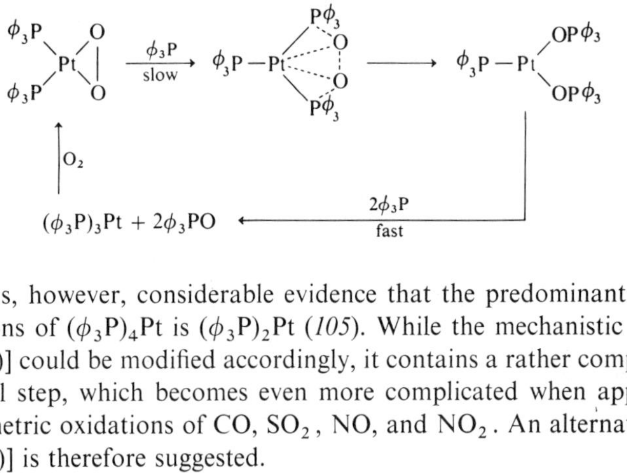

$$(126)$$

There is, however, considerable evidence that the predominant species in solutions of $(\phi_3P)_4Pt$ is $(\phi_3P)_2Pt$ (105). While the mechanistic scheme [Eq. (126)] could be modified accordingly, it contains a rather complicated and novel step, which becomes even more complicated when applied to stoichiometric oxidations of CO, SO_2, NO, and NO_2. An alternate route [Eq. (127)] is therefore suggested.

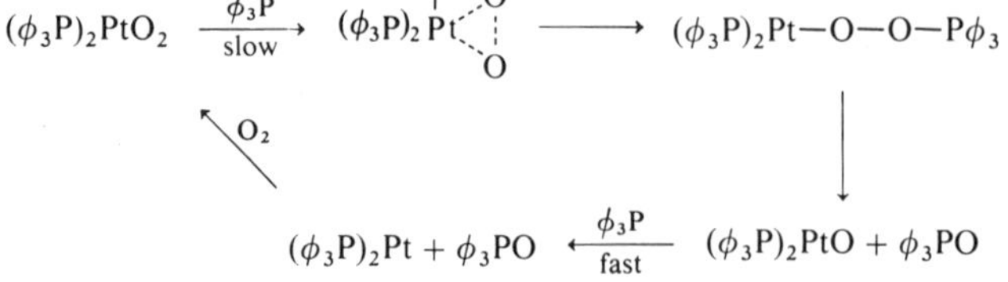

$$(127)$$

This scheme is more nearly in accord with mechanistic ideas concerning insertions of π-bonded ligands, and avoids the necessity of complexing two more ligands other than ϕ_3P in stoichiometric oxidations of NO and NO_2. Facile reduction of freshly prepared platinum and palladium oxides to zerovalent metal complexes by ϕ_3P has been demonstrated (183,184).

Several studies indicate similarities between the catalytic behavior of Co(II) and of the Group VIII mononuclear, diamagnetic oxygen complexes or their precursors. Catalysis of cyclohexene autoxidation by $(\phi_3P)_2PtO_2$, $(\phi_3P)_2IrCOI$, and $(\phi_3P)_3RhCl$ (178,188), of ethyl and propyl benzene and of tetralin autoxidation by $(\phi_3P)_3RhCl$ (167,186,188) and $(\phi_3P)_3RuCl$ (187) and of diphenylmethane autoxidation by $(\phi_3P)_2RhCOCl$ (189) has been reported. All of these reactions have the characteristics of radical chain processes (166,185,188,189). In the case of cyclohexene or ethylbenzene autoxidations catalyzed by $(\phi_3P)_2RhCl$, products and rates were found to be virtually indistinguishable from those obtained with rhodium (III) acetylacetonate or 2-ethylhexanoate or cobalt (II)-2-ethylhexanoate (188). The reactions were inhibited by free radical inhibitors, such as phenols, and could be initiated by peroxides. Similar results were found for the autoxidation of diphenylmethane catalyzed by $(\phi_3P)_2RhCOCl$ (189). Suggested mechanisms, therefore, involve peroxide decomposition catalysis either by a Rh(II)-Rh(III) (188) or by a Rh(I)-Rh(III) (189) couple [Eqs. (128) and (129)].

$$Rh(I) + 2ROOH \longrightarrow Rh(III) + 2RO\cdot + 2OH^- \qquad (128)$$

$$Rh(II) + 2ROOH \longrightarrow Rh(I) + 2RO_2\cdot + 2H^+ \qquad (129)$$

Evidence that hydrocarbon, autoxidations may be initiated in peroxide free systems, at least, by some diamagnetic oxygen complexes, has also been obtained. Thus, the autoxidation of hydroperoxide-free cumene was catalyzed by complexes of Pd(O) in benzene solution at 35°C and one atmosphere oxygen (181a). No hydroperoxide decomposition occurred under these conditions, and no reaction was observed in the absence of catalyst. Initial oxygen absorption rates were influenced by the nature of catalyst ligands, tris-(tri-p-tolyphosphine) palladium (O) displaying considerably greater activity than tetrakis-(triphenylphosphine) palladium (O). Tetrakis-(tri-p-chlorophenylphosphine) palladium (O) was found to be inactive. Other complexes capable of oxygen complex formation, such as $(\phi_3P)_4Pt$ and $(\phi_3P)_2IrCOCl$ showed some activity. However, in these cases, lengthy induction periods followed by autocatalysis were observed indicating that, as described above, these complexes catalyzed hydroperoxide decomposition rather than formation. Moreover, $(\phi_3P)_2$ IrCOCl catalyzed cumene autoxidation without an induction period in the presence of hydroperoxides and reacted rapidly with cumene hydroperoxide under anaerobic conditions with CO_2 evolution.

The results obtained with palladium complexes can be rationalized by assuming that the oxygen complexes formed from Pd(O) complexes are capable of abstracting hydrogen from cumene [Eqs. (130)–(133)].

$$(Ar_3P)_4Pd + 2O_2 \longrightarrow (Ar_3P)_2PdO_2 + 2Ar_3PO \qquad (130)$$

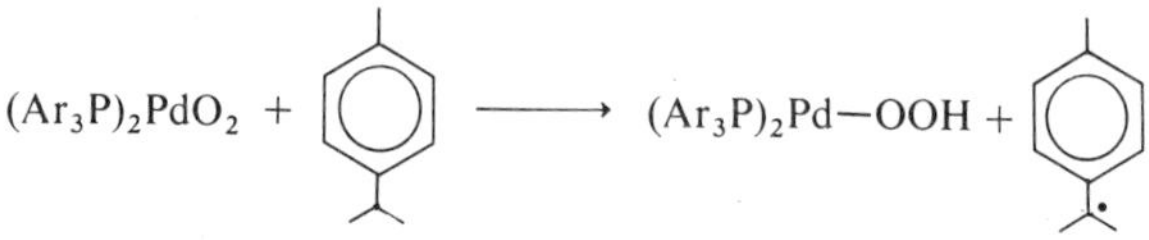

$$(Ar_3P)_2PdO_2 + \text{[cumene]} \longrightarrow (Ar_3P)_2Pd\!-\!OOH + \text{[cumyl radical]} \qquad (131)$$

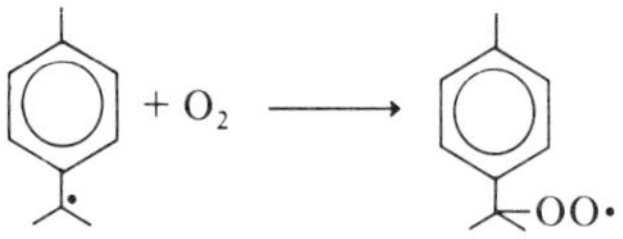

$$\text{[cumyl radical]} + O_2 \longrightarrow \text{[cumylperoxy radical, } -OO\cdot] \qquad (132)$$

$$\text{[cumylperoxy, } -OO\cdot] + \text{[cumene]} \longrightarrow \text{[cumyl hydroperoxide, } -OOH] + \text{[cumyl radical]} \quad \text{etc.} \qquad (133)$$

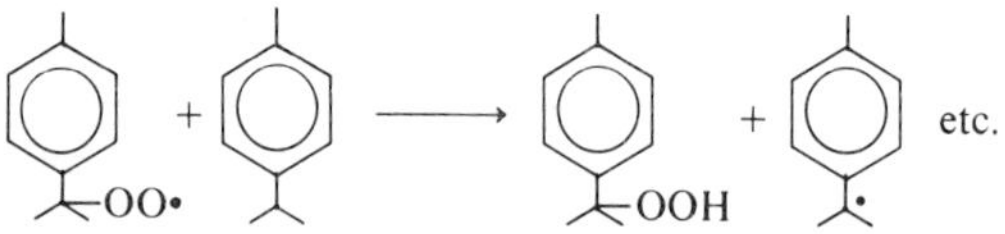

No information concerning the fate of the initiating palladium species is available at this time. Possibilities include decomposition of the metal hydroperoxide either to form $HO_2\cdot$ and regenerate Pd(O) or with formation of $HO\cdot$ and a metal oxide species. Since no insoluble palladium species have been noted, the former path appears more likely.

While it is possible to infer radical character in, at least, those oxygen complexes capable of autoxidation initiation in peroxide free systems, such suggestions appear incompatible with the diamagnetism of these materials. The latter has, as mentioned, been interpreted in terms of a π-bonded oxygen molecule which can be considered as analogous to singlet oxygen. This model is consistent with the long O—O bond lengths found (176) and is not inconsistent with reactions involving either hydrogen abstraction or those in which oxygen behaves as an electrophile (190). Unfortunately, no distinction between a reactive triplet and a reactive singlet species can be made on the basis of products obtained in the reactions discussed. However, in several instances, reactions have been found which strongly resemble those of singlet oxygen. For example, the smooth cleavage of enamines to amides and ketones, which has been observed as a photochemical oxygenation (190,191), has also been catalyzed by CuCl in the absence of light (192). Similarly, ketoperoxychelates formed from $(\phi_3P)_2$

PtO_2 and carbonyl compounds (*193*) [Eq. (134)] resemble intermediates suggested in reactions of singlet oxygen (*190*); $R' = CH_3$, H and $R'' = CH_3$, C_2H_5, C_5H_6

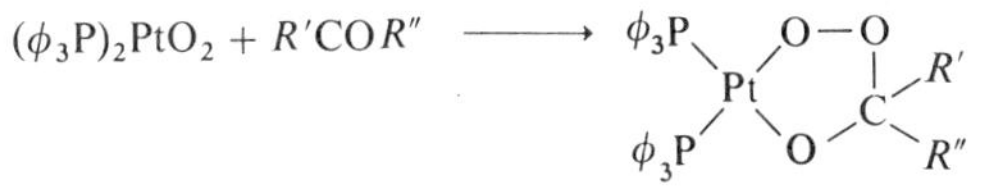

$$(\phi_3P)_2PtO_2 + R'COR'' \longrightarrow \phi_3P \underset{\phi_3P}{\overset{}{\diagdown}} Pt \underset{O}{\overset{O-O}{\diagup}} \underset{R''}{\overset{R'}{\diagdown}} C \tag{134}$$

All of this raises the interesting possibility that both the complexing of oxygen and reactions of complexed oxygen might be further examples of catalysis by symmetry transformation (*194*).

VIII. Conclusions

It can be seen from the foregoing discussion of homogeneous metal catalyzed oxidation reactions that, in many instances, it is possible to rationalize results on the basis of classical organic radical or ionic mechanistic concepts. That is, overall electron transfers can be accounted for in this manner. Actual mechanisms, however, are considerably more complicated, and the precise nature of reactive species is, as yet, unknown. In most of the cases considered, catalysis appears to be a consequence of formation of a metal–substrate bond at some stage of the reaction. The degree to which subsequent reactions of such intermediates resemble radical, ionic, or concerted molecular processes would seem to be a function of the electrophilicity of the metal and the nucleophilicity of the substrate. A better understanding of the variables affecting metal–substrate bonding would be of considerable help in formulating detailed mechanisms.

REFERENCES

1. L. F. Fieser, and M. Fieser, *Reagents for Organic Syntheses*, Wiley, New York, 1967.
2. W. A. Waters, *Mechanisms of Oxidation of Organic Compounds*, Methuen, London, 1964.
3. T. A. Turney, *Oxidation Mechanisms*, Butterworths, London, 1965.
4. J. Halpern, *Quart. Rev.*, **15**, 207 (1961).
5. J. O. Edwards, *Inorganic Reaction Mechanisms*, Benjamin, New York, 1964.
6. W. L. Reynolds and R. W. Lumry, *Mechanisms of Electron Transfer*, Ronald, New York, 1966.
7. F. Basolo and R. G. Pearson, *Mechanisms of Inorganic Reactions*, Wiley, New York, 2nd ed., 1967.
8. J. Smidt, W. Hafner, R. Jira, J. Sedlmeier, R. Sieber, R. Ruttinger, and H. Kojer, *Angew. Chem.*, **71**, 176 (1959).

9. A. Aguiló, *Adv. Organometal. Chem.*, **5**, 321 (1957).
10. E. W. Stern, *Catalysis Rev.*, **1**, 73 (1967).
11. I. I. Moiseev, *Pre-prints, Div. Pet. Chem., Am. Chem. Soc.* **14**, (2), B49 (1969).
12. R. Huttel, J. Kratzer, and M. Bechter, *Ber.*, **94**, 766 (1961).
13. W. Hafner, R. Jira, J. Sedlmeier, and J. Smidt, *Ber.*, **95**, 1575 (1962).
14. W. Clement and C. W. Selwitz, *J. Org. Chem.*, **29**, 241 (1964).
15. H. Okada, T. Noma, Y. Katsuyama, and H. Hashimoto, *Bull. Chem. Soc. Japan*, **41**, 1395 (1968).
16. R. Huttel and H. Christ, *Ber.*, **97**, 1439 (1964).
17. R. Platz and W. Fuchs, *Brit. Pat.*, 1,041,376 (1966).
18. M. N. Vargaftik, I. I. Moiseev, and Ya. K. Syrkin, *Dokl. Akad. Nauk SSSR*, **139**, 1396 (1961).
19. Farbwerke Hoechst, A. G., *Belg. Pat.*, 626,669 (1963).
20. J. Smidt and H. Krekeler, *Erdeol Kohle*, **16**, 560 (1963).
21. J. Smidt, W. Hafner, R. Jira, R. Sieber, J. Sedlmeier, and A. Sabel, *Angew. Chem.*, **74**, 93 (1962).
22. I. I. Moiseev, M. N. Vargaftik, and Ya. K. Syrkin, *Dokl. Akad. Nauk SSSR*, **133**, 377 (1960).
23. E. W. Stern and M. L. Spector, *Proc. Chem. Soc.*, **1961**, 370.
24. E. W. Stern, *Proc. Chem. Soc.*, **1963**, 111.
25. D. Clark, P. Hayden, W. D. Walsh, and W. E. Jones, *Brit. Pat.* 964,001 (1964).
26. A. P. Belov and I. I. Moiseev, *Izv. Akad. Nauk, SSSR Ser. Khim.*, **1966**, 139.
27. W. Kitching, Z. Rappoport, S. Winstein, and W. G. Young, *J. Am. Chem. Soc.*, **88**, 2054 (1966).
28. S. Uemura, T. Okada, and K. Ichikawa, *Nippon Kagaku Zasshi*, **89**, 692 (1968).
29. Imperial Chem. Ind. Ltd., *Belg. Pat.* 635,426 (1964).
30. M. N. Vargaftik, I. I. Moiseev, Ya. K. Syrkin, and V. V. Yashkin, *Izv. Akad. Nauk, SSSR Odt. Khim. Nauk.*, **1962**, 930.
31. R. G. Schultz and D. E. Gross, *Adv. Chem. Ser.*, **70**, 97 (1968).
32. C. B. Anderson and S. Winstein, *J. Org. Chem.*, **28**, 605 (1963).
33. M. Greene, R. N. Haszeldine, and J. Lindley, *J. Organometal. Chem.*, **6**, 107 (1966).
34. W. C. Baird, Jr., *J. Org. Chem.*, **31**, 2411 (1966).
35. A. P. Belov, G. Yu. Pek, and I. I. Moiseev, *Izv. Akad. Nauk SSSR, Ser. Khim.*, **1965**, 2204.
36. D. Clarke and P. Hayden, *Pre-Prints, Petrol. Div.. Am. Chem. Soc.*, **11**, (4) D5 (1966).
37. M. Tamura and T. Yasui, *Chem. Commun.*, **1968**, 1209.
38. R. Van Helden, C. F. Kohll, D. Medema, G. Verberg, and T. Jonkoff, *Rec. Trav. Chim.*, **87**, 961 (1968).
39. R. G. Brown, J. M. Davidson and C. Triggs, *Pre-prints, Petrol, Div. Am. Chem. Soc.*, **14**, (2) B23 (1969).
40. D. Clarke, P. Hayden, and R. D. Smith, *Disc. Farad. Soc.*, **46**, 98 (1969).
41. P. M. Henry, *J. Org. Chem.*, **32**, 2575 (1967).
42. D. Clark, P. Hayden, and R. D. Smith, *Pre-prints, Petrol. Div., Am. Chem. Soc.*, **14**, (2), B10 (1969).
43. P. M. Henry, *J. Am. Chem. Soc.*, **86**, 3246 (1964).
44. P. M. Henry, *J. Am. Chem. Soc.*, **88**, 1595 (1966).

45. T. Dozono and T. Shiba, *Bull. Japan Petrol. Inst.*, **5**, 8 (1963).

46. I. I. Moiseev, M. N. Vargaftik and Ya. K. Syrkin, *Dokl. Akad. Nauk S.S.S.R.*, **130**, 820 (1960).

47. S. V. Pestrikov, I. I. Moiseev, and T. M. Romanova, *Zh. Neorg. Khim.*, **10**, 1199 (1965).

48. R. Jira, J. Sedlmeier, and J. Smidt, *Ann.*, **693**, 99 (1966).

49. I. I. Moiseev, M. N. Vargaftik, and Ya. K. Syrkin, *Izv. Akad. Nauk S.S.S.R.*, *Odt. Khim. Nauk*, **1963**, 1144.

50. M. N. Vargaftik, I. I. Moiseev, and Ya, K. Syrkin, *Dokl. Akad. Nauk. S.S.S.R.*, **147**, 399 (1962).

51. K. I. Matveev, A. M. Osipov, V. F. Odyakov, Yu. V. Suzdal'nitskaya, I. F. Bukhtoyarov, and O. A. Emel'yanova, *Kinetika i Kataliz*, **3**, 661 (1962).

52. K. I. Matveev, I. F. Bukhtoyarov, N. N. Shul'ts, and O. A. Emel'yanova, *Kinetika i Kataliz*, **5**, 649 (1964).

53. R. Ninomiya, M. Sato, and T. Shiba, *Bull. Japan Petrol. Ibst.*, **7**, 31 (1965).

54. A. P. Belov, I. I. Moiseev, and N. G. Uvarova, *Izv. Akad. Nauk S.S.S.R.*, *Ser. Khim.* **1965**, 2224.

55. A. P. Belov, I. I. Moiseev, and N. G, Uvarova, *Izv. Akad. Nauk S.S.S.R.*, *Ser. Khim*, **1966**, 1642.

56. I. I. Moiseev, A. P. Belov, V. A. Igoshin, and Ya. K. Syrkin, *Dokl. Akad. Nauk*, *SSSR*, **173**, 863 (1967).

57. Nippon Gosei, Nagaku Kogyo Kabushiki Kaisha, *Fr. Pat.*, 1,324,029 (1963).

58. M. S. Kharasch, R. C. Seyler, and F. R. Mayo, *J. Am. Chem. Soc.*, **60**, 882 (1938).

59. M. Nakumura and K. Gunji, *J. Japan Petrol. Inst.*, **6**, 191 (1963).

60. I. I. Moiseev, M. N. Vargaftik, and Ya. K. Syrkin, *Dokl. Akad. Nauk SSSR*, **153**, 140 (1963).

61. J. Halpern, *Chem. Eng. News.* **44**, 68 (1966).

62. I. I. Moiseev and M. N. Yargaftik, *Dokl. Akad. Nauk SSSR*, **166**, 370 (1966).

62a. I. I. Moiseev and M. N. Vargaftik, *Izv. Akad. Nauk SSSR*, *Ser. Khim.*, **1965**, 893.

63. W. Kitching, *Organometal. Chem. Rev.*, **3**, 61 (1968).

64. J. Chatt, K. M. Vallarino, and L. M. Venanzi, *J. Chem. Soc.*, **1957**, 3413.

65. J. K. Stille, D. B. Fox, L. F. Hines, R. W. Fries, and R. D. Hughes, *Pre-prints*, *Petrol. Div. Am. Chem. Soc.*, **14**, (2), B149 (1969).

66. P. M. Henry, *J. Am. Chem. Soc.*, **86**, 217 (1964).

67. R. Cramer, *J. Am. Chem. Soc.*, **86**, 217 (1964).

68. C. E. Holloway, G. Hulley, B. F. G. Johnson, and J. Lewis, *J. Chem. Soc.*, *(A)*, **1969**, 53.

69. B. L. Shaw, *Chem. Commun.*, **1968**, 464.

70. S. V. Pestrikov, *Zh. Fiz. Khin.*, **39**, 218 (1965).

71. J. Chatt and B. L. Shaw, *J. Chem. Soc.*, **1962**, 5075.

72. G. E. Coates and G. Calvin, *J. Chem. Soc.*, **1960**, 2008.

73. E. H. Brooks and F. Glockling, *J. Chem. Soc. (A)*, **1967**, 1030.

74. L. G. Cannel and R. W. Taft, Jr., *J. Am. Chem. Soc.*, **78**, 5812 (1956).

75. J. Chatt and B. L. Shaw, *Chem. Ind.*, **1960**, 931.

76. J. Chatt and B. L. Shaw, *Chem. Ind.*, **1961**, 290.

77. L. Vaska and J. W. DiLuzio, *J. Am. Chem. Soc.*, **83**, 1262, 2784 (1961).

78. J. Chatt and I. Leden, *J. Chem. Soc.*, **1955**, 2036.
79. J. R. Joy and M. Orchin, *Z. Anorg. Allgem. Chem.*, **305**, 236 (1960).
80. A. V. Nikiforova, I. I. Moiseev, and Ya. K. Syrkin, *Zh. Obshch. Khim.*, **33**, 3239 (1963).
81. W. G. Lloyd, *J. Org. Chem.*, **32**, 2816 (1967).
82. D. Clark and P. Hayden, *U.S. Pat.*, 3,257,448 (1966).
83. R. Van Helden and G. Verberg, *Rec. Trav. Chim.*, **84**, 1263 (1965).
84. J. M. Davidson and C. Triggs, *Chem. Ind.*, **1966**, 457.
85. J. M. Davidson and C. Triggs, *J. Chem. Soc. (A)*, **1968**, 1324.
86. M. O. Unger and R. A. Fauty, *J. Org. Chem.*, **34**, 18 (1969).
87. I. Moritani and Y. Fujiwara, *Tetrahedron Let.*, **1967**, 1119.
88. Y. Fujiwara, I. Moritani, M. Matsuda, and S. Teranishi, *Tetrahedron Let.*, **1968**, 633.
89. Y. Fujiwara, I. Moritani, and M. Matsuda, *Tetrahedron*, **24**, 4819 (1968).
90. Y. Fujiwara, I. Moritani, M. Matsuda and S. Terranishi, *Tetrahedron Let.*, **1968**, 3863.
91. Y. Fujiwara, I. Moritani, R. Asano, and S. Teranishi, *Tetrahedron Let.*, **1968**, 6015.
92. I. Moritani, Y. Fujiwara, and S. Teranishi, *Pre-prints, Petrol. Div., Am. Chem. Soc.*, **14**(2), B172 (1969).
93. R. F. Heck, *J. Am. Chem. Soc.*, **90**, 5519 (1968).
94. R. F. Heck, *J. Am. Chem. Soc.*, **90**, 5526 (1968).
95. R. F. Heck, *J. Am. Chem. Soc.*, **90**, 5531 (1968).
96. R. F. Heck, *J. Am. Chem. Soc.*, **90**, 5535 (1968).
97. R. F. Heck, *J. Am. Chem. Soc.*, **90**, 5538 (1968).
98. R. F. Heck, *J. Am. Chem. Soc.*, **90**, 5542 (1968).
99. P. M. Henry, *Tetrahedron Let.*, **1968**, 2285.
100. A. N. Nesmeyanov, A. Z. Rubezhov, L. A. Leits, and S. P. Gubin, *J. Organometal. Chem.*, **12**, 187 (1968).
101. T. Saegusa, T. Tsuda, and K. Nishijima, *Tetrahedron Let.*, **1967**, 4255.
102. H. C. Volger, *Rec. Trav. Chim.*, **86**, 677 (1967).
103. C. F. Kohll, and R. Van Helden, *Rec. Trav. Chim.*, **86**, 193 (1967).
104. W. H. Clemment and C. M. Selwitz, *Tetrahedron Let.*, **1962**, 1081.
105. R. Ugo, *Coord. Chem. Rev.*, **3**, 319 (1968).
106. T. J. Wallace, A. Schriesheim, H. Hurwitz, and M. B. Glaser, *Ind. Eng. Chem., Proc. Res. Dev.*, **3**, 237 (1964).
107. J. D. Hopton, C. J. Swan, and D. L. Trim, *Adv. Chem. Ser.*, **75**, 216 (1968).
108. C. F. Cullis and D. L. Trim, *Disc. Farad. Soc.*, **46**, 144 (1969).
109. W. A. Waters, *Disc. Farad. Soc.*, **46**, 185 (1969).
110. W. I. Taylor and A. R. Battersby (eds.), *Oxidative Coupling of Phenols*, Dekker, New York, 1967.
111. G. Eglington and A. R. Galbraith, *Chem. Ind.*, **1956**, 737.
112. A. S. Hay, *J. Org. Chem.*, **25**, 1275 (1960).
113. A. S. Hay, *J. Org. Chem.*, **27**, 3320 (1962).
114. G. F. Endres, A. S. Hay, and J. W. Eustance, *J. Org. Chem.*, **28**, 1300 (1963).
115. A. S. Hay, H. S. Blanchard, G. F. Endres, and J. W. Eustance, *J. Am. Chem. Soc.*, **81**, 6335 (1959).
116. E. Ochiai, *Tetrahedron*, **20**, 1831 (1964).

117. H. Finkbeiner, A. S. Hay, H. S. Blanchard, and G. F. Endres, *J. Org. Chem.*, **31**, 549 (1966).

118. D. R. Bryant, J. E. McKeon, and B. C. Ream, *Tetrahedron Let.*, **1968**, 3371.

119. J. M. Davidson and C. Triggs, *Chem. Ind.*, **1967**, 1361.

120. J. M. Davidson and C. Triggs, *J. Chem. Soc.*, (*A*), **1968**, 1331.

121. D. R. Bryant, J. E. McKeon, and B. C. Ream, *J. Org. Chem.*, **33**, 4123 (1968).

122. D. R. Bryant, J. E. McKeon, and B. C. Ream, *J. Org. Chem.*, **34**, 1106 (1969).

123. C. H. Bushweller, *Tetrahedron Let.*, **1968**, 6123.

124. A. B. Evnin and A. Y. Lam, *Chem. Commun.*, **1968**, 1184.

125. P. Fitton, J. E. McKeon, and B. C. Ream, *Chem. Commun.*, **1969**, 370.

126. L. Reich and S. S. Stivala, *Autoxidation of Hydrocarbons and Polyolefins, Kinetics and Mechanisms*, Dekker, New York 1969.

127. W. O. Lundberg (ed.) *Autoxidation and Antioxidants*, Interscience, New York (1961).

128. K. U. Ingold, *Acc. Chem. Res.*, **2**, 1 (1969).

129. J. K. Kochi, *Science*, **155**, 415 (1967).

130. N. M. Emanuel, Z. K. Maizus, and I. P. Skibida, *Angew. Chem., Internat. Ed.*, **8**, 97 (1969).

131. R. F. Gould (ed.), *Oxidation of Organic Compounds I and II. Adv. Chem. Ser.* **75** and **76**, Am. Chem. Soc., Washington, D.C. (1968).

132. *Disc. Forad. Soc.*, **46** (1969).

133. T. A. Cooper, A. A. Clifford, D. J. Mills, and W. A. Waters, *J. Chem. Soc.* (*B*), **1966**, 793.

134. N. Uri, *Nature*, **177**, 1177 (1956).

135. H. Kropf and H. Hoffman, *Tetrahedron Let.*, **1967**, 659.

136. A. T. Betts and N. Uri, *Adv. Chem. Ser.*, **76**, 160 (1968).

137. H. Kropf, *Ann.*, **637.**, 73 (1960).

138. L. N. Denisova and E. T. Denisov, *Izv. Akad. Nauk SSSR Ser. Khim.*, **1966**, 1095.

139. E. I. Heiba, R. M. Dessau, and W.J. Koehl, Jr., *J. Am. Chem. Soc.*, **90**, 1082 (1968).

140. H. Finkbenier and J. B. Bush, Jr., *Disc. Farad. Soc.*, **46**, 150 (1969).

141. E. I. Heiba and R. M. Dessau, *Disc. Farad. Soc.*, **46**, 189 (1969).

142. F. Haber and J. Weiss, *Naturwiss.* **20**, 948 (1932).

143. Y. Kamiya, S. Beaton, A. Lafortune, and K. U. Ingold, *Can. J. Chem.*, **41**, 2020 (1963).

144. Y. Kamiya, S. Beaton, A. Lafortune, and K. U. Ingold, *Can. J. Chem.*, **41**, 2034 (1963).

145. C. Copping and N. Uri, *Disc. Farad. Soc.*, **46**, 202 (1969).

146. E. Ochiai, *Tetrahedron*, **20**, 1819 (1964).

147. J. K. Kochi and P. E. Mocadlo, *J. Org. Chem.*, **30**, 1134 (1965).

148. W. A. Waters, *Disc. Farad. Soc.*, **46**, 158 (1969).

149. J. K. Kochi and R. V. Subramanian, *J. Am. Chem. Soc.*, **87**, 4855 (1965).

150. J. D. Bacha and J. K. Kochi, *J. Org. Chem.*, **33**, 83 (1968).

151. L. H. Vogt, Jr., H. M. Faigenbaum, and S. E. Wiberley, *Chem. Rev.*, **63**, 269 (1963).

152. M. E. Winfield, *Oxidases and Related Redox Systems* (T. E. King, H. S. Mason, and M. Morrison, eds.), Wiley, New York (1965), p. 115.

153. S. Fallab, *Angew, Chem. Int. Ed.*, **6**, 496 (1967).

154. L. Vaska, *Science*, **140**, 809 (1963).

155. J. H. Bayston, F. D. Looney, and M. E. Winfield, *Aust. J. Chem.* **16**, 557 (1963).
156. J. H. Bayston, N. K. King, F. D. Looney, and M. E. Winfield, *J. Am. Chem. Soc.*, **91**, 2775 (1969).
157. M. Mori, J. A. Weil, and M. Ishiguro, *J. Am. Chem. Soc.*, **90**, 615 (1968).
158. J. H. Bayston and M. E. Winfield, *J. Catalysis*, **3**, 123 (1964).
159. G. Pregaglia, D. Morelli, F. Conti, G. Gregorio, and R. Ugo, *Disc. Farad. Soc.*, **46**, 110 (1969).
160. C. F. Wells, *Disc. Farad. Soc.*, **46**, 137 (1969).
161. S. Takahashi, K. Sonogashira, and N. Hagihara, *Mem. Inst. Sci. Ind. Res. Osaka Univ.*, **23**, 69 (1966).
162. G. Wilke, H. Schott, and P. Heimbach, *Angew. Chem. Internat. Ed.*, **6**, 92 (1967).
163. C. D. Cook and G. S. Jahual, *Inorg. Nucl. Let.*, **3**, 31 (1967).
164. S. Otsuka, A. Nakamura, and Y. Tatsumo, *Chem. Commun.*, **1967**, 836.
165. P. B. Chock and J. Halpern, *J. Am. Chem. Soc.*, **88**, 3511 (1966).
166. J. P. Collman, *Acc. Chem. Res.*, **1**, 136 (1968).
167. J. Blum, H. Roseman, and E. D. Bergmann, *Tetrahedron Let.*, **1967**, 3665.
168. K. R. Laing and W. R. Roper, *Chem. Commun.*, **1968**, 1556.
169. J. A. McGinnety, R. J. Doedens, and J. A. Ibers, *Science*, **155**, 709 (1967).
170. L. Vaska, *Acc. Chem. Res.*, **1**, 335 (1968).
171. K. Hirota and M. Yamamoto, *Chem. Commun.*, **1968**, 533.
172. W. P. Griffith and T. D. Wickins, *J. Chem. Soc.*, (*A*), **1968**, 397.
173. S. J. LaPlaca and J. A. Ibers, *J. Am. Chem. Soc.*, **87**, 2581 (1965).
174. T. Kashiwagi, N. Yasuoka, N. Kasai, and M. Kakudo, *Chem. Commun.*, **1969**, 743.
175. C. D. Cook, P. T. Cheng, and S. C. Nyburg, *J. Am. Chem. Soc.*, **91**, 2123 (1969).
176. R. Mason, *Nature*, **217**, 543 (1968).
177. C. J. Nyman, C. E. Wymore, and G. Wilkinson, *J. Chem. Soc.*, (*A*), **1968**, 561.
178. J. P. Collman, M. Kubota, and J. W. Hosking, *J. Am. Chem. Soc.*, **89**, 4809 (1967).
179. C. D. Cook and G. S. Jauhal, *J. Am. Chem. Soc.*, **89**, 3066 (1967).
180. J. J. Levinson and S. D. Robinson, *Chem. Commun.*, **1967**, 198.
181a. E. W. Stern, *Chem. Commun.*, **1970**, 736.
181. E. W. Stern, unpublished results.
182. J. P. Birk, J. Halpern, and A. L. Pickard, *J. Am. Chem. Soc.*, **90**, 4491 (1968).
183. L. Malatesta and C. Cariello, *J. Chem. Soc.*, **1958**, 2323.
184. L. Malatesta and M. Angoletta, *J. Chem. Soc.*, **1957**, 1187.
185. E. S. Gould and M. Rado, *J. Catalysis*, **13**, 238 (1969).
186. A. J. Birch and G. S. R. Subba Rao, *Tetrahedron Let.*, **1968**, 2917.
187. J. Blum and H. Rosenman, *Israel J. Chem.*, **5**, 69p (1967).
188. V. P. Kurkov, J. Z. Pasky, and J. B. Lavigne, *J. Am. Chem. Soc.*, **90**, 4743 (1968).
189. L. W. Fine, M. Grayson, and V. H. Suggs, *Pre-prints, Petrol. Div. Am. Chem. Soc.*, **14**(4), C98 (1969).
190. C. S. Foote, *Acc. Chem. Res.*, **1**, 104 (1968).
191. J. E. Huber, *Tetrahedron Let.*, **1968**, 3271.
192. R. Van Rheenen, *Chem. Commun.*, **1969**, 314.
193. R. Ugo, F. Conti, S. Cenini, R. Mason, and G. B. Robertson, *Chem. Commun.*, **1968**, 1498.
194. F. D. Mango and J. H. Schachtschnieder, *J. Am. Chem. Soc.*, **89**, 2484 (1967).

5

Carbonylation

D. T. THOMPSON and R. WHYMAN

Imperial Chemical Industries Limited
Petrochemical & Polymer Laboratory
The Heath
Runcorn, Cheshire
England

I. Introduction

Carbonylation and hydroformylation reactions have achieved considerable importance in synthetic organic chemistry for incorporating CO into a wide variety of chemical systems. The conversion of olefins into plasticiser alcohols using the oxo reaction has also achieved wide application in industry. The activation of organic functional groups by transition metals in homogeneous solution is possible in a variety of reactions involving unsaturated hydrocarbons, e.g., hydrogen transfer (*1*), hydrogenation (*2*), oxidation (*3*), isomerization (*4*), and polymerization (*5*), but the most versatile and well-established reaction of this type from the viewpoint of a synthetic organic chemist is undoubtedly carbonylation.

In this article we summarize and discuss the alternative methods for forming organic functional groups using carbonylation reactions. In general, the section headings are for the products, and subsections are given for each type of starting material.

The reaction in which olefins are hydroformylated to give aldehydes and alcohols (the oxo reaction) has been studied in detail from a mechanistic viewpoint; these results are discussed under aldehydes. The reactions involve features which are typical of those homogeneously catalyzed by transition metals. These include the initial dissociation of a metal hydride catalyst to leave a vacant coordination site; the reactant then takes up this site and two successive insertions by olefin and CO into metal-H and metal-*R* bonds respectively give an acyl intermediate which, upon hydrogenolysis, yields both product and the catalyst for recycle (*6*). The conversion of methyl-acetylene to methyl methacrylate is also operated commercially and a mechanistic interpretation of this reaction is given in the ester section. Other reactions have also received mechanistic investigations either from a kinetic viewpoint or by the isolation or identification of intermediates. The more significant developments are described in the appropriate sections.

Most of the reactions described in this chapter are carried out in the presence of transition metal carbonyls and a significant proportion of these are catalytic. Recently, however, H. C. Brown and co-workers (*7*) have introduced an important stoichiometric synthesis of a wide range of organic products which involves the carbonylation of organoborane intermediates obtained from olefinic starting materials. These reactions, which involve intramolecular transfer of alkyl groups from boron to carbon (*8*), proceed in very high yields and are convenient for the synthesis of carbonylated materials in the laboratory under very mild conditions. The carboxylation of olefins using acid catalysts is a well established reaction and the two step modification introduced by Koch (see Section V, A) is used in the industrial synthesis of branched chain carboxylic acids.

The oxo process is based on cobalt carbonyl catalysts and cobalt is the most commonly used transition metal for hydroformylation and carbonylation reactions. Presumably this is partly because of cheapness and availability and recently it has been shown that in certain circumstances some other group VIII metals may be used as catalysts. All the group VIII metals have been used in stoichiometric carbonylation reactions. Other transition metals, such as manganese, are much less effective (9).

In general, olefinic materials are much more reactive under hydroformylation conditions than the corresponding acetylenic materials, but, as is mentioned above, the Markownikoff addition of methanol to methylacetylene in the presence of CO alone is effectively catalyzed by $Ni(CO)_4$. Acetylene may be dicarbonylated to ester in the presence of CO, alcohol, and palladium. The conversion of acetylenes to carbonyl compounds has also been reported using $TlCl_3$ in acetic acid (10). Aromatic hydrocarbons are usually inert and are sometimes used as solvents in carbonylation reactions.

Good carbonylatiom methods are available for the synthesis of aldehydes, alcohols, ketones, esters, carboxylic acids (and derivatives), and quinones. The reactions of nitrogen compounds with carbon monoxide and/or hydrogen either alone or with other substrates has led to the synthesis of carbonylated nitrogen systems such as amides, ureas, imides, and lactams. In general, the incorporation of CO into an organic system leads to the homologated product, e.g., C_6 olefin gives C_7 aldehyde. A number of decarbonylation reactions are also included in the discussion. Alkyl halides and hydrocarbons are sometimes produced under carbonylation conditions and these reactions are also summarized.

The carbonylation reactions described below are generally considered to take place in homogeneous solution, but catalyst deposition on the walls of high pressure vessels suggests that in some cases heterogeneous catalysis may also make a contribution. It is also possible to hydroformylate olefins to aldehydes in the gas phase (11). Pure CO is only rarely available and most CO cylinders supplied commercially contain a small percentage of hydrogen—this may play a significant part in the mechanism of some reactions described as taking place under CO.

Many of the "classic" carbonylation reactions reported in the literature are carried out at high pressures and temperatures. More recently, con-

siderable emphasis has been placed on a search for transition metal catalysts which will enable the same reactions to be performed under much
milder conditions. These efforts have met with some success and the results are included where appropriate. Brown's hydroboration method also
enables stoichiometric carbonylation synthesis to be performed under mild
conditions.

For completeness, this chapter includes reactions in which no CO gas
is used and in which the sole source of CO is the ligand of a metal carbonyl
compound. Some isomerization and hydrogenation reactions which take
place under carbonylation and hydroformylation conditions are also described where these are relevant to the discussion.

II. Aldehydes

The conversion of olefinic functional groups to mixtures of *n*- and *iso*-
aldehydes by hydroformylation at 150°/200 atm in the presence of cobalt
catalysts is a well-established reaction and forms the basis of the indus-
trially important process known as the oxo reaction. Hydroboration may
be used in the laboratory to prepare *n*- or *iso*-aldehydes from olefins under
mild conditions in almost quantitative yields. Carbonylation methods have
also been used to prepare aldehydes from dienes, epoxides, and functionally
substituted olefins including nitriles. Hydroformylation of alkynes is very
difficult but low yields of aldehydes have been obtained from hex-1-yne
using a rhodium catalyst.

A. FROM OLEFINS VIA HYDROBORATION

Olefins react with BH_3 to give trialkylboranes, R_3B. Aldehydes may be
prepared from these by treatment with CO in the presence of $LiAlH(OCH_3)_3$
(7). The lithium hydride promotes the formation of an intermediate in
which only one of the three R groups has been transferred from boron to
carbon; oxidation of this intermediate with hydrogen peroxide gives the
aldehyde in high yields:

$$R_3B + CO + LiAlH(OCH_3)_3 \xrightarrow[25\text{-}45°]{} R_2B-\underset{\underset{OAl(OCH_3)_3Li}{|}}{\overset{\overset{H}{|}}{C}}-R \xrightarrow{[O]} R\text{CHO} \quad (1)$$

e.g.:

$$CH_3(CH_2)_3CH{=}CH_2 \longrightarrow CH_3(CH_2)_5CHO \qquad (2)$$
$$98\%$$

$$\underset{94\%}{CH_3CH{=}CHCH_3 \longrightarrow CH_3CH_2\overset{\overset{\displaystyle CH_3}{|}}{C}HCHO} \qquad (3)$$

In the above procedure only one of the three alkyl groups in BR_3 is used. This disadvantage may be overcome by reacting the olefin with 9-bora-bicyclo[3,3,1]nonane (9-BBN), obtained from 1,5-cyclooctadiene and BH_3 (12). Reaction of these B-R-9-borabicyclo[3,3,1]nonane derivatives with CO and $LiAlH(OCH_3)_3$ results in preferential reaction at the B-alkyl group and high overall yields of aldehyde from olefin, e.g.:

$$CH_3(CH_2)_3CH{=}CH_2 \longrightarrow CH_3(CH_2)_5CHO \qquad (4)$$
$$77\%$$

The introduction of the aldehyde takes place with retention of configuration and the synthesis can be carried out in the presence of nitrile and ester functional groups (7,13), e.g.:

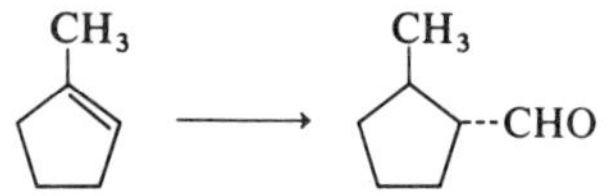

$$\qquad (5)$$

$$H_2C{=}CH(CH_2)_8CN \longrightarrow OHC(CH_2)_{10}CN \qquad (6)$$
$$95\%$$

B. From Olefins Using Transition Metal Catalysts

The reaction which has received most attention as a means for converting olefins into aldehydes has been that of hydroformylation in the presence of cobalt carbonyls. This has achieved considerable industrial importance and is known as the oxo reaction (14):

$$RCH{=}CH_2 + CO + H_2 \longrightarrow RCH_2CH_2CHO + RCH(CHO)CH_3 \qquad (7)$$

There have been many studies of the mechanism of this reaction. The active catalyst is thought to be either $HCo(CO)_3$ or $HCo(CO)_4$ and under typical reaction conditions (ca. 150°, 200 atm) the most significant steps are insertion of an olefin into a cobalt–hydrogen bond, followed by CO insertion into a cobalt–carbon bond (6). The following scheme refers to the formation of n-aldehyde:

$$HCo(CO)_4 \rightleftharpoons HCo(CO)_3 + CO \qquad (8)$$

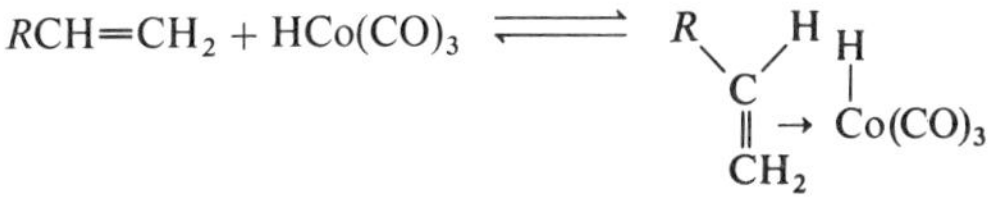

$$RCH=CH_2 + HCo(CO)_3 \rightleftharpoons \begin{array}{c} R \quad H \; H \\ \diagdown \diagup \quad | \\ C \quad \rightarrow Co(CO)_3 \\ \| \\ CH_2 \end{array} \qquad (9)$$

$$\begin{array}{c} R \quad H \; H \\ \diagdown \diagup \quad | \\ C \quad \rightarrow Co(CO)_3 \\ \| \\ CH_2 \end{array} \rightleftharpoons RCH_2CH_2Co(CO)_3 \qquad (10)$$

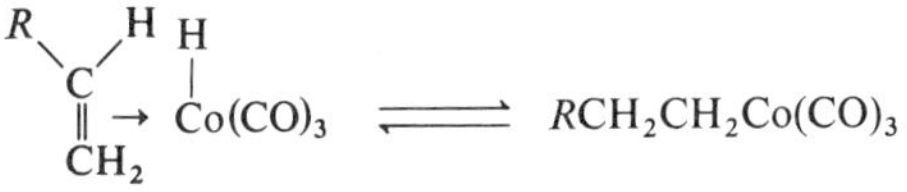

$$RCH_2CH_2Co(CO)_3 + CO \rightleftharpoons RCH_2CH_2COCo(CO)_3 \qquad (11)$$

$$RCH_2CH_2COCo(CO)_3 + H_2 \longrightarrow RCH_2CH_2CHO + HCo(CO)_3 \qquad (12)$$

Various alternative steps may also be involved, e.g.:

$$RCH_2CH_2COCo(CO)_3 + HCo(CO)_4 \longrightarrow RCH_2CH_2CHO + Co_2(CO)_8 \qquad (13)$$

Kinetic evidence for the formation of $HCo(CO)_3$ has recently been obtained (15). By using ^{13}CO, it has been shown (16) that when (triphenylphosphine)benzylcobalt tricarbonyl forms (triphenylphosphine)-phenylacetylcobalt tricarbonyl the acyl group is formed by incorporation of a carbonyl ligand, whereas the CO from the gas phase enters the coordination sphere of the cobalt as a new ligand. The production of the isomeric aldehyde $RCH(CHO)CH_3$ can take place via CO insertion into the corresponding alkyl $RCH(CH_3)Co(CO)_3$ or by isomerization of alkyl or acyl intermediates. Hydrogenation or isomerization of the olefin sometimes takes place in competition with the hydroformylation to produce by-products.

The stoichiometric hydroformylation of norbornene under ambient conditions is largely, if not entirely, a *cis*-process (17):

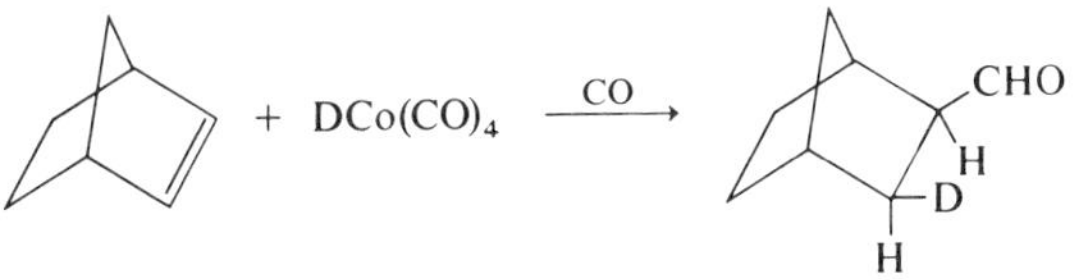

$$(14)$$

In addition, studies of the catalytic high pressure hydroformylation of unsaturated carbohydrates having vinyl ether structures has clearly indicated *cis*-hydroformylation and this stereochemistry may be a general feature of hydroformylation reactions. $HCo(CO)_4$ adds to 1,2-diphenylcyclobutene-1 in a *cis*-fashion (18).

Cobalt may be added in a number of forms but the key catalytic intermediate, $HCo(CO)_3X$ ($X = CO$ or PR_3), is normally formed during the reaction, e.g., $\pi\text{-}C_4H_7Co(CO)_2PBu_3$ gives $HCo(CO)_3PBu_3$ (19).

Stoichiometric hydroformylation occurs with $Co_2(CO)_8$ at room temperature and pressure (20). With catalytic amounts of cobalt compounds, however, a minimal CO partial pressure is needed in order to promote the regeneration of $Co_2(CO)_8$ or $HCo(CO)_4$ and to stabilize these catalysts. A small increase in CO partial pressure above this increases the reaction rate up to a maximum which depends on temperature and type of olefin. A further increase in CO partial pressure causes a reduction in rate. An increase in hydrogen partial pressure always increases reaction rate, although at very high pressures this effect becomes smaller. Natta's equation may be used for the normal ranges of temperature and pressure (100°–160°/50–250 atm)

$$\frac{d\,(\text{aldehyde})}{dt} = k[\text{olefin}][\text{Co}][P_{H_2}][P_{CO}]^{-1} \qquad (15)$$

Other equations have also been proposed (20a,20b) and shown to be useful for certain sets of conditions. When using 1 : 1 CO/H_2, the hydroformylation reaction seems to be independent of pressure over a considerable range—the two pressure effects operate in opposite directions. Increase in temperature usually increases the rate of reaction but may increase the proportion of side reactions. The rate of reaction is increased by the presence of small amounts of nitrogeneous organic bases, e.g., pyridine (21) but is decreased by larger amounts of pyridine. Rates are usually decreased by the presence of phosphines (see Section III). The production of high proportions of n-isomers is favored by high partial pressures of CO (22), the presence of phosphines (22a,22b), and low temperatures. The properties of the solvent are another factor (23), but solvent effects are usually small (24). Catalyst concentration has no effect on product distribution (22). An interesting new development is the use of solid poly-2-vinylpyridine cobalt carbonyls as catalyst reservoirs which reversibly and rapidly release enough $HCo(CO)_4$ both to destroy impurities in olefin feedstocks and to promote reproducible rates of hydroformylation (without an induction period) (24a).

The rate of hydroformylation decreases with increasing alkyl substitution in the olefin (21), i.e., straight chain terminal olefins > straight chain internal olefins > branched chain olefins. The effect is larger the closer the branching is to the double bond. With cyclic olefins the order of reactivity is cyclopentene > cyclohexene < cycloheptene > cyclooctene.

The scope of the hydroformylation of substituted olefins in organic synthesis is wide and a few examples are given below. These include formation of a precursor for a THF derivative (25), direct formylation of a methyl group (26), and hydroformylation of (+)-tris [(S)-1-methylpropoxy] methane (27):

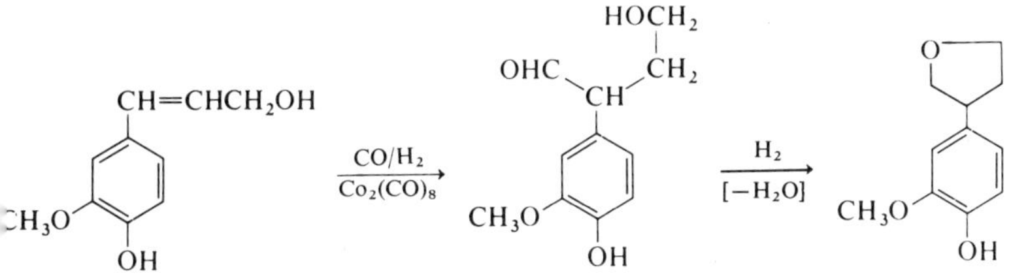

$$
\text{CH=CHCH}_2\text{OH} \xrightarrow[\text{Co}_2(\text{CO})_8]{\text{CO/H}_2} \quad \xrightarrow[\text{[-H}_2\text{O]}]{\text{H}_2} \tag{16}
$$

$$
\begin{array}{c}
\text{CH}_2 \\
\parallel \\
\text{CH} \\
\mid \\
\text{CHCH}_3 \\
\mid \\
\text{C}_3\text{H}_7
\end{array}
\xrightarrow[\text{Co}_2(\text{CO})_8]{\text{CO/H}_2}
\begin{array}{c}
\text{CH}_3 \\
\mid \\
\text{CH}_2 \\
\mid \\
^*\text{CHCH}_2\text{CHO} \\
\mid \\
\text{C}_3\text{H}_7
\end{array}
+
\begin{array}{c}
\text{CH}_3 \\
\mid \\
\text{CH}_2 \\
\mid \\
\text{C=CH}_2 \\
\mid \\
\text{C}_3\text{H}_7
\end{array}
\tag{17}
$$

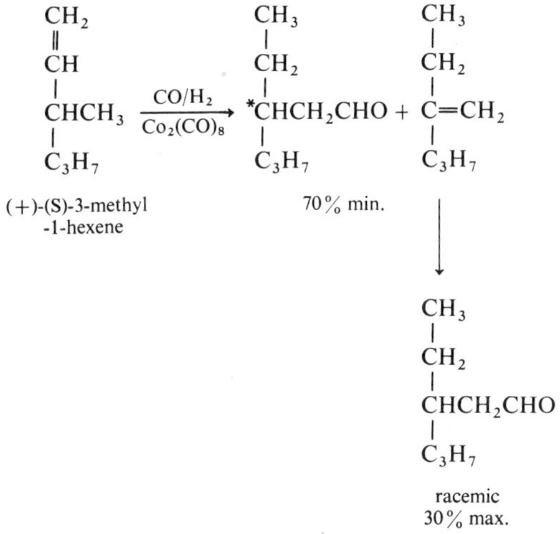

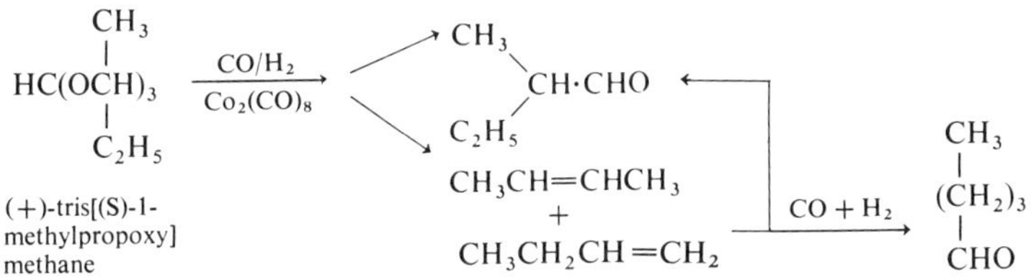

(+)-(S)-3-methyl
-1-hexene   70% min.

$$
\begin{array}{c}
\text{CH}_3 \\
\mid \\
\text{CH}_2 \\
\mid \\
\text{CHCH}_2\text{CHO} \\
\mid \\
\text{C}_3\text{H}_7
\end{array}
$$

racemic
30% max.

$$
\begin{array}{c}
\text{CH}_3 \\
\mid \\
\text{HC(OCH)}_3 \\
\mid \\
\text{C}_2\text{H}_5
\end{array}
\xrightarrow[\text{Co}_2(\text{CO})_8]{\text{CO/H}_2}
\begin{array}{l}
\text{CH}_3 \\
\quad\text{CH·CHO} \\
\text{C}_2\text{H}_5 \\[4pt]
\text{CH}_3\text{CH=CHCH}_3 \\
\quad + \\
\text{CH}_3\text{CH}_2\text{CH=CH}_2
\end{array}
\xrightarrow{\text{CO + H}_2}
\begin{array}{c}
\text{CH}_3 \\
\mid \\
(\text{CH}_2)_3 \\
\mid \\
\text{CHO}
\end{array}
\tag{18}
$$

(+)-tris[(S)-1-
methylpropoxy]
methane

Substituted olefins such as acrylonitrile (120°–130°/200 atm) and vinyl chloride may be hydroformylated in the presence of $\text{Co}_2(\text{CO})_8$ to give high yields of $\text{OHC(CH}_2)_2\text{CN}$ (82%) (28) and $\text{CH}_3\text{CHClCHO}$ (85–90%) (20). In methanol as solvent both α- and β-formylpropionitriles have been isolated from the cobalt catalyzed hydroformylation of acrylonitrile in 8 and 10% yields respectively; the principal product of this reaction is $(\text{CH}_3\text{O})_2\text{CH(CH}_2)_2\text{CN}$ (71%) (28a). α-Formylpropionitrile polymerizes spontaneously at room temperature. N-Alkylaminoolefins give N-alkyl-amino-aldehydes in the presence of cobalt or rhodium carbonyl catalysts (*14*), e.g.,:

$$CH_2=CHCH_2N(CH_3)_2 \xrightarrow[\text{Co}_2(CO)_8]{\text{CO/H}_2(120°/165 \text{ atm})} \underset{87\%}{OHCCH_2CH_2CH_2N(CH_3)_2} \quad (19)$$

For olefins, rhodium complexes achieve rates of hydroformylation comparable with cobalt under milder conditions (*29–32*). The comparative order of activity is Rh > Co > Ir for carbonyls(*33*). $RhH(CO)_2(PPh_3)_2$, formed by the addition of CO to $RhH(CO)(PPh_3)_2$, or from $RhH(CO)(PPh_3)_3$, catalyses the hydroformylation of alkenes at 25°/1 atm to give a high proportion of n-aldehyde. The rhodium reactions obey similar kinetic laws to those catalyzed by cobalt (*35*) although minor variations are possible (*32,36*). It is interesting to note that at ambient conditions the addition of phosphine increases the rate of reaction. As with cobalt, the rhodium reaction involves the formation of acyl-metal intermediates (*37*). It has been shown (*37*) that the hydroformylation of ethylene with $RhH(CO)(PPh_3)_3$ may be inhibited by high pressures of CO. The mechanism involves the formation of a square planar acyl derivative which then undergoes oxidative addition by hydrogen.

New oxo processes using $RhH(CO)(PPh_3)_3$ (*37a*) or $RhCl(CO)(PPh_3)_2$ (*37b*) and operating at *ca.* 100°/30 atm give an aldehyde selectivity of 98 %, and up to 80 % of n-isomer. Since the conditions for the two processes are very similar, it seems likely that the different rhodium compounds are precursors to the same catalytically active species. The highest yields of straight chain aldehydes are favored by the presence of excess phosphine. In the limiting case, i.e. molten PPh_3, yields of 94 % n-aldehyde were obtained but at the expense of reaction rate (*37c*).

Hydroformylation of methyl methacrylate produces a mixture of α- and β-isomers; high proportions of the branched α-isomer being favored by low temperatures and the presence of PR_3 (*23*), factors which favor the formation of straight chain isomers from unsubstituted olefins:

$$\underset{CH_2=CCO_2CH_3}{\overset{\overset{CH_3}{|}}{}} \xrightarrow[\text{RhH(CO)[P(OPh)}_3]_3]{\text{CO/H}_2} \underset{\alpha}{\overset{\overset{CH_3}{\underset{|}{|}}}{CH_3CCO_2CH_3}}_{\underset{CHO}{|}} + \underset{\beta}{\overset{\overset{CH_3}{|}}{OHCCH_2CHCO_2CH_3}} \quad (20)$$

Hydroformylation of ethyl acrylate follows a similar pattern (*38*). 4-Nitrostyrene and 3-nitropropene may be catalytically hydroformylated in the presence of rhodium, but not cobalt, carbonyls to give predominantly branched products (*38a*). The electron withdrawing effects of the

CO_2Me and NO_2 groups therefore seem to favor the formation of branched isomers.

Iron carbonyls are less spectacular carbonylation catalysts than cobalt or rhodium carbonyls but they can be used to prepare certain aldehydes (*39–42*), e.g.:

$$HFe(CO)_2\pi\text{-}C_5H_5 + RCH=CH_2 \longrightarrow RCH_2CH_2Fe(CO)_2\pi\text{-}C_5H_5$$

$$\downarrow CO \;\; 150°$$

$$RCH_2CH_2COFe(CO)_2\pi\text{-}C_5H_5$$

$$\downarrow H_2$$

$$HFe(CO)_2\pi\text{-}C_5H_5 + RCH_2CH_2CHO \tag{21}$$

$$KHFe(CO)_4 + n\text{-}C_3H_7I \xrightarrow[\text{(one mole)}]{CO} n\text{-}C_3H_7\overset{\overset{\displaystyle H}{|}}{C}OFe(CO)_4 \longrightarrow n\text{-}C_3H_7CHO \tag{22}$$

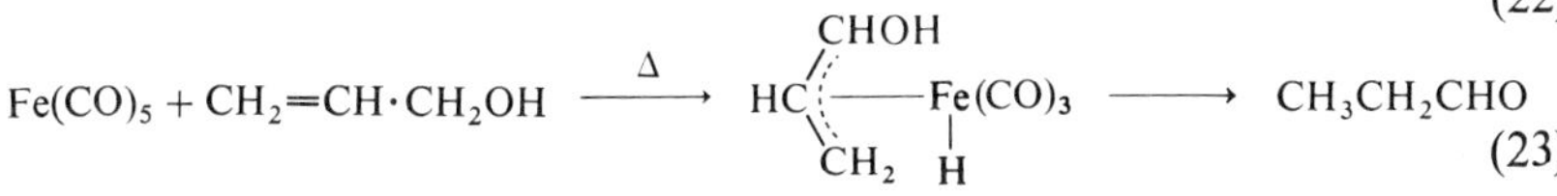

$$Fe(CO)_5 + CH_2=CH\cdot CH_2OH \xrightarrow{\Delta} HC\underset{CH_2}{\overset{CHOH}{\underset{\diagdown}{\diagup}}}\!\!\!-Fe(CO)_3 \longrightarrow CH_3CH_2CHO \tag{23}$$

The first of these reactions is catalytic but the second stoichiometric, and the third is apparently an isomerization rather than a carbonylation. Examples in which ruthenium has been used under hydroformylation conditions to give aldehydes include the following (*43,44*):

$$CH_3(CH_2)_2CH=CH_2 \xrightarrow[Ru(CO)_3(PPh_3)_2]{CO/H_2(100°/100\,atm)} \underset{80\%}{\text{hexaldehydes}} \tag{24}$$

$$C_6H_5CH_2OH \xrightarrow{CO/H_2} C_6H_5CHO + C_6H_5CH_3 + H_2O \tag{25}$$

The benzaldehyde seems to be formed by hydrogen transfer.

Palladium also catalyzes the conversion of olefins into aldehydes under CO and hydrogen but the yields are low since a high proportion of olefin is hydrogenated to hydrocarbon (*45*), e.g.:

$$CH_2=CH_2 + CO + H_2 \longrightarrow CH_3CH_2CHO + CH_3CH_3 \tag{26}$$

Aldehydes can be decarbonylated by heating at $> 180°$ in the presence of catalytic quantities of palladium (*46,47*):

$$RCH_2CH_2CHO \xrightarrow[Pd]{200°} RCH=CH_2 + RCH_2CH_3 + CO + H_2 \tag{27}$$

These results led to a proposed mechanism for the conversion of acid chlorides to aldehydes in the heterogeneous Rosenmund reaction (see Section XI):

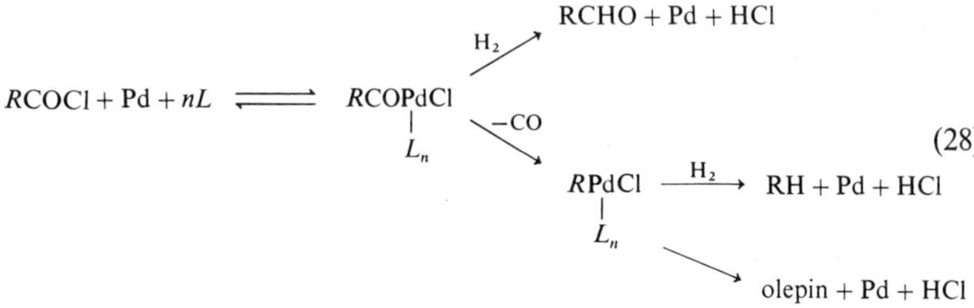

$$RCOCl + Pd + nL \rightleftharpoons \underset{\underset{L_n}{|}}{RCOPdCl} \begin{array}{c} \overset{H_2}{\nearrow} RCHO + Pd + HCl \\ \underset{\searrow}{\overset{-CO}{}} \\ \underset{\underset{L_n}{|}}{RPdCl} \overset{H_2}{\longrightarrow} RH + Pd + HCl \\ \searrow \\ \text{olepin} + Pd + HCl \end{array} \qquad (28)$$

C. FROM ACETYLENES

Hex-1-yne reacts with a 1:4 mixture of $H_2:CO$ at 110°/120 atm in the presence of $RhCl(PPh_3)_3$ in methanol/benzene to give a 15% yield of n-heptaldehyde and 2-methylhexaldehyde in equal amounts (*48,49*). With cobalt catalysts it is necessary to use much more severe conditions to get any reaction and the products then are alcohols, e.g., pent-1-yne gives low yields of n-hexanol (6%) and 2-methylpentanol (5%) at 185°/250 atm (*50*) (see Section III, C.).

D. FROM DIENES

Normally, under hydroformylation conditions, conjugated dienes give only saturated monoaldehydes (e.g., butadiene gives n- and *iso*-valeraldehyde (*51*), but it has recently been shown that the use of a rhodium tributylphosphine catalyst prevents isomerisation of the double bonds and gives 80–90% yields of hydroformylation products, about half of which is $OHC(CH_2)_4CHO$ (*52*).

E. FROM EPOXIDES

Aldehydes are sometimes obtained from epoxides under hydroformylation conditions (*14*), e.g., ethylene oxide gives acetaldehyde by isomerization or β-hydroxypropionaldehyde in low yields, and propylene oxide has been reported to give β-hydroxy-n-butyraldehyde in 42% yield. $HCo(CO)_4$ reacts rapidly with epoxides at 0°C under CO to give β-hydroxyacylcobalt tetracarbonyls which hydrogenolyse to give either aldehydes or alchols (*27*):

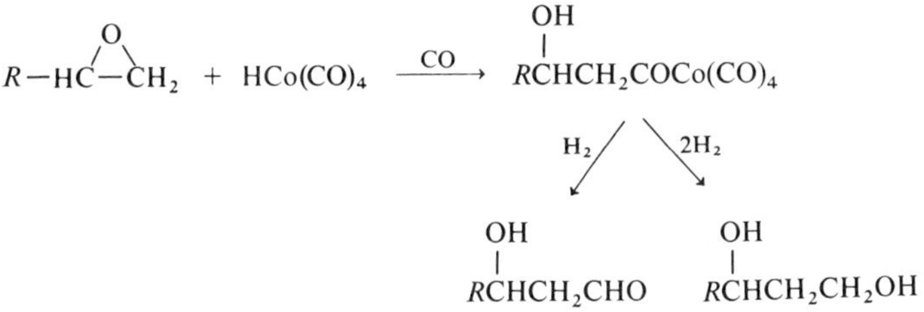

$$R-HC\overset{O}{\underset{\diagdown}{\diagup}}CH_2 + HCo(CO)_4 \xrightarrow{CO} R\overset{OH}{\underset{|}{C}}HCH_2COCo(CO)_4$$

$$H_2 \diagup \quad \diagdown 2H_2 \tag{29}$$

$$R\overset{OH}{\underset{|}{C}}HCH_2CHO \qquad R\overset{OH}{\underset{|}{C}}HCH_2CH_2OH$$

The rate for expoxides is 20–40 times greater than for epichlorhydrins (*53*).

F. FROM α,β UNSATURATED ALIPHATIC NITRILES

Saturated aldehydes may be obtained from α,β unsaturated aliphatic. nitriles (see Section II,B) using the following reaction carried out at room temperature (*54*), e.g.:

$$CH_2=CHCN + Co_2(CO)_8 + 5HCl + H_2O$$
$$\xrightarrow[CH_3OH]{N_2} \underset{85\%}{CH_3CH_2CHO} + 2CoCl_2 + NH_4Cl + 8CO \tag{30}$$

III. Alcohols

As with aldehydes, alcohols may be obtained from olefins by two principal routes—the stoichiometric hydroboration method or the catalytic hydroformylation route. Primary, secondary, and tertiary alcohols may be obtained via organoborane intermediates. Hydrogenation of aldehydes to alcohols occurs under more extreme conditions of hydroformylation than are normally employed. The addition of phosphines to hydroformylation reaction mixtures enables alcohols to be obtained directly particularly with cobalt catalysts. Reaction of olefins with $Fe(CO)_5$ under aqueous alkaline conditions gives alcohols. Acetylenes and epoxides are also possible starting materials for the preparation of alcohols by carbonylation methods.

A. FROM OLEFINS VIA HYDROBORATION

Organoborane intermediates, prepared by reacting olefins with BH_3, may be treated with CO in the presence of lithium trimethoxyalumino-hydride to give species in which only one of the R groups has been transferred from boron to carbon (cf. Sections II,A; IV,A; V,A; VI,A). Alkaline hydrolysis without oxidation produces almost quantitative yields of primary methylol derivatives (*12,55*):

$$R_3B \xrightarrow[\text{LiAlH(OCH}_3)_3]{\text{CO(25-45°)}} \underset{\substack{| \\ \text{OAl(OCH}_3)_3\text{Li}}}{\overset{\substack{\text{H} \\ |}}{R_2B-C-R}} \xrightarrow{\text{OH}^-} RCH_2OH \quad (31)$$

The use of 9-BBN (see Section II,A) can enable the conversion of all the olefin to alcohol, rather than one third (*12*), and also provides a route to 9-alkylbicyclo[3,3,1]nonan-9-ols (*56*) in which both the 9-alkyl and 9-hydroxy groups are bonded axially to a six membered ring:

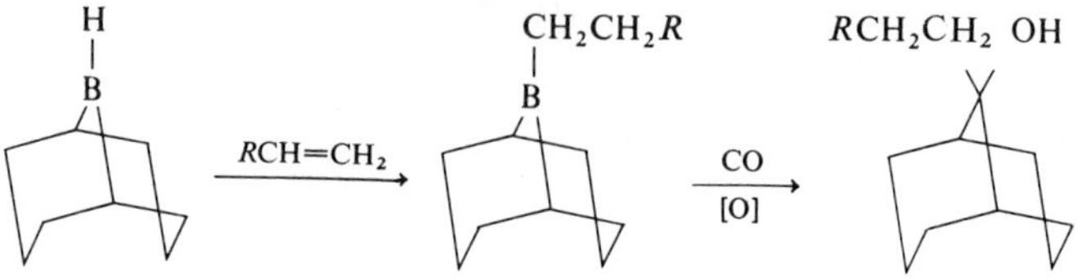

$$(32)$$

9-BBN

As with the synthesis of aldehydes (see Section II,A), this hydroboration route may be used for stereospecific synthesis, e.g. (*57*):

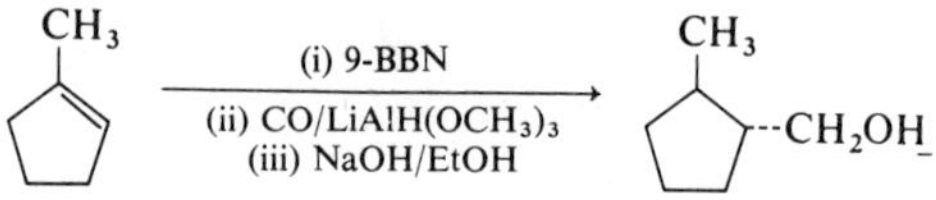

$$(33)$$

Reaction of R_3B with CO in the presence of one mole of water allows two of the three alkyl groups to be transferred from boron to carbon. The water apparently converts the boraepoxide into the corresponding hydrate, and the latter is less susceptible to the transfer of the third alkyl group than the intermediates which are formed under anhydrous conditions (*7*). Alkaline hydrolysis then gives secondary alcohols (*58*):

$$R_3B + CO \longrightarrow \underset{\overset{\diagdown\diagup}{O}}{RB-CR_2} \xrightarrow{\text{H}_2\text{O}} \underset{\substack{| \quad | \\ \text{OH} \quad \text{OH}}}{RB-CR_2}$$

$$\xrightarrow{\text{OH}^-} \qquad\qquad\qquad\qquad (34)$$

$$R_2CHOH + RB(OH)_2$$

Treatment of BR_3 with CO alone allows all three alkyl groups to be transferred (*8,58*) and alkaline oxidation then gives tertiary alcohols:

$$R_3B + CO \xrightarrow{100-125°} (R_3CBO)_x \xrightarrow[\text{NaOH}]{\text{H}_2\text{O}_2} R_3COH \quad (35)$$

$$\text{e.g., } CH_3CH_2CH=CH_2 \longrightarrow \underset{90\%}{(CH_3CH_2CH_2CH_2)_3COH} \quad (36)$$

The following route has been used to prepare the first member of a new class of compounds having a central carbon atom carrying a functional group supported by three rings with common sides (*59*):

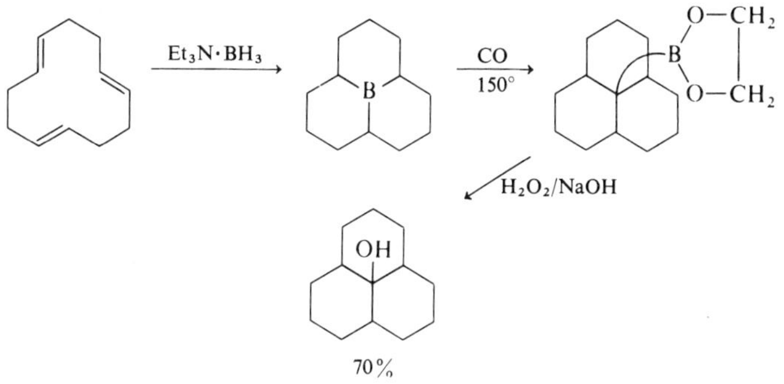

(37)

70%

B. FROM OLEFINS USING TRANSITION METAL CATALYSTS

Dissolution of $Fe(CO)_5$ in aqueous alkali gives salts of $H_2Fe_2(CO)_8$ which are effective catalysts for converting olefins into alcohols (*21,60*) e.g., at $100–150°/160–180$ atm, ethylene gives propanol and propionic acid, propylene gives butanol, and but-1-ene gives *n*-amyl alcohol and 2-methyl-butanol.

In the hydroformylation reaction using cobalt catalysts (see Section II,B) the addition of phosphines produces two effects (*34,61–64*): First the products, under comparable conditions, are alcohols rather than aldehydes, and second, a high proportion of *n*-isomers is obtained. The first effect indicates that phosphine activates the catalyst for hydrogenation and the second is probably steric (*34,63,65*) although inhibition of double-bond isomerization may also play a part (*63,66*). The polymeric compound $[Co(CO)_2PR_3]_n$ has been isolated (*62*) from stoichiometric hydroformylation reactions, e.g.:

$$HCo(CO)_3PR_3 + CH_3CH=CH_2 + H_2 \xrightarrow[120°]{} [Co(CO)_2PR_3]_n + n\text{-}C_4H_9OH$$

(38)

Other cobalt compounds such as $[Co(CO)_3PBu_3]_2$ and $[Co(CO)_3(PBu_3)_2]^+$ $[Co(CO)_4]^-$ have also been isolated from hydroformylation reaction solutions (*61*). All these compounds, however, may be derivatives of the active catalysts.

Addition of PBu_3 to $RhCl_3$ in the presence of sodium acetate (HX acceptor) showed that at $195°/35$ atm ($CO/H_2 = 1/2$) the rhodium catalyst is more selective for the formation of aldehydes than is $HCo(CO)_3PBu_3$ (61).

The conversion of aldehydes to alcohols using hydrogen transfer reactions has been reported in the presence of ruthenium (67,68), cobalt (69), rhodium (70), and iridium (71) catalysts, e.g., $[Ru_2Cl_3(PPhEt_2)_6]^+Cl^-$, $CoH_3(PPh_3)_3$, $RhCl(PPh_3)_3$, and $IrH_3(PPh_3)_3$.

C. FROM ACETYLENES

Acetylenes react with iron carbonyls in alkaline aqueous media to produce complexes of formulas $RC_2R' \cdot Fe_2(CO)_6(COH)_2$ which have the structure ($R=R'=CH_3$) (6):

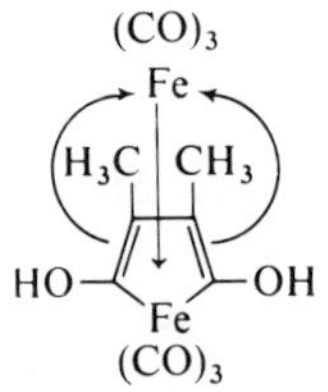

$$
\begin{array}{c}
(CO)_3 \\
Fe \\
H_3C \quad CH_3 \\
HO \quad OH \\
Fe \\
(CO)_3
\end{array}
$$

(1)

Pent-1-yne reacts slowly under hydroformylation conditions [$Co_2(CO)_8$; $185°/250$ atm] to give low yields of 1-hexanol and 2-methylpentanol (50).

D. FROM EPOXIDES

Styrene oxide hydrogenates in the presence of $KHFe(CO)_4$ or $K_2Fe_2(CO)_8$ to give mainly β-phenylethyl alcohol (72) (see Section XI). Although cobalt hydridocarbonyl reacts rapidly with epoxides at 0°C under CO to give β-hydroxyacylcobalt tetracarbonyls, this reaction sometimes gives alcohols by hydrogenolysis (cf. Section II,B), e.g., cyclohexene oxide is partially hydrogenated to cyclohexanol under hydroformylation conditions (14).

IV. Ketones

Brown's hydroboration method may be used to convert olefins into symmetrical, unsymmetrical, or cyclic ketones. Transition metal complexes have been used to convert a whole range of types of organic compounds into ketones by carbonylation—examples of starting materials include olefins, acetylenes, dienes, and alkyl and aryl metal derivatives. Metal carbonyls catalyze the isomerization of epoxides and unsaturated alcohols to aldehydes and ketones, and benzoic anhydride has been decarbonylated to fluorenone using a rhodium catalyst. There is one example where palladium is used as a template for joining allyl to acetylacetonate ligands under CO.

A. FROM OLEFINS VIA HYDROBORATION

As is usual in this general method (see Sections II,A; III,A; V,A; VI,A) the reaction of BH_3 with olefin gives BR_3. The subsequent reaction with CO can be stopped at the stage where two of the three alkyl groups have been transferred from boron to carbon by the addition of one mole of water (see Section III,A). Treatment of the intermediate so formed with alkaline hydrogen peroxide gives the ketone in high yield (treatment with alkali gives the secondary alcohol, see Section III,A) (7), e.g.:

$$CH_3CH_2CH=CH_2 \longrightarrow \underset{85\%}{(CH_3CH_2CH_2CH_2)_2CO} \qquad (39)$$

$$CH_3CH=CHCH_3 \longrightarrow \underset{81\%}{(CH_3CH_2\overset{\overset{\displaystyle CH_3}{\displaystyle |}}{CH})_2CO} \qquad (40)$$

Unsymmetrical ketones may be synthesized if the trialkylborane is synthesized in two stages, e.g. (7,8,73):

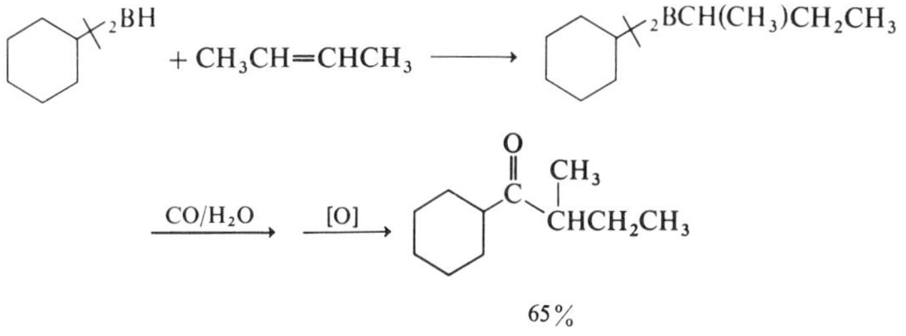

$$(41)$$

65%

Unsymmetrical ketones may also be synthesized using thexyldialkylboranes (thexyl = 2,3-dimethyl-2-butyl); the tertiary thexyl group does not migrate to the carbon as readily as the primary or secondary alkyl groups (74):

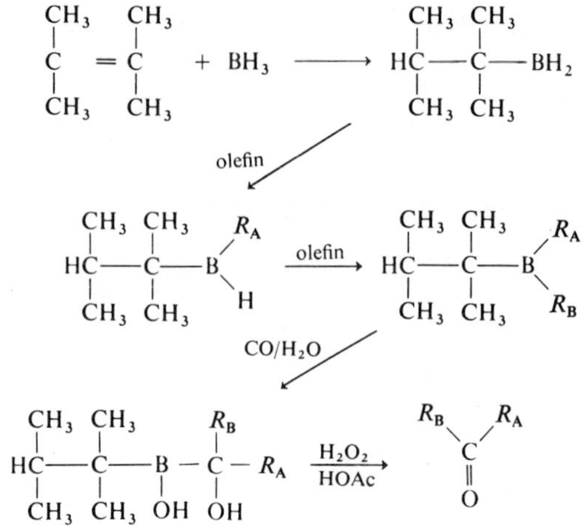

$$\underset{\substack{CH_3 \\ | \\ C \\ | \\ CH_3}}{\overset{CH_3}{|}} = \underset{\substack{| \\ CH_3}}{\overset{CH_3}{C}} + BH_3 \longrightarrow \underset{\substack{| \\ CH_3}}{\overset{CH_3}{HC}} - \underset{\substack{| \\ CH_3}}{\overset{CH_3}{C}} - BH_2$$

olefin

$$\underset{\substack{| \\ CH_3 \; CH_3}}{\overset{CH_3 \; CH_3}{HC - C - B}} \overset{R_A}{\underset{H}{}} \xrightarrow{\text{olefin}} \underset{\substack{| \\ CH_3 \; CH_3}}{\overset{CH_3 \; CH_3}{HC - C - B}} \overset{R_A}{\underset{R_B}{}} \qquad (42)$$

CO/H₂O

$$\underset{\substack{| \quad | \quad | \\ CH_3 \; CH_3 \; OH \; OH}}{\overset{CH_3 \; CH_3 \qquad R_B}{HC - C - B - C - R_A}} \xrightarrow[\text{HOAc}]{H_2O_2} \underset{O}{\overset{R_B \diagdown \diagup R_A}{\underset{\parallel}{C}}}$$

e.g., $\underset{\substack{| \\ CH_3}}{\overset{CH_3}{CH_3C}}$=CH₂ + CH₂=CHCH₂CO₂Et $\longrightarrow$ (CH₃)₂CHCH₂$\underset{\substack{\parallel \\ O}}{C}$(CH₂)₃CO₂Et

$$(43)$$

84%

Two examples of analogous reactions to give cyclic ketones are (75,76):

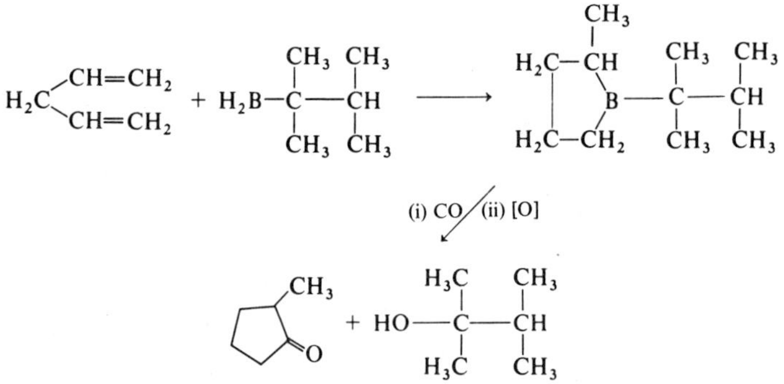

$$\underset{\diagdown CH=CH_2}{\overset{\diagup CH=CH_2}{H_2C}} + \underset{\substack{| \quad | \\ CH_3 \; CH_3}}{\overset{CH_3 \; CH_3}{H_2B - C - CH}} \longrightarrow \underset{\substack{| \quad | \\ H_2C - CH_2}}{\overset{\substack{CH_3 \\ | \\ H_2C - CH}}{\qquad}} \underset{\substack{| \quad | \\ CH_3 \; CH_3}}{\overset{CH_3 \; CH_3}{B - C - CH}}$$

$$(44)$$

(i) CO / (ii) [O]

$$\underset{O}{\overset{CH_3}{\bigpentagon}} + \underset{\substack{| \quad | \\ H_3C \quad CH_3}}{\overset{H_3C \quad CH_3}{HO - C - CH}}$$

$$\underset{H}{\overset{}{\bighexagon}} \xrightarrow[\substack{THF \\ 0°}]{RBH_2} \underset{H}{\overset{H}{\underset{\overset{}{\bigvee}}{}}}\text{-Br} \xrightarrow[\text{(ii) NaOAc/H}_2O_2]{\text{(i) CO(50°/70 atm)H}_2O} \underset{H}{\overset{H \overset{O}{\parallel}}{}}$$

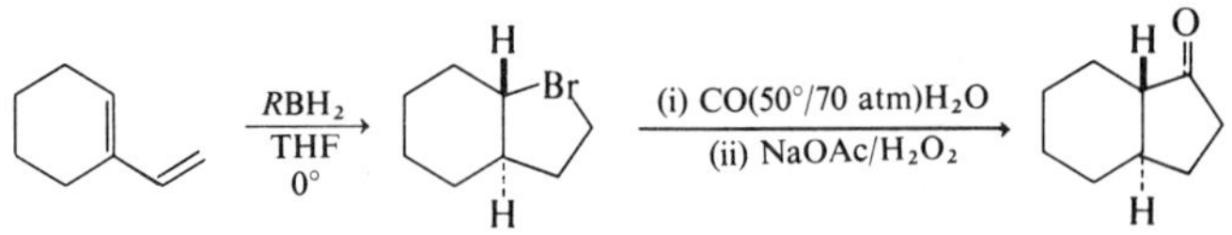

$$(45)$$

66%

This last reaction represents the first preparation of stereochemically pure ketones of this type.

B. FROM OLEFINS USING TRANSITION METAL CATALYSTS

Bicyclo[2,2,1]heptene reacts with $Fe_2(CO)_9$ at room temperature to give the following cyclic ketone (21);

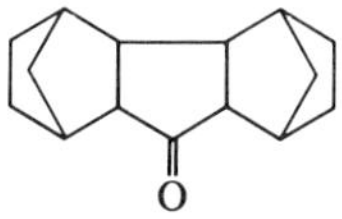

(2)

Treatment of cyclohexene with CO/H_2 at $300°/800$ atm in a stainless steel autoclave gives an aldehyde, alcohol, and ketone (77), the formation of catalytic quantities of iron carbonyls being a distinct possibility under these conditions:

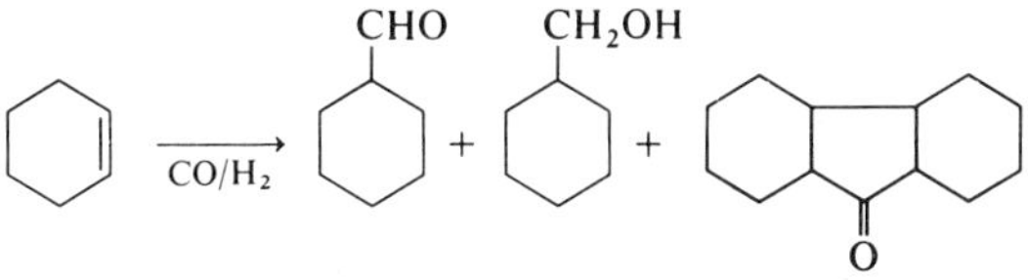

In the presence of ethylene, the oxo synthesis gives ketonic products (14). These could be formed by either or both the following two routes (78,79);

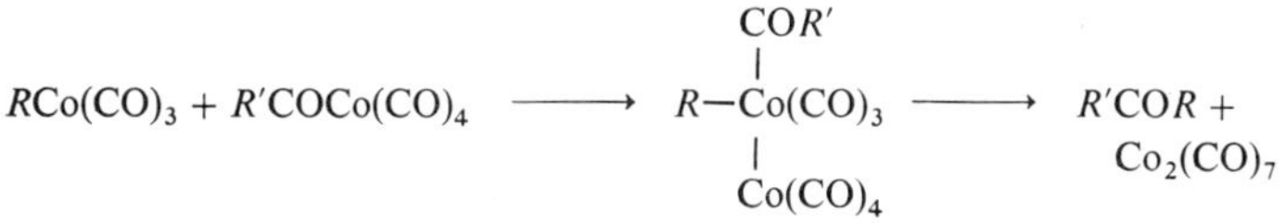

$$RCo(CO)_3 + R'COCo(CO)_4 \longrightarrow R-\underset{\underset{Co(CO)_4}{|}}{\overset{\overset{COR'}{|}}{Co}}(CO)_3 \longrightarrow R'COR + \quad (47)$$
$$Co_2(CO)_7$$

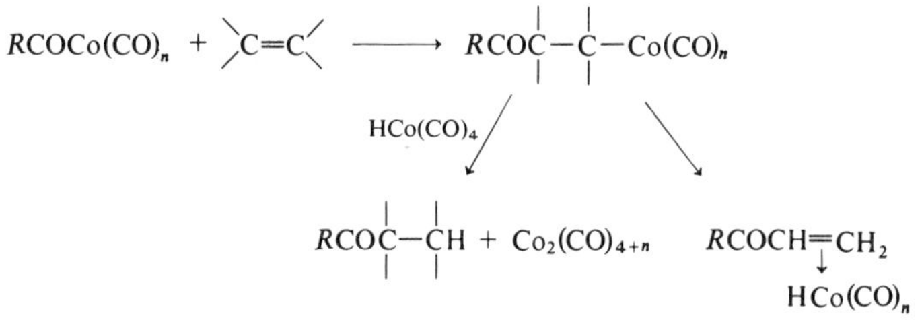

$$RCOCo(CO)_n + {\overset{}{\underset{}{>}}}C{=}C{\overset{'}{\underset{\diagdown}{}}} \longrightarrow RCO\overset{|}{\underset{|}{C}}-\overset{|}{\underset{|}{C}}-Co(CO)_n$$

$$HCo(CO)_4 \diagup \qquad \diagdown$$

$$RCO\overset{|}{\underset{|}{C}}-\overset{|}{\underset{|}{C}}H + Co_2(CO)_{4+n} \qquad RCOCH{=}CH_2 \quad (48)$$
$$\qquad\qquad\qquad\qquad\qquad\qquad\downarrow$$
$$\qquad\qquad\qquad\qquad\qquad HCo(CO)_n$$

Diethyl ketone is formed from ethylene, CO and a limited quantity of hydrogen (79–81). The organic ligand in 5-hexenoylcobalt tetracarbonyl cyclizes to give a mixture of ketonic products (81,82):

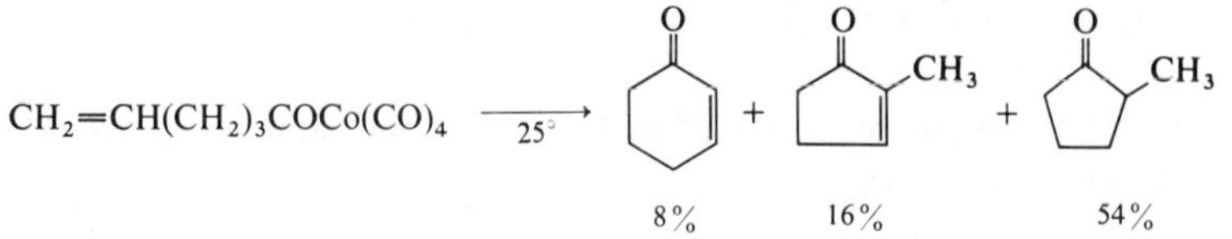

$$CH_2{=}CH(CH_2)_3COCo(CO)_4 \xrightarrow{25°}$$

8%   16%   54%   (49)

Carbonylation of dicyclooctenenickel in the presence of nickel carbonyl or nickel bromide gives a 1:1 mixture of dicycloctenenickel and dicyclo-octenylketone (83).

C. FROM ACETYLENES

The reactions of iron carbonyls with acetylenes in organic solvents yield a number of types of complex having ketonic ligands (6), e.g.

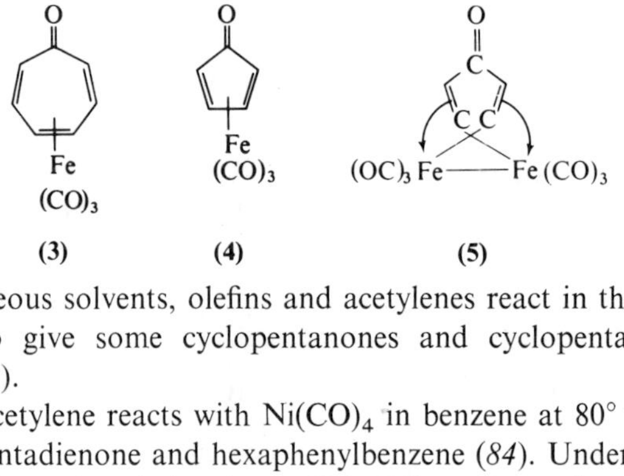

(3)   (4)   (5)

In nonaqueous solvents, olefins and acetylenes react in the presence of $Co_2(CO)_8$ to give some cyclopentanones and cyclopentadienones respectively (21).

Diphenylacetylene reacts with $Ni(CO)_4$ in benzene at 80° to give tetra-phenylcylopentadienone and hexaphenylbenzene (84). Under similar conditions acetylene gives indan-1-one (85). When diphenylacetylene is carboxylated with nickel carbonyl in dioxan/aqueous ethanol in the presence of acid the main product is 1,2,3,4-tetraphenylcyclopent-2-en-1-one and there is a small yield of ethyl-trans-α-phenylcinnamate (21,86).

D. FROM DIENES

Acyldienes may be prepared from dienes using iron carbonyl complexes (87). Bicyclo[2,2,1]heptadiene reacts with iron carbonyls to give cyclic ketones (21):

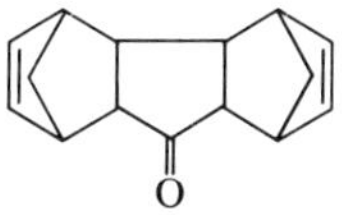

(6) 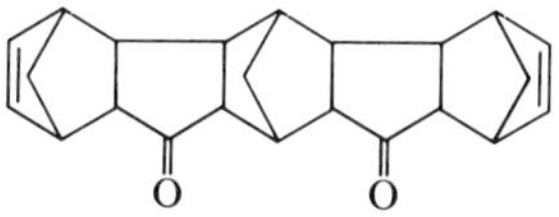   **(7)**

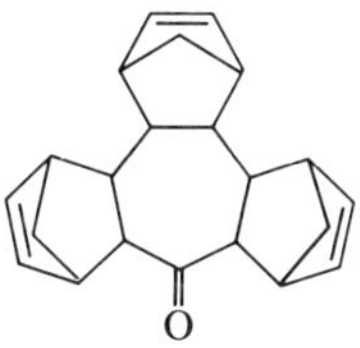

(8)

Synthesis of 1-acyl-1,3-dienes from alkyl or acyl halides, CO and conjugated dienes is a general reaction (*88*):

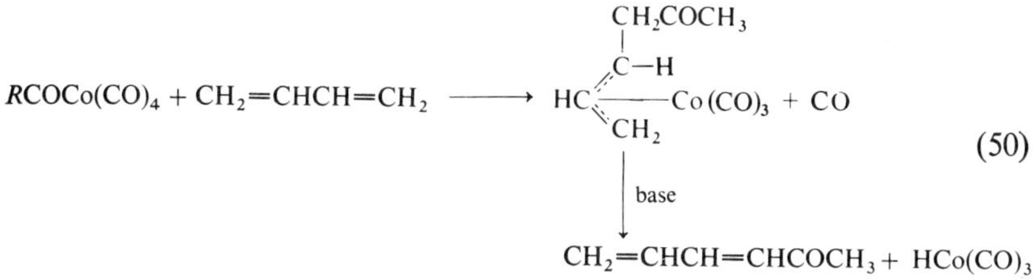

$$RCOCo(CO)_4 + CH_2=CHCH=CH_2 \longrightarrow \begin{array}{c} CH_2COCH_3 \\ | \\ C-H \\ HC\diagdown\diagup \quad Co(CO)_3 + CO \\ CH_2 \end{array}$$

(50)

base ↓

$$CH_2=CHCH=CHCOCH_3 + HCo(CO)_3$$

Bicyclo[2,2,1]heptadiene in water/ethanol/acetic acid with $Ni(CO)_4$ gives a mixture of the following acid, its ester, and the ketone shown:

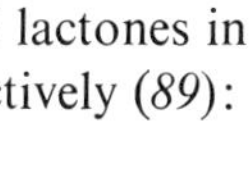

(9) 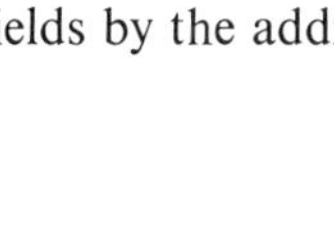   **(10)**

α,ω,dienes are carbonylated at 200°/1000 atm in the presence of palladium to give ketones and lactones in low yields by the addition of one and two moles of CO respectively (*89*):

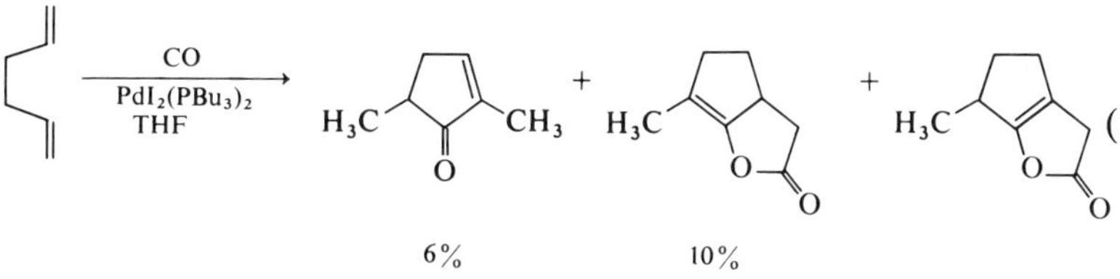

(51)

6%   10%

In methanol, under similar conditions the following transformation takes place (90):

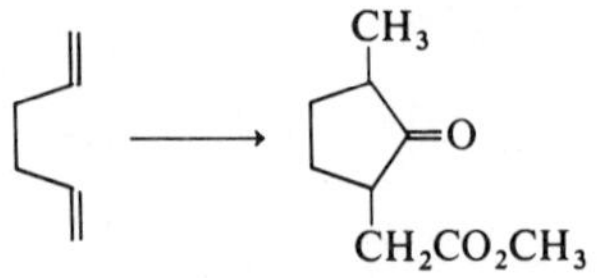

$$(52)$$

The presence of acids has a catalytic effect on these palladium systems. Palladium complexes also catalyze the transannular addition of CO to 1,5-cyclooctadiene (91):

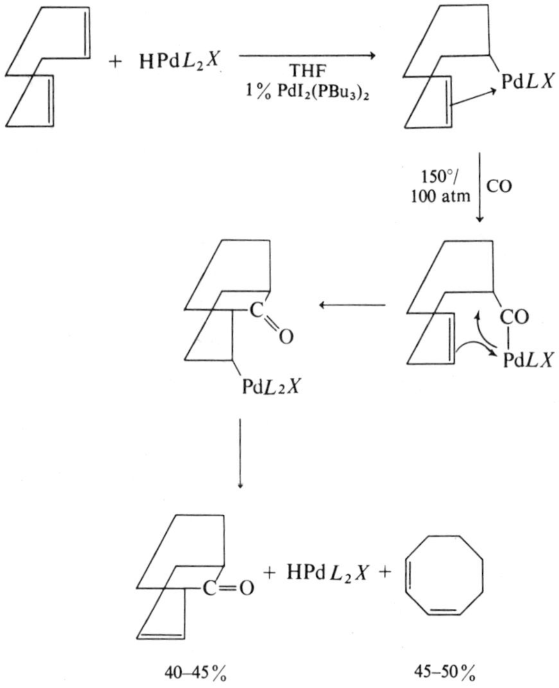

$$(53)$$

40–45%   45–50%

E. FROM EPOXIDES

Cobalt carbonyl systems may convert α-epoxides into ketones by one of the following isomerization mechanisms which involve formation of the enol form of the ketone or an anionic intermediate (82):

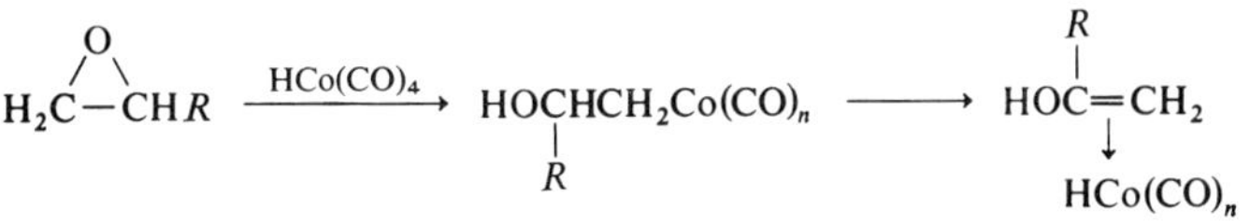

$$\underset{\text{H}_2\text{C}-\text{CH}R}{\overset{\text{O}}{\triangle}} \xrightarrow{\text{HCo(CO)}_4} \underset{R}{\overset{}{\text{HOCHCH}_2\text{Co(CO)}_n}} \longrightarrow \underset{\overset{\downarrow}{\text{HCo(CO)}_n}}{\overset{R}{\text{HOC}=\text{CH}_2}} \quad (54)$$

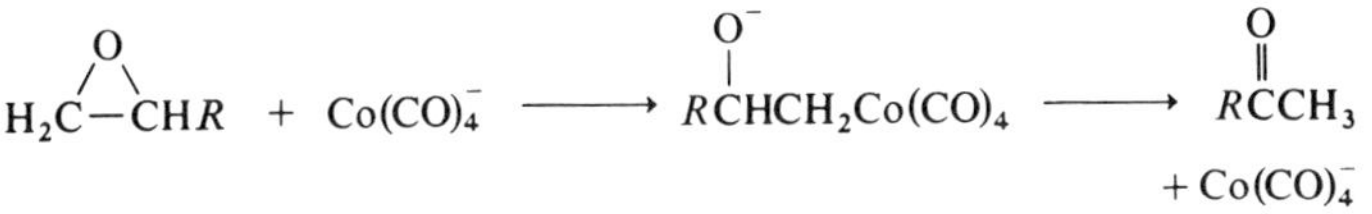

$$\underset{\text{H}_2\text{C}-\text{CH}R}{\overset{\text{O}}{\triangle}} + \text{Co(CO)}_4^- \longrightarrow \overset{\text{O}^-}{R\text{CHCH}_2\text{Co(CO)}_4} \longrightarrow \overset{\text{O}}{\underset{}{RC\text{CH}_3}} \quad (55)$$
$$+ \text{Co(CO)}_4^-$$

F. FROM ALCOHOLS

Heating acyclic secondary unsaturated alcohols either neat or in hydrocarbon solvents with 10–20 mole % of Fe(CO)_5 for 2–6 hours gives 60–80% isomerization to ketones of greater than 95% purity (42), e.g.:

$$\text{CH}_2=\text{CHCH}_2\text{CHOHC}_3\text{H}_7 \xrightarrow[110-125°]{\text{Fe(CO)}_5} \overset{\text{O}}{\underset{}{\text{C}_3\text{H}_7\text{CC}_3\text{H}_7}} \quad (56)$$

Unsaturated primary alcohols and cyclic olefinic alcohols give lower yields due to extensive dimerization, but irradiation assists in the following transformation:

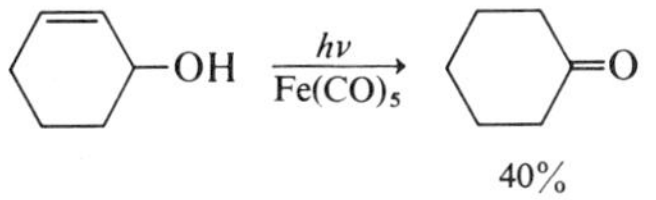

$$\langle\ \rangle\text{-OH} \xrightarrow[\text{Fe(CO)}_5]{hv} \langle\ \rangle\text{=O} \quad (57)$$
$$40\%$$

HCo(CO)_4 has been reported to catalyze isomerization (and carbonylation) of allyl and substituted alkyl allyl alcohols to aldehydes (and lactones) but only in low yields (3–21%) (42).

G. BY DECARBONYLATION OF ANHYDRIDE

$\text{RhCl(PPh}_3)_3$ catalyzes the formation of fluorenone from benzoic anhydride at 240–245° (93):

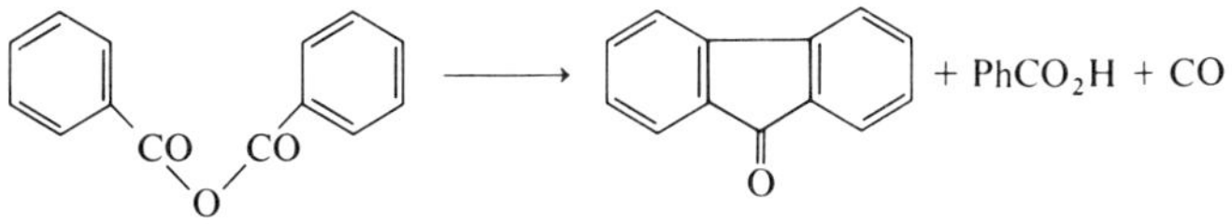

$$\underset{\underset{\text{O}}{\text{CO}\ \text{CO}}}{\bigcirc\ \bigcirc} \longrightarrow \overset{}{\underset{\text{O}}{\bigcirc\bigcirc}} + \text{PhCO}_2\text{H} + \text{CO} \quad (58)$$

H. By Removal of Metal from Ligand

Treatment of the following π-allylacetylacetonate complex with CO under ambient conditions furnishes a diketone as the sole product (94):

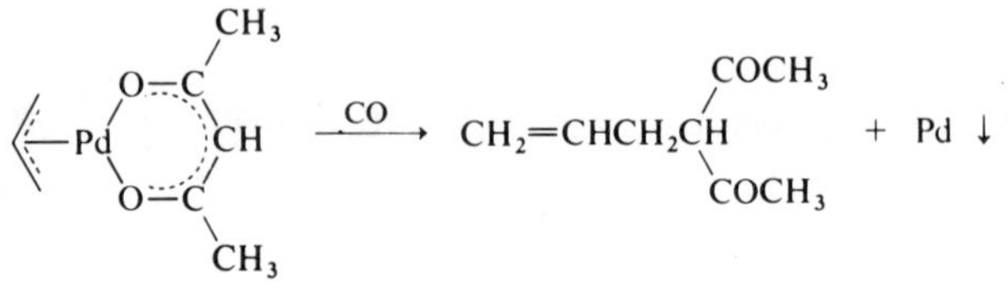

$$
\text{[}\pi\text{-allyl-Pd(acac)]} \xrightarrow{\ \text{CO}\ } \text{CH}_2\!=\!\text{CHCH}_2\underset{\text{COCH}_3}{\overset{\text{COCH}_3}{\text{CH}}} + \text{Pd} \downarrow \qquad (59)
$$

In this reaction the palladium acts as a template for bringing the allyl and acetylacetonate groups together.

I. From Alkyl and Aryl Metal Derivatives

Among the products from carbonylation of dicyclopentadienyldiphenyltitanium is diphenylketone (95).

Both dialkyl and diaryl ketones may be prepared by reacting the mercuric bromide derivative or diorganomercurial with $Co_2(CO)_8$ at room temperature (96,97):

$$
R\text{HgBr} + Co_2(CO)_8 \xrightarrow[\text{THF}]{} \text{Hg[Co(CO)}_4]_2 + R_2\text{CO} + \text{CoBr}_2 \qquad (60)
$$
$$
(83\% \text{ for } R = \text{Ph})
$$

In the case of aryl ketones the reaction can be made catalytic with respect to $Co_2(CO)_8$ by carrying it out in the presence of CO and irradiation, e.g.:

$$
Ph_2\text{Hg} + \text{CO} \xrightarrow[\substack{Co_2(CO)_8 \\ \text{THF}}]{hv} Ph_2\text{CO} + \text{Hg} \qquad (61)
$$
$$
91\%
$$

Treatment of an aryllithium compound with $Ni(CO)_4$ at $-70°$ gives an acyl nickel species which reacts with an acetylene to yield a diketone (98):

$$
R\text{Li} + Ni(CO)_4 \xrightarrow{-70°} 2\text{Li}[R\underset{\text{O}}{\overset{\|}{-}}\text{C}-Ni(CO)_3] \xrightarrow[70°,\,\text{H}^+]{R'\text{C}\equiv\text{CH}} R\underset{\text{O}}{\overset{\|}{-}}\text{C}-\underset{\underset{47-74\%}{}}{\overset{R'}{\underset{|}{\text{CH}}}}\text{CH}_2\underset{\text{O}}{\overset{\|}{-}}\text{C}-R \qquad (62)
$$

n-Butyllithium gives an analogous 1,4-diketone from mesityl oxide using a similar route (*99*):

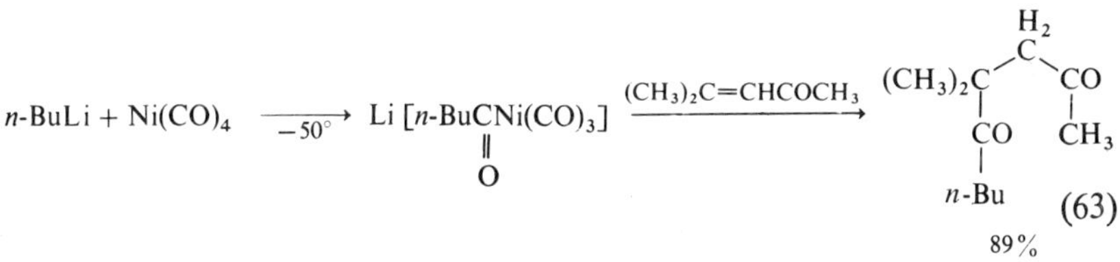

$$n\text{-BuLi} + \text{Ni(CO)}_4 \xrightarrow[-50°]{} \underset{\underset{O}{\parallel}}{\text{Li}[n\text{-BuCNi(CO)}_3]} \xrightarrow{(\text{CH}_3)_2\text{C}=\text{CHCOCH}_3}$$

$$\underset{\underset{89\%}{}}{\underset{n\text{-Bu}}{(\text{CH}_3)_2\text{C}\overset{\overset{\text{H}_2}{\overset{\text{C}}{\diagdown}}}{\underset{\text{CO}}{\diagup}}\overset{\text{CO}}{\underset{\text{CH}_3}{}}}} \quad (63)$$

With nickel carbonyl, benzyl halides give a mixture of dibenzyl and dibenzylketone (*100*):

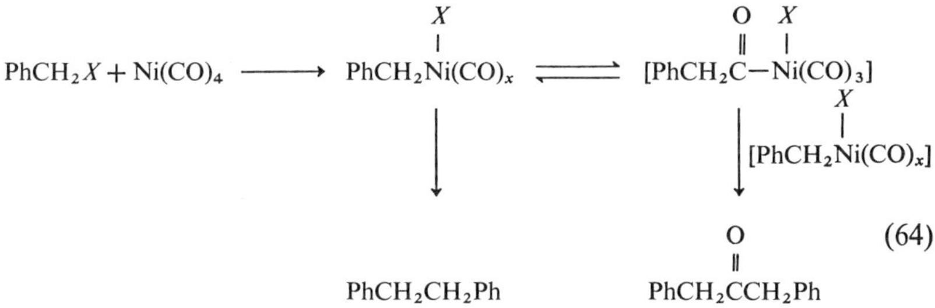

$$\text{PhCH}_2X + \text{Ni(CO)}_4 \longrightarrow \text{PhCH}_2\overset{X}{\underset{}{|}}\text{Ni(CO)}_x \rightleftharpoons [\text{PhCH}_2\overset{O}{\overset{\parallel}{\text{C}}}\overset{X}{\underset{}{|}}-\text{Ni(CO)}_3]$$

$$\downarrow \qquad\qquad\qquad\qquad \downarrow \overset{X}{\underset{}{|}}[\text{PhCH}_2\text{Ni(CO)}_x]$$

$$\text{PhCH}_2\text{CH}_2\text{Ph} \qquad\qquad \text{PhCH}_2\overset{O}{\overset{\parallel}{\text{C}}}\text{CH}_2\text{Ph} \qquad (64)$$

Allyl halides, CO and acetylenes give ketonic products in the presence of Ni(CO)$_4$ (*101*):

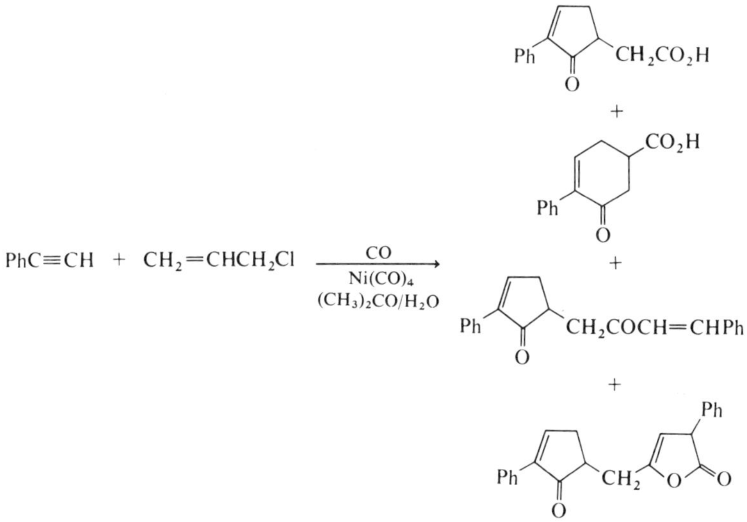

$$\text{PhC}\equiv\text{CH} + \text{CH}_2=\text{CHCH}_2\text{Cl} \xrightarrow[\substack{\text{Ni(CO)}_4 \\ (\text{CH}_3)_2\text{CO/H}_2\text{O}}]{\text{CO}} \quad (65)$$

V. Carboxylic Acids and Anhydrides

The preparation of carboxylic acids by carbonylation techniques may be conveniently divided into reactions yielding either saturated aliphatic, unsaturated, or aryl carboxylic acids. Most of these preparations were initially carried out under extreme conditions of temperature and pressure but recently interest has focussed on the use of milder reaction conditions. When the carbonylation reactions described here are carried out in alcoholic solvents rather than water then esters are usually the final products. Carboxylic acids may be obtained from the hydrolysis of acid halides and anhydrides.

A. Saturated Carboxylic Acids

The most common systhesis of saturated carboxylic acids is by the hydroxycarboxylation of olefins ($21,102$):

$$RCH{=}CH_2 + CO + H_2O \longrightarrow RCH_2CH_2CO_2H + RCHCH_3 \quad (66)$$
$$\underset{\textstyle CO_2H}{|}$$

The conditions generally used involve the reaction of the olefin with carbon monoxide under aqueous acidic conditions in the presence of a nickel salt, e.g., NiI_2, which is presumbably converted into $Ni(CO)_4$ *in situ*. Catalytic reaction conditions are 200 atm CO and 250° but the reaction may also be carried out stoichiometrically under milder conditions (50 atm CO and 160°). Although nickel is the most commonly used catalyst, metal salts of ruthenium, osmium, cobalt, rhodium, palladium, and platinum are also effective in this reaction. Cobalt salts display comparable activity to those of nickel but other metals are less efficient.

Acid catalysts also promote the carboxylation of olefins. Early patents indicated the use of severe conditions ($100-350°/500-1000$ atm) and this did not encourage industrial exploitation (20). However, Koch et al. (20) showed that high olefin conversion can be achieved under mild conditions if the reaction is carried out in two stages. The olefin is first treated with CO and acid (e.g. H_2SO_4, H_3PO_4, HCl, or BF_3) in the absence of water, and the resulting product subsequently hydrolysed; $-20-+80°/$ $1-100$ atm can then be used to give high yields of carboxylic acids:

$$H_2C{=}CHR + H^+ \longrightarrow H_3\overset{+}{C}{-}CHR \xrightarrow{\text{CO}} H_3C{-}\overset{+}{C}HR{-}CO \quad (66a)$$

$$H_3C{-}CHR{-}\overset{+}{C}O + HOH \longrightarrow H_3C{-}CHR{-}CO_2H + H^+ \quad (66b)$$

This reaction is accompanied by double bond isomerization and isomerization of the carbon skeleton, and the addition of the carboxyl group strictly follows the Markownikoff rule to give only branched chain products. Ethylene is consequently the only olefin with which addition to a terminal position occurs. The number of carboxylic acids formed in these reactions are often high since in many cases, especially at low CO pressure, carbonylation occurs only after oligomerization of the starting material has taken place. Lower carbon number acids may also be obtained as a result of catalytic cracking of the starting material. However, in spite of these potential disadvantages, the Shell versatic acid process uses this reaction for converting isobutylene and diisobutylene into pivalic acid and olefin cuts in the range of C_6 to C_{10} into C_7 to C_{11} carboxylic acid mixtures, using CO pressures of *ca.* 70 atm at 70° and H_3PO_4/BF_3 as catalyst. Over 90% tertiary acids are obtained. There have been a number of recent mechanistic studies of the Koch reaction (*103,103a,103b,103c*).

Saturated carboxylic acids may be prepared from olefins under very mild conditions using the hydroboration-carbonylation-oxidation technique (see Sections II,A; III,A; IV,A; VI,A) (*73*). Thus *n*-octene may be converted into cyclohexyl-*n*-octyl ketone (by hydroboration followed by carbonylation) which may be oxidized to nonanoic acid in 79% yield.

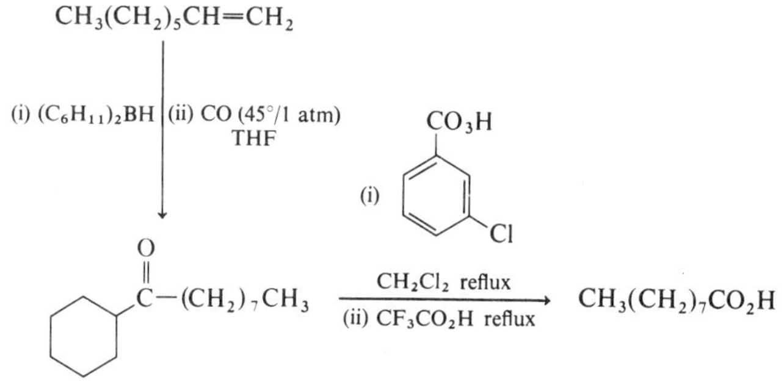

CH3(CH2)5CH=CH2
(i) (C6H11)2BH
(ii) CO (45°/1 atm)
THF
CO3H
(i)
Cl
O
C—(CH2)7CH3
CH2Cl2 reflux
(ii) CF3CO2H reflux
CH3(CH2)7CO2H

(67)

Saturated alcohols may be carbonylated to carboxylic acids but more forcing conditions are needed than those required for olefins (*21,102*). Primary alcohols react at 300°, and secondary and tertiary alcohols at 275°. Considerable amounts of branched chain acids are obtained from primary alcohols and it seems that the reaction may proceed via formation of the olefin. Again cobalt and nickel salts appear to be the most active catalysts for these reactions.

As mentioned earlier, it has been assumed that the actual catalytic species in these reactions are the metal carbonyls, formed *in situ* by reaction of the metal halide salt with carbon monoxide. Mizoroki and Nakayama have studied visible spectral changes under high pressure-high temperature conditions during the carbonylation of methanol to acetic acid with cobalt acetate and an iodide as catalyst (104). These workers conclude that the tetrahedral species $[Co(CH_3CO_2)_{4-n}I_n]^{2-}$ are predominant in the reaction mixture and that most of the cobalt is not present as $Co_2(CO)_8$. However, the limitations of the experimental method do not preclude the formation of some $Co_2(CO)_8$ or $HCo(CO)_4$ which may be present in sufficient quantities to catalyze the reaction. The acceleration of the reaction rate by small quantities of potassium acetate (105) and the effect of the latter on the carbonylation of methyl iodide (106) do, however, strongly suggest that acetate ions play a significant role in the carbonylation of methanol to acetic acid.

The synthesis of carboxylic acids under mild conditions has been of recent interest and novel homogeneous systems have been described for the carbonylation of methanol to acetic acid ($3,107$). Here a solution of a rhodium or iridium salt in the presence of a halogen promoter, e.g., CH_3I, in a common organic solvent catalyzes the carbonylation of methanol at $175°$ and 27 atm total pressure ($P_{CO} < 13$ atm). The selectivity to acetic acid is 99% and significant rates of carbonylation can be observed even at atmospheric pressure. The available evidence suggests that the active catalyst is a d^8 Rh(I) complex and that the reaction is initiated by oxidative addition of CH_3I (107).

Carboxylic acids may be obtained from aliphatic hydrocarbons under very mild conditions. Thus cyclohexane carboxylic acid may be obtained in 90% yield from methylcyclopentane by treatment with carbon monoxide under ambient conditions in the presence of hydrofluoric acid and antimony pentafluoride, followed by hydrolysis (108).

$$\qquad\qquad\qquad (68)$$

This method may also be used for the preparation of ketones. The reaction mechanism is assumed to proceed via carbonium and acylium hexafluoroantimonates. Acetic acid may also be obtained from methane (109).

B. Unsaturated Carboxylic Acids

α,β-Unsaturated carboxylic acids may be prepared by the catalytic carboxylation of acetylenes (21). Thus acrylic acid may be prepared from acetylene by reaction with carbon monoxide in aqueous acid in the presence of $Ni(CO)_4$.

$$HC\equiv CH + H_2O \xrightarrow[\text{Ni(CO)}_4]{\text{CO(150°/300 atm)}} \underset{95\%}{CH_2=CHCO_2H} \qquad (69)$$

The reaction may also be carried out stoichiometrically at 40°C to give the unsaturated acid in yields of *ca.* 50%. Other catalysts, e.g., Fe(CO)$_5$, have been investigated for this reaction, but the yields of acid are usually lower. In the case of substituted acetylenes the direction of addition of the elements of formic acid obeys the Markownikoff rule (*110*), and the general reaction appears considerably more selective than the carboxylation of olefins to produce saturated carboxylic acids. Reactions of this type are frequently carried out in aqueous ethanol when the product is the ethyl ester (see Section VI,C).

The oxidative carbonylation of olefins in the presence of Pd(II)/Cu(II) catalysts has recently been demonstrated to yield α,β-unsaturated carboxylic acids by pyrolysis of the initially formed β-substituted acids (*111,112*).

$$RCH=CH_2 + \tfrac{1}{2}O_2 \xrightarrow[\text{LiCl/Pd(OAc)}_2/\text{CuCl}_2]{\text{CO(125-150°/73 atm)}} \underset{\underset{\text{OAc}}{|}}{R-CH-CH_2-CO_2H} \Big\downarrow \text{pyrolyze} \qquad (70)$$

$$R-CH=CH-CO_2H + AcOH$$

If the oxidative carbonylation of ethylene is carried out in the presence of acetic anhydride, the formation of β-acetoxypropionic acid is suppressed, thus making the pyrolysis step unnecessary. Similarly, butadiene yields vinyl acrylic acid. α,β-Unsaturated acids may also be obtained by the Co$_2$(CO)$_8$ catalyzed carbonylation of epoxides in nonhydroxylic solvents (*88*).

$$\underset{\underset{O}{\diagdown\diagup}}{R-HC-CH} \xrightarrow[\text{Co}_2\text{(CO)}_8]{\text{CO(160°/400 atm)}} RCH=CHCO_2H \qquad (71)$$

In the case of $R=H$, ethylene oxide is converted into acrylic acid in 81% yield.

Unsaturated acids containing two double bonds in the α,β and δ,ε-positions may be prepared under ambient conditions by the carboxylation of allyl halides or allyl alcohol in the presence of acetylene and Ni(CO)$_4$ (*101*).

$$RCH=CHCH_2X + HC\equiv CH + H_2O$$

$$\xrightarrow[\text{Ni(CO)}_4]{\text{CO(20°/1 atm)}} RCH=CHCH_2CH=CHCO_2H + HX \qquad (72)$$

The reaction is commonly carried out in alcoholic solvents when the products are unsaturated esters (see Section VI,G). In ketonic solvents keto-lactones may be formed in addition to the unsaturated acids (113) (see Section VII,D). Higher yields of unsaturated acid are obtained when allylic alcohols are used instead of the halides.

$$RCH{=}CHCH_2OH + HC{\equiv}CH \xrightarrow[\substack{Ni(CO)_4 \\ HCl}]{CO} RCH{=}CHCH_2CH{=}CHCO_2H \atop 80\text{-}85\%} \quad (73)$$

β,γ-Unsaturated acids may be prepared under mild conditions by the carboxylation of allyl halides alone in the presence of Ni(CO)$_4$ (101).

$$RCH{=}CHCH_2X + H_2O \xrightarrow[Ni(CO)_4]{CO(20°/2\text{-}3 \text{ atm})} RCH{=}CHCH_2CO_2H + HX \quad (74)$$

Allenic acids may be obtained by the stoichiometric carbonylation of halogenoacetylenes under aqueous conditions in the presence of Ni(CO)$_4$ but the yields are usually low (114).

$$\underset{\underset{CH_3}{|}}{\overset{\overset{Cl}{|}}{CH_3{-}C{-}C{\equiv}CH}} \longrightarrow \underset{\underset{CH_3}{|}}{CH_3{-}C{=}C{=}CH{-}CO_2H} \atop 45\% \quad (75)$$

Other by-products including substituted maleic anhydrides and ketones are also formed.

C. ARYL CARBOXYLIC ACIDS

Aryl halides react with CO in the presence of Ni(CO)$_4$ and aqueous acid to give the corresponding benzoic acid in ca. 30% yield (21).

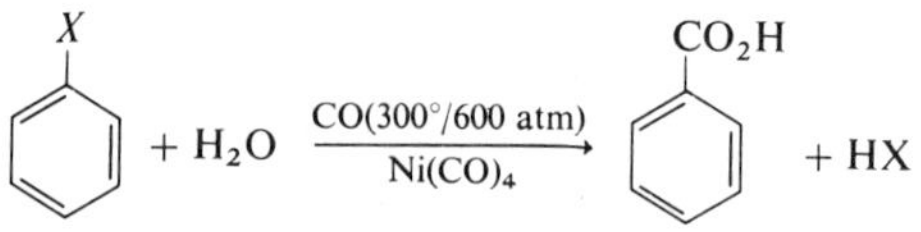

$$\text{(aryl-}X\text{)} + H_2O \xrightarrow[Ni(CO)_4]{CO(300°/600 \text{ atm})} \text{(aryl-}CO_2H\text{)} + HX \quad (76)$$

The use of iron and cobalt carbonyls as catalysts in this reaction has also been reported but the yields of arylcarboxylic acids are generally found to be very dependent on catalyst and solvent. However, when stoichiometric amounts of salts of carboxylic acids, e.g., potassium acetate or sodium benzoate, are added to the starting materials, the yields may be markedly increased under less extreme reaction conditions ($115,116,116a$).

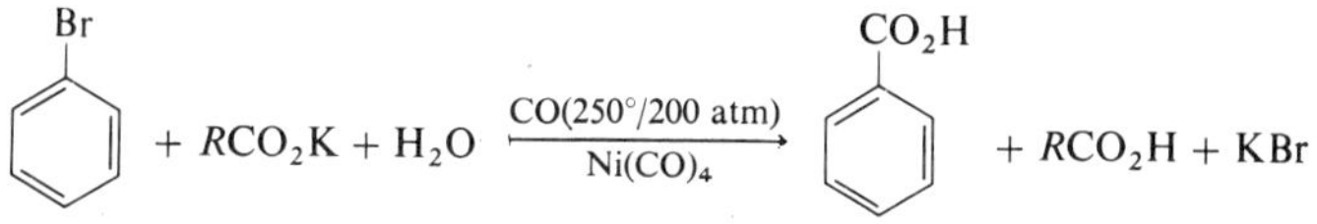

$$\text{(Br-phenyl)} + RCO_2K + H_2O \xrightarrow[\text{Ni(CO)}_4]{\text{CO(250°/200 atm)}} \text{(CO}_2\text{H-phenyl)} + RCO_2H + KBr \qquad (77)$$

The formation of HX in the initial reaction is thought to inhibit the carboxylation reaction and the function of the carboxylic acid salt is to decrease the concentration of HX during the course of the reaction.

$$RCO_2K + HX \longrightarrow RCO_2H + KX \qquad (78)$$

In the presence of carboxylic acid salts the method becomes a general one for the formation of aryl carboxylic acids in high yields (95–100%).

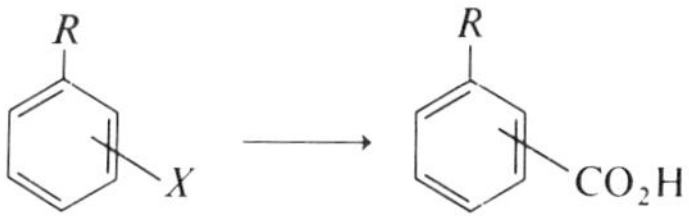

$$\text{(R-phenyl)}X \longrightarrow \text{(R-phenyl)}CO_2H \qquad (79)$$

Many of the reactions of this type do not proceed in the absence of KCO_2CH_3.

Aryl carboxylic acids may also be obtained by the stoichiometric reaction of aryl diazonium salts with metal carbonyls at ambient pressure and temperature. $Fe(CO)_5$, $Co_2(CO)_8$, and $Ni(CO)_4$ all react in ethanol, acetone, or acetic acid solution with the chloride or tetrafluoroborate salts to yield aryl carboxylic acids (117,118). Yields are variable but p-anisic acid and terephthalic acid may be obtained in ca.75% yield from the reaction of the appropriate BF_4^- salt in the presence of $Ni(CO)_4$. This reaction provides a useful single-step alternative to the Sandmeyer reaction.

p-Hydroxybenzoic acid may be conveniently prepared in quantitative yield by acidification of the reaction product from potassium phenoxide, potassium carbonate and CO (119,120):

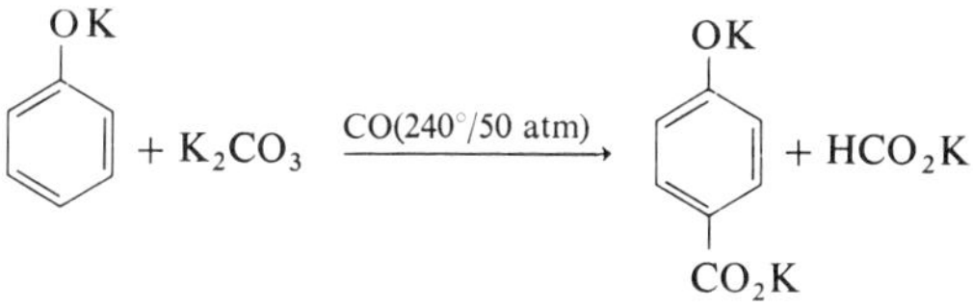

$$\text{(OK-phenyl)} + K_2CO_3 \xrightarrow{\text{CO(240°/50 atm)}} \text{(OK-phenyl-CO}_2\text{K)} + HCO_2K \qquad (80)$$

D. ACID ANHYDRIDES

The formation of anhydrides by carbonylation techniques has been described for various transition metal catalysts, of which the following examples are typical.

Acetylenes react stoichiometrically with $Fe(CO)_5$ in aqueous alkali to yield diol complexes which may be oxidized to maleic anhydrides (21).

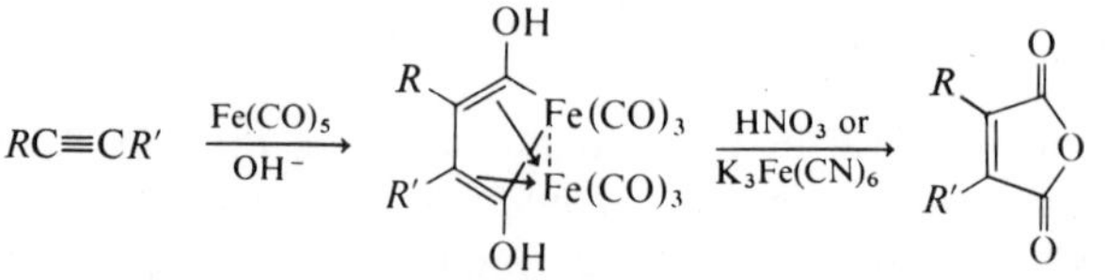

$$RC{\equiv}CR' \xrightarrow[OH^-]{Fe(CO)_5} \quad \xrightarrow[K_3Fe(CN)_6]{HNO_3 \text{ or}} \quad \tag{81}$$

β-Propiolactone may be catalytically carbonylated with $Co_2(CO)_8$ to succinic anhydride and acrylic acid (121).

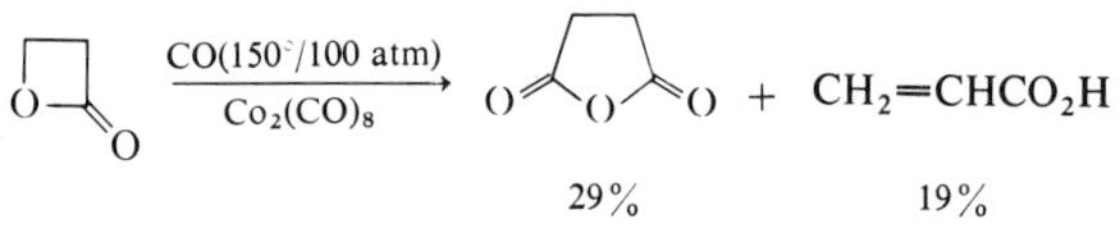

$$\xrightarrow[Co_2(CO)_8]{CO(150^\circ/100 \text{ atm})} \quad + \quad CH_2{=}CHCO_2H \tag{82}$$

$$\qquad\qquad\qquad\qquad\qquad\qquad 29\% \qquad\qquad 19\%$$

Benzoic anhydrides may be disproportionated to phthalic anhydride and benzene at elevated temperatures and pressures of carbon monoxide.

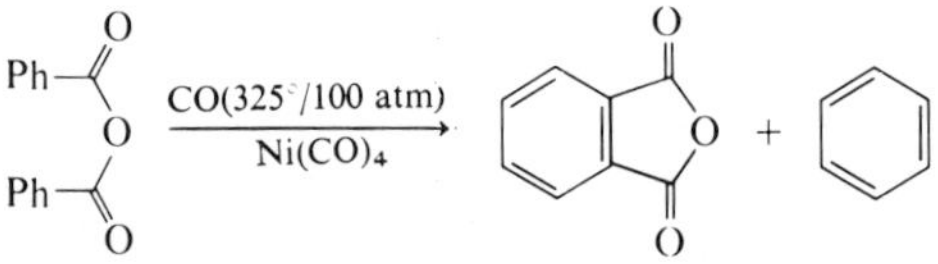

$$\xrightarrow[Ni(CO)_4]{CO(325^\circ/100 \text{ atm})} \quad + \quad \tag{83}$$

$Ni(CO)_4$ may also be used as a catalyst under mild conditions. Thus, on heating a mixture of oct-l-yne, $Ni(CO)_4$, and acetic acid in methanol at 60°, the mixed anhydride of acetic and α-hexylacrylic acid is formed (101,122).

Palladium salts catalyze the carbonylation of allyl ethers and acetylenic alcohols to produce anydrides (123). In the case of allyl ethers, esters are formed by the insertion of one mole of CO, and anhydrides from the further insertion of CO into the ester (124). Thus allyl acetate yields the mixed anhydride of 3-butenoic acid and acetic acid:

$$CH_2{=}CHCH_2OR \xrightarrow[PdCl_2]{CO(80^\circ/150 \text{ atm})} CH_2{=}CHCH_2CO_2R$$

$$\xrightarrow{\hspace{2cm}} CH_2{=}CHCH_2CO_2COR \tag{84}$$

Diallyl ether affords the anhydride of 3-butenoic acid. This is a useful method of synthesis of β,γ-unsaturated acid anhydrides.

$$\begin{array}{c} CH_2{=}CHCH_2 \\ \qquad\qquad\qquad O \\ CH_2{=}CHH_2C \end{array} \longrightarrow \begin{array}{c} CH_2{=}CHCH_2C \\ \qquad\qquad\qquad O \\ CH_2{=}CHH_2C \end{array} \longrightarrow \begin{array}{c} CH_2{=}CHCH_2C \\ \qquad\qquad\qquad O \\ CH_2{=}CHCH_2C \end{array} \tag{85}$$

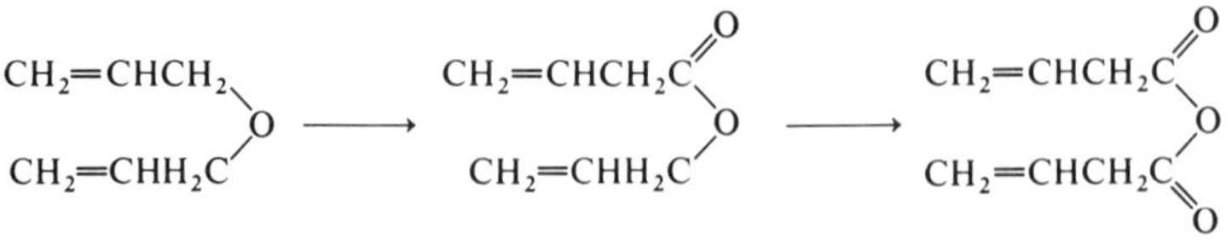

Acetylenic alcohols may also be carbonylated in benzene to yield anhydrides (125); in alcoholic solvents esters are formed.

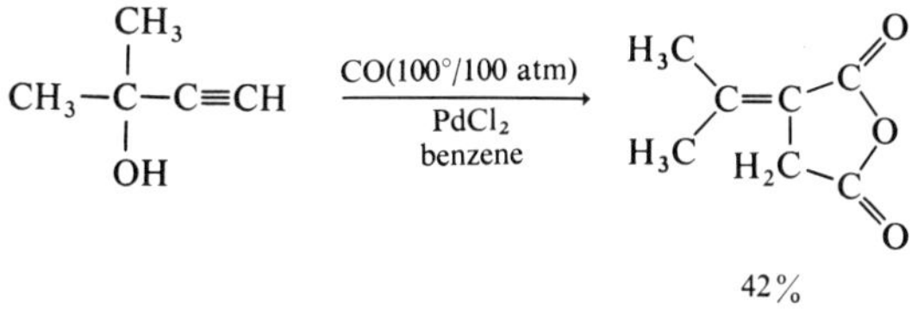

$$
\underset{\underset{\text{OH}}{|}}{\overset{\overset{\text{CH}_3}{|}}{\text{CH}_3-\text{C}-\text{C}\equiv\text{CH}}} \quad \xrightarrow[\substack{\text{PdCl}_2 \\ \text{benzene}}]{\text{CO}(100°/100\ \text{atm})} \quad \text{(structure, 42\%)} \tag{86}
$$

VI. Esters

The hydroboration technique may be used to prepare homologated esters from olefins. Esters may be obtained from olefins, acetylenes, dienes, aldehydes, epoxides, and halogenated hydrocarbons under carbonylation and hydroformylation conditions in the presence of alcohols.

A. FROM OLEFINS VIA HYDROBORATION

Treatment of olefins with BH_3 gives trialkylboranes (see Sections II,A; III,A; IV,A; V,A) which react with ethyl mono and dihaloacetates under the influence of t-BuOK to give esters and α-haloesters (*126,127*).

$$
R_3B + BrCH_2CO_2Et \quad \xrightarrow[t\text{BuOH}]{t\text{BuOK}} \quad \underset{\text{93\% for } R = Bu}{RCH_2CO_2Et} \tag{87}
$$

$$
R_3B + Br_2CHCO_2Et \quad \xrightarrow[t\text{BuOH}]{t\text{BuOK}} \quad \underset{\underset{Br}{|}}{RCHCO_2Et} \tag{88}
$$

This reaction is not a carbonylation in the usual sense but does enable the preparation of monoisomeric esters in high yield—hydroformylation methods generally give mixtures of *n*- and *iso*-esters. Only one-third of the olefin is used in the above reaction but the use of 9-BBN (see Sections II,A; III,A) utilizes all the olefin (*127*);

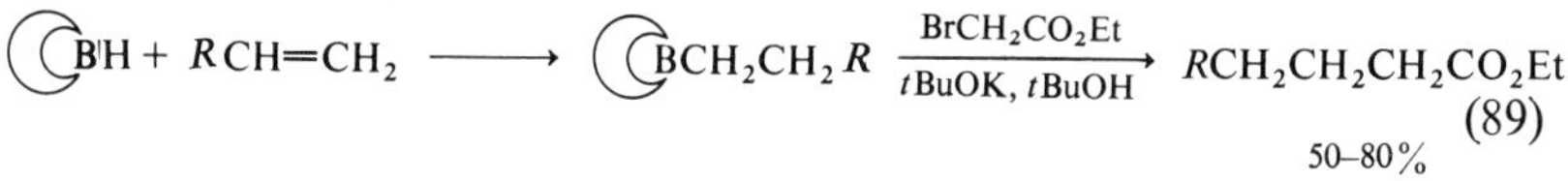

$$
\text{(B)}BH + R\text{CH}=\text{CH}_2 \longrightarrow \text{(B)}BCH_2CH_2R \xrightarrow[t\text{BuOK},\ t\text{BuOH}]{BrCH_2CO_2Et} \underset{50-80\%}{RCH_2CH_2CH_2CO_2Et} \tag{89}
$$

B. FROM OLEFINS USING TRANSITION METAL CATALYSTS

The acylcobalt tetracarbonyls formed from olefins (*22,82*) under hydroformylation conditions react with alcohols to form *n*- and *iso*-esters (see Section II,B):

$$RCOCo(CO)_4 + R'OH \longrightarrow RCO_2R' + HCo(CO)_4 \qquad (90)$$

$$\text{e.g., } CH_3CH=CH_2 \xrightarrow[CH_3OH]{CO} CH_3(CH_2)_2CO_2CH_3 + \underset{H_3C}{\overset{H_3C}{\diagdown}}CHCO_2CH_3 \qquad (91)$$

Functionally substituted olefins behave similarly, but the n-isomer pre-dominates in the first reaction ($128,129$):

$$CH_2=CHCN \xrightarrow[\substack{py/CH_3OH \\ Co_2(CO)_8}]{CO/H_2(100°/190\ atm)} CH_3O_2C(CH_2)_2CN + CH_3\overset{\overset{\displaystyle CO_2CH_3}{|}}{CH}CN \qquad (92)$$

$$CH_3O_2CCH=CH_2 \xrightarrow[120°-160°]{} \underset{CH_2CO_2CH_3}{\overset{CH_2CO_2CH_3}{\underset{|}{\overset{|}{}}}} \qquad (93)$$

In this type of reaction the effectiveness of catalysts falls in the order Co > Rh $\gg$ Ir (33) (see Section II,B).

Carbonylation of olefins in the presence of alcohol, HX, and palladium (130) or nickel (131) catalysts gives saturated esters, e.g., (132):

$$CH_2=CH_2 + CO + ROH$$

$$\xrightarrow[100°/40\ atm]{} CH_3CH_2CO_2R + CH_3CH_2COCH_2CH_2CO_2R \qquad (94)$$

The product incorporating one mole of ethylene is the major one. With higher olefins mixtures of n- and iso-esters are obtained and the iso-ester is usually predominant; Pd(O) (132) or $PdCl_2(PPh_3)_2$ (133) are suitable catalysts. The phosphine complex, which may be reduced to Pd(O) under the reaction conditions, has very high catalytic activity and this is particularly so for substituted olefins. Tsuji has proposed the following mechanism for this reaction (123):

$$HX + Pd + nL \rightleftharpoons \underset{L_n}{\overset{CH_2=CH_2}{\underset{|}{\overset{|}{HPdX}}}} \Bigg\} \rightleftharpoons \underset{CH_2=CH_2}{\overset{L_n}{\underset{|}{\overset{|}{HPdX}}}} \rightleftharpoons$$

$$\underset{CH_3CH_2PdX}{\overset{L_n}{\underset{|}{}}} \underset{-CO}{\overset{CO}{\rightleftharpoons}} \underset{CH_3CH_2COPdX}{\overset{L_n}{\underset{|}{}}} \rightleftharpoons Pd + nL +$$

$$CH_3CH_2COX \xrightarrow{ROH} CH_3CH_2CO_2R + HX$$

$$(95)$$

This mechanism accounts for the regeneration of the catalytic species Pd and HX.

A reaction in which the n-ester is the predominant product involves the use of H_2PtCl_6—$SnCl_2$ at 90°/200 atm. to carboxylate dodecene-1 in methanol to 85% linear ester ($133a$). Since a tin : platinum ratio of 5 : 1 is used, the formation of the straight chain ester could be the result of steric crowding around the catalyst.

C. From Acetylenes

Spectroscopic evidence indicates that $HCo(CO)_4$ reacts with acetylene to give $(C_2H_2)Co_2(CO)_6$ (*133b*); treatment of this complex with ethanol gives stoichiometric quantities of ethyl acrylate. $CH_3CCo_3(CO)_9$, obtained by treating the product of the reaction between acetylene and $Co_2(CO)_8$ with strong acid, reacts stoichiometrically with methanol to give an 80% yield of $CH_3CH(CO_2CH_3)_2$ (*133c*). In the presence of $PBu_3{}^n$ methanolysis gives $EtCO_2CH_3$, $CH_3CH(CO_2CH_3)_2$, and $(CH_3O_2CCH_2)_2$.

Allyl compounds react with CO and acetylene in alcohol in the presence of $Ni(CO)_4$ catalyst to give unsaturated esters (*101*) (see Section VI,G). In the presence of an alcohol acetylene is carbonylated to acrylic esters (*20*):

$$HC{\equiv}CH + CO + HOR \xrightarrow{\ Ni(CO)_4\ } H_2C{=}CHCO_2R \qquad (96)$$

The analogous conversion of methylacetylenes to methacrylic esters is possible by the following route (*135*):

$$Ni(CO)_4 + HX \rightleftharpoons HNi(CO)_{4-n}X + nCO \qquad (97)$$

$$CH{\equiv}CR + HNi(CO)_{4-n}X$$

$$\longrightarrow \underset{\displaystyle |}{\overset{\displaystyle R}{CH_2{=}C}}Ni(CO)_{4-n}X + RCH{=}CHNi(CO)_{4-n}X$$

$$\Big\downarrow CO$$

$$\underset{\displaystyle |}{\overset{\displaystyle R}{CH_2{=}C}}CONi(CO)_{4-n}X \qquad (98)$$

$$\Big\downarrow R'OH$$

$$\underset{\displaystyle |}{\overset{\displaystyle R}{CH_2{=}C}}CO_2R' + HNi(CO)_{4-n}X$$

This reaction is the basis for the commercial manufacture of methyl methacrylate from methyl acetylene. In the palladium catalyzed carbonylation of acetylenic bonds extensive dicarbonylation takes place (*134,136*), e.g.:

$$HC{\equiv}CH + 2CO + 2CH_3OH \longrightarrow CH_3O_2CCH{=}CHCO_2CH_3 + H_2 \qquad (99)$$

$$2HC{\equiv}CH + 2CO + 2CH_3OH \longrightarrow (CH_3O_2CCH{=}CH)_2 + H_2 \qquad (100)$$

The yield of the first diester is 50 moles per mole of palladium. These reactions take place under ambient conditions in the presence of palladium halides and thiourea plus a little oxygen to prevent hydrogenation of the acetylene and the products. Propargyl chloride and propargyl alcohol react with CO in methanol as indicated below (123,125) (see Section VI,G).

$$HC{\equiv}CCH_2Cl + CO + CH_3OH$$

$$\xrightarrow[\text{PdCl}_2]{} \underset{\overset{|}{Cl}}{H_2C{=}CCH_2CO_2CH_3} + \underset{\overset{|}{CH_2CO_2CH_3}}{CH_2{=}CCO_2CH_3} \qquad (101)$$

$$HC{\equiv}CCH_2OH + CO + CH_3OH$$

$$\xrightarrow[\text{PdCl}_2]{} \underset{\overset{|}{CH_2{=}CCO_2CH_3}}{CH_2OCH_3} + \underset{\overset{|}{CH_2CO_2CH_3}}{CH_2{=}CCO_2CH_3} + \underset{\overset{\overset{|}{CCO_2CH_3}}{\|}}{\underset{CHCO_2CH_3}{CH_2CO_2CH_3}} \qquad (102)$$

An acetylenic ester is produced in reactions recently described by Tsutsumi et al. (136a). Lithium phenyl acetylide reacts with $Fe(CO)_5$ or $Ni(CO)_4$ at $-15°$ or $-30°$ to give lithium phenylethynylmetal carboxylates. Decomposition of these intermediates with iodine in methanol under reflux, or at $-30°$ respectively gives $PhC{\equiv}CCO_2CH_3$ in both cases but $PhC{\equiv}CC{\equiv}CPh$ in only the nickel case.

D. From Dienes

$Ni(CO)_4$ catalyzes the formation of monocarboxylic esters from allene (137–139):

$$CH_2{=}C{-}CH_2 \xrightarrow[\text{Ni(CO)}_4]{\text{CO/CH}_3\text{OH}} \underset{62\%}{CH_3\overset{\overset{CH_2}{\|}}{C}CO_2CH_3} + CH_2{=}CHCH_2CO_2CH_3 \quad (103)$$

Ruthenium and platinum/tin catalysts have also been used to prepare methyl methacrylate from allene in 50% and 40% yields respectively (140,141). Carbonylation of conjugated dienes is catalyzed by palladium and gives β,γ-unsaturated esters (142):

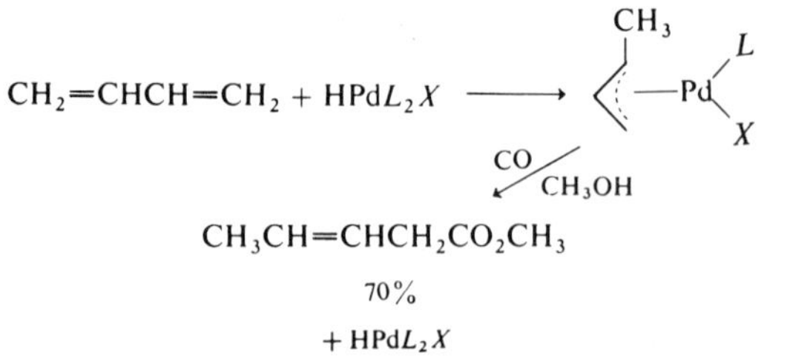

$$CH_2{=}CHCH{=}CH_2 + HPdL_2X \longrightarrow$$

$$CH_3CH{=}CHCH_2CO_2CH_3$$

$$70\%$$

$$+ HPdL_2X$$

where $L = PBu_3$ and $X = I$. Nonconjugated dienes may give cyclic keto esters under the same conditions (90):

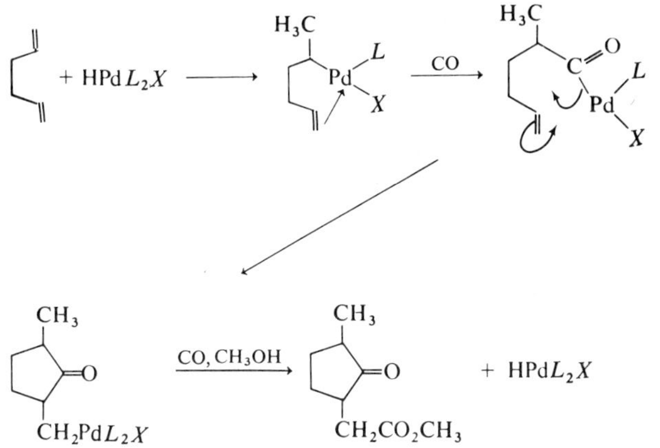

$$(105)$$

Cyclooctadiene reacts under similar conditions to give cyclooct-4-ene-1-carboxylates and under more severe conditions to give saturated dicarboxylic esters ($133,142a$).

E. FROM ALDEHYDES

Hydroformylation of propionaldehyde and hexahydrobenzaldehyde gives propyl formate and cyclohexyl formate respectively in yields of 35% (143).

F. FROM EPOXIDES

Catalytic carboxylation of epoxides with CO and an alcohol gives 3-hydroxypropionate esters (82), e.g. (144):

$$CH_3-HC-CH_2 + CO + CH_3OH \xrightarrow{Co_2(CO)_8} CH_3CHCH_2CO_2CH_3 \quad (106)$$

with O bridging HC–CH$_2$, OH on the product, 40%

Terminal olefin oxides are converted to esters in low yields under ambient conditions using the iron carbonylates $KHFe(CO)_4$, $K_2Fe(CO)_4$, or $K_2Fe_2(CO)_8$ ($40,145$):

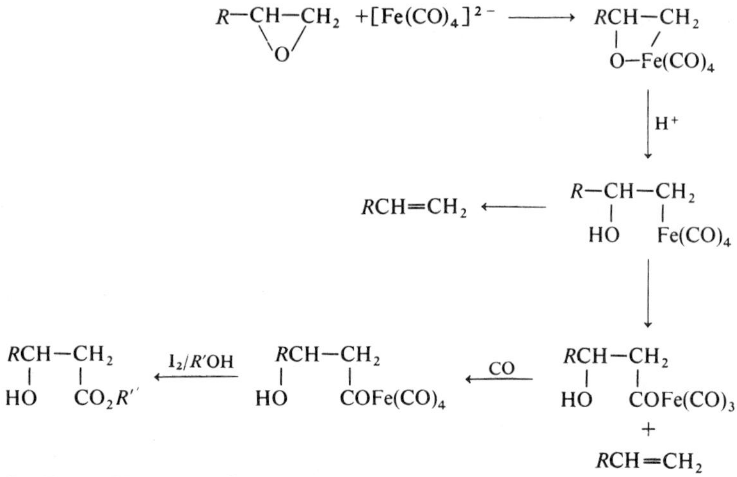

$$R-\underset{\underset{O}{\diagdown\diagup}}{CH-CH_2} + [Fe(CO)_4]^{2-} \longrightarrow \underset{\underset{O-Fe(CO)_4}{|\diagup}}{RCH-CH_2}$$

$$\downarrow H^+$$

$$RCH=CH_2 \longleftarrow \underset{\underset{HO \quad Fe(CO)_4}{|\quad\quad|}}{R-CH-CH_2} \quad (107)$$

$$\downarrow$$

$$\underset{\underset{HO \quad CO_2R''}{|\quad\quad|}}{RCH-CH_2} \xleftarrow{\text{I}_2/R'OH} \underset{\underset{HO \quad COFe(CO)_4}{|\quad\quad|}}{RCH-CH_2} \xleftarrow{\text{CO}} \underset{\underset{HO \quad COFe(CO)_3}{|\quad\quad|}}{RCH-CH_2}$$
$$+$$
$$RCH=CH_2$$

G. From Halogen Compounds

Alkyl halides undergo alkoxycarbonylation in the presence of base and cobalt carbonyl (82):

$$RX + Co(CO)_4^- + CO \longrightarrow RCOCo(CO)_4 + X^- \quad (108)$$

$$RCOCo(CO)_4 + R'OH \longrightarrow RCO_2R' + HCo(CO)_4 \quad (109)$$

$$HCo(CO)_4 + base \longrightarrow [\text{H-base}]^+[Co(CO)_4]^- \quad (110)$$

Epichlorhydrin may be converted to ester in the presence of $Co_2(CO)_8$ and glycidyl trimethylammonium chloride under CO pressure (146):

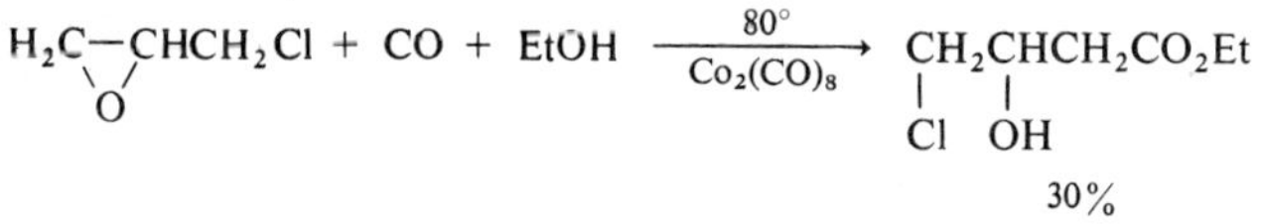

$$\underset{\underset{O}{\diagdown\diagup}}{H_2C-CHCH_2Cl} + CO + EtOH \xrightarrow[\text{Co}_2(\text{CO})_8]{80^\circ} \underset{\underset{Cl \quad OH}{|\quad\quad|}}{CH_2CHCH_2CO_2Et} \quad (111)$$
$$30\%$$

Methyl iodide has been converted to methyl formate as follows (69):

$$CoHN_2(PPh_3)_3 \xrightarrow[\text{PhCH}_3,0^\circ]{CO} \text{"Co formate"} \xrightarrow{CH_3I} HCO_2CH_3 \quad (112)$$

There are two general methods for the preparation of unsaturated esters by carbonylation with $Ni(CO)_4$ as catalyst (101,147–149):

$$RCH=CHCH_2X + CO + R'OH \xrightarrow{\text{Ni(CO)}_4} RCH=CHCH_2CO_2R' + HX \quad (113)$$

$$RCH=CHCH_2X + HC\equiv CH + CO + R'OH$$

$$\xrightarrow{\text{Ni(CO)}_4} RCH=CHCH_2CH=CHCO_2R' + HX \quad (114)$$

The mechanism of these reactions involves the formation of π-allyl intermediates. An extension of this type of reaction is the following (150):

$ClCH_2CH=CHCH_2Cl$

$$\xrightarrow[\substack{CH_3OH \\ HOAc/LiCl}]{\substack{CO/HC\equiv CH \\ Ni(CO)_4}} \quad ClCH_2CH=CHCH_2CH=CHCO_2CH_3$$

$$+ CH_3CO_2CH=CHCH_2CH=CHCH_2CH=CHCO_2CH_3$$

$$(115)$$

Treatment of a wide variety of organic halides, RX, with several equivalents of $Ni(CO)_4$ in alcoholic media ($R'OH$) containing two to three equivalents of alkoxide gives esters (151):

$$\text{e.g., } RX \xrightarrow[CH_3OH, CH_3O]{Ni(CO)_4} RCO_2CH_3 \qquad (116)$$

Benzyl halide reacts with $Ni(CO)_4$ in ethanol to yield an ester product (100):

$$PhCH_2X + EtOH \xrightarrow{Ni(CO)_4} PhCH_2CO_2Et + HX + [Ni(CO)_3] \qquad (117)$$

Allyl chloride also reacts in the presence of palladium chloride to give allylic esters (152):

$$CH_2=CHCH_2Cl \xrightarrow[\substack{PdCl_2 \\ EtOH}]{CO} CH_2=CHCH_2CO_2Et$$

$$\xrightarrow{} EtO_2CCH=CHCH_2CO_2Et$$

$$(118)$$

This carbonylation is best carried out in THF in order to minimize the concentration of free HCl—the ring opening reaction of HCl with the THF removes most of it. The carbonylation of 1,4-dichloro-2-butene can be made catalytic in the presence of $PdCl_2$ and $CuCl_2$ (123):

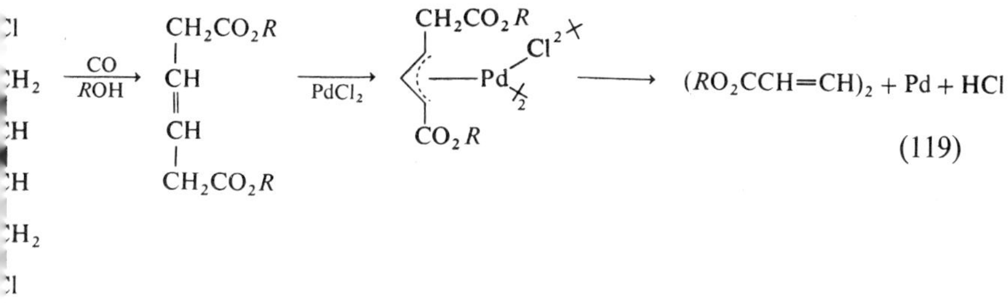

$$(119)$$

Pd(II) is both a good carbonylation catalyst and readily forms π-allyls by hydrogen abstraction. Allyl chloride reacts with butadiene (but not with acetylene as in the nickel case) as follows:

$$CH_2=CHCH_2Cl + CH_2=CHCH=CH_2$$

$$\xrightarrow[Pd]{CO} CH_2=CHCH_2CH_2CH=CHCH_2COCl$$

$$(120)$$

In alcohol this product would presumably be converted to ester. Propargyl chloride carbonylates as follows (via $H_2C=C=CHCO_2CH_3$) (*125*):

$$HC{\equiv}C{-}CH_2{-}Cl + CO + CH_3OH$$

$$\xrightarrow[\text{PdCl}_2 \text{ or Pd/C}]{20{-}100°/100 \text{ atm}}$$

$$H_2C=CCH_2CO_2CH_3 + CH_2=CCO_2CH_3$$
$$\quad\;\; | \qquad\qquad\qquad\qquad\quad |$$
$$\quad\;\; Cl \qquad\qquad\qquad\qquad CH_2CO_2CH_3$$

$$(121)$$

In dilute solutions the itaconate is the main product from propargyl alcohol:

$$HC{\equiv}CCH_2OH + CO + CH_3OH$$

$$\xrightarrow[\text{PdCl}_2 \text{ or Pd/C}]{100°/100 \text{ atm}} \quad \underset{\text{2-methoxymethyl-acrylate}}{\overset{\displaystyle CH_2OCH_3}{\underset{\displaystyle CH_2=CCO_2CH_3}{|}}} + \underset{\text{itaconate}}{\overset{\displaystyle CH_2=CCO_2CH_3}{\underset{\displaystyle CH_2CO_2CH_3}{|}}} + \underset{\text{aconitate}}{\overset{\displaystyle CH_2CO_2CH_3}{\underset{\displaystyle CHCO_2CH_3}{\overset{\displaystyle CCO_2CH_3}{\|}}}}$$

$$(122)$$

The relative proportions of the three products depend on the solvent, catalyst and concentration of HCl. Substituted propargyl alcohols may also be carbonylated to esters in alcohol and anhydrides in benzene.

H. From Orthoformates

Ethyl formate is formed together with acetals under oxo conditions (*14,153*):

$$2HC(OEt)_3 + CO + H_2 \xrightarrow[\text{Co}_2\text{(CO)}_8]{} EtCH(OEt)_2 + 2HCO_2Et + EtOH \quad (123)$$

I. From Mercury Derivatives

Mercuric nitrates carbonylate to give esters at 25 atm and mainly acyl esters at higher pressures (*154*):

$$RHgNO_3 + CO + CH_3 \xrightarrow[25 \text{ atm}]{} RCO_2CH_3 + Hg + HNO_3 + RNO_2$$

$$\xrightarrow[200{-}300 \text{ atm}]{} RCOCO_2CH_3 \qquad\qquad (124)$$

The mechanism does not involve the formation of carbonium ions and it must, therefore, proceed either by a concerted S_N2 or an alkyl migration route.

VII. Lactones

Lactones may be synthesized by five general routes employing carbonylation or hydroformylation techniques; a discussion of each of these is given in Sections A–E. In most cases these methods are of limited generality due to the occurrence of competing side-reactions which can result in low yields of the lactone. However, different methods of synthesis are frequently complementary to one another. All the reactions involve complexes of the group VIII transition metals as catalysts and the majority include $Co_2(CO)_8$, possibly since this compound is the most extensively studied catalyst for carbonylation reactions.

A. RING CLOSURE REACTIONS OF UNSATURATED ALCOHOLS WITH CARBON MONOXIDE

Treatment of unsaturated alcohols with carbon monoxide in the presence of $Co_2(CO)_8$ as catalyst affords lactone ring systems in varying yields (20,155).

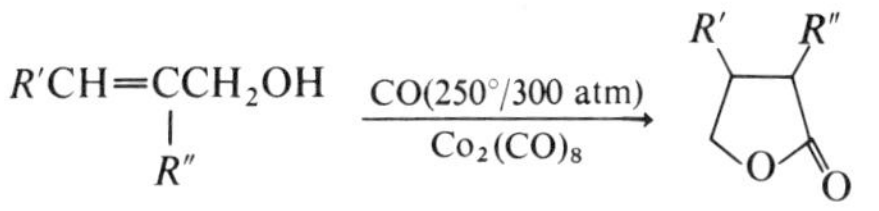

$$R'CH{=}CCH_2OH \underset{R''}{} \xrightarrow[\text{Co}_2\text{(CO)}_8]{\text{CO(250°/300 atm)}} \quad \text{(lactone)} \tag{125}$$

A competing reaction is that of isomerization of the unsaturated alcohol to aldehyde, and in the case of allyl alcohol ($R'{=}R''{=}$H) this is the predominant reaction. Here only ca. 2% of γ-butyrolactone is produced and ca. 50% of the starting alcohol is isomerized to aldehyde. In the presence of acetonitrile as solvent, and by the addition of a catalytic amount of an organic base such as pyridine, aldehyde formation may be minimized and γ-butyrolactone obtained in a maximum yield of 60% (156).

In general, however, the reaction is limited to cases where isomerization of the unsaturated alcohol cannot occur, i.e. when the 2-position is blocked by substituents.

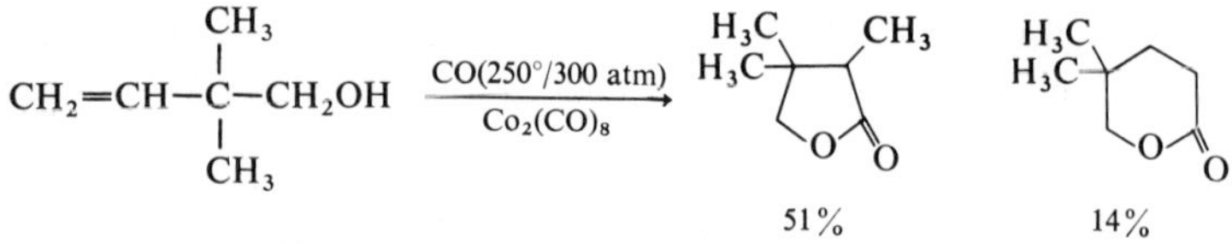

$$\underset{\underset{CH_3}{\overset{CH_3}{|}}}{CH_2{=}CH{-}\overset{\overset{CH_3}{|}}{C}{-}CH_2OH} \xrightarrow[Co_2(CO)_8]{CO(250°/300\ atm)} \quad \text{51\%} \qquad \text{14\%} \qquad (126)$$

The relative proportion of the two products varies with the alkyl substituents but generally the 5-membered ring system predominates. The formation of a 6-membered ring system is favored by the presence of an alkyl substituent in the 3-position.

With γ-unsaturated secondary alcohols, lactones are produced in small yields and the predominant reaction is isomerization to the ketone. Higher yields of lactone are obtained from γ-unsaturated tertiary alcohols, although a competing reaction here is the formation of olefins by elimination of water, followed by hydrogenation of the resulting diene by hydrogen transfer (20,155).

B. Reactions of Unsaturated Carboxylic Acid Esters with Carbon Monoxide and Hydrogen

A convenient preparation of lactones which are not readily accessible by Method A involves the catalytic hydroformylation of esters of unsaturated carboxylic acids, usually in the presence of $Co_2(CO)_8$ as catalyst (14). The reaction conditions are generally more extreme than those commonly used in the hydroformylation of olefins. The α,β-unsaturated carboxylic acid esters may be converted into γ-(and δ-) lactones, where β,γ-unsaturated carboxylic acid esters yield δ-lactones. Using this method, ethyl acrylate may be converted into γ-butyrolactone in 88% yield:

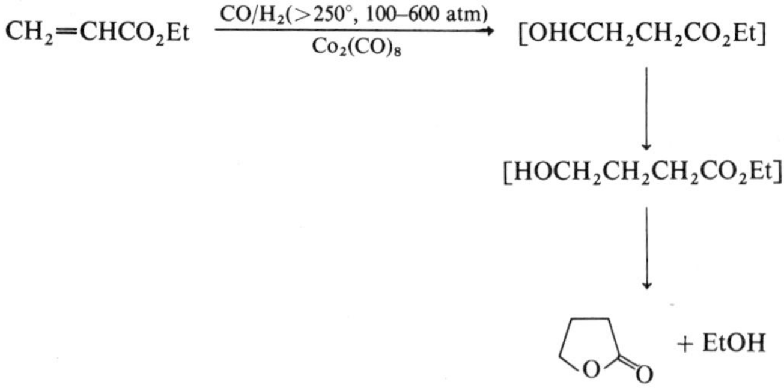

$$CH_2{=}CHCO_2Et \xrightarrow[Co_2(CO)_8]{CO/H_2(>250°,\ 100{-}600\ atm)} [OHCCH_2CH_2CO_2Et]$$

$$\downarrow$$

$$[HOCH_2CH_2CH_2CO_2Et] \qquad (127)$$

$$\downarrow$$

$$+\ EtOH$$

Similarly, methyl crotonate yields δ-valerolactone and β-methyl-γ-butyrolactone.

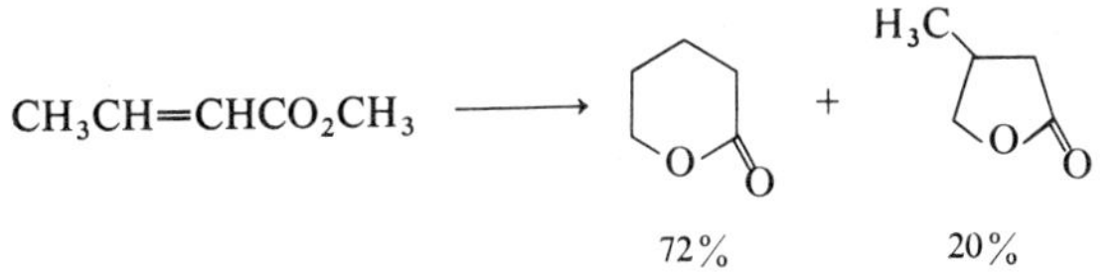

$$CH_3CH{=}CHCO_2CH_3 \longrightarrow \underset{72\%}{\text{[structure]}} + \underset{20\%}{\text{[structure]}} \qquad (128)$$

C. Reactions of Oxo-Ring Systems with Carbon Monoxide

Trimethylene oxide reacts with carbon monoxide in the presence of a $Co_2(CO)_8$ catalyst to yield γ-butyrolactone in 55% yield (88). Similarly β,β-dimethyloxetane yields β,β-dimethyl-γ-butyrolactone.

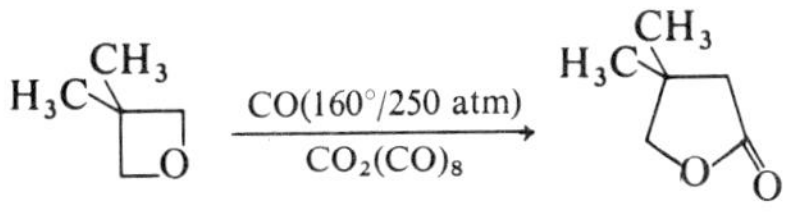

$$\underset{}{\text{[structure]}} \xrightarrow[CO_2(CO)_8]{CO(160^\circ/250\ atm)} \underset{}{\text{[structure]}} \qquad (129)$$

These reactions may also be carried out *stoichiometrically* under mild conditions. Thus trimethylene oxide reacts with $HCo(CO)_4$ with the uptake of one mole of carbon monoxide to give 4-hydroxybutyryl cobalt tetracarbonyl. The organic moiety may then be cyclized by the action of a hindered base (88).

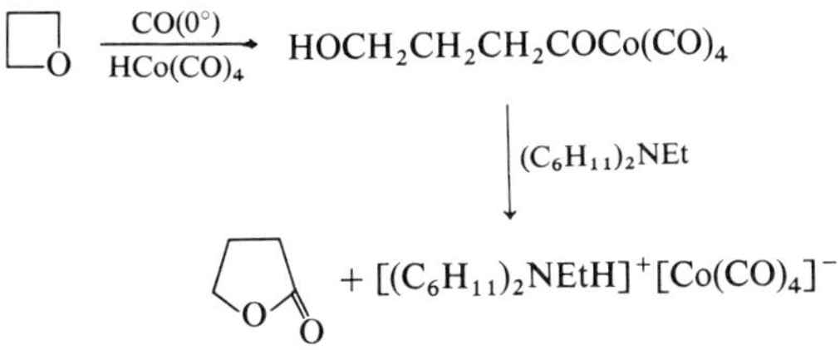

$$\underset{}{\text{[structure]}} \xrightarrow[HCo(CO)_4]{CO(0^\circ)} HOCH_2CH_2CH_2COCo(CO)_4$$

$$\Big\downarrow (C_6H_{11})_2NEt \qquad (130)$$

$$\underset{}{\text{[structure]}} + [(C_6H_{11})_2NEtH]^+[Co(CO)_4]^-$$

D. REACTIONS OF ALLYL, ALKYL, OR ACYL HALIDES WITH ACETYLENES
AND CARBON MONOXIDE

This method of synthesis is a useful route to unsaturated lactone ring
systems and has the advantage of very mild reaction conditions. The reaction may be catalyzed by either cobalt or nickel carbonyls; the use of the
former has been reviewed by Heck (88).

$$RCH_2X + R'C{\equiv}CR' + 2CO \xrightarrow[\text{base}]{\text{Co(CO)}_4^-} \quad + [HX] \qquad (131)$$

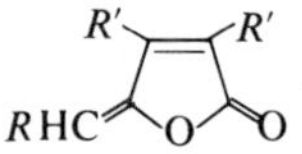

The scope of this reaction is somewhat limited since the lactone is initially
formed as a cobalt carbonyl complex and the latter is not susceptible to
base catalyzed elimination under mild conditions unless R is an activating
substituent, e.g., CN. The reaction is only partially catalytic since a
stable acetylene dicobalt hexacarbonyl complex is also formed and this
tends to consume the catalyst. The yield of lactone depends markedly on
the acetylene used but the reaction works fairly well with disubstituted
acetylenes.

Chiusoli and co-workers (101) have produced detailed studies of the
reactions of allyl and acyl halides with acetylenes and carbon monoxide in
the presence of nickel carbonyl. A variety of organic molecules may be
obtained and it is found that the solvent and presence or absence of water
are critical factors controlling the course of these reactions. In solvents such
as esters or ethers, acyl halides react with a stoichiometric amount of
$Ni(CO)_4$ and a slight excess of acetylene to produce anhydrides as the
major products and only small amounts of γ-lactone, whereas in alcoholic
solvents allyl halides react to form hexadienoic acids and esters. These
reactions have been discussed where appropriate in other parts of this
chapter (see Sections V,B; VI,G).

The reaction between allyl halides and substituted acetylenes in ketonic
solvents generally yields keto-acids and keto-lactones (113). Thus, in the
presence of small amounts of water ($<0.5\%$) allyl chloride and acetylene
react with carbon monoxide and $Ni(CO)_4$ to yield

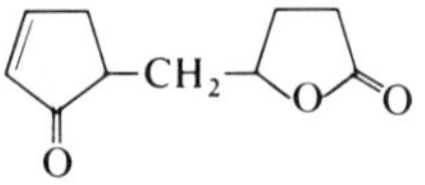

(11)

(see Section VI,G). At very low water concentrations secondary reactions with the solvent can occur yielding molecules such as

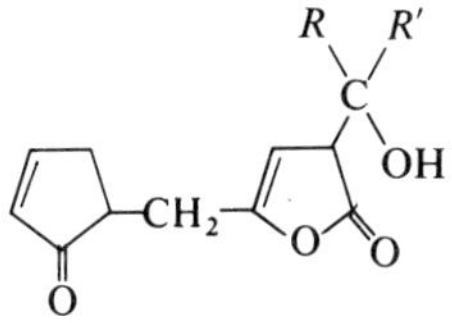

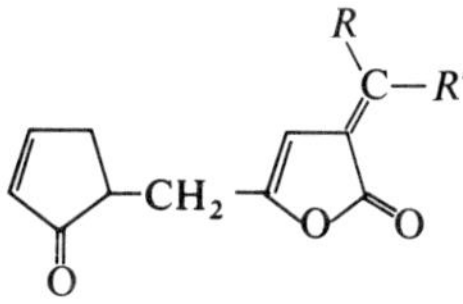

(12)   and   (13)

Acyl halides $RCOCl$ react in a similar manner to yield β,γ-unsaturated γ-lactones of the following type (157):

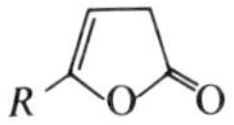

(14)

ε-Lactones containing two double bonds in the ring may be obtained when the reaction conditions are such as to minimize the carbon monoxide concentration in solution, i.e., high flow rates of acetylene (158, 159).

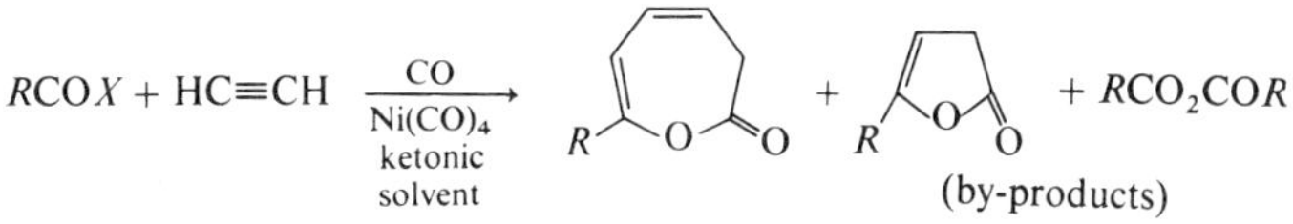

$$RCOX + HC{\equiv}CH \xrightarrow[\substack{Ni(CO)_4 \\ ketonic \\ solvent}]{CO}$$ $+$ $+ RCO_2COR$ (132)

(by-products)

Keto-lactones of the type

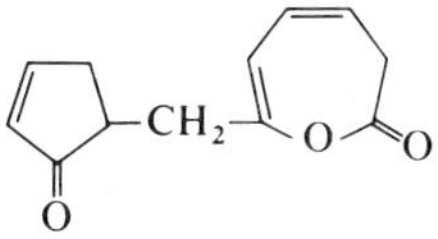

(15)

may be formed in low yield when allyl halides are used under these conditions.

Unsaturated γ-lactones may also be obtained in low yields (2–24%) from the reactions of acetylenes with lithium aroyltricarbonylnickelates at $-30°C$ (98) (see Section IV,I).

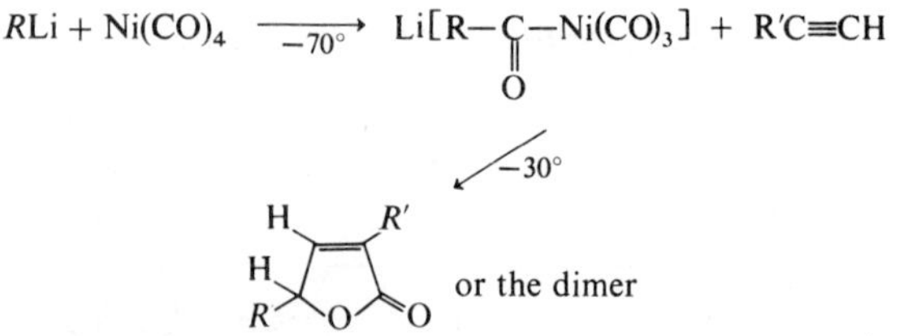

$$RLi + Ni(CO)_4 \xrightarrow{-70°} Li[R-\underset{\underset{O}{\|}}{C}-Ni(CO)_3] + R'C{\equiv}CH$$

$$(133)$$

or the dimer

The normal reaction, however, is carried out at $-70°$ and is a useful synthetic route to 1,4-diketones. These are obtained in 47–74% yields and the γ-lactones are only produced as by-products.

Lactones and olefins are the predominant products from the reaction of olefins in the presence of iodobenzene and nickel carbonyl (*160*):

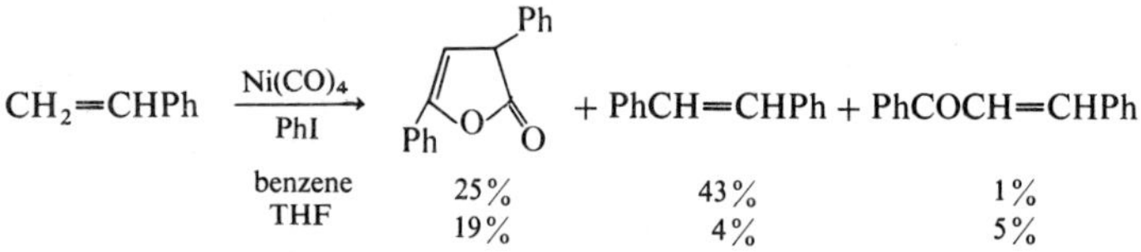

$$CH_2{=}CHPh \xrightarrow[PhI]{Ni(CO)_4} \quad \text{(lactone)} \quad + PhCH{=}CHPh + PhCOCH{=}CHPh \quad (134)$$

| | benzene | 25% | 43% | 1% |
| | THF | 19% | 4% | 5% |

The suggested mechanism involves a benzoyl nickel species $PhCONi(CO)_nI$ as an intermediate which is too reactive to be isolated. This is the first report of the formation of a lactone directly from a mono-olefin.

In summary, the reactions of allyl, alkyl, and acyl halides with acetylenes and carbon monoxide afford extremely versatile syntheses of unsaturated lactones under very mild conditions.

E. Reactions of Unsaturated Hydrocarbons with Carbon Monoxide

Isolated reactions of acetylenes with carbon monoxide to yield lactone systems have been reported. Thus propyne reacts in the presence of $Co_2(CO)_8$ yielding 53% *trans*- and 38% *cis*- 2,6-dimethyl-2,4,6-octatriene-1,4:5,8-diolide and 9% of the 2,7-dimethyl analogue (*14,161*):

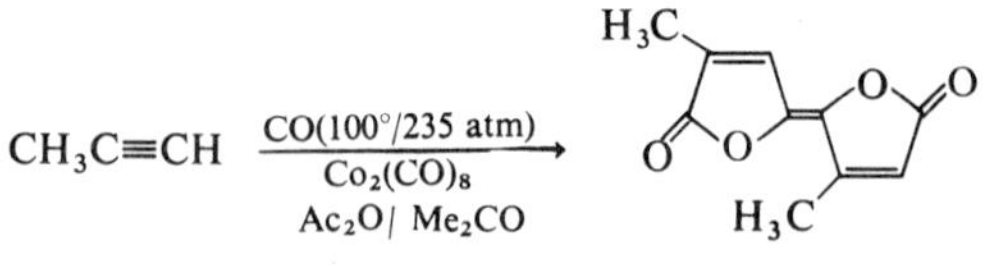

$$CH_3C{\equiv}CH \xrightarrow[\substack{Co_2(CO)_8 \\ Ac_2O/\ Me_2CO}]{CO(100°/235\ atm)} \quad \text{(diolide structure)} \quad (135)$$

Diphenylacetylene and carbon monoxide react in ethanol in the presence of palladium dichloride and HCl to form α,β-diphenyl-γ-crotonolactone (*123,162*).

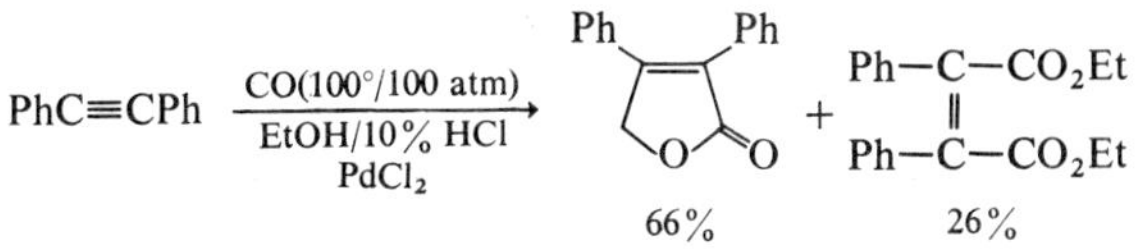

$$PhC\equiv CPh \xrightarrow[\substack{EtOH/10\% \ HCl \\ PdCl_2}]{CO(100°/100 \ atm)} \quad \underset{66\%}{\text{[lactone]}} \quad + \quad \underset{26\%}{\substack{Ph-C-CO_2Et \\ \| \\ Ph-C-CO_2Et}} \qquad (136)$$

The reaction is solvent dependent and no carbonylation occurs in benzene. The lactone is thought to be formed by simultaneous attack of 2 moles of carbon monoxide on the triple bond, followed by reduction of one of the carbonyl functions. This is atypical of the usual carbonylation reactions of acetylenes catalyzed by palladium (see Section VI,C).

Dienes may be carbonylated under forcing conditions in the presence of palladium catalysts to give lactones in low yield (*89*).

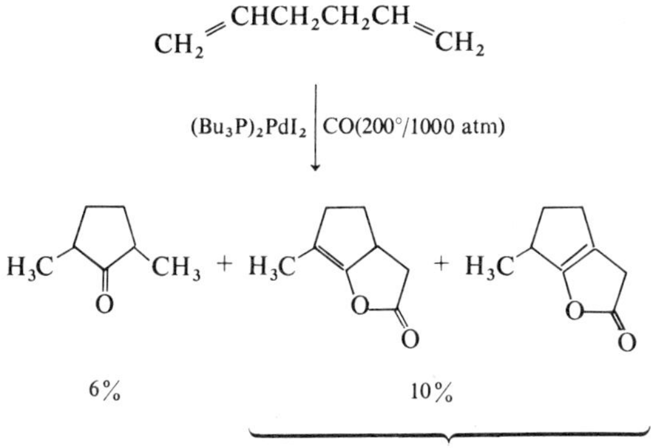

$$CH_2 \overset{CHCH_2CH_2CH}{\diagup \diagdown} CH_2$$

$$(Bu_3P)_2PdI_2 \Big| CO(200°/1000 \ atm)$$

$$(137)$$

A useful extension of the base reaction of cobalt carbonyl to the reaction of *cis*-1-chloro-4-hydroxy-2-butene produces a 6-membered unsaturated lactone ring in good yield (*88,163*).

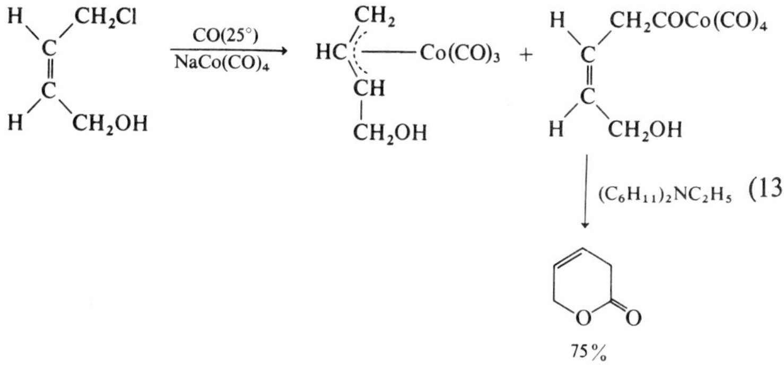

$$(138)$$

75%

VIII. Quinones and Hydroquinones

Stoichiometric reactions of acetylenes with metal carbonyls have been studied in considerable detail and a wide range of complexes including quinone and hydroquinone derivatives may be produced (*21,102,164*). Thus, irradiation of but-2-yne in $Fe(CO)_5$ yields the 2,3,5,6-tetramethyl benzoquinoneiron tricarbonyl complex (*165*):

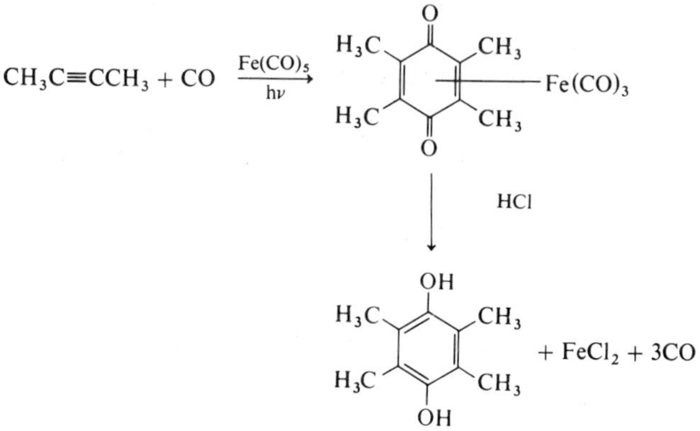
CH3C≡CCH3 + CO
Fe(CO)5
hν
O
H3C
CH3
Fe(CO)3
H3C
CH3
O
HCl
OH
H3C
CH3
+ FeCl2 + 3CO
H3C
CH3
OH

(139)

Decomposition of complexes of this type by air or dilute acids produces quinones or hydroquinones respectively.

Perhaps the most convenient synthesis of hydroquinone is that discovered by Reppe and co-workers several years ago. Acetylene reacts with $Fe(CO)_5$ in aqueous alcohol at low pressures (20–25 atm) and temperatures (50–80°) to form hydroquinone in *ca.* 30% yield (*85*). This reaction is promoted in the presence of bases but is stoichiometric with respect to $Fe(CO)_5$. Similarly mono- or di-substituted acetylenes react to give di- or tetra-substituted hydroquinones.

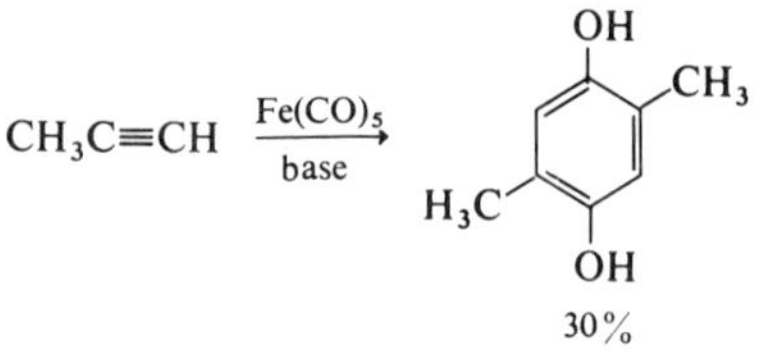
CH3C≡CH
Fe(CO)5
base
OH
CH3
H3C
OH
30%

(140)

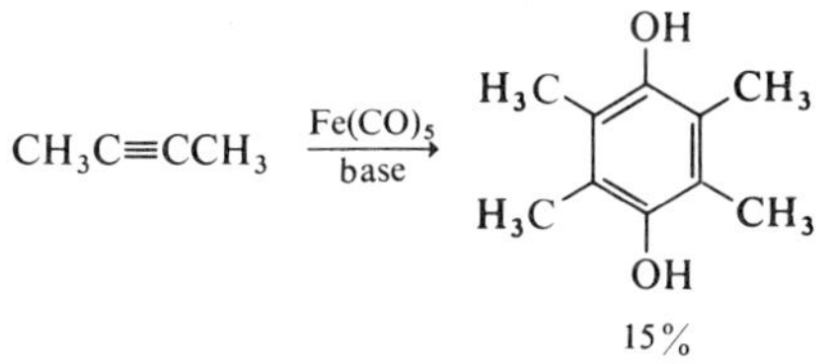
CH3C≡CCH3
Fe(CO)5
base
OH
H3C
CH3
H3C
CH3
OH
15%

(141)

Extension of this synthesis to produce a catalytic reaction has met with partial success either by the use of more extreme but carefully controlled conditions or by the use of catalysts such as $[Fe(NH_3)_6][Co(CO)_4]_2$ and $[Co(NH_3)_6][Co(CO)_4]_2$ (*166*). Yields of up to 70% based on acetylene may be obtained by these methods. Similarly, trimethylhydroquinone may be obtained in 55% yield from propyne, but-2-yne, carbon monoxide and water in the presence of $Fe(CO)_5$. Additional hydroquinones are produced in lower yields.

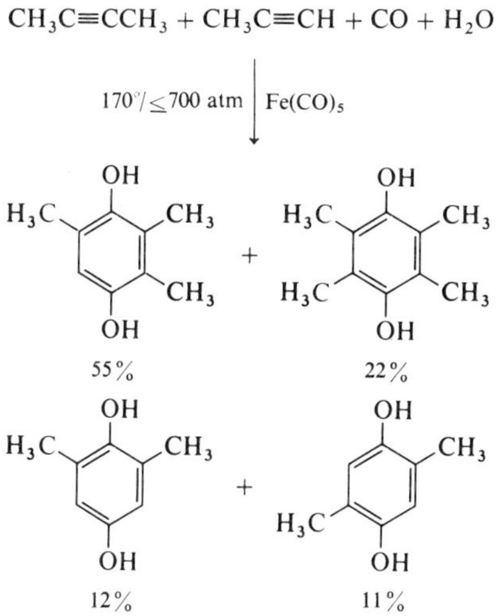
CH3C≡CCH3 + CH3C≡CH + CO + H2O
170°/≤700 atm
Fe(CO)5
55%
22%
12%
11%

(142)

Other metal complexes including halides and carbonyls of ruthenium and rhodium have been investigated as potential catalysts (*166,166a*). Thus hydroquinone may be obtained from the reaction of acetylene, CO and H_2 in the presence of $Ru_3(CO)_{12}$ in THF or dioxan under anhydrous conditions (*167*). Highest yields (55%) are obtained at low partial pressures of hydrogen (5–10 atm):

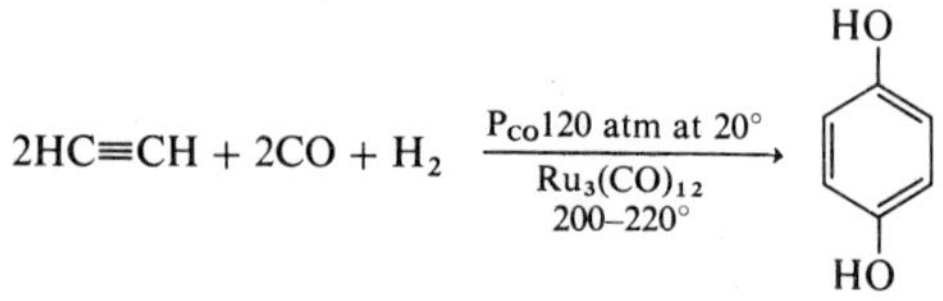

$$2HC{\equiv}CH + 2CO + H_2 \xrightarrow[\substack{Ru_3(CO)_{12} \\ 200-220°}]{P_{CO}120 \text{ atm at } 20°}$$

(143)

Yields of up to 65% may be obtained in the presence of water or alcohol under relatively low partial pressures of CO at 150–250°.

IX. Acyl and Alkyl Halides

Acyl and alkyl halides are best considered in the same section since both species may frequently be obtained by carbonylation techniques in competing reactions. The two are also related by virtue of decarbonylation. Halides may be obtained by halocarbonylation reactions of olefins, acetylenes, allylic compounds, and paraffin hydrocarbons. Tsuji and co-workers (*123,168*) have found that palladium complexes are particularly active catalysts for the formation of acyl and alkyl halides.

A. FROM OLEFINS

The ethylene-palladium chloride complex may be stoichiometrically carbonylated under mild conditions—slowly even under ambient conditions—to give β-chloropropionyl chloride in quantitative yield (*111,169, 170*).

Prior isolation of the ethylene complex is unnecessary and other olefins react directly with carbon monoxide in the presence of $PdCl_2$ to produce β-chloroacyl chlorides. Thus, β-chlorobutyryl chloride and β,β-dichloropropionyl chloride are obtained in good yield from propylene and vinyl chloride, respectively, with carbon monoxide at 90 atm and 90° (*111*).

$$\underset{R'}{\overset{R}{\diagdown}}C=C\underset{R''}{\overset{R'''}{\diagup}} \xrightarrow[\substack{PdCl_2 \\ \text{benzene}}]{CO(20-100°/40-100 \text{ atm})} R'-\underset{\underset{Cl}{|}}{\overset{\overset{R}{|}}{C}}-\underset{\underset{R''}{|}}{\overset{\overset{R'''}{|}}{C}}-COCl$$

(144)

$R =$ H, alkyl, halogen

The olefin may be catalytically carbonylated in alcoholic solvents when an ester is the final product (*132*). The reaction mechanism is thought to involve the intermediate formation of an acyl chloride which is then esterified by the alcohol.

$$CH_2{=}CH_2 \xrightarrow[\substack{50 \text{ atm} \quad 15\%HCl/EtOH}]{\underset{PdCl_2 \text{ or } Pd/C}{CO(80°/50 \text{ atm})}} CH_3CH_2CO_2Et \qquad (145)$$

The HCl plays a definite part in the reaction and the mechanism is thought to be as follows:

$$HX + Pd + nL \;\rightleftharpoons\; \overset{\overset{L_n}{|}}{HPd}{-}X \;\underset{RCH{=}CH_2}{\rightleftharpoons}\; \overset{\overset{L_n}{|}}{RCH_2CH_2Pd}{-}X$$

$$\Big\Vert CO$$

$$RCH_2CH_2COX + Pd + nL \;\rightleftharpoons\; \overset{\overset{L_n}{|}}{RCH_2CH_2COPd}{-}X \quad (146)$$

$$\Big\vert R'OH$$

$$\downarrow$$

$$RCH_2CH_2CO_2R' + HX$$

Studies on the stereochemistry of carbonylation of palladium bicyclo [2,2,1]-heptadiene complexes are consistent with this proposed mechanism (*170a*). Further details on the formation of esters by this method are included in the section concerning esters (see Section VI,B).

Olefins may also be catalytically carbonylated in carbon tetrachloride (*171*). With ethylene, 4,4,4-trichlorobutanoyl chloride is the sole product but higher olefins yield mixtures of acyl and alkyl halides, with high carbon monoxide pressures favoring the former. An interesting feature of this reaction is that it is catalyzed by dinuclear complexes e.g., $[\pi{-}C_5H_5Fe(CO)_2]_2$, $[\pi{-}C_5H_5Mo(CO)_3]_2$, and $Co_2(CO)_8$; mononuclear complexes are ineffective. The cyclopentadienyl complexes catalyze the reaction at $<100°$ where $Co_2(CO)_8$ is effective at $>100°$.

$$RCH{=}CH_2 + CO + CCl_4 \xrightarrow[170{-}200 \text{ atm}]{50{-}140°} \underset{\underset{COCl}{|}}{RCHCH_2CCl_3} + \underset{\underset{Cl}{|}}{RCHCH_2CH_3} \qquad (147)$$

The 2-alkyl-4,4,4-trichlorobutanoyl chlorides are useful precursors for other organic molecules.

B. FROM ACETYLENES

A mixture of muconyl, maleyl, and fumaryl chlorides may be obtained from the carbonylation of acetylene in benzene in the presence of $PdCl_2$ (*172*). Muconyl chloride is a major product in this reaction.

$$HC \equiv CH \xrightarrow[\substack{PdCl_2 \\ benzene}]{CO(100^\circ/100\ atm)} \begin{array}{c} CH=CHCOCl \\ | \\ CH=CHCOCl \end{array} + \begin{array}{c} CHCOCl \\ \| \\ CHCOCl \end{array} + \begin{array}{c} CHCOCl \\ \| \\ ClCOCH \end{array} \quad (148)$$

By a modification of this reaction methyl maleate and methyl muconate can be obtained under very mild conditions (from the reaction of acetylene and CO catalyzed by $PdCl_2$ in methanol and thiourea) (*134*).

C. FROM ALLYIC COMPOUNDS

π-Allylpalladium chloride is carbonylated stoichiometrically in benzene to yield 3-butenoyl chloride (*173*).

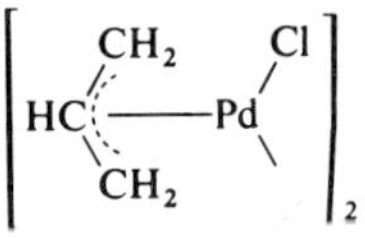

$$\left[\begin{array}{c} CH_2 \quad Cl \\ HC\!\!-\!\!-\!\!Pd \\ CH_2 \end{array} \right]_2 \xrightarrow[benzene]{CO(80^\circ/100\ atm)} CH_2=CHCH_2COCl + Pd \quad (149)$$

The carbonylation of allyl chloride may also be carried out catalytically, preferably in benzene or ether solvents (*111,174*). Thus, allyl chloride reacts with carbon monoxide (90°, 85 atm) in dimethoxyethane in the presence of $(\pi-C_3H_5PdCl)_2$ as catalyst to yield 3-butenoyl chloride selectively and in high conversion. This is found to be a general reaction.

$$R-CH=C-CH_2-X \xrightarrow{CO} H-C=C-CH_2COX \quad (150)$$
$$\overset{|}{R'} \phantom{-CH_2-X \xrightarrow{CO} H-} \overset{|}{R}\ \overset{|}{R'}$$

$R = Cl$, alkyl, H and $R' = $ alkyl, H. The corresponding π-allyl palladium chloride complexes are used as catalysts. For $R = R' = H$, $[(\pi-C_3H_5)RhCl_2]$ and $[(\pi-C_3H_5)_2RhCl]$ may be used as catalysts but their activities are lower than that of the palladium system. The use of non π-allylic catalysts, e.g., Pd, $PdCl_2$, incurs induction periods, presumably during which the π-allylic system is formed, and generally lower rates of reaction (*111*). With π-allylpalladium catalysts, group V ligands may be added, to give a maximum rate when the ligand to palladium ratio is 0.5 (*174a*); a trinuclear species is thought to be the active catalyst in this reaction. When this reaction is carried out in alcoholic solvents esters are again the products.

Conjugated dienes may also be carbonylated in the presence of $PdCl_2$ to yield alkyl and acyl halides (*175*). Presumably the mechanism is similar, a π-allyl complex being formed on initial reaction of the diene with the palladium salt.

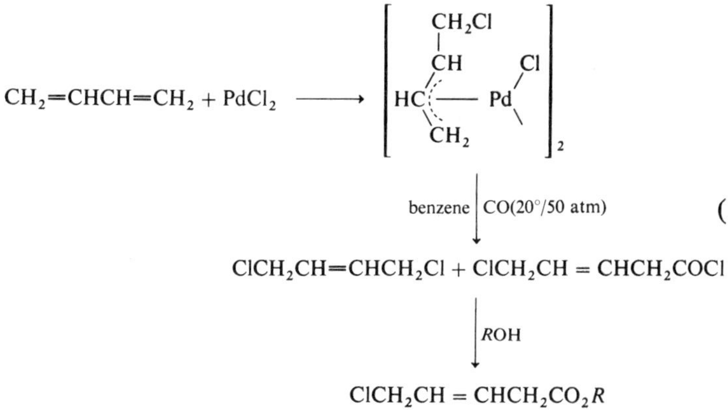

$$CH_2=CHCH=CH_2 + PdCl_2 \longrightarrow \left[\begin{array}{c} CH_2Cl \\ | \\ CH \\ HC \diagdown \diagup \quad Cl \\ \quad \diagup Pd \diagdown \\ CH_2 \end{array} \right]_2$$

$$\text{benzene} \Big| CO(20°/50\ atm) \qquad (151)$$

$$ClCH_2CH=CHCH_2Cl + ClCH_2CH = CHCH_2COCl$$

$$\Big| ROH$$

$$ClCH_2CH = CHCH_2CO_2R$$

In the presence of an alcoholic solvent a mixture of the unsaturated dichloride and an ester are produced.

In an analogous manner, allyl chloride reacts with butadiene in the presence of carbon monoxide and $(\pi-C_3H_5PdCl)_2$ as catalyst to yield 3,7-octadienyl chloride (*111*).

$$CH_2=CHCH_2X + CH_2=CH-CH=CH_2$$

$$\xrightarrow[\substack{(\pi-C_3H_5PdCl)_2 \\ benzene}]{CO(90°/60\ atm)} CH_2=CHCH_2CH_2CH=CHCH_2COX$$

$$(152)$$

This type of reaction offers a potential synthesis of long-chain unsaturated carboxylic acids. Allyl chloride reacts with acetylene and carbon monoxide in the presence of $Ni(CO)_4$ as catalyst to produce 2-*cis*-5-hexadienoyl chloride (*101,176*). However, this product cyclizes readily due to further reaction with acetylene, and may only be isolated in small quantities:

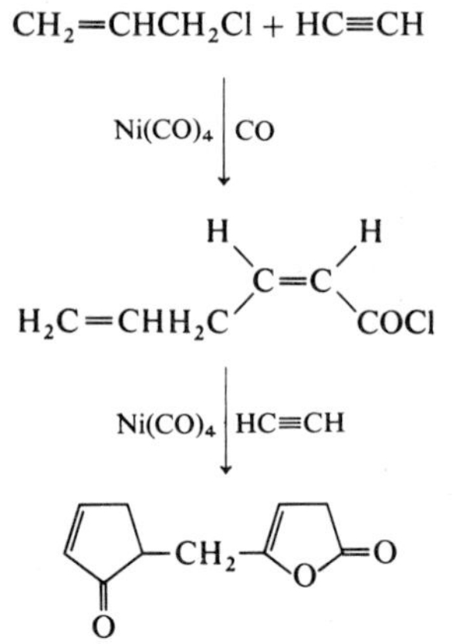

$$CH_2{=}CHCH_2Cl + HC{\equiv}CH$$

$$\Big\downarrow Ni(CO)_4 \;\; CO$$

$$\underset{H_2C=CHH_2C}{\overset{H}{\diagdown}}C{=}C\underset{COCl}{\overset{H}{\diagup}}$$ (153)

$$\Big\downarrow Ni(CO)_4 \;\; HC{\equiv}CH$$

D. CHLOROCARBONYLATION OF SATURATED HYDROCARBONS

Although probably unlikely to provide a general synthetic method, it is worth mentioning that alkyl and acyl halides may be obtained by the γ-radiation initiated chlorocarbonylation of hydrocarbons (177).

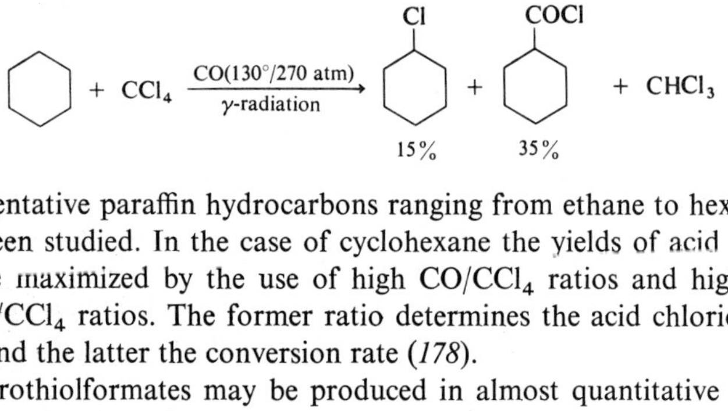

$$+ \;CCl_4 \;\xrightarrow[\gamma\text{-radiation}]{CO(130°/270\;atm)}\; + \; + \; CHCl_3$$ (154)

Representative paraffin hydrocarbons ranging from ethane to hexadecane have been studied. In the case of cyclohexane the yields of acid chloride may be maximized by the use of high CO/CCl_4 ratios and high cyclohexane/CCl_4 ratios. The former ratio determines the acid chloride selectivity and the latter the conversion rate (178).

Chlorothiolformates may be produced in almost quantitative yield by the direct reaction of sulphenyl chlorides with carbon monoxide (79).

$$RSCl \;\xrightarrow[CCl_4]{CO(25°/ \leqslant 400\;atm)}\; RSCOCl$$ (155)

$R{=}CH_3$, Ph.

E. DECARBONYLATION OF ACYL HALIDES

Many of the transition metal catalyzed decarbonylation reactions of acyl halides yields olefins rather than alkyl halides (46), e.g.:

$$RCH_2CH_2COX \;\xrightarrow[200°]{Pd\;or\;PdCl_2}\; RCH{=}CH_2 + CO + HX$$ (156)

However, in the absence of a hydrogen atom in the β-position elimination of HX is not possible and the product is a halide:

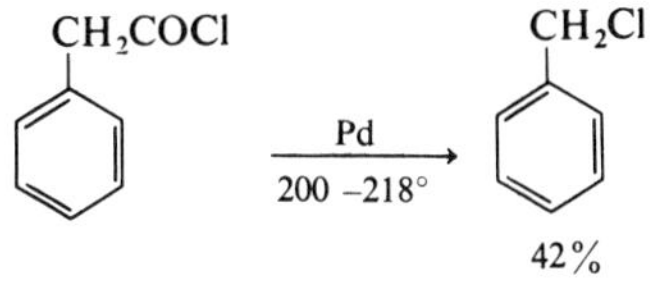

$$(157)$$

$(Ph_3P)_3RhCl$ is an extremely effective catalyst for the decarbonylation of aroyl halides to aryl halides at temperatures above 200° $(46,180–182)$. This reaction is very versatile and is virtually unaffected by electronic or steric influences of substituents. The following mechanism for decarbonylation has been proposed:

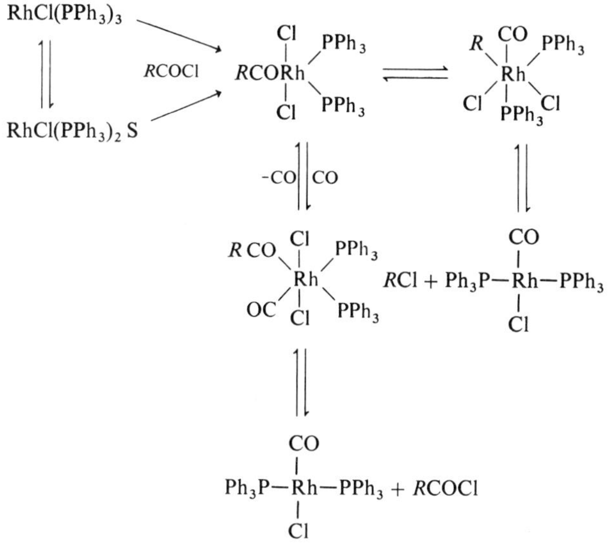

$$(158)$$

In support of this mechanism the acyl complex $RCORhCl_2(PPh_3)_2$ has been isolated (46).

The nature of the ligand does, however, appear critical. Thus, in the decarbonylation of benzoyl chloride, replacement of Ph_3P in RhCl-$(CO)(PPh_3)_2$ by Ph_3As and Ph_3Sb causes the 90% conversion to drop to 20% and 30% respectively (183).

The decarbonylation reaction is reversible and benzyl halides may be catalytically carbonylated to phenylacetyl halides in the presence of $RhCl(CO)(PPh_3)_2$ (46).

X. Organonitrogen Compounds

Carbonylation reactions involving the use of nitrogen containing compounds have been reviewed and classified according to both starting materials (*184*) and final products (*20,155*). Many of these reactions involve ring closure with carbon monoxide, and most are catalyzed by $Co_2(CO)_8$ or $Ni(CO)_4$. Earlier reviews (*20,155,184*) have covered the literature up to 1966 and this section summarizes and up-dates the previous contributions. The types of organonitrogen molecules which can be obtained by carbonylation or hydroformylation techniques are amides and ureas, imides, lactams, phthalimidines, indazolones and quinazolines, and amino acids.

A. Amides and Ureas from Amines and Their Precursors

The reaction of primary and secondary amines with carbon monoxide in the presence of metal carbonyls to yield substituted ureas and N-formyl derivatives is now well known (*21,102,184*).

$$2RNH_2 + CO \longrightarrow RNHCONHR \qquad (159)$$

$$R_2NH + CO \longrightarrow HCONR_2 \qquad (160)$$

These reactions are important not only for their synthetic value but also from the standpoint of mechanistic considerations. Although the initial catalysts used were powdered metals or metal salts, metal carbonyls are probably the true catalytic species under normal reaction conditions (150–270°, 100–300 atm). Aliphatic and aryl amines react differently and this effect has been interpreted in terms of the relative basicities of the amines (*53*). Thus, aliphatic amines react in the presence of metal carbonyls to yield mainly N-formyl derivatives, whereas aniline reacts to form diphenylurea and smaller amounts of formanilide.

$$BuNH_2 \xrightarrow[\text{Ni(CO)}_4]{\text{CO(200°)}} \underset{\text{40–45\%}}{HCONHBu} + \underset{\text{40–45\%}}{BuNHCONHBu} \qquad (161)$$

$$PhNH_2 \xrightarrow[\text{Ni(CO)}_4]{\text{CO(200°)}} \underset{\text{75\%}}{PhNHCONHPh} \qquad (162)$$

The presence of phosphines in the $[Rh(CO)_2Cl]_2$ catalyzed carbonylation of $BuNH_2$ has a dramatic effect upon the selectivity to formamide formation (*185*).

In the presence of $Mn_2(CO)_{10}$ as catalyst, primary aliphatic amines yield 1,3-dialkylureas as the major products (186). Disubstituted formamides may be produced in high selectivity by the copper catalyzed carbonylation of secondary amines (187,188), e.g.:

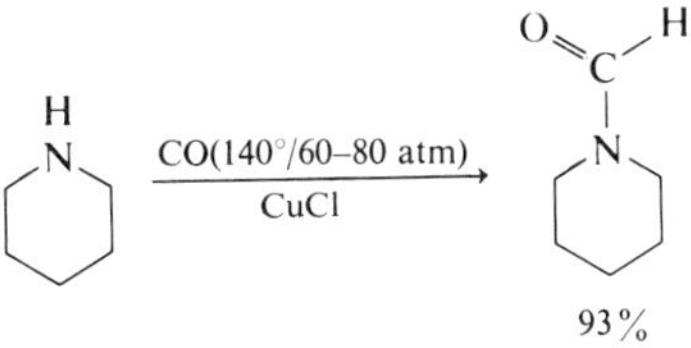

$$(163)$$

93%

Tertiary amines may also be carbonylated and again aliphatic amines react differently from aryl amines:

$$Bu_3N + CO \xrightarrow{Ni(CO)_4} HCONBu_2 \qquad (164)$$

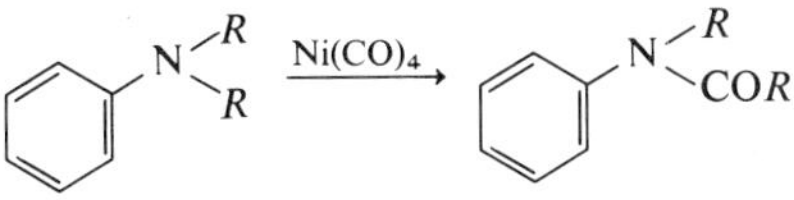

$$(165)$$

The latter reaction has only been studied with N,N-diethylaniline and N,N-dimethyl-β-naphthylamine. Anilides may be prepared in high yield by the reduction of nitrobenzene with carbon monoxide in a carboxylic acid solvent catalyzed by $Fe(CO)_5$, $Co_2(CO)_8$, or $Ni(CO)_4$ (189). The suggested mechanism involves the initial formation of a nitrene which is then carbonylated to produce phenyl isocyanate.

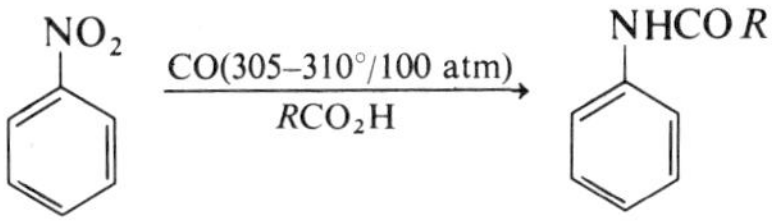

$$(166)$$

$R = CH_3$, Et. Diphenylureas may be obtained in low yield by treatment of nitrobenzene, azobenzene, hydrazobenzene, and substituted azobenzenes with CO/H_2 (1 : 1) at 200 atm and 120–150° in the presence of catalytic amounts of $Co_2(CO)_8$ (190).

The mechanisms of the reaction of amines with carbon monoxide catalyzed by metal carbonyls are thought to involve intermediates containing the 31771-5-88a grouping. This suggestion is supported by the isolation of isocyanates from the stoichiometric carbonylation of primary amines in the presence of $PdCl_2$ under mild conditions (191); $PdCl_2$ is reduced to metallic palladium during the reaction.

$$RNH_2 \xrightarrow[\text{PdCl}_2]{\text{CO}(65\text{-}85°/1.4\text{ atm})} RNCO \tag{167}$$

$R = \text{Bu } 49\%$ and $R = \text{Ph } 68\%$.

Under more vigorous conditions, with metallic palladium as catalyst, n-decylamine yields 1,3-didecylurea and N,N-didecyloxamide (192).

$$RNH_2 \xrightarrow[\substack{\text{Pd} \\ \text{benzene}}]{\text{CO}(180°/100\text{ atm})} RNHCONHR + RNHCOCONHR + H_2 \tag{168}$$

It is only recently that the first intermediates in these reactions of amines with carbon monoxide have been isolated. Secondary amines have been found to react with $Co_2(CO)_8$ to yield N-dialkylcarbamoylcobalt tetracarbonyls which may be isolated as their triphenylphosphine derivatives (193).

$$R_2NH + Co_2(CO)_8 \longrightarrow R_2NCOCo(CO)_4 \xrightarrow{\text{PPh}_3} R_2NCOCo(CO)_3PPh_3$$

$$\tag{169}$$

$R_2NH = \text{di-}n\text{-propylamine and piperidine.}$

The formylation reactions may also be carried out stoichiometrically, e.g.,

$$Re(CO)_5Cl + 3NH_3 \longrightarrow [Re(CO_4)(NH_3)_2]Cl + HCONH_2 \tag{170}$$

$$Ru(CO)_2I_2 + 2NH_3 \xrightarrow{< -30°} Ru(CO)_2(NH_3)I_2$$

$$\Big\downarrow {}^{> -30°} \, NH_3$$

$$[Ru(NH_3)_n(HCONH_2)_{4-n}]I_2 \tag{171}$$

Alkali metal amides may be quantitatively converted into alkali-metal formamides by reaction with carbon monoxide at $-45°$ C (194).

$$MNH_2 \xrightarrow[\text{liq. NH}_3]{\text{CO}(-45°)} M(HCONH) \tag{172}$$

where $M = $ Na, K, Rb, Cs. These compounds are very hygroscopic and are hydrolyzed by water to formamide and by an excess of caustic alkali to formates.

B. Amides from Olefins and Halogenated Hydrocarbons

Olefins may be catalytically carbonylated in the presence of amines to yield amides. The isomeric product distribution of anilides obtained from propylene and aniline has been studied as part of a comparison with the hydroformylation reaction (*22*).

$$CH_3CH{=}CH_2 + PhNH_2$$

$$\xrightarrow[\text{CO}_2(\text{CO})_8]{\text{CO}(180°/150\ \text{atm})}$$

(173)

$$CH_3CH_2CONHPh + \underset{H_3C}{\overset{H_3C}{\diagdown}}CHCOHNPh$$

The stoichiometric reaction of benzyl chloride with aniline in the presence of carbon monoxide and $NaCo(CO)_4$ yields phenylacetanilide (*195*)

$$PhCH_2Cl + PhNH_2 \xrightarrow[\substack{NaCo(CO)_4 \\ THF}]{CO(35°)} \underset{47\%}{PhCH_2CONHPh} \qquad (174)$$

Similarly, amides may be obtained by the stoichiometric reaction of amines with acyl cobalt tetracarbonyls under mild conditions.

$$RCOCo(CO)_4 + 2R'NH_2 \longrightarrow RCONHR' + [R'NH_3]^+[Co(CO)_4]^- \quad (175)$$

Unsaturated amide derivatives may be obtained from an amino-carbonylation reaction of halogenated olefins with amines in the presence of $Ni(CO)_4$ (*151*).

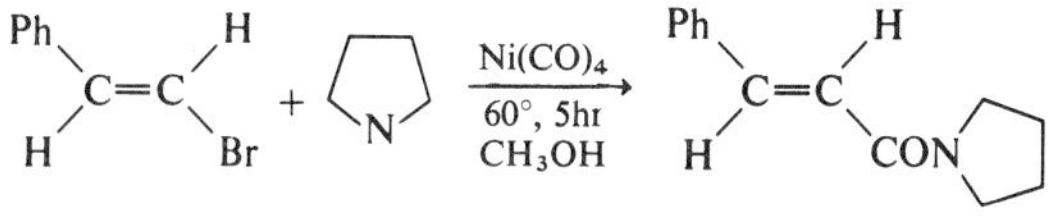

$$\underset{H}{\overset{Ph}{\diagdown}}C{=}C\underset{Br}{\overset{H}{\diagup}} + \left\langle\!\!{N}\right\rangle \xrightarrow[\substack{60°,\ 5hr \\ CH_3OH}]{Ni(CO)_4} \underset{H}{\overset{Ph}{\diagdown}}C{=}C\underset{CON}{\overset{H}{\diagup}}\!\!\left\langle\right\rangle \qquad (176)$$

$$82\%$$

The reaction offers a useful potential route to compounds of this type and occurs under very mild conditions.

A related reaction is the production of substituted amides from the treatment of isocyanides with CX_4 in the presence of $Co_2(CO)_8$ (*196*):

$$RNC + CX_4 + Co_2(CO)_8 \xrightarrow[\substack{benzene \\ H_2O}]{room\ temperature} RNHCOCHX_2 + Co(CNR)_4X_2 + CO$$

$$(177)$$

The proposed reaction mechanism involves a carbene intermediate.

C. IMIDES

Imides may be prepared from α,β-unsaturated carboxylic acid amides by reaction with carbon monoxide in the presence $Co_2(CO)_8$ as catalyst (20,155,184):

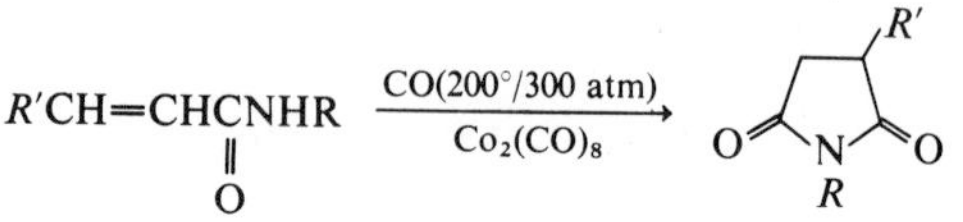

$$R'CH=CHCNHR \xrightarrow[\text{Co}_2\text{(CO)}_8]{\text{CO(200°/300 atm)}} \quad \text{[imide ring with } R' \text{ and } R\text{]} \qquad (178)$$
$$\underset{O}{\|}$$

$$\underset{O\ \ O}{\underset{\|\ \ \|}{PhCNHCPh}} \xrightarrow[\text{Ni(CO)}_4]{\text{CO(325°/200 atm)}} \quad \text{[phthalimide ring NH]} \qquad (179)$$

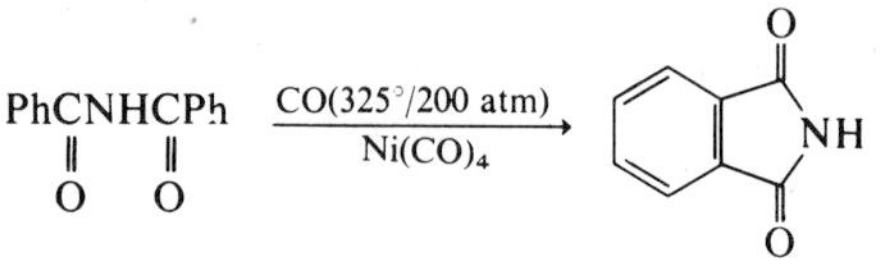

The reaction is very versatile and can be applied with various types of R and R' groups. Best yields (72–94%) are obtained when R = alkyl and R' = H and the product is an N-alkylsuccinimide. Six-membered ring systems may be prepared from β,γ-unsaturated carboxylic acid amides or the α,β-analogues which undergo isomerization prior to cyclization.

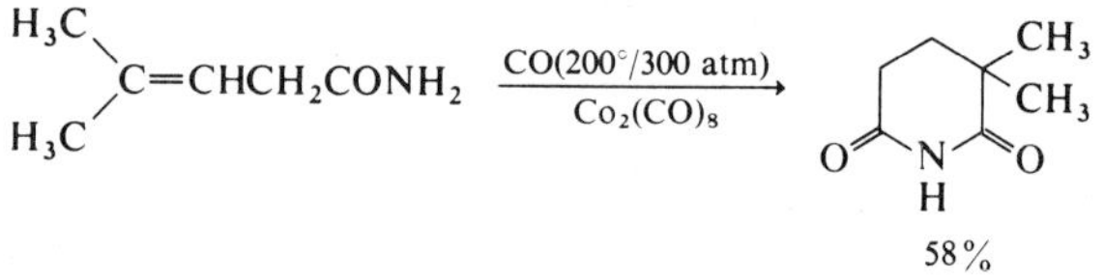

$$\underset{H_3C}{\overset{H_3C}{>}}C=CHCH_2CONH_2 \xrightarrow[\text{Co}_2\text{(CO)}_8]{\text{CO(200°/300 atm)}} \quad \text{[six-membered ring with CH}_3\text{, CH}_3\text{, N-H]} \qquad (180)$$
$$58\%$$

D. LACTAMS

The carbonylation of unsaturated amines affords lactams (20). Thus, N-alkylallylamines react with carbon monoxide in the presence of $Co_2(CO)_8$ to give N-alkyl-2-pyrrolidones in yields varying up to a maximum of 87% when R = H.

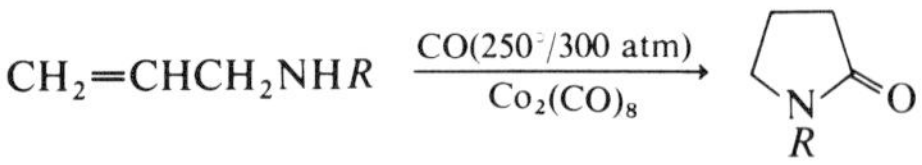

$$CH_2=CHCH_2NHR \xrightarrow[\text{Co}_2\text{(CO)}_8]{\text{CO(250°/300 atm)}} \quad \text{[pyrrolidone ring with N-}R\text{, =O]} \qquad (181)$$

Six-membered ring systems may also be formed, presumably by isomerization of the unsaturated amine; thus crotylamine yields 3-methyl-2-pyrrolidone and 2-piperidone.

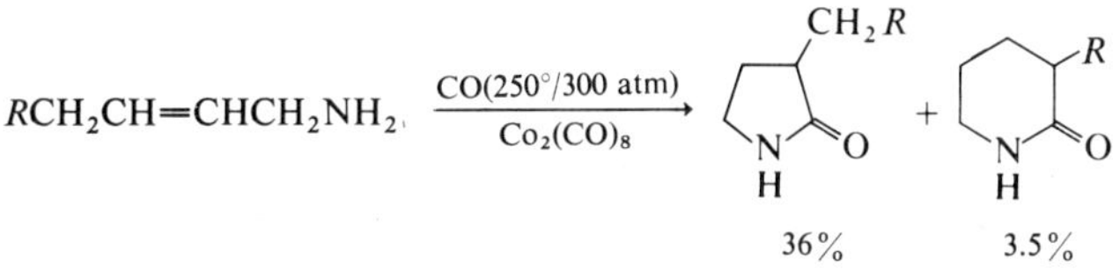

$$RCH_2CH{=}CHCH_2NH_2 \xrightarrow[\text{Co}_2(\text{CO})_8]{\text{CO}(250°/300 \text{ atm})}$$

36%   3.5%   (182)

As with the imide preparations, $Co_2(CO)_8$ shows the highest catalytic activity of the metal carbonyls tested. Thus rhodium carbonyl shows only 27% conversion of allylamine to 2-pyrrolidone and $Fe(CO)_5$ is inactive in this type of reaction.

E. PHTHALIMIDINES

Phthalimidines may be synthesized from a variety of starting compounds including Schiff bases, aromatic nitriles, aromatic ketoximes, phenyl-hydrazones, semicarbazones, and azines (20,155,184). The reactions and conditions are summarized in the following reaction sequences:

1. *Schiff Bases*

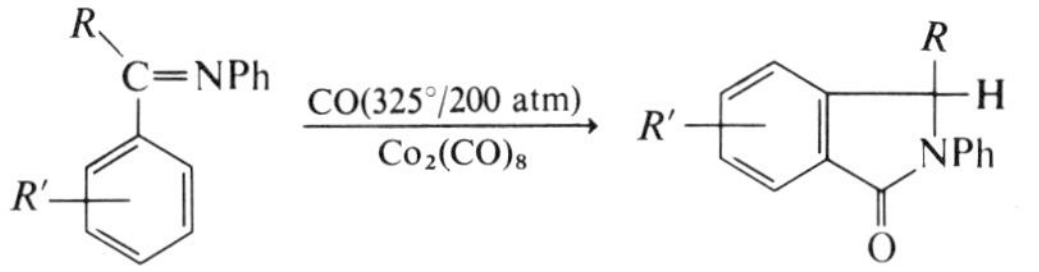

$$\xrightarrow[\text{Co}_2(\text{CO})_8]{\text{CO}(325°/200 \text{ atm})}$$

(183)

This reaction is very versatile and good yields can be obtained for a variety of R and R' substituents. This reaction was the first example of a $Co_2(CO)_8$ catalyzed reaction in which the CO became attached to an aromatic nucleus (for $R = R' = H$). $Fe(CO)_5$ is a less effective catalyst and $Ni(CO)_4$ is inactive. Schiff bases of aliphatic amines will also give phthalimidines.

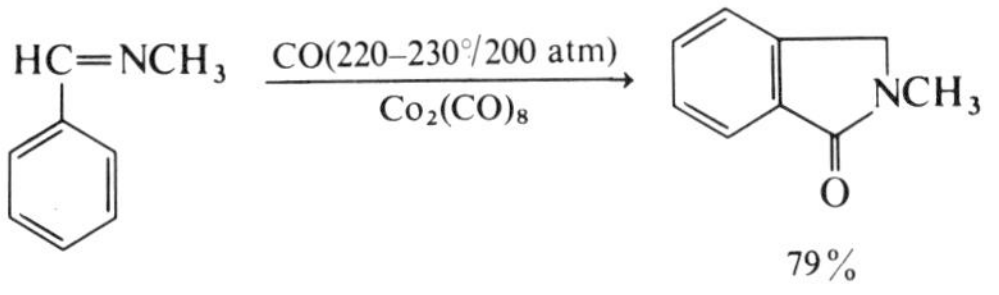

$$HC{=}NCH_3 \xrightarrow[\text{Co}_2(\text{CO})_8]{\text{CO}(220–230°/200 \text{ atm})}$$

79%   (184)

2. Aromatic Nitriles

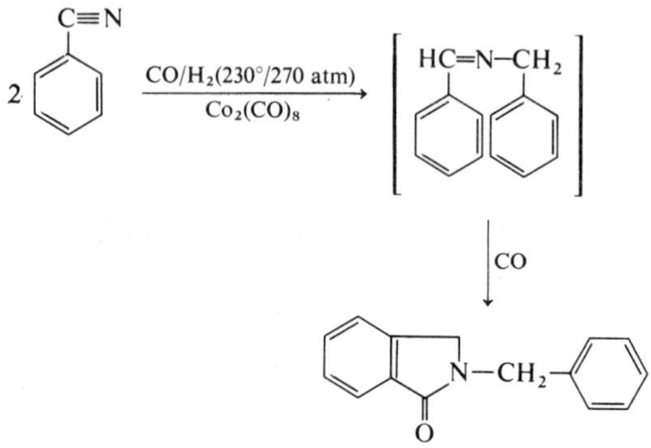

(185)

22%

This reaction proceeds in the presence of a small quantity of pyridine and under a pressure of carbon monoxide containing ca. 2% hydrogen. The mechanism is presumed to involve a Schiff base intermediate and is strictly analogous to the previous reaction.

3. Aromatic Ketoximes

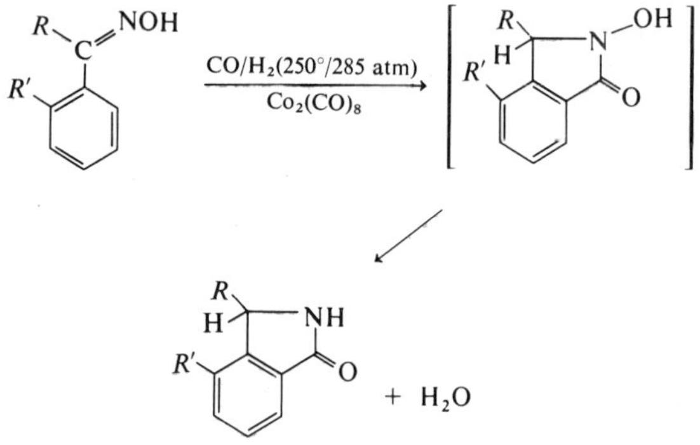

(186)

$+ H_2O$

80% $(R = Ph, R' = H)$

Aromatic ketoximes react with a mixture of carbon monoxide and hydrogen (98.5 : 1.5%) to give phthalimidines. In this reaction it is thought that the initial products are N-hydroxy-phthalimidines which are then hydrogenated. Diaryl ketoximes show greater selectivity than alkyl aryl ketox-

imes and in the latter reaction other by-products may be formed. The course
of the reaction is also dependent upon the amount of hydrogen present in
the carbon monoxide.

4. Phenylhydrazines of Aromatic Ketones

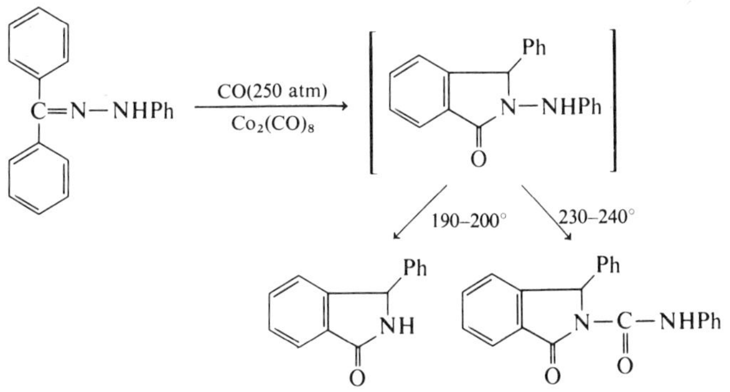

$$(187)$$

Hydrogenolysis of the N—N bond occurs at 190–200°C yielding the
phthalimidine but at higher temperatures insertion of CO into the N—N
bond occurs with the formation of N-(N′-phenylcarbamoyl)-phthalimi-
dine.

5. Semicarbazones of Aromatic Ketones

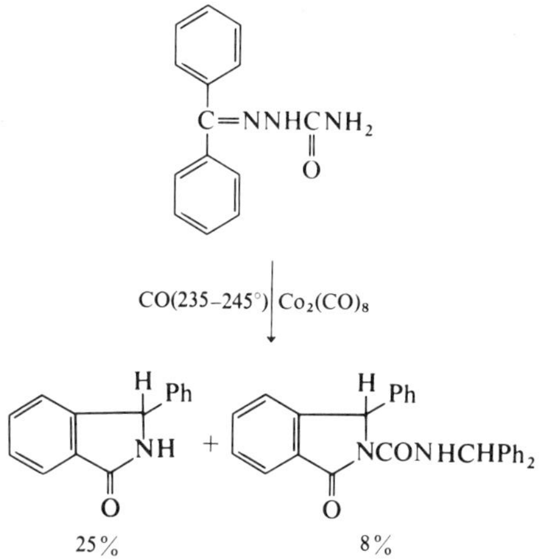

$$(188)$$

Here again the intermediate N-ureidophthalimidines cannot be isolated, presumably due to hydrogenolysis. Other by-products may also be formed in this reaction. In general, the N-phenylphthalimidines can be prepared in higher yield from the Schiff bases than from the aldehyde phenyl-hydrazones.

F. INDAZOLONES AND QUINAZOLINES

Aromatic azocompounds react with carbon monoxide to form inda-zolones at 180°–190°C and above this temperature insertion of CO into the N—N bond occurs with the formation of 2,4-dioxo-1,2,3,4,-tetrahydro-quinazolines (184).

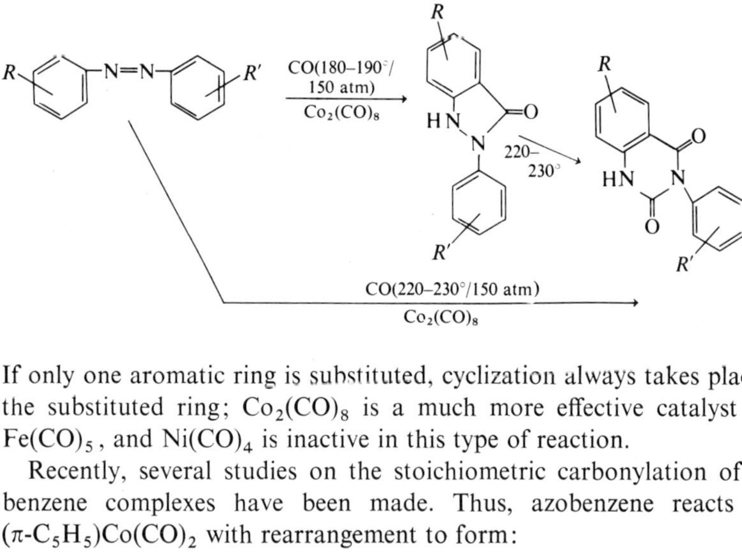

$$(189)$$

If only one aromatic ring is substituted, cyclization always takes place on the substituted ring; $Co_2(CO)_8$ is a much more effective catalyst than $Fe(CO)_5$, and $Ni(CO)_4$ is inactive in this type of reaction.

Recently, several studies on the stoichiometric carbonylation of azo-benzene complexes have been made. Thus, azobenzene reacts with $(\pi\text{-}C_5H_5)Co(CO)_2$ with rearrangement to form:

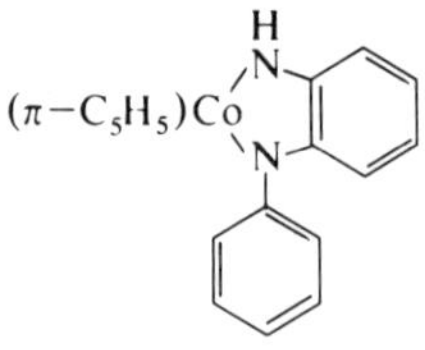

(16)

This complex may be carbonylated (200°, 100 atm) to give N-phenyl-benzimidazolone in 60 % yield (197).

Similarly, Tsuji and co-workers (123,198) have studied the carbonylation of azobenzene-palladium dichloride complexes in protic solvents (H_2O or EtOH). 2-Aryl-3-indazolines are produced in good yield. Degradative work on the carbonylated products suggests that unsymmetrically substituted azobenzenes react with $PdCl_2$ to form a Pd—C σ-bond preferentially with the benzene ring having an electron donating group, i.e. σ-bond formation is an electrophilic substitution on the benzene ring.

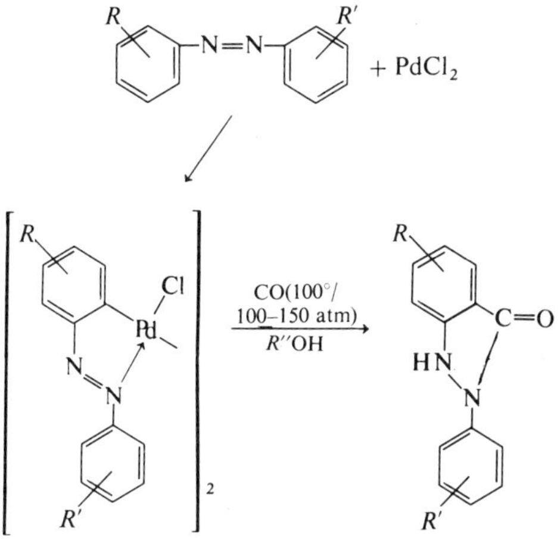

(190)

Recent work by Heck (199) has thrown some light upon the mechanism of these azobenzene carbonylation reactions. By ligand exchange reactions of palladium complexes, 2-(phenylazo)phenylmetal derivatives of cobalt, manganese, and rhenium have been prepared.

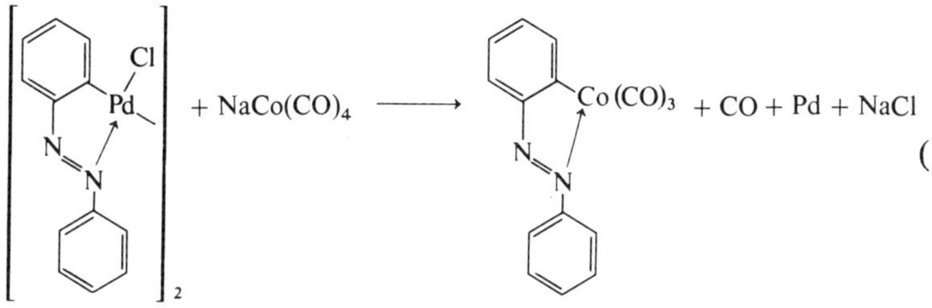

(191)

Carbonylation under mild conditions leads to the observation of the following sequence of reactions:

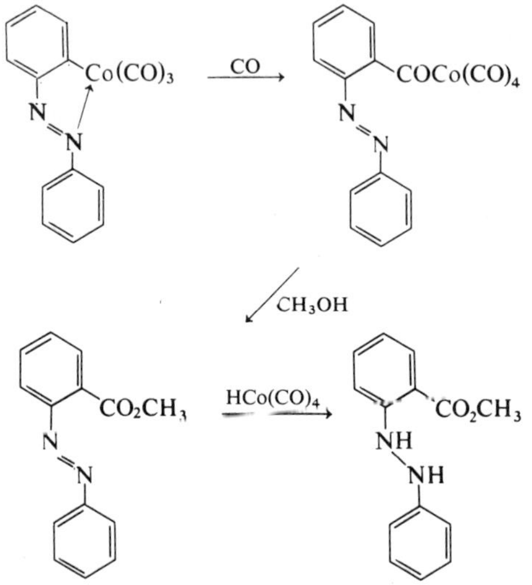

(192)

The final product is 2-carbomethoxyhydrazobenzene and under normal carbonylation conditions this would cyclize to produce the indazolone. These results support the postulate that the cobalt tricarbonyl complex is an intermediate in the $Co_2(CO)_8$ catalyzed carbonylation of azobenzene.

G. AMINO ACIDS

The palladium catalyzed carbonylation of tertiary allylamines offers a potential synthesis of amino acids under mild conditions (*111*). Thus, the complex formed between $PdCl_2$ and N,N-dimethylallylamine in methanol may be carbonylated in benzene solution at atmospheric pressure to give an ester of β-methoxy-γ-N, N-dimethylaminobutyric acid in 50% yield.

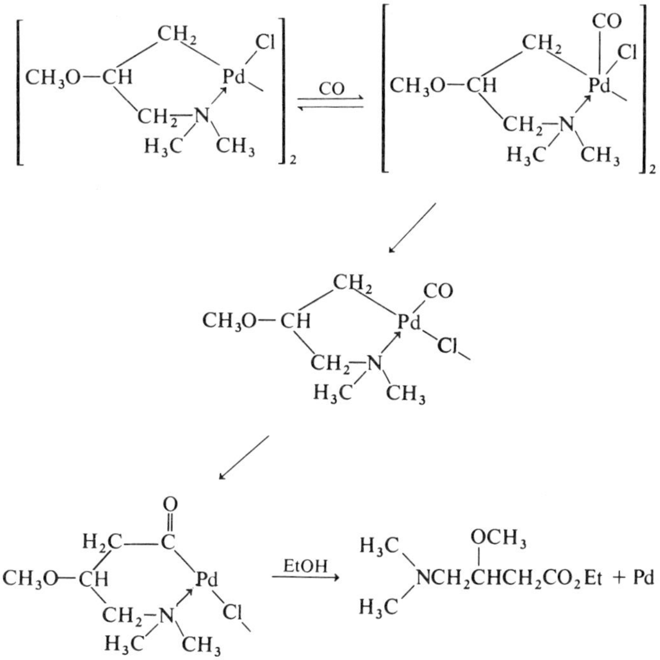

$$(193)$$

β-Cyanopropionaldehyde, obtained from the hydroformylation of acrylonitrile in Lewis base solvents, may be converted into glutamic acid by treatment with HCN in 20% NH_4OH followed by hydrolysis with sodium hydroxide (14).

$$CH_2=CHCN \xrightarrow[Co_2(CO)_8]{CO/H_2(120\text{-}130°/200\ atm)} OHCCH_2CH_2CN$$

$$\xrightarrow[\substack{ii.\ NaOH \\ iii.\ HCl}]{i.\ HCN\ in\ 20\%\ NH_4OH} HO_2C(CH_2)_2\underset{\underset{CO_2H}{|}}{CH}NH_2 \qquad (194)$$

XI. Hydrocarbons

Hydrocarbons may be formed as by-products in carbonylation reactions, and since their formation does not involve the net incorporation of carbon monoxide, only those reactions where there are mechanistic implications, e.g., decarbonylation, are discussed in any detail here.

Saturated hydrocarbons may be formed under hydroformylation conditions by reactions of the following type, which compete with the carbonyl insertion reaction:

$$RCo(CO)_4 + HCo(CO)_4 \longrightarrow RH + Co_2(CO)_8 \qquad (195)$$

This is illustrated by the hydroformylation of nuclear-substituted benzyl alcohols where electron withdrawing substituents favor hydrocarbon formation (200,201).

Probably the most relevant aspect of hydrocarbon formation to this chapter is that concerning decarbonylation, i.e., the reverse of the carbonylation reaction. Decarbonylation reactions can yield both saturated and unsaturated hydrocarbons. Thus the stoichiometric reaction of acid chlorides with a cobalt-nitrogen complex has recently been found to give the corresponding saturated hydrocarbon in 80% yield (202)

$$RCOCl \xrightarrow{(Ph_3P)_3CoN_2H \cdot Et_2O} \underset{80\% \quad 2-3\%}{RH + RR} \qquad (196)$$

$R = CH_3$, Ph, PhCH$_2$, PhOCH$_2$, and (CH$_3$)$_2$CH.

In general, however, decarbonylation reactions of acid chlorides and aldehydes yield olefins (102).

$$RCH_2CH_2COCl \xrightarrow{-CO} RCH_2CH_2Cl \xrightarrow{-HCl} RCH = CH_2 \qquad (197)$$

$$RCH_2CH_2CHO \longrightarrow RCH = CH_2 \qquad (198)$$

In the case of acyl chlorides containing no β-hydrogen atom then, as mentioned above (see Section IX,E), the final product is an alkyl chloride, since elimination of HCl to form the olefin is not possible. Saturated hydrocarbons are also formed in small amounts during the decarbonylation of aldehydes and this is presumed to arise from hydrogenation of the olefin.

Decarbonylation reactions of this type were initially carried out heterogeneously but recently the emphasis has been placed on homogeneous catalysts. The most widely studied homogeneous system is that containing RhCl(PPh$_3$)$_3$. The decarbonylation reaction becomes catalytic at 200°C (181) but most of the mechanistic studies have been carried out stoichiometrically under considerably milder conditions. Tsuji and Ohno (168,203) have proposed the following mechanism for the decarbonylation of acyl halides:

$$(Ph_3P)_3RhCl + RCH_2CH_2COCl \longrightarrow RCH_2CH_2CORhCl_2(PPh_3)_2$$

$$RCH_2CH_2Rh(CO)Cl_2(PPh_3)_2$$

$$RCH=CH_2 + HCl + Rh(CO)Cl(PPh_3)_2$$

$$(199)$$

In support of this mechanism the intermediate acyl complex has been isolated.

The kinetics of the decarbonylation of aldehydes is consistent with an analogous mechanism although here no intermediates could be detected (*200*).

$$(Ph_3P)_3RhCl + RCH_2CH_2CHO \longrightarrow RCH_2CH_2CORhHCl(PPh_3)_2$$

$$RCH_2CH_2Rh(CO)HCl(PPh_3)_2$$

$$RCH=CH_2 + H_2 + RhCl(CO)(PPh_3)_2$$

$$(200)$$

Although $RhCl(PPh_3)_3$ is by far the most extensively studied catalyst, other systems affording decarbonylation have been described. Thus aldehydes may be stoichiometrically decarbonylated to olefins in the presence of $[Ru_2Cl_3(Et_2PhP)_6]Cl$ (*67,68*). Hexanoic acid is converted into pent-2-ene, and butyraldehyde yields propene and a small amount of propane. The relative proportions of alkene and alkane are governed by the reaction temperature.

The palladium catalyzed decarbonylation of acyl chlorides and aldehydes to form olefins has been studied and related with the Rosenmund reduction (*46*) (see Section II,B); this is presumably a heterogeneous reaction.

Hydrocarbons may be formed under mild conditions by coupling reactions of allyl complexes or their precursors in the presence of $Ni(CO)_4$ (83).

$$ (201) $$

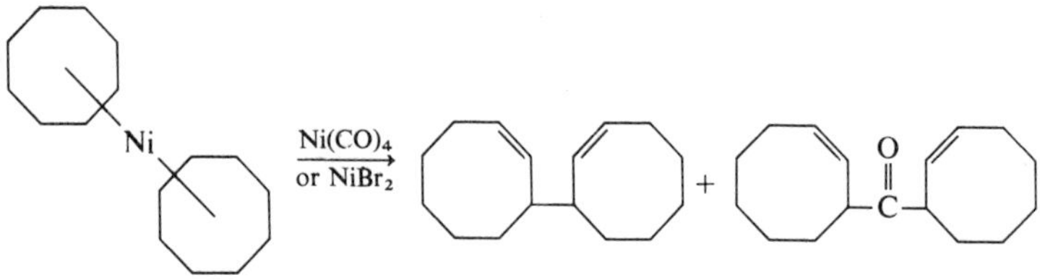

$$ (202) $$

In the latter reaction the two products are obtained in 1 : 1 mole ratio.

Olefin oxides and allylic chlorides react with carbon monoxide in the presence of potassium iron carbonylates [$KHFe(CO)_4$, $K_2Fe(CO)_4$, $K_2Fe_2(CO)_8$] under ambient conditions to produce olefins (72,145,205). The carbon monoxide is oxidized to carbon dioxide.

$$ R-HC-CH_2 + CO \longrightarrow RCH=CH_2 + CO_2 $$
$$ \underset{O}{\diagdown\diagup} $$

$$ (203) $$

1,2-Olefin oxides generally yield the terminal olefin with a small amount of isomerized product and internal olefin oxides do not react. Styrene oxide reacts to give ethylbenzene and α- and β-phenylethyl alcohols in addition to styrene (see Section III,D).

The cyclic trimerization of acetylenes is catalyzed by metal carbonyls and their derivatives and although this does not involve carbonylation in the usual sense it is included here for completeness (102,206). These reactions frequently afford substituted benzenes which are difficult to obtain by other synthetic methods. Thus, the $Co_2(CO)_8$ catalyzed trimerization of t-butylacetylene yields 1,2,4-tri-t-butylbenzene (207). Various cobalt carbonyl complexes have been isolated as intermediates in this reaction.

REFERENCES

1. Y. M. Y. Haddad, H. B. Henbest, J. Husbands, and T. R. B. Mitchell, *Proc. Chem. Soc.*, 361 (1964).
2. J. A. Osborn, F. H. Jardine, J. F. Young, and G. Wilkinson, *J. Chem. Soc. (A)*, 1711 (1966).
3. J. Smidt, *Chem. and Ind.*, 54 (1962).

4. J. F. Harrod and A. J. Chalk, *J. Amer. Chem. Soc.*, **86**, 1776 (1964).

5. G. Natta, *Science*, **147**, 261 (1965).

6. J. P. Candlin, K. A. Taylor, and D. T. Thompson, *Reactions of Transition Metal Complexes*, Elsevier, Amsterdam, 1968.

7. H. C. Brown, *Acc. Chem. Res.*, **2**, 65 (1969).

8. H. C. Brown and M. W. Rathke, *J. Amer. Chem. Soc.*, **89**, 4528 (1967).

9. N. S. Imjanitov and D. M. Rudkoskij, *J. Prakt. Chem.*, **311**, 712 (1969).

10. S. Uemura, R. Kitoh, K. Fujita, and K. Ichikawa, *Bull Chem. Soc. Japan*, **40**, 1499 (1967).

11. *Chem. Process. Eng.*, **50**, 9 (1969).

12. H. C. Brown, E. F. Knights, and R. A. Coleman, *J. Amer. Chem. Soc.*, **91**, 2144 (1969).

13. H. C. Brown and R. A. Coleman, *J. Amer. Chem. Soc.*, **91**, 4606 (1969).

14. A. J. Chalk and J. F. Harrod, *Adv. Organomet. Chem.*, **6**, 119 (1968).

15. F. Ungváry and L. Markó, *J. Organomet. Chem.*, **20**, 205 (1969).

16. Z. Nagy-Magos, G. Bor, and L. Markó, *J. Organomet. Chem.*, **14**, 205 (1968).

17. W. E. Fichteman and M. Orchin, *J. Org. Chem.*, **34**, 2790 (1969).

18. W. L. Fichteman and M. Orchin, *J. Org. Chem.*, **33**, 1281 (1968).

19. W. W. Spooncer, A. C. Jones, and L. H. Slaugh, *J. Organomet. Chem.*, **18**, 327 (1969).

20. J. Falbe, *Carbon Monoxide in Organic Synthesis*, Springer-Verlag, Berlin, 1970.

20a. R. Martin, *Chem. and Ind.*, 1536 (1954).

20b. R. Iwanaga, *Bull. Chem. Soc. Japan*, **35**, 778 (1962).

21. C. W. Bird, *Chem. Rev.*, **62**, 283 (1962).

22. P. Pino, F. Piacenti, M. Bianchi, and R. Lazzaroni, *Chim. Ind. (Milan)*, **50**, 106 (1968).

22a. F. Piacenti, M. Bianchi, E. Benedetti, and P. Frediani, *J. Organomet. Chem.*, **23**, 257 (1970).

22b. E. R. Tucci, *Ind. Eng. Chem. Prod. Res. Dev.*, **8**, 286 (1969).

23. R. L. Pruett and J. A. Smith, *J. Org. Chem.*, **34**, 327 (1969).

24. Y. Takegami, Y. Watanabe, H. Masada, and T. Mitsudo, *Bull. Chem. Soc. Japan*, **42**, 206 (1969).

24a. A. J. Moffat, *J. Catalysis*, **18**, 193 (1970).

25. L. S. Nahum, *J. Org. Chem.*, **33**, 3601 (1968).

26. F. Piacenti, S. Pucci, M. Bianchi, R. Lazzaroni, and P. Pino, *J. Amer. Chem. Soc.*, **90**, 6847 (1968).

27. F. Piacenti, M. Bianchi, and P. Pino, *J. Org. Chem.*, **33**, 3653 (1968).

28. G. Gut, M. H. El-Makhzangi, and A. Guyer, *Helv. Chim. Acta.*, **48**, 1151 (1965).

28a. Y. Ono, S. Sato, M. Takesada, and H. Wakamatsu, *Chem. Comm.*, 1255 (1970).

29. P. S. Hallman, D. Evans, J. A. Osborn, and G. Wilkinson, *Chem. Comm.*, 305 (1967).

30. J. A. Osborn, G. Wilkinson, and J. F. Young, *Chem. Comm.* 17 (1965).

31. P. Legzdins, G. L. Rempel, and G. Wilkinson, *Chem. Comm*, 825 (1969).

32. K. L. Olivier and F. B. Booth, *Amer. Chem. Soc. Preprints, Div. Petr. Chem.*, **14**, A7-A11 (1969).

33. N. S. Imyanitov and D. M. Rudkovskii, *Kin. and Cat.*, **8**, 1051 (1967).

34. D. Evans, J. A. Osborn, and G. Wilkinson, *J. Chem. Soc. (A)*, 3133 (1968).

35. B. Heil and L. Markó, *Chem. Ber.*, **102**, 2238 (1969).

36. M. Yamaguchi, *J. Chem. Soc. Japan Ind. Chem. Sect.*, **72**, 671 (1969).
37. G. Yagupsky, C. K. Brown, and G. Wilkinson, *Chem. Comm.*, 1244 (1969); *J. Chem. Soc. (A)*, 1392 (1970).
37a. K. L. Olivier and F. B. Booth, *Hydrocarbon Processing*, **49** (4), 112 (1970).
37b. J. H. Craddock, H. Hershman, F. E. Paulik, and J. F. Roth, *Ind. Eng. Chem. Prod. Res. Dev.*, **8**, 291 (1969).
37c. C. K. Brown and G. Wilkinson, *J. Chem. Soc. (A)*, 2753 (1970).
38. Y. Takegami, Y. Watanabe, and H. Masada, *Bull. Chem. Soc. Japan*, **40**, 1459 (1967).
38a. M. Takesada and H. Wakamatsu, *Bull. Chem. Soc. Japan*, **43**, 2192 (1970).
39. J. Tsuji and Y. Mori, *Bull. Chem. Soc. Japan*, **42**, 527 (1969).
40. Y. Takegami, Y. Watanabe, H. Masada, and I. Kanaya, *Bull. Chem. Soc. Japan*, **40**, 1456 (1967).
41. W. T. Hendrix, F. G. Cowherd, and J. L. von Rosenberg, *Chem. Comm.*, 97 (1968).
42. R. Damico and T. J. Logan, *J. Org. Chem.*, **32**, 2356 (1967).
43. D. Evans, J. A. Osborn, F. H. Jardine, and G. Wilkinson, *Nature*, **208**, 1203 (1965).
44. G. Braca, G. Sbrana, and P. Pino, *Chim. Ind. (Milan)*, **50**, 121 (1968).
45. J. Tsuji, N. Iwamoto, and M. Morikawa, *Bull. Chem. Soc. Japan*, **38**, 2213 (1965).
46. J. Tsuji and K. Ohno, *J. Amer. Chem. Soc.*, **90**, 94, 99 (1968).
47. J. Tsuji, K. Ohno, and T. Kujimoto, *Tetrahedron Letters*, 4565 (1965).
48. F. H. Jardine, J. A. Osborn, G. Wilkinson, and J. F. Young, *Chem. and Ind.* 560 (1965).
49. C. W. Bird, E. M. Briggs, and J. Hudec, *J. Chem. Soc. (C)*, 1862 (1967).
50. H. Greenfield, J. H. Wotiz, and I. Wender, *J. Org. Chem.*, **22**, 542 (1957).
51. H. Adkins and J. L. R. Williams, *J. Org. Chem.*, **17**, 980 (1952).
52. B. Fell and W. Rupelius, *Tetrahedron Letters*, 2721 (1969).
53. N. S. Imyanitov and D. M. Rudkovskii, *Kin. and Cat.*, **9**, 859 (1968).
54. H. Wakamatsu and K. Sakamaki, *Chem. Comm.*, 1140 (1967).
55. M. W. Rathke and H. C. Brown, *J. Amer. Chem. Soc.*, **89**, 2740 (1967).
56. E. F. Knights and H. C. Brown, *J. Amer. Chem. Soc.*, **90**, 5283 (1968).
57. H. C. Brown, M. M. Rogic, M. W. Rathke, and G. W. Kabalka, *J. Amer. Chem. Soc.*, **91**, 2150 (1969).
58. H. C. Brown and M. W. Rathke, *J. Amer. Chem. Soc.*, **89**, 2737, 2738 (1967).
59. H. C. Brown and E. Negishi, *J. Amer. Chem. Soc.*, **89**, 5478 (1967).
60. H. W. Sternberg, R. Markby, and I. Wender, *J. Amer. Chem. Soc.*, **78**, 5704 (1956).
61. L. H. Slaugh and R. D. Mullineaux, *J. Organomet. Chem.*, **13**, 469 (1968).
62. A. Pregaglia, A. Andreeta, G. Ferrari, and R. Ugo, *Chem. Comm.*, 590 (1969).
63. A. Hershman and J. H. Craddock, *Ind. Eng. Chem. Product Res. and Dev.*, **7**, 226 (1968).
64. M. Bianchi, E. Benedetti, and F. Piacenti, *Chim. Ind. (Milan)*, **51**, 613 (1969).
65. W. Kniese, J. H. Nienburg, and R. Fischer, *J. Organomet. Chem.*, **17**, 133 (1969).
66. F. Asinger, B. Fell, and W. Rupilius, *Ind. Eng. Chem. Product Res and Dev.*, **8**, 214 (1969).
67. R. H. Prince and K. A. Raspin, *J. Chem. Soc. (A)*, 612 (1969).
68. R. H. Prince and K. A. Raspin, *Chem. Comm.*, 156 (1966).
69. A. Misono, Y. Uchida, M. Hidai, and T. Kuse, *Chem. Comm.*, 981 (1968).
70. F. H. Jardine and G. Wilkinson, *J. Chem. Soc. (C)*, 270 (1967).

71. R. S. Coffey, *Chem. Comm.*, 923 (1967).
72. Y. Takegami, Y. Watanabe, T. Titsudo, I. Kanaya, and H. Masada, *Bull. Chem. Soc. Japan*, **41**, 158 (1968).
73. H. C. Brown, G. W. Kabalka, and M. W. Rathke, *J. Amer. Chem. Soc.*, **89**, 4530 (1967).
74. H. C. Brown and E. Negishi, *J. Amer. Chem. Soc.*, **89**, 5285 (1967).
75. H. C. Brown and E. Negishi, *J. Amer. Chem. Soc.*, **89**, 5477 (1967).
76. H. C. Brown and E. Negishi, *Chem. Comm.*, 594 (1968).
77. P. T. Lansbury and R. W. Meschke, *J. Org. Chem.*, **24**, 104 (1959).
78. J. A. Bertrand, C. L. Aldridge, S. Husebye, and H. B. Jonassen, *J. Org. Chem.*, **29**, 790 (1964).
79. R. F. Heck, *J. Amer. Chem. Soc.*, **85**, 3116 (1963).
80. J. H. Staib, W. R. F. Guyer, and O. C. Slotterbeck, U.S. Patent 2,864,864 (1958).
81. R. F. Heck, *Adv. Organomet. Chem.*, **4**, 243 (1966).
82. R. F. Heck, *J. Amer. Chem. Soc.*, **90**, 5518 (1968).
83. E. J. Corey, M. F. Semmelhack, and L. S. Hegedus, *J. Amer. Chem. Soc.*, **90**, 2416 (1968).
84. C. W. Bird and J. Hudec, *Chem. and Ind.*, 570 (1959).
85. W. Reppe and H. Vetter, *Annalen*, **582**, 133 (1953).
86. G. P. Mueller and F. L. MacArtor, *J. Amer. Chem. Soc.*, **76**, 4621 (1954).
87. F. M. Chaudhari, G. R. Knox, and P. L. Pauson, *J. Chem. Soc. (C)*, 2255 (1967).
88. R. F. Heck, *Organic Syntheses via Metal Carbonyls* (I. Wender and P. Pino, eds.), Interscience, New York, Vol. 1, p. 373 (1968).
89. S. Brewis and P. R. Hughes, *Chem. Comm.*, 71 (1967).
90. S. Brewis and P. R. Hughes, *Chem. Comm.*, 489 (1965).
91. S. Brewis and P. R. Hughes, *Chem. Comm.*, 6 (1966).
92. R. W. Goetz and M. Orchin, *J. Amer. Chem. Soc.*, **85**, 1549 (1963).
93. J. Blum and Z. Lipshes, *Progress in Co-ordination Chemistry*, Elsevier, Amsterdam, 1968, A17.
94. Y. Takahashi, S. Sakai, and Y. Ishii, *Chem. Comm.*, 1092 (1967).
95. H. Masai, K. Sonogashira, and N. Hagihara, *Bull. Chem. Soc., Japan*, **41**, 750 (1968).
96. D. Seyferth and R. T. Spohn, *J. Amer. Chem. Soc.*, **90**, 540 (1968).
97. D. Seyferth and R. T. Spohn, *J. Amer. Chem. Soc.*, **91**, 3037 (1969).
98. Y. Sawa, I. Hashimoto, M. Ryang, and S. Tsutsumi, *J. Org. Chem.*, **33**, 2159 (1968).
99. E. J. Corey and L. S. Hegedus, *J. Amer. Chem. Soc.*, **91**, 4926 (1969).
100. E. Yoshisato and S. Tsutsumi, *J. Org. Chem.*, **33**, 869 (1968).
101. G. P. Chiusoli and L. Cassar, *Angew. Chem. Internat. Edit.*, **6**, 124 (1967).
102. C. W. Bird, *Transition Metal Intermediates in Organic Synthesis*, Logos Press, London, 1967.
103. Ya. T. Éidus, S. D. Pirozhkov, and K. V. Puzitskii, *J. Org. Chem. USSR*, **4**, 369 (1968).
103a. Y.-P. Yang, K. V. Puzitskii, and Ya. T. Eidus, *Izv. Akad. Nauk. SSSR Ser. Khim.* 424 (1970).
103b. H. Hogeveen, F. Baardman, and C. F. Roobeek, *Rec. Trav. Chim.*, **89**, 227 (1970).
103c. H. Hogeveen and C. J. Gaasbeek, *Rec. Trav. Chim.*, **89**, 395 (1970).
104. T. Mizoroki and M. Nakayama, *Bull Chem. Soc. Japan*, **38**, 1876 (1965).
105. T. Mizoroki and M. Nakayama, *Bull. Chem. Soc. Japan*, **39**, 1477 (1966).

106. T. Mizoroki and M. Nakayama, *Bull. Chem. Soc. Japan*, **41**, 1628 (1968).

107. F. E. Paulik and J. F. Roth, *Chem. Comm.*, 1578 (1968).

108. R. Paatz and G. Weisgerber, *Chem. Ber.*, **100**, 984 (1967).

109. H. Hogeveen, J. Lukas, and C. F. Roobeek, *Chem. Comm.*, 920 (1969).

110. C. W. Bird and E. M. Briggs *J. Chem. Soc. (C)*, 1265 (1967).

111. D. Medema, R. van Helden, and C. F. Kohll, *Inorg. Chim. Acta*, **3**, 255 (1969).

112. D. M. Fenton, K. L. Oliver, and G. Biale, *Amer. Chem. Soc., Prepr. Div. Pet. Chem.* **14**, C77 (1969).

113. L. Cassar and G. P. Chiusoli, *Chim. Ind. (Milan)*, **48**, 323 (1966).

114. E. R. H. Jones, G. H. Whitham, and M. C. Whiting, *J. Chem. Soc.*, 4628 (1957).

115. T. Mizoroki and M. Nakayama, *Bull. Chem. Soc. Japan*, **40**, 2203 (1967).

116. M. Nakayama and T. Mizoroki, *Bull. Chem. Soc. Japan*, **42**, 1124 (1969).

116a. M. Nakayama and T. Mizoroki, *Bull. Chem. Soc. Japan*, **43**, 569 (1970).

117. G. N. Schrauzer, *Chem. Ber.*, **94**, 1891 (1961).

118. J. C. Clark and R. C. Cookson, *J. Chem. Soc.*, 686 (1962).

119. Y. Yashuhara and T. Nogi, *Chem. and Ind.*, 229 (1967).

120. Y. Yashuhara, T. Nogi, and H. Saisho, *Bull. Chem. Soc. Japan*, **42**, 2070 (1969).

121. Y. Mori and J. Tsuji, *Bull. Chem. Soc. Japan*, **42**, 777 (1969).

122. G. P. Chiusoli and A. Cameroni, *Chim. Ind. (Milan)*, **46**, 1063 (1964).

123. J. Tsuji, *Acc. Chem. Research*, **2**, 144 (1969); *Adv. in Org. Chem. Methods and Results*, Vol. 6, Ed. E. C. Taylor and H. Wynberg, Interscience, New York, 1969.

124. J. Tsuji, J. Kiji, S. Imamura, and M. Morikawa, *J. Amer. Chem. Soc.*, **86**, 4350 (1964).

125. T. Nogi and J. Tsuji, *Tetrahedron*, 4099 (1969).

126. H. C. Brown, M. M. Rogic, M. W. Rathke, and G. W. Kabalka, *J. Amer. Chem. Soc.*, **90**, 818 (1968).

127. H. C. Brown and M. M. Rogic, *J. Amer. Chem. Soc.*, **91**, 2146 (1969).

128. A. Matsuda, *Bull. Chem. Soc. Japan*, **40**, 135 (1967).

129. A. Matsuda, *Bull. Chem. Soc. Japan*, **42**, 571 (1969).

130. A. Matsuda, *Bull. Chem. Soc. Japan*, **42**, 2596 (1969).

131. G. P. Chiusoli and S. Merzoni, *Chim. & Ind. (Milan)*, **51**, 612 (1969).

132. J. Tsuji, M. Morikawa, and J. Kiji, *Tetrahedron Letters*, 1437 (1963).

133. K. Bittler, N. v. Kutepow, D. Neubauer, and H. Ries, *Angew. Chem. Internat. Edit.*, **7**, 329 (1968).

133a. L. J. Kehoe and R. A. Schell, *J. Org. Chem.*, **35**, 2846 (1970).

133b. Y. Iwashita, F. Tamura, and H. Wakamatsu, *Bull. Chem. Soc. Japan*, **43**, 1520 (1970).

133c. K. Tominaga, N. Yamagami, and H. Wakamatsu, *Tetrahedron Letters*, 2217 (1970).

134. G. P. Chiusoli, C. Venturello, and S. Merzoni, *Chem. and Ind.*, 977 (1968).

135. S. Kunichika, Y. Sakakibara, and T. Nakamura, *Bull. Chem. Soc. Japan*, **41**, 390 (1968).

136. J. Tsuji and T. Nogi, *J. Org. Chem.*, **31**, 2641 (1966).

136a. I. Rhee, M. Ryang, and S. Tsutsumi, *Tetrahedron Letters*, 4593 (1969).

137. S. Kunichika, Y. Sakakibara, and T. Okamoto, *Bull. Chem. Soc. Japan*, **40**, 885 (1967).

138. J. Tsuji and T. Susuki, *Tetrahedron Letters*, 3027 (1965).

139. T. Susuki and J. Tsuji, *Bull. Chem. Soc. Japan*, **41**, 1954 (1968).

140. T. J. Kealy and R. E. Benson, *J. Org. Chem.*, **26**, 3126 (1961).
141. E. L. Jenner and R. V. Lindsey, *U.S. Patent*, 2,876,254 (1959).
142. S. Brewis and P. R. Hughes, *Chem. Comm.*, 157 (1965).
142a. J. Tsuji, S. Hosaka, J. Kiji, and T. Susuki, *Bull. Chem. Soc. Japan*, **39**, 141 (1966).
143. L. Markó and P. Szabo, *Chem. Tech. (Berlin)*, **13**, 482 (1961).
144. J. Eisenmann, R. L. Yamartino, and J. F. Howard, *J. Org. Chem.*, **26**, 2102 (1961).
145. Y. Takegami, Y. Watanabe, T. Mitsudo, and H. Masada, *Bull. Chem. Soc. Japan*, **42**, 202 (1969).
146. J. D. McClure, *J. Org. Chem.*, **32**, 3888 (1967).
147. F. Guerrieri and G. P. Chiusoli, *Chem. Comm.*, 781 (1967).
148. G. P. Chiusoli, D. Dubini, M. Ferraris, S. Merzoni, and G. Mondelli, *J. Chem. Soc. (C)*, 2889 (1968).
149. F. Guerrieri and G. P. Chiusoli, *J. Organomet. Chem.*, **15**, 209 (1968).
150. J. B. Mettalia and E. H. Specht, *J. Org. Chem.*, **32**, 3941 (1967).
151. E. J. Corey and L. S. Hegedus, *J. Amer. Chem. Soc.*, **91**, 1233 (1969).
152. J. Tsuji and S. Imamura, *Bull. Chem. Soc. Japan*, **40**, 197 (1967).
153. F. Piacenti, C. Cioni, and P. Pino, *Chem. and Ind.*, 1240 (1960).
154. J. M. Davidson, *J. Chem. Soc. (A)*, 193 (1969).
155. J. Falbe, *Angew. Chem. Internat. Edit.*, **5**, 435 (1966).
156. A. Matsuda, *Bull. Chem. Soc. Japan*, **41**, 1876 (1968).
157. L. Cassar, G. P. Chiusoli, and M. Foa, *Chim. Ind. (Milan)*, **50**, 515 (1968).
158. M. Foa, L. Cassar, and M. Tacchi Venturi, *Tetrahedron Letters*, 1357 (1968).
159. L. Cassar and M. Foa, *Chim. Ind. (Milan)*, **51**, 673 (1969).
160. E. Yoshisato, M. Ryang, and S. Tsutsumi, *J. Org. Chem.*, **34**, 1500 (1969).
161. G. Albanesi, *Chim. Ind. (Milan)*, **46**, 1169 (1964).
162. J. Tsuji and T. Nogi, *J. Amer. Chem. Soc.*, **88**, 1289 (1966).
163. R. F. Heck, *J. Amer. Chem. Soc.*, **85**, 1460 (1963).
164. W. Hubel, *Organic Syntheses via Metal Carbonyls*, (I. Wender and P. Pino, eds.), Interscience, New York, 1968, Vol. 1, p. 273.
165. H. W. Sternberg, R. Markby, and I. Wender, *J. Amer. Chem. Soc.*, **80**, 1009 (1958).
166. W. Reppe, N. v. Kutepow, and A. Magin, *Angew. Chem. Internat. Edit.*, **8**, 727 (1969).
166a. *Canadian Patents*, 847924, 848479.
167. P. Pino, G. Braca, C. Sbrana, A. Cuccuru, *Chem. and Ind.*, 1732 (1968).
168. J. Tsuji and K. Ohno, Homogeneous Catalysis (Industrial Applications and Implications) *Adv. in Chem. Ser.*, **70**, Ed. R. F. Gould, *Amer. Chem. Soc. Washington D.C.*, 155 (1968); *Synthesis*, 157 (1969).
169. J. Tsuji, M. Morikawa, and J. Kiji, *Tetrahedron Letters*, 1061 (1963).
170. J. Tsuji, M. Morikawa, and J. Kiji, *J. Amer. Chem. Soc.*, **86**, 4851 (1964).
170a. J. K. Stille and L. F. Hines, *J. Amer. Chem. Soc.*, **92**, 1798 (1970).
171. T. Susuki and J. Tsuji, *Tetrahedron Letters*, 913 (1968); *J. Org. Chem.* **35**, 2982 (1970).
172. J. Tsuji, M. Morikawa, and N. Iwamoto, *J. Amer. Chem. Soc.*, **86**, 2095 (1964).
173. J. Tsuji, J. Kiji, M. Morikawa, and S. Imamura, *J. Amer. Chem. Soc.*, **86**, 4350 (1964).
174. W. T. Dent, R. Long, and G. H. Whitfield, *J. Chem. Soc.*, 1588 (1964).
174a. H. C. Volger, K. Vrieze, J. W. F. M. Lemmers, A. P. Praat, and P. W. N. M. van Leeuwen, *Inorg. Chim. Acta.*, **4**, 435 (1970).

175. J. Tsuji, J. Kiji, and S. Hosaka, *Tetrahedron Letters*, 605 (1964).
176. R. F. Heck, *J. Amer. Chem. Soc.*, **85**, 2013 (1963).
177. W. A. Thaler, *J. Amer. Chem. Soc.*, **89**, 1902 (1967).
178. W. A. Thaler, *J. Amer. Chem. Soc.*, **90**, 4370 (1968).
179. W. A. Thaler, *Chem. Comm.*, 527 (1968).
180. J. Tsuji and K. Ohno, *J. Amer. Chem. Soc.*, **88**, 3452 (1966).
181. J. Blum, *Tetrahedron Letters*, 1605 (1966).
182. J. Blum, E. Oppenheimer, and E. D. Bargmann, *J. Amer. Chem. Soc.*, **89**, 2338 (1967).
183. C. Lassau, Y. Chauvin, and G. Le Febvre, *Progress in Coordination Chemistry*, Elsevier, Amsterdam, 1968, A16.
184. A. Rosenthal and I. Wender, *Organic Syntheses via Metal Carbonyls* (eds. I. Wender and P. Pino), Interscience, New York, 1968, Vol. 1, p. 405.
185. D. Durand and C. Lassau, *Tetrahedron Letters*, 2329 (1969).
186. F. Calderazzo, *Inorg. Chem.*, 4, 293 (1965).
187. T. Saegusa, S. Kobayashi, K. Hirota, and Y. Ito, *Bull. Chem. Soc.*, *Japan*, **42**, 2610 (1969).
188. T. Saegusa, S. Kobayashi, K. Hirota and Y. Ho, *Tetrahedron Letters*, 6125 (1966).
189. T. Kajimoto and J. Tsuji ,*Bull. Chem. Soc. Japan*, **42**, 827 (1969).
190. S. Murahashi and S. Horiie, *Bull. Chem. Soc. Japan*, **33**, 78 (1960).
191. E. W. Stern and M. Spector, *J. Org. Chem.*, **31**, 596 (1966).
192. T. Tsuji and N. Iwamoto, *Chem. Comm.*, 380 (1966).
193. J. Palágyi and L. Markó, *J. Organomet. Chem.*, **17**, 453 (1969).
194. R. Nast and P. Dilly, *Angew. Chem. Internatl. Edit.*, **6**, 357 (1967).
195. R. F. Heck and D. S. Breslow, *J. Amer. Chem. Soc.* **85**, 2779 (1963).
196. Y. Yamamoto and N. Hagihara, *Bull. Chem. Soc. Japan*, **42**, 2077 (1969).
197. T. Joh, N. Hagihara, and S. Murahashi, *Bull. Chem. Soc. Japan*, **40**, 661 (1967).
198. H. Takahashi and J. Tsuji, *J. Organometal. Chem.*, **10**, 511 (1967).
199. R. F. Heck, *J. Amer. Chem. Soc.*, **90**, 313 (1968).
200. I. Wender, H. Greenfield, S. Metlin, and M. Orchin, *J. Amer. Chem. Soc.*, **74**, 4079 (1952).
201. Y. C. Fu, H. Greenfield, S. J. Metlin, and I. Wender, *J. Org. Chem.*, **32**, 2837 (1967).
202. S. Tyrliek and H. Stepowska, *Tetrahedron Letters*, 3593 (1969).
203. K. Ohno and J. Tsuji, *J. Amer. Chem. Soc.*, **90**, 99 (1968).
204. M. C. Baird, C. J. Nyman, and G. Wilkinson, *J. Chem. Soc.(A)*, 348 (1968).
205. Y. Takegami, Y. Watanabe, T. Mitsudo, and T. Okajima, *Bull. Chem. Soc. Japan*, **42**, 1992 (1969).
206. C. Hoogzand and W. Hübel, *Organic Syntheses via Metal Carbonyls* (I. Wender and P. Pino, eds.), Interscience, New York, 1968, Vol. 1, p. 343.
207. V. Krüerke, C. Hoogzand, and W. Hübel, *Chem. Ber.*, **94**, 2817 (1961).

6

Catalysis of Symmetry Forbidden Reactions

FRANK D. MANGO

and

J. H. SCHACHTSCHNEIDER

Shell Development Company
Emeryville, California

I. Introduction

Molecules tend to undergo transformation from one bonding configuration to another in an ordered fashion and simple molecular orbital theory provides a surprisingly clear picture of this process. This is due, to a large extent, to the critical role orbital symmetry plays in bond transformation. The ordered transformation of a molecular system can be approximated

223

by considering the transformation of the molecular orbitals corresponding to the bonds undergoing change. The symmetries of the molecular orbitals will tend to be preserved throughout the transformation. Orbitals in the starting system will thus correlate with orbitals in the product system of the same symmetry. When all bonding orbitals of one system correlate with bonding orbitals of the product system, orbital symmetry restrictions to reaction are minimal. When bonding orbitals correlate with antibonding orbitals, however, symmetry restraints to molecular transformation exist and reaction along that path experiences an energy barrier reflecting the energy of the indicated excited state.

The significance of orbital symmetry controls in reaction chemistry is best illustrated by the Woodward-Hoffmann rules (1). Essentially all concerted organic reactions can be divided into symmetry allowed and symmetry forbidden categories by the symmetry rules. Predictability is remarkable. When alternative symmetry allowed reaction paths exist, transformation along forbidden paths are essentially unobserved. Since the first rule was described in 1965 (2), a massive research effort has been directed toward testing the basic concept. The chemical information accumulated since then (3) provides powerful experimental evidence supporting the Woodward-Hoffmann theory. Indeed, there can be little doubt at this time that their basic ideas are correct and that the symmetry rules are sound.

The orbital symmetry treatments of Woodward and Hoffmann have been used by Eaton to similarly describe the transformation of inorganic metal complexes (4). This approach, however, deals with systems in which energy difference between ground states and excited states can be quite small. The transformation restraints due to orbital symmetry conservation are thus not comparable to those encountered in organic systems where these differences are substantial. Eaton's approach has been questioned by Whitesides (5), who offers a somewhat better description of inorganic transformations based on state correlations, an approach first introduced by Longuet-Higgins and Abrahamson (6).

Pearson has recently refocused attention on Baders' rule (7) for predicting molecular structure (8). Based on the second-order Jahn-Teller effect, the symmetry rules were able to predict the stable structures of molecules XY_n for $n = 2$–7. This approach also has predictive power regarding molecular transformations of transition metal complexes. The extension of Baders' symmetry rules into this area gives predictions which differ from those of Eaton (4) in some cases. Pearson's treatment has broad application in inorganic systems and again points attention to the power of simple orbital symmetry descriptions of transforming molecular systems.

We are concerned here with a different application of orbital symmetry, one involving both organic transformations and the rearrangements of ligand systems on the coordination spheres of transition metals. We address ourselves specifically to the reactions of transition metal coordinated-organic ligands which, in the absence of metals, would be symmetry forbidden. This process, "forbidden-to-allowed catalysis," requires special operations on the part of the transition metal system which can, conceivably, introduce additional restraints to reaction. A surprisingly detailed picture of this catalytic interaction can be given which essentially defines model systems for this kind of catalysis. Both of the above aspects of orbital symmetry are involved. The metal provides a ground-state, symmetry allowed path for the organic ligands through specific electronic interactions, and, in the process, can experience concomitant energy restraints due to negative interactions with the new ligand systems. Forbidden-to-allowed catalysis, as a transformation proceeding on the coordination sphere of a transition metal, differs from any other metal-coordinated transformation. In this chapter we will describe this hypothetical process, its contrast to similar metal-assisted organic transformations and discuss the broad body of catalytic chemistry believed to be a consequence of this special kind of catalysis. Preliminary notes on this topic have appeared. The forbidden-to-allowed concept was first introduced as it applied to [2 + 2] and [2 + 2 + 2 + 2] cycloaddition reactions (9). The same approach was then applied to symmetry forbidden electrocyclic reactions by two separate groups of workers, Volger and Hogeveen (10), and Merk and Pettit (11). Acetylene cycloaddition [2 + 2] to cyclobutadiene was next described, and in this case the possibility of significant orbital symmetry restraints attending the forbidden-to-allowed process was introduced (12). Metal-catalyzed cyclobutanation of dienes has also been discussed (13) and a brief review covering some of the above subjects and metal-assisted sigmatropic transformations has appeared (14).

II. The [2 + 2] Cycloadditions

A. ROLE OF THE TRANSITION METAL

The concerted, suprafacial fusion of two olefin π bonds to two cyclobutane σ bonds (**1** → **2**) is a symmetry forbidden process (15).

This symmetry restriction is general and holds for all bond transformations involving the spatial relocalization (i.e., suprafacial, suprafacial) of carbon–carbon bonds illustrated in Eq. (1) (16).

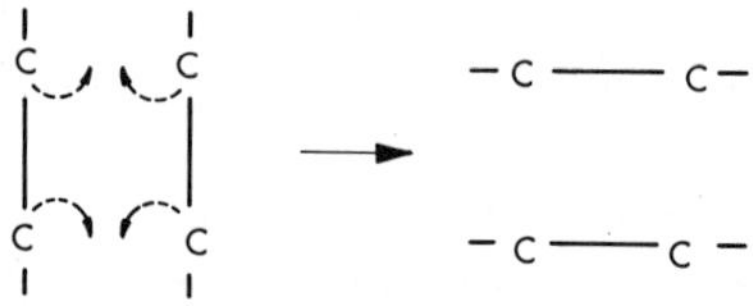

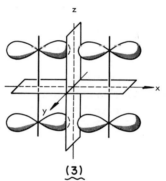

$$\tag{1}$$

For the fusion of two olefin π bonds, the symmetry restrictions are easily described by considering the π and π^* interactions of approaching olefins. In (3), two approaching olefins (the σ bond networks rest in parallel planes, i.e., with the ZY plane) interact along p orbitals sharing common axes.

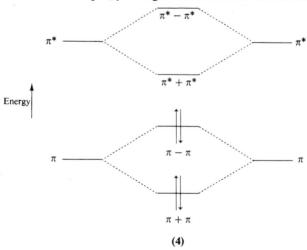

(3)

When the olefins approach within van der Waals distance, the energy levels of the π and π^* orbitals split, yielding a new molecular orbital description.

(4)

The electrons of the two π bonds remain in molecular orbitals constructed of π orbitals [i.e., the $(\pi + \pi)$ and $(\pi - \pi)$]. If the approaching olefins proceed along the reaction coordinate leading to a cyclobutane ring, the energy of the system will climb sharply upward since the symmetry of the $\pi - \pi$ combination correlates with that of a σ^* orbital in cyclobutane. The appropriate symmetry assignments are illustrated in the correlation diagram in Fig. 1. The symmetry symbols in Fig. 1 refer to orbital sign symmetries about the ZY and XY planes in (3). These are the elements of symmetry preserved by all transforming molecular orbitals across the reaction coordinate. The symbols signify symmetric (S) or antisymmetric (A) about

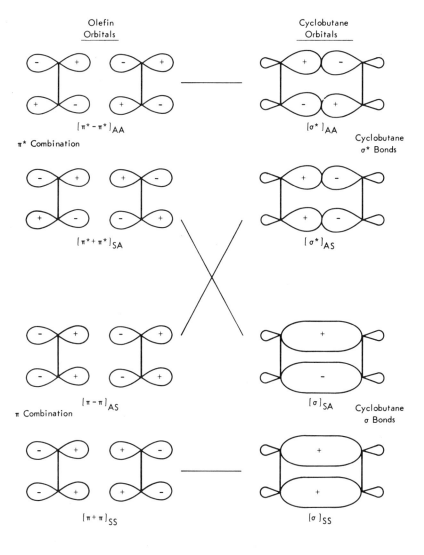

Fig 1. Correlation diagram for the cyclobutanation of two olefin π bonds.

the respective planes, e.g., the $[\pi - \pi]$ combination, assigned AS, is antisymmetric about ZY and symmetric about XY.

The symmetry restrictions to this reaction—and thus the symmetry forbidden assignment—stem from the fact that the ground state of the interacting olefin system $[(\pi + \pi)_{SS}^2, (\pi - \pi)_{AS}^2]$ correlates with a high energy excited state of cyclobutane, $[(\sigma)_{SS}^2, (\sigma^*)_{AS}^2]$. This essentially guarantees a high energy barrier to reaction and significantly restrains molecular systems in these bonding configurations from interconversion.

The symmetry restrictions just described can, in theory, be totally removed by introducing into the system the appropriate transition metal complex (9). The reactants, resting on the coordination sphere of the metal center, are thus free to transform, pass from one bonding configuration to the other $[(1) \rightleftarrows (2)]$, while remaining in the ground state. The metal, in this process, carries out two critical operations. It injects an electron pair into an antibonding ligand combination and removes an electron pair from a bonding combination using metal nonbonding atomic orbitals of the appropriate symmetry. For the interconversion $(1) \rightleftarrows (2)$, the critical ligand orbitals from Fig. 1 are

$$\uparrow\downarrow (\text{metal}) + [\pi^* + \pi^*]_{SA} \quad \rightleftarrows \quad [\sigma]_{SA}^2 + \quad (\text{metal}) \qquad (2)$$

$$(\text{metal}) + [\pi - \pi]_{AS}^2 \quad \rightleftarrows \quad [\sigma^*]_{AS} + \uparrow\downarrow (\text{metal}) \qquad (3)$$

The metal can use d orbitals and a pair of nonbonding d electrons for this purpose. The exchange of electron pairs between ligand system and metal system is illustrated, graphically, in Fig. 2 (electron density, denoted by shading, is localized in orbitals of origin and termination for conceptual clarity). This process—the exchange of electron pairs between metal and ligand system—is best envisaged as proceeding smoothly and essentially simultaneously across the reaction coordinate. For two olefin ligands occupying adjacent ligand position in the metal complex (5), the d_{YZ} metal atomic orbital has SA symmetry and the d_{ZX} AS symmetry.

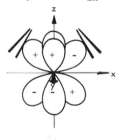

(5)

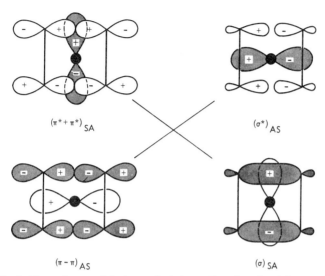

$(\pi^* + \pi^*)_{SA}$ $(\sigma^*)_{AS}$

$(\pi - \pi)_{AS}$ $(\sigma)_{SA}$

Fig. 2. The exchange of electron pairs between ligands and metal proceeding through orbitals of SA and AS symmetries.

The olefins have their p orbitals directed along axes running parallel to the ZX plane and normal to the Y axis. The ligand field provided by the olefin π orbitals in this configuration splits the two metal d orbitals in (**5**), the d_{ZX} rising in energy relative to d_{YZ}. If the ligand field provided by the two olefin ligands—and neglecting for the moment that provided by other metal ligands—is sufficiently strong, two metal valence electrons to be distributed between d_{YZ} and d_{ZX} will occupy d_{YZ} leaving d_{ZX} free of metal electrons. If these metal orbitals are essentially nonbonding in the metal complex (exclusive of the olefin ligands), i.e., they are not part of a bonding network, electron density is free to pass in and out of these orbitals. Cyclobutanation of the coordinated ligands, therefore, can proceed with electron density in d_{ZY} flowing from the metal to the ligand while olefin electron density flows into d_{ZX}. This process need not disrupt the metal system. Significant charge differential between metal and ligand systems would not be generated if the exchange process proceeds symmetrically. Summarizing, the olefin ligands impress a ligand field upon the metal which can order the metal valence electrons as indicated in Fig. 2, thereby providing the prerequisite distribution of electron density for a ground state, π bond fusion to a cyclobutane ring [i.e., (**6**) → (**7**)]. The cost to the metal system, in this simplified hypothetical case, is a reordering of its d electrons, i.e. $[d_{YZ}^2, d_{ZX}] \to [d_{YZ}, d_{ZX}^2]$.

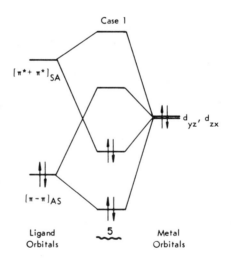

The energy and overlap properties of the transition metal d orbitals play an important role in the forbidden-to-allowed process. The correlation diagram in Fig. 2 is a good framework in which to view this aspect of the catalysis. The propensity of two interacting olefins to transform along the reaction coordinate to a cyclobutane ring will be enhanced with increasing electron density in $[\pi^* + \pi^*]_{SA}$ and decreasing electron density in $[\pi - \pi]_{AS}$. The metal, therefore, in addition to providing the orbital symmetry prerequisites outlined above, can further promote molecular transformation through orbital interactions which enhance this distribution of electron density. The extent of interaction of two orbitals of the appropriate symmetry increases as the orbital overlap increases and their energy differential decreases. Assuming adequate orbital overlaps, the metal d_{yz} and d_{zx} orbitals should best be positioned in energy between the $[\pi^* + \pi^*]_{SA}$ and $[\pi - \pi]_{AS}$ ligand combinations. If the two metal d orbitals are of the same energy (exclusive of the ligand field provided by the olefin ligands), the following energy diagram is obtained (Fig. 3). Shifting the d energy

Fig. 3. Energy level diagrams for transition metal complex (**5**); the metal d_{YZ} and d_{ZX} orbitals are degenerate.

band upwards will increase $d_{YZ} + [\pi^* + \pi^*]_{SA}$ metal-ligand interaction but decrease $d_{ZX} + [\pi - \pi]_{AS}$ interaction. The electron density in these two ligand combinations will therefore increase relative to Case 1. Lowering the d energy band from the midpoint will decrease the interaction between d_{YZ} and $[\pi^* + \pi^*]_{SA}$ and increase that between d_{ZX} and $[\pi - \pi]_{AS}$ thereby decreasing electron density in both ligand combinations. The energy distribution in Case 1 seems to better fulfill both requirements, namely, increasing electron density in $[\pi^* + \pi^*]_{SA}$ and decreasing electron density in $[\pi - \pi]_{AS}$.

If the two d orbitals are not degenerate, two additional possibilities exist. Energy level diagrams for these cases (Cases 2 and 3) are illustrated in Fig. 4. Of the two energy distributions Case 3 better provides the desired

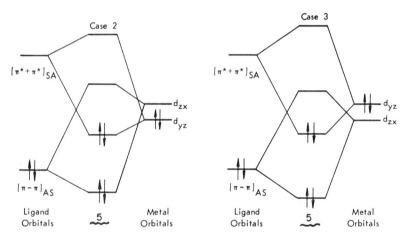

Fig. 4. Energy level diagrams for transition metal complex (**5**); the metal d_{YZ} and d_{ZX} orbitals are split.

distribution of electrons in the two critical ligand orbitals. Since the energy differential between the metal d levels and the ligand combinations is smaller in Case 3 than in Cases 1 and 2, metal-ligand interaction is greater, and the desired distribution of electron density is enhanced. Considering only this aspect of the metal catalytic role, then, the metal systems should promote the forbidden-to-allowed process (**1**) $\rightarrow$ (**2**) in the following order of increasing activity, Case 2 < Case 1 < Case 3.

There is a second element of the forbidden-to-allowed process worth discussing. This is the reordering of metal d electrons which proceeds with ligand transformation. In Case 1, the reordering process (i.e., $[d_{YZ}^2, d_{ZX}] \rightarrow$

$[d_{YZ}, d_{ZX}^2]$) is of little consequence with respect to the metal complex since these atomic orbitals are of the same energy. For Case 3, the reordering process is energetically favorable since d electron density is refocused into a d orbital of lower energy (Fig. 5). The energy diagram in the figure is not a correlation diagram and thus does not reflect the energy paths of the various molecular orbitals. The ligand combinations do, in fact, transform along the sharp energy slopes indicated by dotted lines in the figure. However, since they are mixed with the indicated d orbitals, the electron pairs ride a lower energy path more directly to the energy levels (heavy lines) of complex (**7**). The main point of the diagram is the reordering of metal d electrons. Assuming the low energy path of orbitals (SA, AS) (**6**) to (AS, SA)(**7**), the reordering process—a consequence of the SA/AS crossing —is illustrated in the diagram by dividing the orbitals of (**6**) and (**7**) into

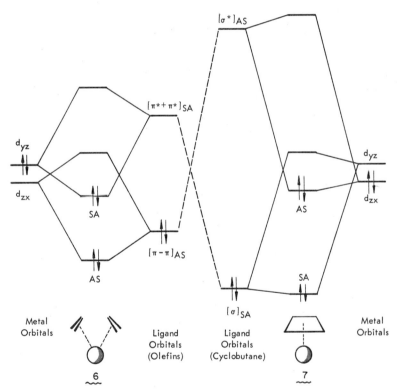

Fig. 5. Energy level diagrams for the metal-catalyzed cyclobutanation of coordinated olefin ligands; the reordering of metal d electrons for Case 3.

their metal d and ligand component parts. The concept of d orbital re-ordering can also be illustrated in terms of the exchange of electron pairs between metal and ligand system discussed earlier (cf. Fig. 2).

The situation is quite different for Case 2, however. Here the ordering of d electrons in the product complex is not ground state. Thus, the interesting possibility of generating an electronically excited metal complex through the forbidden-to-allowed catalytic process is raised. The likelihood of chemiluminescence would have to be weighed against the energetics of the forbidden-to-allowed process and the energy differential separating the critical d levels. The more favorable situations would seem to be those involving energetically favorable ligand transformations. The energy gained through ligand transformation could thus compensate for negative energetics associated with the generation of an electronically excited metal complex. The transformation of quadricyclene (**8**) to norbornadiene (**9**) is a good example of an energetically favorable transformation. Quadricy-

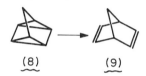

$$\underline{(8)} \qquad \underline{(9)}$$

clene is a highly strained molecule relative to norbornadiene. Numerous examples of this catalytic process [i.e., (**8**) → (**9**)] have been reported and will be treated later in the chapter when valence isomerization is discussed. For molecular transformations which are not themselves significantly exothermic, however, the energy split in d levels for Case 2 could provide a substantial energy barrier to reaction. The activities, then, of the three possible metal systems on the bases of d electron reordering alone should increase in the order, Case 2 < Case 1 < Case 3; the same order noted above.

Our treatment to this point has focused primarily on the ligand transformation and the metals role in assisting this process. The ligand combinations and metal systems have essentially been treated as separate molecular units. This approach was adopted to better illustrate the metals operations in the forbidden-to-allowed process. But the metal complex is a molecular unit in itself and the changes in bond character within it which necessarily accompany ligand transformation can play an important role in the catalytic process. As ligands transform from one bonding configuration to another [e.g., (**1**) → (**2**)], the bonding relationship with the co-ordinated metal changes. The ligand-to-metal bond change can be one of degree (either becoming stronger or weaker) or one of kind (e.g., from bonding to antibonding). This latter case has been treated before (*12*), and will not be discussed in this chapter.

The process of ligand-to-metal bond change with ligand transformation will first be discussed, qualitatively, as it applies to simple olefin cyclobutanation (**6**) $\rightleftarrows$ (**7**). There are four important bonding interactions between the metal and ligand system in (**6**). The first two arise from olefin π combinations (SS and AS) and the second two from olefin π^* combinations (SA and AA). The metal couples with these orbital combinations using atomic orbital combinations of the appropriate symmetries. For transformation (**6**) $\rightleftarrows$ (**7**), metal atomic orbitals fall into the following symmetry groups: SS (s, p_Z, d_{Z^2}, $d_{X^2-Y^2}$); SA (p_Y, d_{YZ}); AS (p_X, d_{ZX}); AA (d_{XY}). The actual mixture of atomic orbitals used by a particular metal will differ with each metal systems depending on energy and overlap properties. For simplicity, we will use only the metal d orbitals. The ligand combinations and the metal d orbitals are coupled in a pictorial representation of the four bonding combinations in Fig. 6.

The four bonding combinations can be looked upon as ligand-to-metal donor bonds (SS and AS) and metal-to-ligand backbonds (SA and AA). The donor bonds are populated by ligand π electrons while the backbonds result from metal valence electrons. For most complexes, the energies of the four orbitals will fall in the increasing order, SS < AS < SA < AA. The SS and AS bonding orbitals will always be populated since they are constructed from π combinations. The SA and AA combinations, in contrast, must compete for metal electron density with the rest of the metal system. For metal systems rich in valence electrons, the SA and AA combinations will very likely both be populated, thereby making the maximum contribution to the overall ligand-metal bond. For metal systems with fewer valence electrons, these orbitals could only be partially populated; in d^2 systems, for example, the AA combination would very likely be empty, making insignificant contributions to the metal-olefin bond energy.

As the ligands undergo transformation to a cyclobutane ring in (**6**) → (**7**), the π and π^* orbital combinations transform to cyclobutane orbitals of the same symmetry. The new ligand orbitals interact with the same set of metal orbitals generating the four combinations in Fig. 7. The pictorial representation in the figure has been simplified for clarity. It does not contain the complete set of cyclobutane σ and σ^* orbitals. There are two degenerate σ orbitals and two degenerate σ^* orbitals in the C-4 ring system which render the metal d_{ZX} and d_{YZ} identical with respect to ligand-metal interaction (Fig. 8). Since for a given molecular transformation, we need consider only the bonds undergoing change, the remaining portion of the

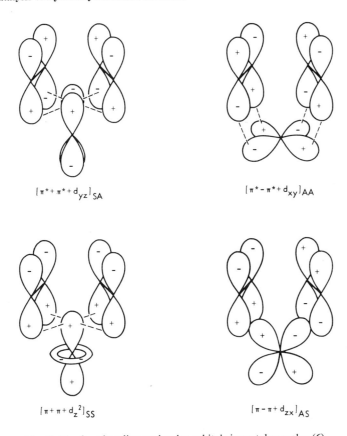

$[\pi^{*}+\pi^{*}+d_{yz}]_{SA}$

$[\pi^{*}-\pi^{*}+d_{xy}]_{AA}$

$[\pi+\pi+d_{z^{2}}]_{SS}$

$[\pi-\pi+d_{zx}]_{AS}$

Fig. 6. The four bonding molecular orbitals in metal complex (**6**).

orbital network can be ignored. The total representation of the cyclic C-4 bonding network, however, becomes important when considering alternative paths of reaction for (**7**) → (**6**). In the present case, though, the bonding assignments of d_{zx} to the σ^{*} and d_{yz} to the σ were arbitrary, reflecting this specific course of molecular transformation.

The metal-cyclobutane bonding may differ significantly in energy from the corresponding metal-olefin bond. The differences are mainly due to changes in energy and overlap properties of the cyclobutane orbitals. The σ orbitals of simple (and unstrained) cyclobutanes are of lower energy than the π orbitals of the corresponding olefins; similarly, the σ^{*} are higher in energy than the π^{*} combinations. For metal systems in which the d orbitals are positioned in energy approximately midway between π and π^{*}

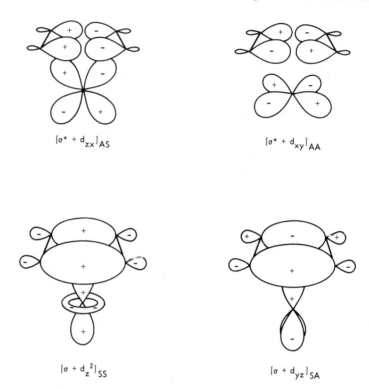

$[\sigma^* + d_{zx}]_{AS}$　　　　$[\sigma^* + d_{xy}]_{AA}$

$[\sigma + d_{z^2}]_{SS}$　　　　$[\sigma + d_{yz}]_{SA}$

Fig. 7. The four bonding molecular orbitals in metal complex (7) formed from transformation of (6).

combinations in (6) (cf. Fig. 3), transformation to (7) means a loss in ligand-to-metal bonding energy. In the catalytic process, this loss in bonding energy must be balanced against the energy gained in π bond fusion.

For a given olefin system, the facility of transformation (6) → (7) should be enhanced with decreasing ligand-to-metal bond energy in (6) relative to (7), other factors remaining constant. For metals with fewer valence electrons, a differential in bonding energy in the desired direction can conceivably be realized. The forbidden-to-allowed process (6) → (7) requires electronic population of $[\pi^* + \pi^* + d_{yz}]_{SA}$. It does not, however, require a metal electron pair in $[\pi^* - \pi^* + d_{XY}]_{AA}$, the metal-ligand second backbond (cf. Fig. 6). Since the bonding energy associated with the molecular orbital of this symmetry in (7) (i.e., $[\sigma^* + d_{XY}]_{AA}$) should be significantly less than $[\pi^* + \pi^* + d_{yz}]_{SA}$ in (6), transformation (6) → (7) will realize some loss in bonding energy, approximately equal to that separating these

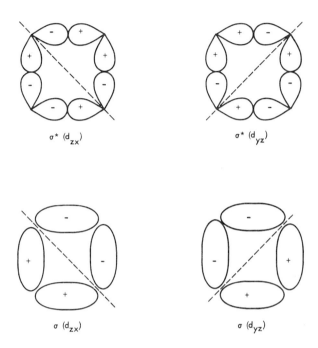

Fig. 8. The degenerate σ and σ^* molecular orbitals in the C–4 ring system.

two orbitals. When this orbital is not occupied in (6), however, this bonding energy is not sacrificed with ligand transformation. For transformation (6) → (7), therefore, two valence electrons are prerequisite for the forbidden-to-allowed process and any additional valence electrons would seem to retard the process since they would tend to contribute ligand-to-metal bonding energy preferentially to (6) relative to (7).

   In this section we have discussed, qualitatively, factors which contribute to the catalytic process (6) → (7). Some of these factors are associated with the forbidden-to-allowed process itself and, therefore, play a critical role in this catalytic transformation. Others are associated primarily with the energetics of the process and need not be significant in this chemistry. These latter factors, therefore, should not be looked upon as selection rules defining which metal systems will or will not have catalytic activity. Rather, they should be viewed as contributors to a framework within which a clearer picture of an otherwise obscure phenomenon is visible. Metals with two valence electrons, for example, should better effect cyclobutanation of olefin ligands than metals with more than two, other factors being equal. This last qualification is important. Given this situation, and assuming a smooth reaction path is available, olefin ligands occupying adjacent ligand positions should have a greater propensity to transform on a d^2 metal system than on a d^{2+}. This does not mean that transition metals rich in valence electrons will not have catalytic activity. There are numerous examples in which d electron-rich metal systems appear to have activity (9), and some will be discussed further in this chapter. The factors associated with the forbidden-to-allowed process, then, can be viewed as prerequisites to this catalytic reaction and the remaining factors as contributors to a clearer understanding of the chemistry stemming from this process.

B. Olefin Reactions

There are, to our knowledge, no reports of a catalytic system which will transform simple olefins to their corresponding cyclobutane derivatives. A number of transition metal systems, however, catalyze the dimerization of bicyclic olefins to cyclobutanes. Norbornadiene (9) undergoes facile dimerization to the cyclobutane dimer (10) in the presence of zero-valent Fe (17), Ni (18,19) and Co (20) catalysts. This chemistry has been reviewed

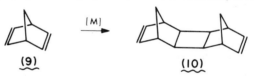

(9)   (10)

by G. N. Schrauzer who first postulated a concerted mechanism (π complex multicenter reaction) for these catalytic processes (21,22). The question of concertedness in these reactions remains essentially open. If in fact these products do arise through the concerted fusion of coordinated olefin ligands, as Schrauzer first postulated, then these are genuine examples of the forbidden-to-allowed process and the role of the transition metal described in the previous section applies. The symmetry arguments themselves, however, do not provide supporting evidence for the concerted path. The absence of symmetry restrictions to a transformation does not imply a likelihood for reaction along that reaction path. In the absence of sound experimental evidence for the concerted transformation, the case for the forbidden-to-allowed process must rest on the broad body of diverse chemistry currently emerging which is consistent with this hypothesis. We turn our attention now to this chemistry.

Green and Wood have reported a number of reactions of tricarbonyl-cyclooctatetraene ruthenium and iron with electrophilic dieneophiles (23). They report the smooth, room temperature reaction of the ruthenium complex with hexafluoroacetone, 1,1-dicyano-2,2-bis-(trifluoromethyl)ethylene, trans-1,2-dicyano-1,2-bis(trifluoromethyl)ethylene, and tetracyanoethylene yielding crystalline 1:1 adducts (11), (12), (13), and (14), respectively. The similar reaction with the iron complex and 1,1-dicyano-2,2-bis-(trifluoromethyl)ethylene gave crystalline (15).

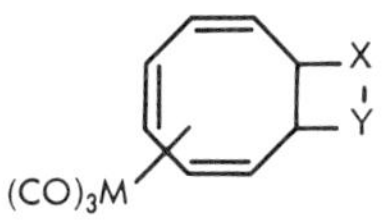

(11) $M = Ru,$ $X = 0,$ $Y = C(CF_3)_2$

(12) $M = Ru,$ $X = C(CN)_2,$ $Y = C(CF_3)_2$

(13) $M = Ru,$ $X = Y = C(CN) CF_3$

(14) $M = Ru,$ $X = Y = C(CN)_2$

(15) $M = Fe,$ $X = C(CN)_2, = C(CF_3)_2$

This particular cycloaddition is of special interest since these olefins normally undergo 1,4-cycloaddition. Involvement of the transition metal, however, may open up the interesting option of 1,2-addition, an otherwise symmetry-forbidden process.

Although simple olefins are not known to undergo direct catalytic cyclobutanation, they do undergo a remarkable catalytic transformation which appears to be related. We refer here to olefin disproportionation (24), certainly the most intriguing catalytic reaction to be discovered in this decade. The basic olefin transformation is illustrated in Eq. (4).

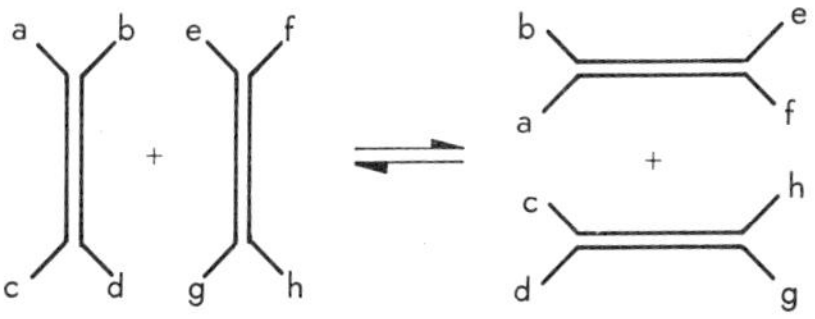

$$(4)$$

In this catalytic process, carbon-carbon double bonds are broken and made with striking ease allowing a spectrum of smooth olefin interconversions. This chemistry has been studied by a number of workers and a review on the subject has recently appeared (25). The name "olefin disproportionation," first applied to this chemistry by Banks and Bailey, may have been an unfortunate choice. The term implies a process involving the formation of dissimilar parts. This is true in only a limited number of cases involving unsymmetrical open-chain olefins. Other workers reporting this chemistry have used different names. Bradshaw, Howman, and Turner have called the reaction "dismutation," of marginal improvement over disproportionation (26). A more descriptive name is "olefin metathesis" which was used by Calderon, Chen, and Scott, describing a homogeneous catalytic system (27). Metathesis, we feel, better reflects the novel nature of the catalytic transformation and more completely embraces the total scope of this chemistry.

Cyclic olefins also undergo transformation along reaction (4) (28). Here, the products are large ring polyenes, of the general structure (16). Cyclic olefins reacting with open-chain olefins open a path to a variety of non-conjugated dienes (29).

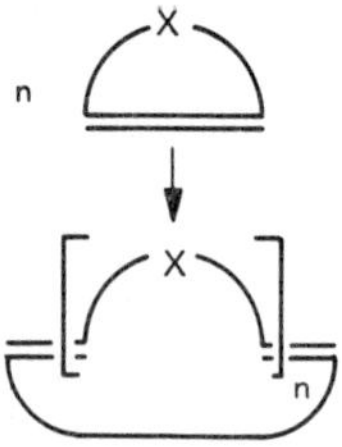

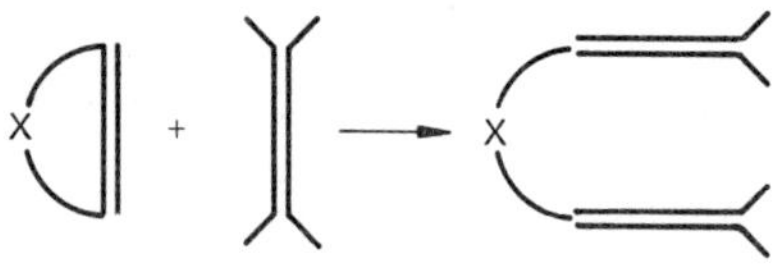

$$(5)$$

(16)

The mechanism for olefin disproportionation has been studied by a number of groups and an account of the work is collected in the review (25). All studies point to a "four-center" type intermediate formed from

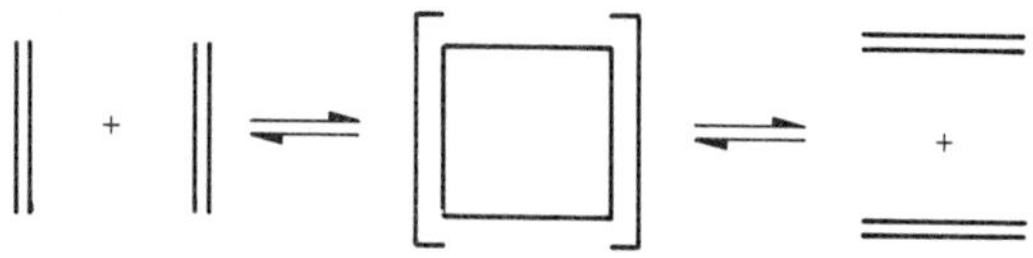

$$(6)$$

fusion of the olefin π bonds [Eq. (6)]. The mechanistic studies, however, have focused primarily on the nature of the olefin transformation and have given less attention to the course of the critical catalytic step itself. The results nevertheless point directly to the interchange of olefin alkylidene groups very likely through the generation of a symmetrical, four-centered intermediate somewhere along the reaction coordinate. From a careful study of the system, Calderon, et al. first proposed a mechanism for olefin metathesis involving the concerted fusion of transition metal-coordinated olefin ligands to the symmetrical cyclic intermediate (17) (30). The body of evidence supports this proposal.

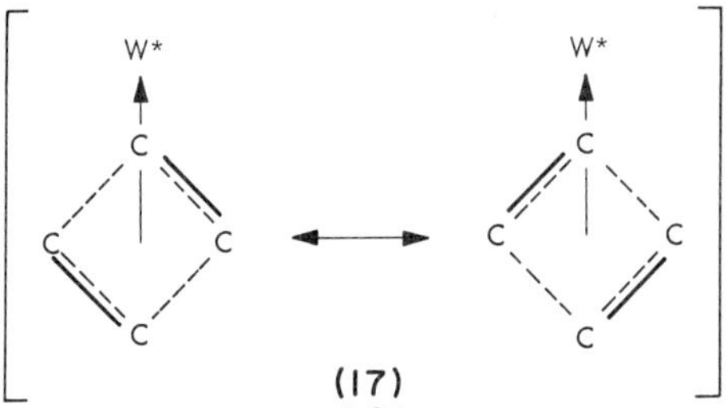

$$(17)$$

Olefin disproportionation is dominated by two transition metals, molybdenum and tungsten. Other transition metals have shown activity, but with the exception of rhenium (31), the activity levels have been low and their scope of application, limited. Smooth, clean reactions can be carried out on heterogeneous metal oxide catalysts (25) or homogeneously using a reduced tungsten chloride system (30) or nitrosyl complexes of tungsten and molybdenum obtained from the addition of aluminum alkyls to $L_2Cl_2(NO)_2M$ (M = Mo, W; L = Ph$_3$P, C$_5$H$_5$N, Ph$_3$PO, etc.) (32). The catalytic transformations are, in either case, essentially identical. The homogeneous systems are distinguished by remarkably high levels of activity. The tungsten halide system is active at room temperature; 2-pentene is converted to the equilibrium mixture of 2-butene, 2-pentene, and 3-hexene in a few seconds. The nitrosyl systems exhibits similar levels of activity in the temperature range 0° to 50°. Both systems have been studied for reaction mechanism. Wang and Menapase studied the metathesis of pure *cis*-, pure *trans*- and mixtures of *cis* and *trans*-2-pentene using a tungsten hexachloride–*n*-butyllithium system (33). The *cis* isomer and the mixed isomers were found to be considerably more reactive than the pure *trans*-2-pentene. In all cases, however, the reaction, at room temperature, was clean, yielding 2-butene and 3-hexene in 100 mole %. These results, coupled with the fact that the optimum mole ratio of *n*-Bu Li per WCl$_6$ was 2, led the authors to conclude the following: (1) the displacement of 2Cl$^-$ by 2R^-, probably occurring in rapid succession [Eq. (7)], where the observed order of efficiency, *n*-butyl > *sec*-butyl > *t*-butyl, roughly parallels the nucleophilicity of these anions; (2) unimolecular or

$$W(VI)Cl_6 + 2R^-Li^+ \longrightarrow \overset{\displaystyle R}{\underset{\displaystyle R}{\mid}} W(VI)Cl_4 + 2LiCl \qquad (7)$$

bimolecular reduction of W, and coordination with 2-pentene as in Eq. (8); and (3) formation

$$
\begin{array}{c}
R \\
| \\
W(VI)Cl_4 \xrightarrow{\;-2R\cdot\;} W(IV)Cl_4 \\
| \\
R
\end{array}
\qquad\qquad (8)
$$

2 olefins fast

$$W(IV)Cl_4 \cdot 2\ \text{olefins}$$

2 olefins $-2R\cdot$

of products and regeneration of catalyst as in Eq. (9).

$$W(IV)Cl_4 \cdot 2\ \text{olefins} \;\rightleftarrows\; W(IV)Cl_4 + \text{products} \qquad (9)$$

two new olefins

They further rationalized the stereochemical and other aspects of the system in terms of the reaction path outlined in Fig. 9.

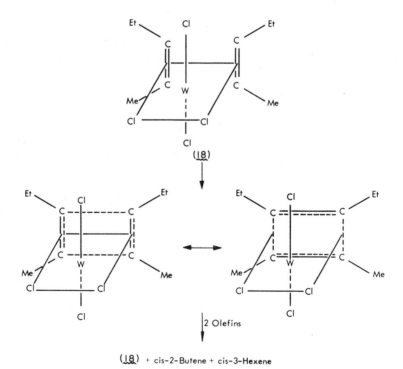

(18) + cis-2-Butene + cis-3-Hexene

Fig. 9. Reaction scheme proposed by Wang and Menopace (*33*) for the metathesis of 2-pentene.

Hughes came to similar conclusions studying the nitrosyl molybdenum system $[(C_5H_5N)_2Mo(NO)_2] \cdot Cl_2\text{-EtAlCl}_2$ (*34*). This study further revealed the striking stereoselectivity of the critical catalytic step. By studying the reaction at low temperature (0°) and at the early stages where kinetic control predominates, it was shown that *cis*-2-pentene reacts preferentially to *cis*-2-butene and *cis*-3-hexene, and that *trans*-2-pentene reacts with equal selectivity to *trans*-2-butene and *trans*-3-hexene. The study further showed that the rate of disproportionation and the rate of isomerization decreased simultaneously, implying that the two processes are intimately connected and probably occur on the same catalyst. The author proposed an octahedral bisolefin complex similar to (**18**) (Fig. 9). This study, however, was not directed to the oxidation state of the molybdenum system and the exact nature of the other four ligands was not specified. The stereoselectivity of the catalyst was explained through steric effects associated with the *cis*-positioning of the two olefins on the octahedral complex. The observed selectivity, then, is nicely explained by assuming a disrotatory twist

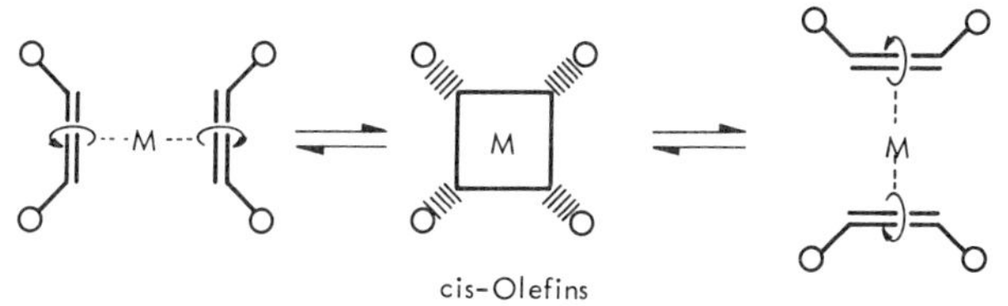

cis–Olefins
Disrotatory Twist

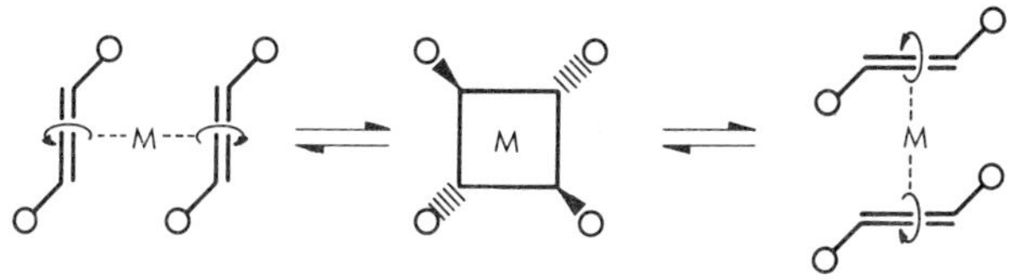

(10)

trans–Olefin
Disrotatory Twist

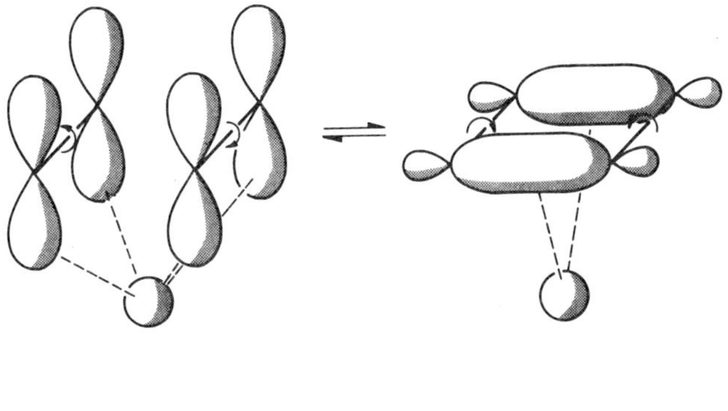

(11)

(12)

about the carbon-carbon double bonds for both the bis(*cis*-2-pentene) and bis(*trans*-2-pentene) metal complexes. The disrotatory mode of twist [Eq. (12)] maintains maximum ligand-metal orbital overlap and, therefore, is almost essential to any mechanism incorporating metal-assisted cyclobutanation of coordinated olefin ligands.

The above results suggest the forbidden-to-allowed process. Indeed this description has been used to define the role of the transition metal in these catalytic reactions (*30,35*). The fact that essentially identical chemistry is observed in relatively simple homogeneous systems (*30,32*) as well as in a spectrum of heterogeneous systems (*25*) points to a common mechanism, one that is not complex, and very likely involves a single transition metal center. The observed stereoselectivity in the homogeneous systems (*33,34*) is consistent with one metal center with the reactants coordinated at adjacent ligand sites. The stereoselectivity in the course of reaction with preferential *cis* to *cis* and *trans* to *trans* (*34*) further supports this picture and provides evidence for the concerted, [2 + 2] cycloaddition of metal-coordinated olefins to a metal-coordinated cyclobutane ring system. The proposed oxidation state for the homogeneous tungsten system (*33*) supports the forbidden-to-allowed role for the metal. Tungsten(IV) has two valence electrons, the minimum number prerequisite to the ground state ligand transformation, and the suggested number to equalize the ligand-to-metal bonding relationship for that transformation.

Alternative mechanisms are difficult to draw. It is possible that the catalytic process [Eq. (4)] proceeds through generation of carbenoid species through the breaking of carbon-carbon double bonds; products would then arise through the recoupling of intermediates. Carbene chemistry, however, does not emerge from these systems. Further, it is doubtful that metal-coordinated carbenes could be generated with such striking ease (between 0° and room temperature, for example) through the breaking of carbon-carbon double bonds. The low activation energies suggested for this chemistry indicate a mechanism in which the bond order of the transforming olefin system is maximized in the critical transformation.

Mechanisms involving metalocyclo intermediates (**19**) can also be drawn [Eq. (**13**)]. The generation of species like (**19**) is not at all unlikely in tran-

(13)

(19)

sition metal systems. Transfer of alkylident groups, however, requires either the generation of a symmetrical intermediate (path 1), a sigmatropic shift of the metal center within the C-4 ring system (path 2), or the direct breakdown of (**19**) to the observed products (path 3).

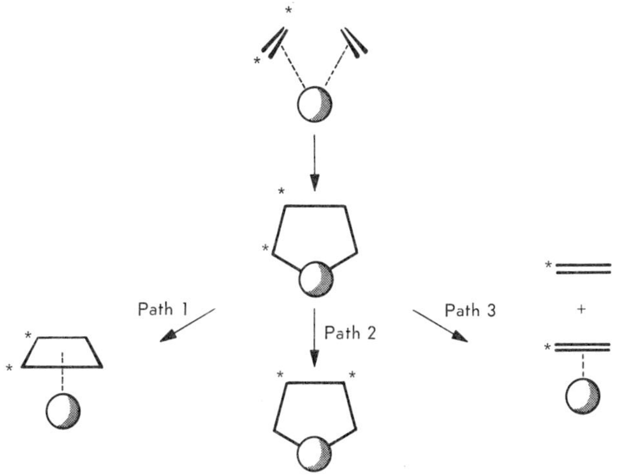

(14)

Path 1 involves the reversible insertion of the metal center into the cyclobutane ring. Further, the metal center must preserve bonding with the ring system throughout this process, since distinct cyclobutane products are apparently not produced in this chemistry, and from the thermodynamics of the system, cyclobutane products should be observed if they are free intermediates. Assuming the metal occupies a near normal ring position in (**19**), it seems unlikely that the rearrangement in path 1 could proceed with the retention of significant cyclobutane-to-metal bond energy, particularly in the early stage of the molecular rearrangement. Path 2 has similar problems. In most reasonable ring geometries, the metal in (**19**)—which should be near planar with the other four ring atoms—could only experience minor interaction with the 3-carbon atom; thus, the [1, 3] shift required in path 2 would seem to require higher activation energies than is indicated for the metathesis process. Path 3 also suffers from energy deficiencies. Although a second metal center could conceivably aid the process, this molecular transformation further encounters orbital symmetry restriction: $[M—C\,\sigma_S^2,\ M—C\,\sigma_A^2,\ C—C\sigma_S^2$ and $C—C\,\sigma_A^2] \rightarrow [M\text{-Olefin }\pi_S^2,\ M\text{-Olefin }\pi_A^{*2},\ \text{Olefin }\pi_S^2,\ \text{and }M\text{-Olefin }\sigma_S^2]$. It would be difficult for the metal in (**19**) to remove these restrictions and essentially impossible for a metal system initially d^2 since it would be d° in (**19**).

The forbidden-to-allowed mechanism suffers from none of these deficiencies. The metal removes the symmetry restraints and opens the path to a concerted [2 + 2] cycloaddition. The ligand-to-metal bond energy which may be lost in the process is balanced against that gained through fusing two π bonds into two σ bonds. The cyclobutane ring system formed need not, in fact, be a bad ligand. It would very likely be planar, and somewhat stretched from its normal configuration. The molecular orbitals of the C-4 ring system have the proper symmetry to interact with the metal system and in the configuration described could conceivably form a sufficiently strong bond to hold the ring system to the metal. In this system, the metal-coordinated cyclobutane ring can be viewed as the valence isomer of the bisolefin system. The transformation from one bonding configuration (cyclobutane $\cdot$ metal) to another [bis(olefin) $\cdot$ metal] may reflect no more than a molecular vibration along a different mode. Thus, the cyclobutane is not formed as a distinct, free intermediate, but instead exists, in this view, as a short-lived species fixed to the metal center responsible for its formation.

In the forbidden-to-allowed process, the transition metal experiences a reordering of its d electrons (i.e., $d_{YZ}^2, d_{ZX} \rightarrow d_{YZ}, d_{ZX}^2$) with cyclobutanation of its olefin ligands. In order to reopen the ring in the other direction [Eq. (6)], the two d electrons must first be relocalized ($d_{YZ}, d_{ZX}^2 \rightarrow d_{YZ}^2, d_{ZX}$), or the ring must rotate 90°. It would seem that the reordering process would proceed best when the d_{YZ} and d_{ZX} orbitals are near degenerate; thus a molecular vibration of the C-4 ring system in one direction or another would again split the orbitals allowing the reordering for the ring scission process. Thus an attractive model for the metathesis process would place the metal center in a ligand field which leaves the d_{YZ} and d_{ZY} orbitals degenerate, exclusive of the bis(olefin) system. The octahedral structure discussed above could undergo rearrangement with olefin transformation to a molecular geometry in which the four ligands have C_{4v} symmetry. The actual catalyst structure could, for example, be a trigonal prism. The octahedral structure, however, offers an alternate possibility. If the splitting of the two d orbitals is not too large, reaction could proceed. The reordering of d electrons at the cyclobutane stage of reaction, then, would be insured since the product complex would be formally in an excited state [cf. Fig. 5; these assignments refer to the coordinate system in (5)].

These comments on the actual geometry of the catalytic system and the ligand field splittings are speculative. There is no evidence supporting any one structure. Our purpose here is to frame the proper questions regarding conceivable structures and attractive splitting patterns. More definitive statements await harder experimental facts or reliable molecular orbital studies based on sound molecular orbital calculations. Until then, we must rely on the conceptual models stemming from the qualitative orbital symmetry picture described earlier.

C. Valence Isomerization

"Valence isomerization" here refers to molecular transformations in which bonds undergo a change in configuration while the nuclei do not. We shall focus on symmetry forbidden valence isomerizations and the role of the transition metal in assisting them. Metal-catalyzed valence isomerizations of this sort are a comparatively new catalytic process. The first example was the smooth ring opening of quadricyclane (8) to norbornadiene (9) (page 233) reported by Hogeveen and Volger (36). Some of this chemistry has recently been reviewed (14); it will, therefore, be discussed here only where relevant to more recent disclosures in this category.

Two striking features associated with metal-catalyzed valence isomerization (8) → (9) are its ease and specificity. At $-26°$, $t_{1/2}$ for (8) is 45 minutes in the presence of 2 mole % di-μ-chloro-bis (bicyclo[2.2.1]hepta-2,5-diene)dirhodium(II). In the absence of a transition metal $t_{1/2}$ for (8) is more than 14 hours at 140° (37). The remarkable specificity of this system is reflected in the catalytic valence isomerization of (20) to hexamethyl-Dewarbenzene (21) (38). Thermal rearrangement (90°–120°) gives primarily

(20)   (21)

the thermodynamically more stable product, hexamethylbenzene. The transition metal, in contrast, at $-30°$ operates cleanly and specifically on the C-4 ring giving mainly (21) (ca 97%), the remaining product being the benzene derivative.

A large part of the facility with which these ring systems undergo rearrangement in the presence of transition metals lies in the nature of the ring systems. Both systems [8] and [20] are highly strained relative to their valence isomers. However, rearrangement paths to greater stability are blocked by orbital symmetry restraints. These ring systems can, in fact, be looked upon as being locked in their respective bonding configurations. The appropriate transition metal system can, in theory, "unlock" these systems allowing bonds to flow into energetically more preferred regions of space. The ease and specificity noted above suggests that this is in fact occurring. We turn now to a more detailed examination of this catalytic process.

Metal-catalyzed valence isomerization of (**8**) requires that the C-4 ring first coordinate to the transition metal, for example, (**22**). Edgewise coordination of cyclopropane rings to transition metals has been reported (*39*). More important, the platinum complex (**23**) has been prepared (*40*)

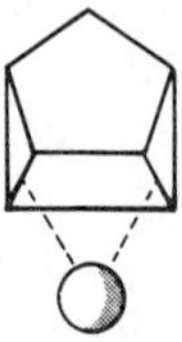

(**22**) 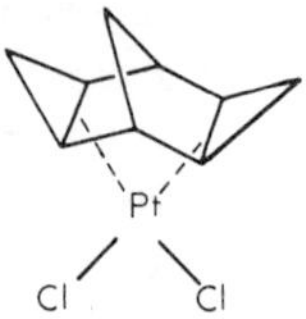   (**23**)

and the tricyclic ring ligand *exo,exo*-tetracyclo[3.3.1.0^{2,4}.0^{6,8}]-nonane (**24**) in (**23**) can be easily displaced with norbornadiene and recovered unchanged. The coordination properties of (**24**) are very likely associated with the π character of the cyclopropane bonds common to 3- and 4-membered ring systems (*41*). The π character of the ring systems would be expected to increase with ring strain reflecting greater *p* character and higher energy. It should not be surprising, therefore, that quadricyclene (**8**), which should contain significantly more strain than (**24**), would possess a propensity to coordinate to transition metal centers.

The cyclopropane σ bonds in (**8**) which rest in the cyclobutane base should differ in π character from the other two σ bonds in the C-4 ring system. They should possess greater *p* character, protruding downward out of the cyclobutane base, as illustrated in (**25**). This dissimilarity be-

(**25**)

tween the two sets of bonds would alter the coordination character of this ring system in an important way. Ligand-to-metal bonding should focus at the cyclopropane centers where the greater π character rests [cf. (**22**)]. The quadricyclane ring system (**8**) would, in this view, resemble norbornadiene (**9**), adopting the role of a bidentate ligand. Coordinating to the metal, then, the cyclopropane centers would interact with the metal orbitals in the manner described for olefins earlier, i.e., as in (**26**) and (**27**).

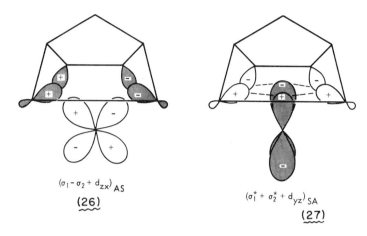

$(\sigma_1 - \sigma_2 + d_{zx})_{AS}$

(26)

$(\sigma_1^* + \sigma_2^* + d_{yz})_{SA}$

(27)

In this electronic configuration, the ligand is free to isomerize to (8) in the ground state, generating, with the concomitant exchange of electron density, the norbornadiene complex (29) in its bonding configuration.

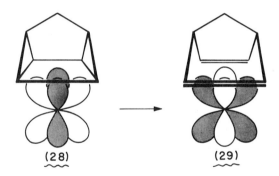

(28)   (29)

We suggest, then, that quadricyclane (8) should be a sufficiently strong ligand to impress upon certain transition metal systems a ligand field which orders the d electrons as in (28) which opens a ground state transformation path to norbornadiene (29).

We have focused our major attention on the theoretical model for forbidden-to-allowed valence isomerization. Our description of this hypothetical process stems from simple molecular orbital symmetry principles which, we feel, are fundamentally sound. The pertinence of these principles to real chemistry has been convincingly demonstrated in a broad body of organic chemistry designed specifically to establish the extent of control of the Woodward-Hoffmann rules (3). But the application of orbital symmetry principles to metals in the special capacity referred to here is a new direction, and these arguments alone will not bear on the breadth of forbidden-to-allowed chemistry. The molecular orbital symmetry picture states that molecular systems locked in bonding configurations by symmetry restraints can be unlocked in a special way by metal centers. A single, concerted molecular transformation defined forbidden becomes allowed on the coordination spheres of certain transition metals. To assign a given molecular transformation symmetry allowed, however, does not necessarily imply a propensity to react along that path. Indeed, symmetry allowed valence isomerizations can be energetically quite unfavorable. The actual role of a catalyst, then, in any chemical transformation which would be otherwise symmetry restricted cannot be defined by orbital symmetry arguments alone. There are frequently alternative paths to products which are equally allowed, and which do not involve forbidden-to-allowed processes as defined above.

To establish unequivocally that the forbidden-to-allowed catalytic process in fact occurs in a given reaction can be a difficult and complex experimental task. For [2 + 2] cycloaddition processes, for example, it requires evidence for a single step, concerted ligand transformation. The orbital symmetry principles supporting the Woodward-Hoffmann rules do not rest solely on this kind of evidence, but rather on the entire spectrum of organic chemistry consistent with the rules. We believe that the forbidden-to-allowed process has broad application in transition metal chemistry and we base this on the nature and breadth of chemistry currently associated with metal-assisted symmetry restricted reactions.

For valence isomerizations which are energetically very favorable [e.g., (20) → (21)], low energies of activation are to be anticipated. In this forbidden-to-allowed process, ligand transformation could involve little more than a molecular vibration along a primary mode previously blocked by symmetry restrictions. Activation energies for valence isomerizations which are symmetry allowed are remarkably low. Bullvalene is an excellent example (42), particularly since each valence isomer is energetically identical to all others. Forbidden-to-allowed processes involving energetically favorable valence isomerization, therefore, should be observable at lower temperatures than some other catalytic processes operating along different reaction coordinates. Further, under these conditions where activation energies of other processes may be sufficiently high to "screen out" competing reactions, high degrees of reaction specificity should persist. Low activation energies and high reaction specificity, then, can be reasonably associated with forbidden-to-allowed catalysis. We have just discussed two catalytic processes (36,38) exhibiting both features, and now we turn to some additional examples.

The *exo*-tricyclo[3.2.1.0^{2,4}]octene (30) is converted smoothly to its valence isomer tetracyclo[3.3.0.0^{2,8}.0^{4,6}]octane (31) at room temperature

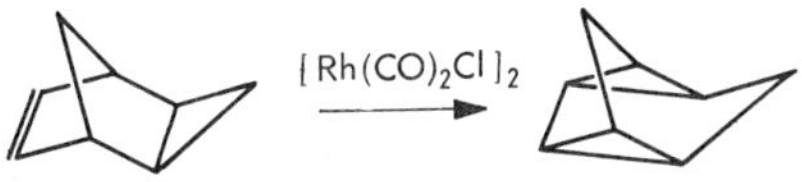

[Rh(CO)$_2$Cl]$_2$

(30)   (31)

in a homogeneous catalytic process using [Rh(CO)$_2$Cl]$_2$ (43). This re-action is clean, the valence isomer (31) being the only product observed. In sharp contrast, the *endo*-isomer (32) remains unchanged even at 100° in the presence of the same rhodium catalyst. These results are consistent

(32)

with the ligand-to-metal coordination prerequisites discussed earlier in this section. This forbidden-to-allowed process would best proceed with the ligand coordinated to the metal in a bidentate manner as in (33).

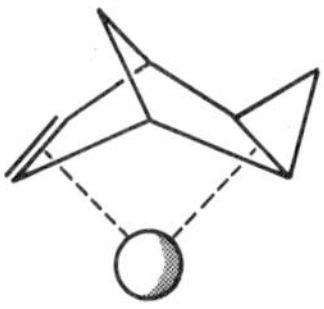

(33)

Bidentate coordination of this sort is more difficult for the *endo*-isomer (32) due to the steric factors associated with *endo* positioning of the cyclo-propane methylene group.

The high reaction specificity of the metal-catalyzed reaction should be contrasted to the thermal isomerization; at 200°, (30) rearranges to

(34)

tricyclo[3.2.1.0^{2,7}]oct-3-ene (34) (44). Photochemical isomerization (45) of either the *endo* (32) or *exo* (30) isomer yields (31) (20%) indicating that the spatial configuration of the fused cyclopropane ring plays no critical role in the valence isomerization, at least in the excited state. It is interesting that the photochemical path parallels that of the rhodium-catalyzed reaction. The photon, in the Woodward-Hoffmann description, plays a forbidden-to-allowed role conceptually similar to that of the transition metal discussed above. The photon, however, experiences no steric restraints in its interaction with a molecular system; it operates equally well with both *endo* and *exo* species.

It is instructive to speculate on the nature of the actual catalyst in the rhodium system. The rhodium in $[Rh(CO)_2Cl]_2$ is d^8 and 4-coordinate in the dimeric form. It is unlikely that rhodium in this metal system can provide the two coordination sites preferred for valence isomerization. The metal complex more reasonably sheds one or both of the carbonyl ligands thereby providing the required degree of coordinate unsaturation. Rhodium(I) chloride does coordinate to bisolefin systems and di-μ-chloro-bis (norbornadiene)-dirhodium(I) is a good example of one such metal complex with valence isomerization activity (36). The existence of di-μ-chloro-bis(diolefin)dirhodium(I) complexes supports the bidentate ligand-to-metal coordination indicated in (33). Bidentate coordination of either a bisolefin system or the olefin cyclopropane-system in (33) implies the d electron ordering prerequisite to valence isomerization. One important question, then, is whether or not a catalyst system in question can provide two coordination sites and the ligand field to stabilize the d electron configuration for the kind of bidentate coordination illustrated in (33). The existence of bis(olefin)rhodium(I)chloride systems suggests that this metal system can.

There is one other aspect to this catalytic process, however, which can play a significant role in the forbidden-to-allowed transformation. We refer here to the additional symmetry restraints which can arise due to the redistribution of d electron density within the metal complex. A ground-state ligand transformation can be visualized in which d electrons are placed in a spatial configuration within the complex which is energetically unfavorable, creating transformation restraints and introducing the possibility of generating a complex in an excited state. It is difficult to generalize this process, however, for it is a function of the relative ligand field strengths of the various ligands in addition to other factors. Each case is best treated separately. Moreover, it should be kept in mind that most ligand systems are labile, and can conceivably rearrange concomitantly with the forbidden-to-allowed ligand reaction, adopting configurations which minimize unfavorable interactions with the new spatial distribution of d electron density. A catalytic transformation, therefore, need not reflect the structure of the starting metal complex with the coordinated ligand reactant(s). Nonreacting ligands can move on and off the primary coordination sphere or undergo structural rearrangements in direct response to the growing redistribution of d electron density.

Transition metal complexes are capable of a number of distinct catalytic transformations and a large number of these involve the movement of hydrogen atoms. A number of processes, then, involving molecular rearrangements can compete with the forbidden-to-allowed process, particularly in those cases where the valence isomerization of interest is not energetically very favorable. In those cases where, at moderate temperatures, metal complexes are not able to shed the required ligands to open the needed coordination positions, or where the ligand fields surrounding the metal centers constitute significant barriers to the redistribution of d electron density or do not order the d electron density properly for the required substrate-metal interaction, competing catalytic processes can dominate the chemistry. This situation is illustrated in the reactions of the *exo*-tricyclooctene (30) with a variety of other transition metal complexes (43).

At 130° in benzene, (30) rearranges quantitatively to 5-methylenebicyclo[2.2.1]hep-2-ene (35) using $IrCl(CO)(PPh)_2$ as catalyst. The re-

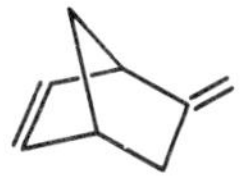

(35)

action was remarkably selective, no traces of (31) or (34) were detected. Similarly, *exo,exo*-tetracyclononane (26) undergoes the same selective catalytic isomerization to *exo*-6-methylenetricyclo[3.2.1.0^{2,4}]octane (36).

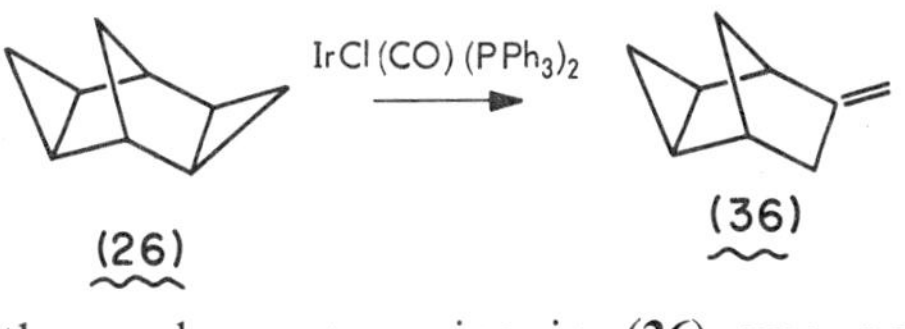

$$ (14) $$

(26)   (36)

Opening of the cyclopropane ring in (36) was not observed. The authors attributed the inactivity of (36) to its inability to coordinate as a bidentate ligand. These rearrangements involve a movement of a hydrogen atom from one carbon atom in the multicyclic ring system to another. This particular iridium complex, discovered by Vaska (46), is known for its facile ability to undergo oxidative addition. In this capacity, the iridium, initially d^8, inserts into a sigma bond of the substrate thereby becoming six-coordinate and d^6. It is possible that a critical step in the rearrangement is a kind of oxidative addition involving the insertion of iridium into the cyclopropane methynyl C—H bond. The multicyclic ring moiety formed could conceivably enjoy some additional stability through homoallylic interaction with the double bond in (30) and the cyclopropane sigma bond in (31).

The catalytic activity of a number of additional metal complexes was also examined by these same workers; $Mo(CO)_6$, Cu_2Cl_2, $RhCl(CO)(PPh_3)_2$, $(CH_6C_5N)_2PdCl_2$, $IrHCl_2(PPh_3)_3$, and $Rh_2(norbornadiene)_2Cl_2$ were all found to be inactive. With $[(C_2H_4)_2PtCl_2]_2$ and $K[(C_2H_4)PtCl_3]$, (30) and (26) give insoluble, stable complexes (37) and (23) respectively.

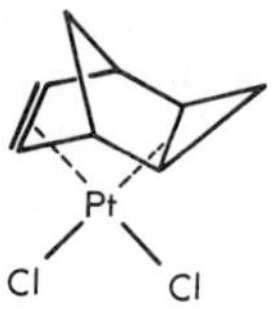

(37)

Katz and Cerefice (47) have also examined the isomerization of *exo*-tricyclooctene (30) using $RhCl(PPh_3)_3$ under different conditions (90°, 2 hours). In this system, (30) yields (34) (62 %), (31) (32 %), and (38) (6 %); with 5 % rhodium on carbon, (130°, 2 days) a similar mixture forms: 25 % (34), 55 % (31) and 20 % (38) (47). Using deuterium labeling, the authors

(38)

were able to show that the rearrangements were primarily intramolecular (and stereospecific in the case of (34)). Katz interprets his results in terms of the following intermediates.

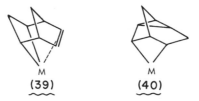

M   M
(39)   (40)

The bonding in (39) (and (40)) should not be considered distinctly different from that in (33). The degree of metal "insertion" into the cyclopropane

σ bond (or the π bonds in the case of olefins) will depend on the overlap and energy properties of the orbitals on the metal center. The ligand will assume a geometrical configuration providing the best mix of carbon s and p orbitals to match that provided by the metal system. The coordinate bond picture (33) differs from the covalent picture (39) more in degree than in kind. Both bonding descriptions are constructed of ligand-to-metal donor bonds (a σ and π bond) and metal-to-ligand backbonds. Increasing backbonding further populates the ligand antibonding orbitals thereby altering the ligand structure in the direction of (39). Real distinctions between the two representations, however, would seem to exist only when the number of metal-to-ligand electrons provided by the metal differs. In the molecular orbital treatment, the two representations differ only in the degree of orbital hybridization; in (39), carbon sp^3 hybrids are directed toward the metal. The carbon-metal bonds are assumed more covalent and stronger. In (33), the two carbon hybrids are directed toward each other, and the bond to the metal is weaker. But the total number of ligand and metal electrons involved in the respective bonds are the same and the symmetries of the respective molecular orbitals identical. Representations (39) and (33), therefore, are best considered limiting structures, and a spectrum of intermediate bonding configurations conceivably exist between them. This description of the metal-bonded cyclopropane ring system is consistent with a study of some reactions of cyclopropane·PtCl$_2$ complexes recently reported (48). The reformation of significant quantities of the cyclopropane compounds upon treatment of a variety of complexes with cyanide, water, lithium aluminum hydride, or hydrogen reduction suggested that appreciable bonding existed between the separated $-CH_2$ groups in the cyclopropane complexes. The authors concluded that the two limiting representations of the coordinated cyclopropane are essentially equivalent. For orbital symmetry considerations, the two representations are of course identical.

Katz interprets his results through mechanisms exclusive of forbidden-to-allowed processes; for two of the products (34) and (38) this is clearly the case. The authors propose that (30) reacts with the rhodium complex giving either (39) or (40) or both, that these give (41), and that (41) yields

HM   (41)

(34). They further propose that (31) is formed from (40) through "extrusion" of the rhodium catalyst. The intermediacy of (40) along a path exclusive of orbital symmetry factors is questionable. The results are just as easily interpreted in terms of intermediate (39) transforming directly to a complex of (31) coordinated to the metal [cf. Eq. (15)]. This species differs from (40) only in being bidentate rather than monodentate. The covalent bonds used by Katz in describing his intermediates in no way alter the orbital symmetry factors. The transformation indicated in Eq. (15) is essentially the same as that in Eq. (16) in this respect.

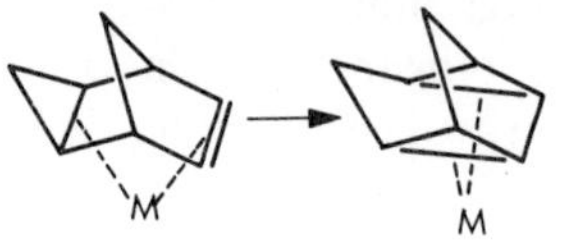

$$(15)$$

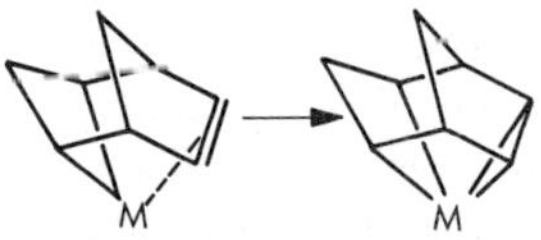

$$(16)$$

Katz and Cerefice have published another communication addressing the mechanism of metal-catalyzed cycloaddition reactions(49). They focus attention on the same catalytic system, tris(triphenylphosphine) rhodium(I) chloride + (30). On that basis of a deuterium labeling study, they were able to demonstrate that both products (31) and (34) are formed from a common intermediate " X ", i.e.,

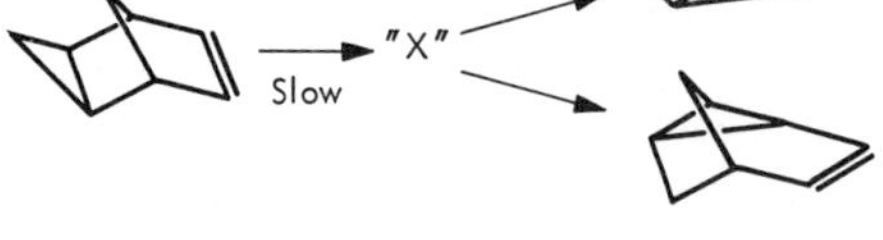

$$(17)$$

The authors concluded that the metal-catalyzed transformation of (30) to (31) is a stepwise process and not a simple metal assisted forbidden-to-allowed valence isomerization. Complex (40) was suggested as a possible structure for " X." The reaction scheme proposed is consistent with the isotope effects noted. This reaction path, however, is not inconsistent with the metal-catalyzed concerted valence isomerization of (30) to (31) as Katz and Cerefice imply. The reaction kinetics reported bear only on the existence of some common intermediate " X," not on its nature. Species " X " could conceivably be a complex of (30) and some rhodium compound in which the tricyclooctene ligand (30) assumes a bidentate bonding configuration [cf. (33)]. Such a bidentate ligand could reasonably be expected to be undergoing rapid transformation from one bonding configuration to another, including one corresponding to the structure suggested [i.e., (40)]. The mechanism the authors eliminate, the independent metal-catalyzed isomerization of (30) to (31), is in any case an unlikely one. This metal system is a complex one in which a broad variety of separate molecular transformations are conceivable. There can be little doubt that stepwise catalytic processes occur as Katz and Cerefice contend, but their relevance to the catalytic processes we address here is unclear. The orbital symmetry factors referred to are not meant to support the forbidden-to-allowed process over other catalytic operations. It is quite possible that in the catalytic system in question the metal operates on the cyclopropane ring in a manner suggested by Katz with (39) transforming directly to (40). The metal system represented by (39) has a number of paths open to it, among them the one represented in Eq. (15). Katz, however, has used his results to question the interpretations of previous workers (*36,38,43*) examining different systems. His study shows that Rh(PPh)$_3$Cl and supported rhodium are capable of carrying out a variety of molecular rearrangements, some involving processes other than "concerted electrocyclic" transformations. These kinds of catalytic processes—and particularly those involving hydrogen shifts—are well known in catalysis, however, and their existence in a particular catalytic reaction does not necessarily bear on mechanisms of different catalytic reactions proceeding in other catalytic systems.

Katz and Cerefice have also reported the smooth rhodium-catalyzed valence isomerization of the tricyclononadiene (42) to the bicyclononatriene (43) (*50*). As in the work of Volger, Hoogeveen, and Gaasbeck (*43*),

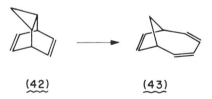

(42)   (43)

[Rh(CO)$_2$Cl]$_2$ was found to be the better catalyst, operating at 25° and yielding (43) essentially quantitatively. RhCl(PPh$_3$)$_3$, in contrast, was active at 130° and yielded the cyclopropane ring-opened product (44) in addition to (43).

(44)

The metal-catalyzed valence isomerization of (42) to (43) could proceed
by the same mechanism involved in the rhodium-catalyzed isomerization
of (30) to (31) [cf. Eq. (15)]. This process is best envisaged with (42) coordinated in a bidentate manner, with the cyclopropane ring occupying
one ligand position (45). Forbidden-to-allowed valence isomerization yields
(46) which yields product via a simple retro Diels-Alder rearrangement
to the observed product (43).

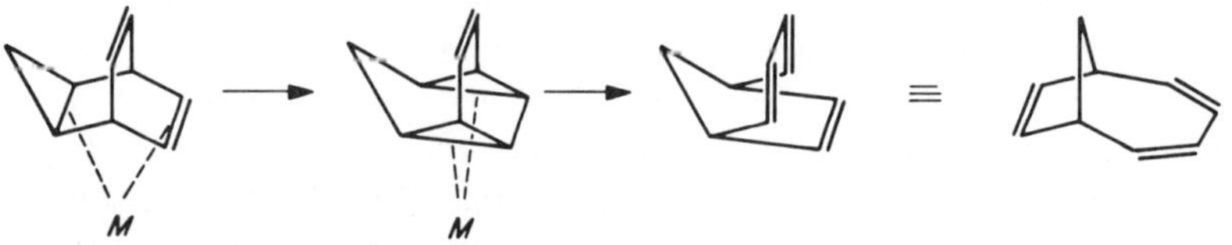

(18)

M M
(45)   (46)   (43)

The sensitivity of highly strained ring systems to metal-catalyzed valence isomerization can be remarkable. Quadricyclanone dimethyl ketal
(47), for example, has been thought to be inherently unstable, rapidly
isomerizing to its valence tautomer (48) under mild conditions (51).

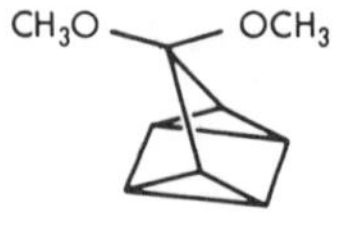

CH₃O ___ OCH₃   CH₃O ___ OCH₃

(47) 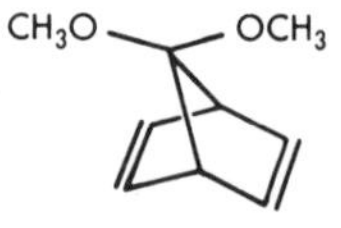   (48)

Gassman and Patton (52) have noted that the instability of (47) is actually related to the "history of the sample," the critical factor being
whether or not the diene precursor of (47) had been exposed to metal
surfaces at elevated temperatures. The dienes apparently complex with
trace metals during purification procedures. After photolysis to the quadricyclane ring structure, the entrained metals, present in trace concentration,

are apparently sufficiently active to catalyze valence isomerization of (47) to (48). When synthesis is carried out with special care, excluding contact with metal surfaces, the quadricyclene obtained is quite stable. Gassman, in fact, reports that the material prepared in this way resisted numerous attempts to bring about thermal isomerization to (48) (52). Indeed, the stability of (47) was such that dichloro(nornornadiene)palladium was required to catalyze the transformation to (48) (room temperature, 8 hour). Gassman reports similar behavior for the dimethyl derivative, (49) (53).

(49)   (50)

The stability of (49) is very dependent on the purity and history of (50), its precursor. The procedure for the purification of (50) included distillation through a spinning-band column. When a stainless steel spinning-band was used, the quadricyclane derivative (49) was unstable and rapidly isomerized to (50), even at $-12°$. When a spinning-band coated with Teflon was used, however, the quadricyclane product showed no detectable change after 4 months at $-12°$. If metal entrainment is responsible for the observed valence isomerization as proposed by Gassman, these results are truly striking. One can only speculate on the quantities of metal extracted by a diene from a stainless steel surface at distillation temperatures. At best, these can be only trace amounts suggesting levels of catalytic activity and catalyst life impressively high.

The importance of ligands and the availability of ligand positions in a metal complex have already been briefly discussed. Manassen has addressed this subject specifically in a careful study on the effects of metal d electron configuration and ligand structure on metal catalyzed valence isomerization (54). Using the isomerization of quadricyclane (8) to norbornadiene (9) as a probe, he examined a number of square planar complexes [phthalocyanine, tetraphenyl porphyrin, and N,N-ethylene-(salicylideneiminato)] containing transition metals of varying d electron configurations, e.g., $d^5(Mn^{+2})$, $d^6(Fe^{+2})$, $d^7(Co^{+2})$, $d^8(Ni^{+2}, Pt^{+2})$, $d^9(Cu^{+2}, Ag^{+2})$, and $d^{10}(Zn^{+2})$. Of the systems examined, the d^6, d^7, and d^8 configurations were the most frequently active. The importance of ligand configuration was found to be particularly interesting. Using N,N-bis(salicylidene)polymethylene diamino cobalt (51) as a model catalyst system, Manassen was

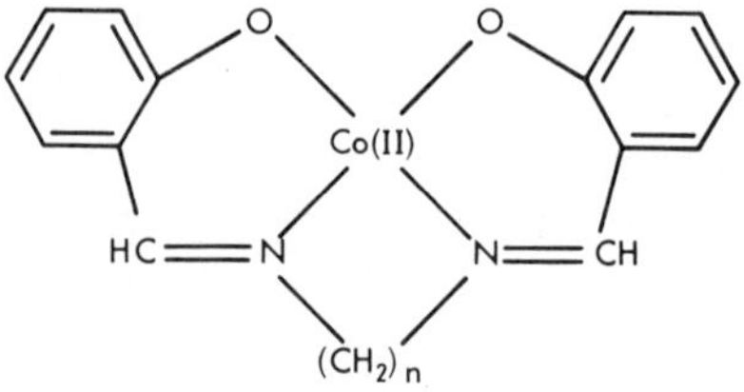

(51)

able to alter ligand geometry about the cobalt going from square planar to the preferred tetrahedral by varying the number (n) of methylenes in the chain linking the nitrogen atoms. Reactivity for (8) → (9) (at 150°) could be switched from its highest point ($n = 2$, square planar) to an intermediate point, ($n = 3$) to zero ($n = 5$, tetrahedral). Thus catalytic activity at the cobalt center is directly related to ligand configuration. It could be attenuated downward by twisting the ligand from its square planar configuration to its tetrahedral configuration where activity vanishes.

Different square planar cobalt(II) complexes were also compared for (8) → (9) activity. At 35°, solutions (1 ml $CHCl_3$, 3 m mole (8) and 0.01 m mole catalyst) of (51) isomerized (8) with a $t_{1/2} = 21$ minutes. Cobalt(II) tetraphenyl porphyrin (CoTPP), under the same conditions, isomerized quadricyclane vigorously, causing the chloroform solution to boil from the rapid release of heat of reaction. The high activity of the porphyrin complex could be controlled somewhat using amine bases (4-picoline and pyridene), known to coordinate at the octahedral sites of metalloporphyrins (55). Mole ratios of amine base to catalyst were varied from 50 to 300, altering the $t_{1/2}$ of quadricyclane from approximately 10 to 30 minutes. The appreciable excess of base over catalyst required to bring the rate of reaction to levels noted was attributed to high levels of CoTPP activity.

Manassen, noting the absence of an empty d orbital of AS symmetry in the CoTPP complex, offers the interesting possibility of using a low energy, empty ligand orbital of that symmetry to accept the quadricyclane electron pair. Ligand orbital energies for metal porphyrins, calculated by Zerner and Gouterman (56), are shown by the authors to rest just above the filled d orbitals of the metal. Of the transition elements Mn through Zn, Co shows the smallest energy separation between the filled d shell and the empty ligand orbital, offering a possible explanation for the exceptional activity of the cobalt porphyrin.

III. The [1,3] Sigmatropic Transformation—A Molecular Orbital Treatment

Olefins are constrained to rearrange along paths dictated by orbital symmetry conservation. One path not open to simple olefins involves the movement of a hydrogen atom from the α carbon, across the face of the allyl moiety to the terminal carbon of the allylic system [Eq. (19)]. This

 (19)

constitutes a suprafacial [1,3] hydrogen shift, a symmetry-forbidden process (57). A transition metal can, in theory, remove the orbital symmetry restrictions, opening the suprafacial corridor for α hydrogen atom migration. Since a fairly detailed description of the metal-catalyzed, forbidden-to-allowed process has recently appeared (14), only a brief treatment of the subject will be attempted here. This section will deal mainly with the results of extended Hückel molecular orbital calculations carried out on simple model systems designed to approximate this catalytic process. The details of the calculations appear in the Appendix.

The essential element in the molecular orbital picture restricting the movement of the α hydrogen atom is the symmetry of the highest occupied molecular orbital of the π-allyl radical, ψ_2. In the [1,3] suprafacial shift,

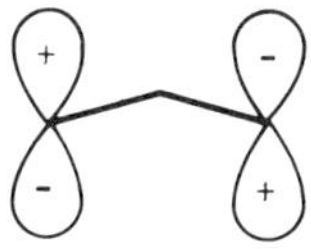

$$\psi_2$$

ψ_2 constitutes the highest occupied molecular orbital in the rearranging system. Because of its symmetry, bonding between the migrating hydrogen atom and allylic moiety beneath it is significantly reduced. The ψ_2 is somewhat antibonding at the transition state, contributing nothing to hydrogen-allyl bonding. A clearer understanding of the nature of the symmetry restrictions is best achieved by examining all the contributing molecular orbitals in the rearranging species and their symmetry-correlated partners in the parent olefin. These are illustrated in Fig. 10. The molecular orbitals in the Figure are displayed in their approximate energy order. Since we are concerned here only with their symmetries with respect to the bonding contributions across the suprafacial arch, no attempt is made to

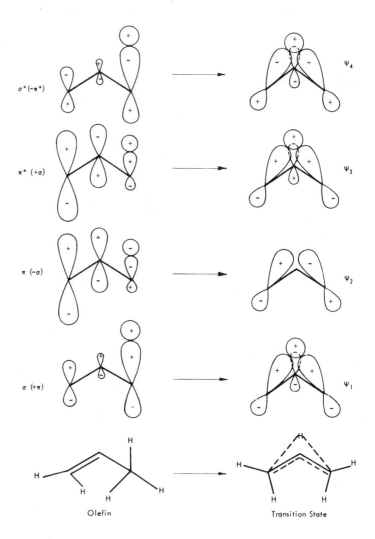

Fig. 10. The molecular orbitals of the transition state of the suprafacial [1,3] sigmatropic transformation correlated with the orbitals in the starting olefin.

reflect actual relative energy levels. The molecular orbitals of the olefin are "mixed" with other orbitals of the same symmetry (noted in parentheses). The extent of mixing in the actual molecular orbital is not implied by the relative sizes of the included orbitals in Fig. 10. These particular combinations were selected to illustrate the symmetry match between correlated partners and to reflect the nature of change in the orbitals as the nuclei proceed along the indicated reaction coordinate. In the π orbital of the olefin, for example, the p orbital on the central carbon and the s orbital of the hydrogen vanish from the molecular orbital as the system approaches the transition state. This process is easily illustrated by carrying out a series of molecular orbital calculations on models approximating the system at various points along the reaction coordinate. The contributions of the s and p orbitals would gradually diminish in the π molecular orbital until a full plane of symmetry was achieved at the transition state where their contributions would be zero.

Two bonds in the olefin are significantly changed approaching the transition state, the π and C—H σ bonds. The molecular orbitals representing these bonds, $\pi(-\sigma)$ and $\sigma(+\pi)$, transform into ψ_2 and ψ_1, respectively. While ψ_1 is a bonding molecular orbital, ψ_2 is not. In fact, ψ_2 is slightly antibonding at the transition state. In other words, the cost of moving the α hydrogen atom across the suprafacial corridor is the loss of one bond in the system. It is this that accounts for the high energy transition state and the symmetry-forbidden assignment.

The role of the transition metal in the forbidden-to-allowed process can best be visualized by first focusing on those molecular orbitals which contribute to the bonding between the allyl moiety and the migrating hydrogen atom. Since we are only considering here the suprafacial mode of rearrangement, it is the bonding along this corridor which is of interest. Of the four molecular orbitals in the transition state, two are bonding across the suprafacial arch (ψ_1 and ψ_3) one is nonbonding (ψ_2) and the other is antibonding (ψ_4). In the metal-free system, only ψ_1 and ψ_2 are populated with electrons; ψ_3 and ψ_4 are empty. Although ψ_3 remains essentially an antibonding molecular orbital (i.e., between the terminal and central carbons and the central carbon and hydrogen) it is substantially bonding across the path of rearrangement. Electronic population of this orbital, then, would restore bonding character between the terminal carbons and the migrating hydrogen atom. A catalyst can conceivably support the [1,3] suprafacial transformation by injecting electron density into ψ_3, thereby providing additional bonding within the corridor of migration.

The nature of interaction between the metal center and the transforming olefin ligand is easily visualized. Since ψ_3 correlates with π^* (cf. Fig. 10), metal electrons, through d orbital interaction with π^*, are provided a smooth avenue of entry into ψ_3. Another dimension to the catalytic role of the metal involves its interaction with the rearranging olefin at the transition state. The ligand system at the transition state resembles the π-allyl ligand, a molecular configuration known to bond nicely to transition metal centers. The catalyst, therefore, can support rearrangement by injecting electron density into ψ_3 through π^* and, in addition, lower the energy of the transformation through π-allyl-like bonding at the transition state. The important interactions here are metal-to-ligand (ψ_3) and ligand (ψ_2)-to-metal; these are illustrated in Fig. 11. The shading is meant to re-

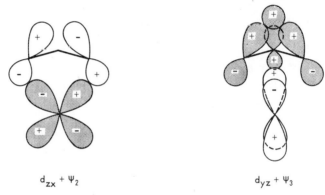

$$d_{zx} + \Psi_2 \qquad\qquad d_{yz} + \Psi_3$$

Fig. 11. The interaction between catalyst d orbitals and the allyl ψ_2 and ψ_3 molecular orbitals at the transition state of the [1,3] suprafacial transformation.

flect a relocalization of electron density from metal-to-ligand (i.e., d_{YZ} to ψ_3) and from ligand-to-metal (i.e., ψ_2 to d_{ZX}).

Extended Hückel (58) molecular orbital calculations have been carried out on model systems to better illustrate the role of the transition metal in this forbidden-to-allowed process. The details of the calculations and the assumptions made are given at the end of this section. Our concern here focuses on the nature of the interactions described above. We, of course, made no attempt to approximate the activation energy for the hypothetical catalytic process. Since we are only interested in the orbital interactions, realistic models of metal systems were not attempted. A simple, naked metal coordinated approximated 2 Å from the ligand model was used in all calculations. Details of the ligand geometries, including bond distances

and angles are also reserved for the Appendix. Two ligand configurations
for the transition state were used. In the first [configuration (A)], the allyl

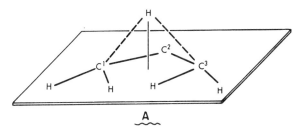

A

ligand was planar, with all bonded hydrogens and carbons occupying the
same plane. The C—C—C angle was 110°; the migrating hydrogen atom
was positioned 1.2 Å above the plane and 1.64 Å from the terminal
methylene carbons (this placed the hydrogen atom 1.54 Å from the central
carbon atom). The second configuration (configuration (B)) was essentially

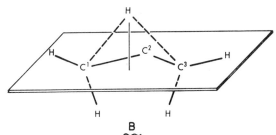

B

the same as A except that the methylene carbons were rotated about the
C—C bonds moving the hydrogen atoms out of the plane and bent slightly
backwards approximating, in space, positions midway between tetrahedral
(sp^3) and planar (sp^2). The hydrogens were thus symmetrically positioned
about axes passing through the terminal carbons and intersecting at the
migrating hydrogen. Configuration (B) is, of course, a better approximation of the transition state than is (A). Configuration (A), however, offers
the opportunity to reduce a rather complex molecular orbital picture to its
contributing components. In (A), for example, the critical molecular orbitals stand out more clearly with a minimum amount of mixing from
other orbitals of the same symmetry. (A) can be looked upon as a π-allyl
system perturbed by a hydrogen atom.
    The molecular orbital calculations on model (A) illustrates an interesting
feature associated with the bonding configuration of the transition state,
namely the polarization of the allyl π orbitals. This is shown in Table 1

TABLE 1

THE ALLYLIC MOLECULAR ORBITALS FOR CONFIGURATION A

Orbital	Energy	Occupation	Atomic orbital coefficients			
			C_1P_Z	C_2P_Z	C_3P_Z	Hs
ψ_1^a	−14.169	↑↓	0.340	0.445	0.340	0.285
	−15.918	↑↓	−0.155	−0.248	−0.155	−0.517
ψ_2	−8.721	↑↓	0.665	0.0	−0.665	0.0
ψ_3	−6.264		0.269	−0.705	0.269	0.423
ψ_4	−4.299		−0.510	0.035	−0.510	0.597

^{a} ψ_1 is significantly mixed with σ molecular orbitals of the same symmetry. The two orbitals with the greatest contribution from ψ_1 are therefore given.

which lists the coefficients of the carbon p_Z orbitals and the s orbital of the migrating hydrogen atom in the four molecular orbitals displayed in Fig. 10. In the ψ_1 molecular orbitals, the hydrogen s orbital is more strongly bonded to the central carbon (C_2) than to the terminal carbons. Since these are the only occupied molecular orbitals contributing to hydrogen–allyl bonding, this means that at the transition state, the hydrogen is more strongly attracted to the central carbon than to either methylene. Although the molecular orbitals of configuration (B) do not illustrate this feature as clearly because of mixing, the direction of bonding is nevertheless the same. The bonding relationship between the migrating hydrogen atom and the three allylic carbons is reflected in the Mulliken population analysis (59), an approximation of the electron density focused in the bonding regions between nuclei.

MULLIKEN POPULATION ANALYSIS

	C_1—H	C_2—H	C_3—H
Configuration A	0.056	0.162	0.056
Configuration B	0.064	0.147	0.064

Although the extent of bonding is slightly shifted toward the terminal methylenes in B, the bonding character is still significantly polarized in the direction of the central carbon. Electronic population of ψ_3 (cf. Table 1) would tend to reverse the direction of bonding, spreading it more symmetrically over the three carbon atoms. The interaction between C_2 and H is strongly antibonding in ψ_3 (i.e., the coefficients are large and have opposite signs). Transition metal interaction with ψ_3 should therefore refocus the H-to-allyl bonding electron density from the plane bisecting the allylic carbon chain into the suprafacial arch.

Molecular orbital calculations were carried out on models of iron complexes with ligands in configurations A and B, as well as simple propylene; ψ_3 mixed significantly with the d orbitals of the transition metal, refocusing the bonding electron density in the expected directions. The extent of interaction is reflected in the Mulliken population analysis for the iron-configuration B complex.

C_1-H	C_2-H	C_3-H
0.092	0.081	0.092

The pattern of interaction and the degree of d orbital mixing can be seen in the actual molecular orbitals, tabulated in the Appendix. The increase in bonding electron density along the suprafacial corridor quite clearly stems from strong metal d orbital-ψ_3 interaction. The important orbital interactions have been selected from the total orbital description and are tabulated in Table 2. The backbonding character in the complex is reflected in the ψ_3 molecular orbitals; ψ_3 correlates with the π^* molecular orbital in the olefin (cf. Fig. 10), the focal point of backbonding in metal-olefin complexes. Thus, some indication of the degree of backbonding maintained through the transition state can be seen in a comparison of ψ_3 and π^* in the molecular orbitals of the appropriate metal complexes. The π^* molecular orbital in the iron-propylene complex (cf. Appendix) compares reasonably well with ψ_3, indicating that no significant change in the degree of metal-to-ligand backbonding occurs with the change in ligand molecular configuration indicated in this approximation of the transition state. The absolute degree of backbonding indicated in these calculations, of course, is not meaningful. The calculations themselves are semiempirical and only approximations of real wave functions. Then the models are purely fictitious created only to reflect the general changes in the composite of ligand molecular orbitals that occur upon introducing a transition metal at a reasonable bonding distance. Although molecular orbital calculations of this kind may be useful in describing activity trends, that is, which metal systems (various metals in a variety of oxidation states and ligand environments) can best carry out the catalytic molecular rearrangement, their value for indicating the likelihood that a given metal system will catalyze

TABLE 2

THE ALLYLIC MOLECULAR ORBITALS FOR IRON-LIGAND CONFIGURATION A COMPLEX[a]

Orbital	Energy	Occupation	Atomic orbital coefficients								
			C_1P_x	C_2P_x	C_3P_x	Hs	$Fe d_{z^2}$	$Fe d_{xz}$	$Fe d_{x^2-y^2}$	$Fe d_{yz}$	$Fe d_{xy}$
ψ_1	−15.65	⇅	−0.091	−0.148	−0.091	−0.445	0.023		0.004		0.022
ψ_1	−14.09	⇅	−0.310	−0.267	−0.310	−0.328	−0.021	−0.026	0.018	0.027	−0.037
ψ_2	−9.22	⇅	0.603	0.0	−0.603	0.0		0.40		0.049	
ψ_3	−7.50	⇅	−0.164	0.332	−0.164	−0.119	0.625		0.318		−0.529
ψ_3	−7.06	⇅	0.040	−0.076	0.040	0.037	−0.134		0.907		0.226
ψ_2	−6.28		0.260	0.0	−0.260	0.0		−0.890		0.058	
ψ_3	−5.49		−0.182	0.626	−0.182	−0.265	−0.286		0.059		0.395
ψ_4	−4.04		0.459	−0.118	0.459	−0.345	0.142		0.107		−0.016

[a] The iron is positioned 2.19 Å from C_1 and C_3, and 2.24 Å from C_2. To be consistent with the complete calculations given in the Appendix, these results refer to a coordinate system which differs from that used in the previous section of this Chapter. In this coordinate system, the X axis passes through the allyl plane with the XY plane besecting the allyl carbon chain.

the rearrangement in question is very limited. For the crude model used here, the calculations clearly show that the transition metal provides bonding along the suprafacial arch as first postulated on the basis of a simple orbital symmetry picture. The calculations further highlight an interesting refocusing of bonding electron density on the suprafacial face. They also illustrate the backbonding flow of metal valence electrons from the starting olefin π^* orbital into the allylic ψ_3, providing bonding along the suprafacial corridor. The extent of bonding between the migrating hydrogen atom and the terminal methylene carbons, approximated in the Mulliken population analysis, was shown to increase by about 44% upon introducing the iron atom. It should be noted, however, that no significant attempt was made to locate the energy minimum for the transition state. For the kinds of questions addressed here, a better approximation of the transition state would probably have been of limited value. The increase in bonding noted could undoubtedly be improved with a better approximation of the transition state. This, however, would be of little additional significance because of the fictitious nature of the model system chosen and the approximate nature of semiempircal calculations of this kind.

The molecular orbital treatment described here supports the general contention that bonding along the suprafacial corridor of the [1,3] sigmatropic path is increased upon coordination of the olefin to a transition metal. It is further shown that this bonding is generated through the introduction of metal valence electrons into the backbonding molecular orbitals of the transforming olefin system. The likelihood that a transition metal complex will, in fact, catalyze a suprafacial [1,3] sigmatropic transformations remains essentially unknown, the rightful subject of experimental chemistry.

The question of an experimental verification of the metal-catalyzed suprafacial [1,3] sigmatropic hydrogen migration has been treated earlier (14). At that time there were few examples in the literature (60) which suggested this kind of catalytic intervention. Interesting cases, however, have subsequently been reported suggesting metal-assisted [1,3] hydrogen shifts. Abley and McQuillin have proposed this kind of metal intervention in the metal-catalyzed reduction of diphenylacetylene to trans-stilbene (61). Using deuterium gas, these authors were able to show that one atom of deuterium is incorporated at the ortho-position of the phenyl group of the product, trans-stilbene. It thus appears that the trans-orientation of the product double bond arises from a suprafacial [1,3] hydrogen shift from the ortho-phenyl position to the olefinic carbon atom. Abley and McQuillin compare this catalytic transformation to analogous systems where benzyl centers undergo palladium-catalyzed hydrogenolysis with stereochemical inversion (*62*). It is interesting to note that in these examples the [1,3]

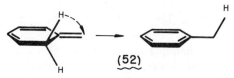

(52)

migration postulated is one that is energetically quite favorable. This transformation is not unlike the valence isomerization of quadricyclane to norbornadiene in that both reactions contain significant driving force. In the [1,3] hydrogen shift, the energy gained through the generation of aromaticity in the incipient benzene ring, could reasonably complement the metals function in providing bonding character across the suprafacial corridor.

Appendix A: Iterative Extended Hückel Calculations as an Aid in Applying Orbital Symmetry Rules in Transition Metal Catalysis

As we have pointed out, the ordering of the *d* levels is crucial in deciding the effect of a given metal complex on a forbidden reaction. It is therefore advantageous to have a reliable method of predicting the ordering of the molecular orbitals in the metal complex. *Ab initio* molecular orbital calculations with large Gaussian basis sets (*63*) could in principle give good results; however, these calculations are far too expensive (hundreds of hours of computer time) for even the simplest transition metal complexes. The only type of calculation which can be routinely applied to transition metal complexes at this time are the so-called extended Hückel molecular orbital methods as proposed by Wolfsberg and Helmholtz (*64*) and extensively applied to organic molecules by Hoffmann (*58*). Similar methods have been used on transition metal complexes by Ballhausen and Gray (*65*) and by Fenske and co-workers (*66*). These methods are empirical and open to many criticisms since they have no sound basis in theory; however, we have found these calculations to be quite useful as long as one is aware of their limitations and does not overinterpret the results.

In particular we have found the self-consistent charge and configuration (SCCC) variation of the iterative extended Hückel method (*65*) useful for model calculations in which we vary the metal, the ligands, and the geometry to obtain some feeling for the effect of the metal and the "symmetry" of the ligand field on the reacting ligands (9). Results of these calculations should be considered qualitative, indicating trends within a series of related molecules, and not quantitative. We shall outline the details of the method we use since it has proved useful when applied in a consistent fashion.

Molecular orbitals, φ_i, are expressed by the well known LCAO method, as a linear combination of atomic orbitals, χ_μ

$$\varphi_i = \sum_\mu C_{\mu i} \chi_\mu \tag{14}$$

where the $C_{\mu i}$ are elements of a matrix C which simultaneously diagonalizes the Hamiltonian matrix H and the overlap matrix S; i.e.

$$C'HC = \varepsilon \tag{15}$$

$$C'SC = I. \tag{16}$$

ε is a diagonal matrix of the molecular orbital energies and I is the unit matrix. The matrix elements $H_{\mu\nu}$ and $S_{\mu\nu}$ are the integrals

$$H_{\mu\nu} = \int \chi_\mu \mathscr{H} \chi_\nu \, d\tau \tag{17}$$

and

$$S_{\mu\nu} = \int \chi_\mu \chi_\nu \, d\tau \tag{18}$$

where $\mathscr{H}$ is the Hamiltonian operator for the molecule. In extended Hückel theory the overlap integral, $S_{\mu\nu}$, is computed exactly from the assumed basis functions χ_μ, and $H_{\mu\nu}$ is approximated empirically as we shall describe below. The basis set is taken to include all the valence atomic orbitals of the atoms in the molecule. For nontransition metals, except hydrogen, we use the s and p valence orbitals for the "best" single zeta Slater SCF functions given by Clementi and Raimondi (67). For the first row transition metals we take the $3d$, $4s$, and $4p$ functions given by Richardson and co-workers (68), and for the second and third row transition metals the valence d, s, and p functions given by Basch and Gray (69). Hydrogen is given a $1s$ Slater orbital with an exponent of 1.2.

Approximation of H_{ii}

The diagonal elements of H are approximated by the valence orbital ionization potential (VOIP) of the orbital in the isolated atom as a function of the charge on the atom

$$H_{ii} = \text{VOIP}_i(q) = -I_i^\circ - A_i q - B_i q^2 \tag{19}$$

where I_i° is the VOIP for the ith orbital on the neutral atom for the weighted average of the energies of all the multiplet terms arising from a given configuration (70). A_i and B_i are empirical parameters obtained by fitting the VOIP as a quadratic function of the charge on the atom. q is the net charge on the atom as computed by a population analysis using Lowdin's orthogonalized atomic orbitals (71). In transition metals three orbital energies are required: d, s, and p. These orbital energies depend strongly on the configuration; e.g., the VOIP for the d orbital in Ni is 5.901 ev for the configuration $d^n \to d^{n-1}$, 10.03 ev for $d^{n-1}s \to d^{n-2}s$ and 11.89 ev for $d^{n-1}p \to d^{n-2}p$. In order to approximate the configuration of the atom in the molecule, the VOIP for each metal orbital is computed as a linear combination of three different configurations according to the following equations:

$$\text{VOIP}(d) = (1.0 - n_s - n_p)\,\text{VOIP}(d:d^n)$$

$$+ n_s\text{VOIP}(d:d^{n-1}s) + n_p(d:d^{n-1})p \tag{20}$$

$$\text{VOIP}(s) = (2.0 - n_s - n_p)\,\text{VOIP}(s:d^{n-1}s)$$

$$+ (n_s - 1.0)\,\text{VOIP}(s:d^{n-2}s^2) + n_p\text{VOIP}(s:d^{n-2}sp) \tag{21}$$

$$\text{VOIP}(p) = (2.0 - n_s - n_p)\,\text{VOIP}(p:d^{n-1}p)$$

$$+ (n_p - 1.0)\,\text{VOIP}(p:d^{n-2}p^2) + n_s\text{VOIP}(p:d^{n-2}sp) \tag{22}$$

where n is the total number of valence electrons, n_s is the number of s electrons, and n_p is the number of p electrons. n_s and n_p are obtained from a Lowdin population analysis. The symbol $\text{VOIP}(d:d^{n-1}p)$ means the ionization potential of a d electron for the configuration $d^{n-1}p$ (i.e.. $d^{n-1} \to d^{n-2}p$).

For nonmetallic atoms the VOIP's do not depend as strongly on the configuration and we use the VOIP for the configuration sp^{n-1} (e.g., for $C:sp^3$ which is not to be confused with the notation for an sp^3 hybrid) which by our experience is very close to the final configuration obtained in most molecules. Table 3 gives the VOIP parameters for the elements through argon obtained from the data of Hinze and Jaffe (72).

The n_s, n_p, and q depend on the molecular orbital coefficients, $C_{\mu i}$, and are iterated to self-consistency of 0.005 units of charge and configuration.

It should be pointed out that the procedure given above neglects the effect of the field of the neighboring atoms and should not be used on highly ionic or charged species. For charged species the diagonal terms should be corrected for the potential due to the charge distributions on neighboring atoms*.

APPROXIMATION OF H_{ij}

In the application of the extended Hückel method there are three commonly used relations for approximating the off-diagonal elements of H in which H_{ij} is taken as some function of S_{ij} and the diagonal elements of H. The first and most widely used approximation suggested by Wolfsberg and Helmholtz (64) uses the relation

$$H_{ij} = kS_{ij}(H_{ij} + H_{jj})/2 \qquad (23)$$

Ballhausen and Gray (65) have suggested the relation

$$H_{ij} = -kS_{ij}(H_{ii}H_{jj})^{1/2} \qquad (24)$$

and a third approximation due to Cusachs (73) uses the equation

$$H_{ij} = S_{ij}(k - |S_{ij}|)(H_{ii} + H_{jj})/2 \qquad (25)$$

where the value of k usually has values in the range 1.5 to 2.2. Care must be used in applying Eq. (25) to preserve the invariance to rotation of the coordinate system. The quantities $S_{ij}(k - |S_{ij}|)$ for all orbitals on a pair of atoms must be evaluated in a local coordinate system and then rotated in the same fashion as the overlap matrix (74). Since none of the relations given above have a sound theoretical justification and since comparative calculations using the three methods show no clear-cut preference for one over the others we have arbitrarily chosen to use Eq. (23) with $k = 1.80$.

In our opinion the iterative extended Hückel method in any of its variations gives at best a crude description of the ground state of the

---

* See, e.g., Fenske et al. (66) in which the correction for H_{ii} on atom a is taken to be

$$\sum_{b \neq a} \beta_b \frac{q_b}{r_{ab}}$$

where q_b is the net charge on atom b, r_{ab} is the distance from a to b, and β_b is a correction for the interpenetration of the electronic charge distributions.

TABLE 3

VOID PARAMETERS (sp^{n-1}) (eV)

Orbital	$I_i°$	A_i	B_i
H(1s)	13.60	16.65	3.80
He(1s)	24.55	20.00	—
Li(2s)	5.37	4.50	—
Li(2p)	3.54	2.98	—
Be(2s)	9.32	7.52	0.82
Be(2p)	5.31	7.17	0.75
B(2s)	14.04	9.37	1.37
B(2p)	7.37	8.78	0.91
C(2s)	19.52	11.75	1.15
C(2p)	9.75	10.86	1.55
N(2s)	25.58	13.31	1.78
N(2p)	12.38	13.09	1.54
O(2s)	32.30	15.35	1.49
O(2p)	14.61	14.77	2.17
F(2s)	40.20	17.02	1.16
F(2p)	17.42	15.08	1.16
Ne(2s)	48.00	14.00	0.80
Ne(2p)	20.40	16.00	0.5
Na(3s)	4.67	4.20	—
Na(3p)	3.04	2.96	—
Mg(3s)	7.64	6.12	0.72
Mg(3p)	4.67	5.27	0.66
Al(3s)	11.32	7.10	0.40
Al(3p)	6.00	5.99	0.10
Si(3s)	15.03	8.38	0.41
Si(3p)	7.07	7.46	1.23
P(3s)	18.16	10.12	—
P(3p)	9.85	7.93	0.35
S(3s)	21.13	10.83	1.24
S(3p)	11.07	10.37	1.68
Cl(3s)	25.23	11.48	0.70
Cl(3p)	13.92	10.44	0.24
A(3s)	29.50	13.00	—
A(3p)	16.80	10.50	—

molecule and hopefully the correct ordering of the highest few filled and lowest few empty molecular orbitals. It should be used only for comparative calculations in which one tries to establish trends in a series of related molecules or configurations or to obtain a crude picture of the ground state molecular orbitals.

Appendix B: Iterative Extended Huckel Calculations on the [1, 3] Sigmatropic Transformation

In order to investigate the feasibility of a metal-catalyzed [1,3] sigmatropic transformation, SCCC-MO calculations of the type described in Appendix A have been carried out on six model compounds.

(I) Propylene with the methyl group rotated so that one CH bond is perpendicular to the $C=C-C$ plane.

(II) Propylene complexed to Fe with the carbons about 2.1Å from the metal atom.

(III) A model intermediate for the [1,3] sigmatropic transformation with the migrating hydrogen atom above the $C=C-C$ plane 1.64Å from the terminal methylene carbons and with the CH_2 groups in the $C=C-C$ plane. (Configuration A.)

(IV) Configuration A complexed to Fe with the carbons about 2.1Å from the metal atom.

(V) The same as (II) except with the CH_2 groups rotated and bent backward approximating a geometry midway between tetrahedral (sp^3) and planar (sp^2). (Configuration B.)

(VI) Configuration B complexed to Fe with the carbons about 2.1Å from the metal atom.

Atom numbers, Cartesian coordinates, and more detailed geometry for each model are given in Table 4. Figure 12 shows the projections of the model compounds in the yz plane. Basis functions for H, C, and Fe are tabulated in Table 5, and the VOIP parameters are given in Table 6. The calculations were iterated to self-consistency of <0.005 ev in charge and configuration. Net charges computed by a Löwdin population analysis are listed in Table 7, and the Mulliken overlap populations in Table 8. Molecular orbital coefficients and orbital energies are given in Table 9.

TABLE 4

Geometry for Model Compounds

(a) Propylene (I and II)[a]

Cartesians[b]

Atom	Number	x	y	z
C	1	1.82242	−0.21316	−1.19903
C	2	2.03681	0.37588	0.0
C	3	1.80102	−0.27193	1.31867
H	4	2.01851	0.32560	−2.10246
H	5	1.45678	−1.21775	−1.24384
H	6	2.40277	1.38135	0.0
H	7	0.99086	−0.9951	1.22833
H	8	1.53036	0.48677	2.05298
H	9	2.70914	−0.78159	1.64066
Fe	10	0.0	0.0	0.0

Bond distances (Å)	Bond angles
$R(1-2) = 1.353$	$\angle 1{-}2{-}3 = 124.8°$
$R(2-3) = 1.488$	$\angle 4{-}1{-}2 = 120.0°$
$R(1-4) = 1.07$ (CH_2)	$\angle 6{-}2{-}1 = 117.6°$
$R(2-6) = 1.07$ (CH)	CH_3 tetrahedral
$R(3-7) = 1.09$ (CH_3)	H(9) up

(b) Configuration A (III and IV)

Cartesians[c]

Atom	Number	x	y	z
C	1	1.76216	−0.37870	−1.14681
C	2	2.03681	0.37588	0.0
C	3	1.76216	−0.37870	1.14681
H	4	1.39957	−1.37492	−0.94074
H	5	1.95241	0.14399	−2.07255
H	6	2.40277	1.38135	0.0
H	7	1.39957	−1.37492	0.94074
H	8	1.95241	0.14399	2.07255
H	9	2.78038	−0.97020	0.0
Fe	10	0.0	0.0	0.0

Bond distances (Å) Angles

$R(1-2) = 1.40$ $\angle 1{-}2{-}3 = 110°$[e]

$R(1-4) = 1.08$ $\angle 4{-}1{-}2 = 114°$

$R(2-6) = 1.07$ $\angle 6{-}2{-}1 = 125°$

$R(Fe-1) = 2.1362$ Dihedral angles

$R(Fe-2) = 2.0719$ CH_2 group in C–C–C plane

$R(4-9) = 1.6437$ The C–C–C plane is tilted 20° from

$R(2-9) = 1.5378$ the plane perpendicular to the x axis.

(c) Configuration B (V and VI)

Cartesians[d]				
Atom	Number	x	y	z
H	4	0.75798	−0.77465	−1.18206
H	5	2.25284	−0.01722	−2.03843
H	7	0.75798	−0.77465	1.18206
H	8	2.25284	−0.01722	2.03843

Other atoms the same as Configuration A.

Dihedral angles
$\angle 5\text{–}1\text{–}2\text{–}3 = 160°$
$\angle 4\text{–}1\text{–}2\text{–}3 = -55°$
$\angle 8\text{–}3\text{–}2\text{–}1 = -160°$
$\angle 7\text{–}3\text{–}2\text{–}1 = 55°$

[a] For comparison of MO coefficients propylene coordinates are in the same coordinate system as models A and B.
[b] See Fig. 12a for projection in yz plane.
[c] See Fig. 12b for projection in yz plane.
[d] See Fig. 12c for projection in yz plane.

TABLE 5

BASIS FUNCTIONS[a]

$\chi(H_{1s}) = 1.00\ (1, 0, 1.2)$
$\chi(C_{2s}) = -0.235078\ (1, 0, 5.6727) + 1.024702$
$\qquad (2, 0, 1.6083)$
$\chi(C_{2p}) = 1.00\ (2, 1, 1.5679)$
$\chi(Fe_{3d}) = 0.5366\ (3, 5, 5.35) + 0.6678\ (3, 2, 1.80)$
$\chi(Fe_{4s}) = -0.02078\ (1, 0, 25.38) + 0.07052\ (2, 0, 9.75)$
$\qquad -0.1744\ (3, 0, 4.48) + 1.0125\ (4, 0, 1.40)$
$\chi(Fe_{4p}) = 0.01118\ (2, 1, 10.60) - 0.03833\ (3, 1, 4.17)$
$\qquad + 1.00067\ (4, 1, 0.80)$

[a] Basis functions are given as linear combinations of Slater functions. Quantities in parentheses give principal quantum number, n, azimuthal quantum number, l, and Slater exponent, ζ, i.e. (n, l, ζ).
Fe functions from Ref. *68*.
C functions from Ref. *67*.

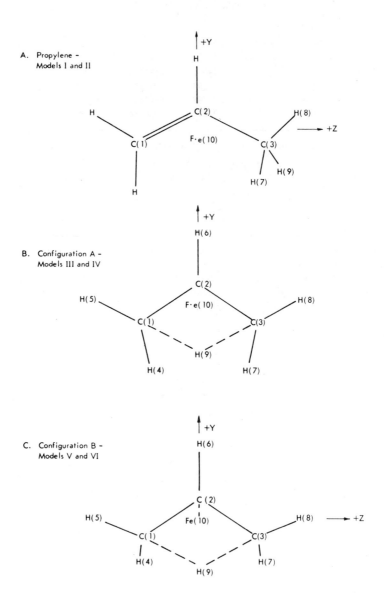

Fig. 12. Model compounds for SCCC-MO calculations.

TABLE 6

Voip Parameters for Fe^a (eV)

Ionization			I°	A	B
Fe(3d)	$d^n \longrightarrow$	d^{n-1}	5.195	10.687	1.711
	$d^{n-1}s \longrightarrow$	$d^{n-2}s$	8.195	12.584	1.711
	$d^{n-1}p \longrightarrow$	d^{n-2}	10.067	12.633	1.711
Fe(4s)	$d^{n-1}s \longrightarrow$	d^{n-1}	7.104	7.916	0.911
	$d^{n-2}s^2 \longrightarrow$	$d^{n-2}s$	8.468	9.057	0.911
	$d^{n-2}sp \longrightarrow$	$d^{n-2}p$	10.092	8.350	0.911
Fe(4p)	$d^{n-1}p \longrightarrow$	d^{n-1}	3.707	6.298	0.905
	$d^{n-2}p^2 \longrightarrow$	$d^{n-2}p$	4.922	7.166	0.905
	$d^{n-2}sp \longrightarrow$	$d^{n-2}s$	4.996	7.166	0.905

^{a} From Ref. 70. See Table 3 for C and H.

TABLE 7

Net Charges (Lowdin)

	I	II	III	IV	V	VI
C(1)	0.0049	−0.0533	−0.0578	−0.0411	−0.0560	−0.0736
C(2)	0.0419	−0.0141	0.1368	0.0547	0.1615	—0.0676
C(3)	0.0520	0.0545	−0.0578	⁺0.0411	−0.0560	−0.0736
H(4)	−0.0207	−0.0390	−0.0278	−0.0085	−0.0305	0.0065
H(5)	−0.0209	−0.0242	−0.0369	−0.0315	−0.0445	−0.0522
H(b)	−0.0198	−0.0339	0.0134	−0.0073	0.0222	−0.0075
H(7)	−0.0143	0.0124	−0.0278	−0.0085	−0.0305	0.0065
H(8)	−0.0145	−0.0102	−0.0369	−0.0315	−0.0445	−0.0522
H(9)	−0.0087	−0.0124	0.0956	0.0469	0.0772	0.0154
Fe(10)	—	0.1203	—	0.0677	—	0.1632

TABLE 8

MULLIKEN OVERLAP POPULATIONS

	I	II	III	IV	V	VI
C(1)—C(2)	0.6521	0.5774	0.4556	0.4356	0.4367	0.4368
C(2)—C(3)	0.3921	0.4222	0.4556	0.4356	0.4367	0.4368
C(1)—C(4)	0.4125	0.4363	0.4108	0.4423	0.4031	0.3890
C(1)—H(5)	0.4123	0.4428	0.4050	0.4349	0.4067	0.4290
C(2)—H(6)	0.4181	0.4467	0.4182	0.4480	0.4154	0.4429
C(3)—H(7)	0.4078	0.4203	0.4108	0.4423	0.4031	0.3890
C(3)—H(8)	0.4069	0.4371	0.4050	0.4349	0.4067	0.4290
C(3)—H(9)	0.4021	0.4209	0.0563	0.0802	0.0635	0.0920
C(1)—H(9)	−0.0039	−0.0023	0.0563	0.0802	0.0635	0.0920
C(2)—H(9)	−0.0294	−0.0289	0.1619	0.1025	0.1466	0.0811
C(1)—Fe	—	−0.0340	—	−0.0571	—	−0.0930
C(2)—Fe	—	−0.0445	—	−0.0612	—	−0.0655
C(3)—Fe	—	−0.1333	—	−0.0571	—	−0.0930
H(4)—Fe	—	−0.0310	—	−0.0374	—	0.0334
H(5)—Fe	—	−0.0360	—	−0.0323	—	−0.0298
H(6)—Fe	—	−0.0337	—	−0.0328	—	−0.0223
H(7)—Fe	—	−0.0077	—	−0.0374	—	0.0334
H(8)—Fe	—	−0.0384	—	−0.0323	—	−0.0298
H(9)—Fe	—	−0.0264	—	−0.0575	—	−0.0304

TABLE 9

MOLECULAR ORBITAL COEFFICIENTS AND ORBITAL ENERGIES*

MODEL I

IGINAL BASIS

, NO. ERGY CUPATION	1 -26,487 2,000	2 -22,722 2,000	3 -18,555 2,000	4 -15,285 2,000	5 -15,004 2,000	6 -14,529 2,000	7 -13,201 2,000	8 -12,598 2,000
(1)2S	-,339	,429	,243	-,023	,001	-,014	,056	-,082
(1)2PZ	-,006	,001	-,156	,259	,012	-,104	,150	-,390
(1)2PX	-,001	,000	-,036	-,066	,052	,106	,103	-,033
(1)2PY	-,004	,001	-,099	-,167	-,003	,320	,284	-,083
(2)2S	-,456	,089	-,386	,101	,010	-,009	,097	,002
(2)2PZ	,001	-,099	-,046	,015	-,009	-,114	-,016	,523
(2)2PX	,003	-,000	-,052	-,080	,151	-,002	-,165	-,003
(2)2PY	,009	-,001	-,144	-,191	-,074	,039	-,430	-,004
(3)2S	-,365	-,440	,143	-,011	-,003	,005	,046	,074
(3)2PZ	,013	-,024	,136	-,281	-,027	-,084	-,153	-,429
(3)2PX	-,002	,003	-,036	-,101	,460	-,164	,080	,040
(3)2PY	-,007	,010	-,100	-,190	-,204	-,341	,263	,112
(4)1S	-,056	,170	,196	-,340	-,002	,266	,112	,267
(5)1S	-,059	,158	,241	,161	,018	-,353	-,319	,034
(6)1S	-,084	,038	-,361	-,121	-,014	,029	-,432	-,002
(7)1S	-,068	-,162	,153	,225	-,212	,365	-,226	-,001
(8)1S	-,065	-,172	,099	-,312	-,283	-,257	,117	-,191
(9)1S	-,069	-,173	,129	-,092	,487	,009	-,062	-,094

FE(10)3DX2Y2 means the Fe $3d_{x^2-y^2}$ orbital on atom number 10.

TABLE 9 (Continued)
MODEL I

ORIGINAL BASIS

MO. NO.	9	10	11	12	13	14	15	16
ENERGY	-11.473	-6.540	6.199	8.168	9.439	13.064	16.543	22.212
OCCUPATION	2.000	.000	.000	.000	.000	.000	.000	.000
C (1)2S	.001	-.000	-.021	-.029	-.022	-.143	.302	.706
C (1)2PZ	.002	.002	.169	-.526	.003	.336	-1.005	-.250
C (1)2PX	-.602	-.770	-.175	-.021	.086	.295	.185	-.205
C (1)2PY	.214	.284	-.461	-.012	.031	.840	.502	-.560
C (2)2S	.001	-.001	.061	-.146	-.037	.177	.225	.746
C (2)2PZ	-.002	.000	.384	-.600	-.019	.725	.050	.095
C (2)2PX	-.560	.793	-.231	.028	-.211	.124	-.269	.247
C (2)2PY	.206	-.291	-.626	.027	.101	.312	-.728	.693
C (3)2S	-.002	-.001	-.034	-.003	-.000	.095	.173	.230
C (3)2PZ	.004	.004	.386	-.960	-.071	.177	.518	.160
C (3)2PX	.127	-.019	-.219	-.227	1.146	-.299	.035	-.083
C (3)2PY	-.039	.019	-.700	-.447	-.436	-.705	.029	-.349
H (4)1S	-.004	.002	.449	-.405	-.008	-.051	-1.079	-.334
H (5)1S	.004	-.002	.483	-.007	.047	.887	.160	-.817
H (6)1S	.000	-.002	.652	.117	-.011	-.441	.452	-.934
H (7)1S	-.090	.094	-.588	-.516	.503	-.631	-.050	-.277
H (8)1S	-.087	.088	.213	.869	.574	.163	-.379	-.053
H (9)1S	.179	-.180	-.228	.249	-1.023	-.180	-.246	-.227

MO. NO.	17	18
ENERGY	40.674	52.708
OCCUPATION	.000	.000
C (1)2S	.781	-1.292
C (1)2PZ	.119	-.618
C (1)2PX	.098	-.094
C (1)2PY	.261	-.264
C (2)2S	-.009	1.504
C (2)2PZ	.875	-.432
C (2)2PX	-.029	-.126
C (2)2PY	-.077	-.339
C (3)2S	-1.538	-.764
C (3)2PZ	.004	.272
C (3)2PX	-.036	.004
C (3)2PY	-.090	-.030
H (4)1S	-.349	.293
H (5)1S	-.189	.317
H (6)1S	.034	-.383
H (7)1S	.578	.265
H (8)1S	.634	.203
H (9)1S	.610	.208

TABLE 9 (Continued)
MODEL II

IGINAL BASIS

NO.	1	2	3	4	5	6	7	8
ERGY	-26.917	-22.645	-18.184	-15.182	-14.998	-14.278	-12.880	-12.130
OCCUPATION	2.000	2.000	2.000	2.000	2.000	2.000	2.000	2.000
1)2S	-.329	.458	.252	-.042	.009	-.002	.048	.094
1)2PZ	-.014	.001	-.139	.150	-.069	-.192	.074	.413
1)2PX	.004	-.007	-.034	-.031	-.015	.116	.094	.032
1)2PY	-.002	.004	-.094	-.052	.034	.325	.269	.133
2)2S	-.465	.153	-.391	.082	-.039	-.043	.094	.036
2)2PZ	-.002	-.089	-.053	-.003	.001	-.110	.055	-.507
2)2PX	.014	-.000	-.047	-.079	-.089	.049	-.172	-.042
2)2PY	.010	-.000	-.133	-.087	.106	.129	-.417	-.072
3)2S	-.442	-.411	.114	-.019	-.005	-.009	.068	-.072
3)2PZ	.030	-.039	.138	-.277	.127	.042	-.241	.379
3)2PX	.004	.015	.046	-.286	-.394	-.388	.055	-.039
3)2PY	-.009	.014	-.123	-.235	.283	-.263	.228	-.098
4)1S	-.076	.187	.199	-.208	.084	.379	.166	-.259
5)1S	-.096	.182	:264	.039	-.020	-.379	-.327	-.097
6)1S	-.120	.063	-.363	-.052	.042	.113	-.440	-.068
7)1S	-.142	-.188	.179	.411	.099	.247	-.171	-.014
8)1S	-.121	-.187	.085	-.305	.388	-.146	.048	.172
9)1S	-.118	-.171	.116	-.234	-.443	.066	-.076	.076
10)4S	.123	.001	-.004	.040	.030	-.004	.024	-.001
10)4PZ	.028	.116	.011	.000	-.006	.012	.001	.009
10)4PX	.211	-.018	-.011	.005	.001	-.004	.010	.000
10)4PY	-.012	-.008	.045	.019	-.017	-.014	.014	.005
10)3DZ2	-.005	.001	.016	-.007	.016	.013	-.026	-.006
10)3DX7	.005	.004	-.004	.024	.019	.023	.015	-.080
10)3DX2Y2	.021	-.001	-.011	.006	-.002	-.015	.076	.019
10)3DY7	-.003	-.004	.001	-.028	.001	-.045	.002	-.007
10)3DXY	-.004	.000	-.013	-.022	.003	.001	.031	.013

NO.	9	10	11	12	13	14	15	16
ERGY	-10.969	-8.206	-7.815	-7.735	-7.692	-7.142	-5.340	-4.664
OCCUPATION	2.000	2.000	2.000	2.000	2.000	.000	.000	.000
1)2S	.015	.002	-.005	.019	-.001	.018	-.017	.021
1)2PZ	.038	.025	.003	-.004	.048	.088	-.021	.044
1)2PX	.555	-.330	-.031	.066	-.025	-.140	-.617	-.365
1)2PY	-.188	.164	.025	-.101	.079	.077	.238	.088
2)2S	.013	.001	.005	.007	.019	.000	.002	.027
2)2PZ	-.034	.049	.039	.007	-.007	-.121	-.017	.003
2)2PX	.533	.312	.011	-.003	.033	-.133	.746	-.025
2)2PY	-.219	-.123	-.035	.000	-.061	.033	-.232	-.076
3)2S	.002	.002	-.013	-.005	.000	-.028	-.010	.052
3)2PZ	-.004	-.067	-.017	.000	-.070	.023	.073	-.036
3)2PX	-.100	.035	.072	.019	.008	.115	-.002	-.163
3)2PY	.047	.073	.092	.028	-.003	.089	.011	-.059
4)1S	-.001	-.004	.011	-.046	-.016	-.031	-.003	.011
5)1S	.006	-.057	-.000	.043	-.012	.041	.023	.028
6)1S	-.000	.004	-.037	-.018	=.062	-.018	-.013	.028
7)1S	.067	.041	.038	.001	.023	.087	.068	.008
8)1S	.091	.032	.043	-.020	-.018	-.023	.068	.043
9)1S	-.152	-.080	-.049	.010	-.008	.089	-.133	-.076
10)4S	-.098	.053	.095	-.029	.183	.085	.141	-.796
10)4PZ	.013	.013	.012	-.018	.004	-.128	-.085	-.001
10)4PX	.017	-.014	-.011	.014	-.011	-.017	-.138	.583
10)4PY	-.006	.019	.005	-.038	-.024	.049	-.143	-.084
10)3DZ2	.041	.590	-.498	.409	.277	.219	-.340	-.027
10)3DX7	.119	-.398	-.283	-.001	-.016	.020	.259	.063
10)3DX2Y2	-.088	-.060	-.572	.161	-.689	-.279	.201	-.181
10)3DY7	-.015	-.119	-.556	-.649	.434	-.249	.019	.010
10)3DXY	-.093	-.439	-.096	.615	.473	-.319	.281	.055

TABLE 9 (Continued)
MODEL II

ORIGINAL BASIS

MO. NO.	17	18	19	20	21	22	23	24
ENERGY	-2.918	-.855	7.276	8.522	9.538	11.323	13.493	16.1
OCCUPATION	.000	.000	.000	.000	.000	.000	.000	.0
C (1)2S	.022	-.090	.048	-.034	.116	-.132	-.154	.3
C (1)2PZ	-.035	.172	.376	.323	.174	-.348	.296	-.9
C (1)2PX	-.162	.188	-.179	.046	-.115	-.053	.278	.1
C (1)2PY	-.217	.010	-.368	-.010	-.013	-.317	.784	.5
C (2)2S	-.026	.012	.142	.041	.110	-.152	.148	.2
C (2)2PZ	.016	.241	.464	.467	.078	-.135	.750	.0
C (2)2PX	.136	-.024	.313	.045	.181	.004	.133	-.2
C (2)2PY	-.240	.013	.522	-.012	.021	-.482	.199	-.7
C (3)2S	.017	.106	.025	-.143	.079	-.291	-.027	.2
C (3)2PZ	.008	.229	.517	.843	.186	-.020	.203	.5
C (3)2PX	-.127	-.185	.001	.364	.927	-.514	-.429	.0
C (3)2PY	-.281	-.039	.437	.372	.648	-.076	-.710	.0
H (4)1S	.024	-.046	.046	.195	.165	-.135	-.014	-1.0
H (5)1S	-.061	.101	-.305	-.110	.091	.586	.789	.2
H (6)1S	.065	-.003	.664	-.275	-.052	.358	-.365	.4
H (7)1S	-.075	.040	-.245	.428	-.106	.675	-.805	.0
H (8)1S	.066	.007	.039	-.800	-.741	-.001	.111	-.3
H (9)1S	-.022	.032	-.293	-.355	.921	.400	-.054	.2
FE(10)4S	.012	.018	-.275	.427	-.326	.764	.215	-.1
FE(10)4PZ	-.009	-1.090	.112	.166	.110	.051	.188	.0
FE(10)4PX	.057	.051	-.508	.628	-.385	.731	.140	-.1
FE(10)4PY	.979	-.012	-.231	.121	.086	-.305	-.010	.0
FE(10)3DZ2	-.048	.006	.081	-.013	-.029	-.030	.020	.0
FE(10)3DXZ	.019	-.203	.073	.060	-.052	-.003	.028	-.0
FE(10)3DX2Y2	.006	.004	-.147	.123	.102	.063	.012	-.0
FE(10)3DYZ	-.011	.040	-.036	.042	-.018	-.061	-.103	.0
FE(10)3DXY	.138	-.006	-.015	.006	.009	-.058	-.051	-.0

MO. NO.	25	26	27
ENERGY	22.155	41.928	52.342
OCCUPATION	.000	.000	.000
C (1)2S	.712	.960	-1.167
C (1)2PZ	-.272	.219	-.581
C (1)2PX	-.207	.090	.077
C (1)2PY	-.555	.294	-.216
C (2)2S	.763	-.193	1.517
C (2)2PZ	.126	.896	-.317
C (2)2PX	.216	-.032	-.157
C (2)2PY	.654	-.035	-.343
C (3)2S	.181	-1.401	-.932
C (3)2PZ	.174	-.062	.253
C (3)2PX	-.115	-.011	.005
C (3)2PY	-.391	-.047	-.022
H (4)1S	-.346	-.375	.252
H (5)1S	-.859	-.188	.336
H (6)1S	-.884	.124	-.355
H (7)1S	-.363	.670	.418
H (8)1S	-.015	.651	.304
H (9)1S	-.212	.598	.302
FE(10)4S	.046	-.197	-.146
FE(10)4PZ	-.001	-.133	-.065
FE(10)4PX	.041	-.136	-.103
FE(10)4PY	-.116	.046	.066
FE(10)3DZ2	.005	-.016	.067
FE(10)3DXZ	.047	.047	-.038
FE(10)3DX2Y2	-.066	.010	-.013
FE(10)3DYZ	-.017	.040	.023
FE(10)3DXY	-.080	.029	-.039

TABLE 9 (Continued)
MODEL III

GINAL BASIS

NO.	1	2	3	4	5	6	7	8
RGY	-26.828	-21.291	-18.897	-15.919	-14.171	-13.794	-13.671	-12.142
UPATION	2.000	2.000	2.000	2.000	2.000	2.000	2.000	2.000
1)2S	-.305	.439	.201	.135	.001	.047	-.033	-.052
1)2PZ	.009	-.018	-.078	-.185	-.167	.080	-.165	-.379
1)2PX	.005	-.004	-.023	-.094	.231	-.118	.012	-.032
1)2PY	-.007	-.010	-.087	.080	.094	-.325	-.269	-.089
2)2S	-.509	-.000	-.340	-.067	-.068	.000	-.163	.000
2)2PZ	-.000	-.184	.000	.000	.000	.183	.000	.570
2)2PX	.008	-.000	-.067	-.193	.366	-.000	.358	.000
2)2PY	.008	-.000	-.244	.185	-.130	.000	.346	-.000
3)2S	-.305	-.439	.201	.135	.001	-.047	-.033	.052
3)2PZ	-.009	-.018	.078	.185	.167	.080	.165	-.379
3)2PX	.005	.004	-.023	-.094	.231	.118	.012	.032
3)2PY	-.007	.010	-.087	.080	.094	.325	-.269	.089
4)1S	-.043	.175	.202	.009	-.248	.390	.271	-.020
5)1S	-.035	.205	.105	.311	.257	-.223	-.012	.305
6)1S	-.109	-.000	-.417	.060	-.083	.000	.339	-.000
7)1S	-.043	-.175	.202	.009	-.248	-.390	.271	.020
8)1S	-.035	-.205	.105	.311	.257	.223	-.012	-.305
9)1S	-.132	.000	.123	-.484	.289	-.000	.103	.000

NO.	9	10	11	12	13	14	15	16
RGY	-8.720	-6.266	-4.299	7.881	8.346	14.108	16.663	21.942
UPATION	2.000	.000	.000	.000	.000	.000	.000	.000
1)2S	-.000	.059	.085	-.153	.063	-.479	-.161	-.334
1)2PZ	.000	.096	.139	.036	-.864	.749	-.300	-.266
1)2PX	.683	-.325	.611	.210	-.076	-.055	.315	.159
1)2PY	-.248	.144	-.170	.681	-.208	-.130	.865	.484
2)2S	.000	.082	.115	-.001	-.000	-.367	-.000	-.604
2)2PZ	-.000	.000	.000	-.000	-.795	.000	.370	-.000
2)2PX	-.000	.876	-.091	.204	-.000	.074	.000	-.384
2)2PY	.000	-.420	-.102	.657	-.000	.223	.000	-1.005
3)2S	-.000	.059	.085	-.153	-.063	-.479	.161	.334
3)2PZ	.000	-.096	-.139	-.036	-.864	-.749	-.300	.266
3)2PX	-.683	-.325	.611	.210	.076	-.055	-.315	.159
3)2PY	.248	.144	-.170	.681	.208	-.130	-.865	.484
4)1S	-.000	.087	.143	.679	-.196	.090	.906	.484
5)1S	-.000	-.020	-.022	-.163	-.663	.898	-.410	-.131
6)1S	-.000	-.018	-.025	-.700	.000	.065	.000	1.151
7)1S	-.000	.087	.143	.679	.196	.090	-.906	.484
8)1S	.000	-.020	-.022	-.163	.663	.898	.410	-.131
9)1S	.000	-.561	-.861	.138	-.000	.031	.000	.077

TABLE 9 (Continued)
MODEL III

ORIGINAL BASIS

MO. NO.	17	18
ENERGY	31,378	54,382
OCCUPATION	,000	,000
C (1)2S	1,127	-1,015
C (1)2PZ	,104	-,527
C (1)2PX	,103	-,094
C (1)2PY	,282	-,262
C (2)2S	,000	1,574
C (2)2PZ	,985	-,000
C (2)2PX	-,000	-,211
C (2)2PY	-,000	-,584
C (3)2S	-1,127	-1,015
C (3)2PZ	,104	,527
C (3)2PX	-,103	-,094
C (3)2PY	-,282	-,262
H (4)1S	-,411	,263
H (5)1S	-,542	,212
H (6)1S	,000	-,286
H (7)1S	,411	,263
H (8)1S	,542	,212
H (9)1S	,000	-,009

TABLE 9 (Continued)
MODEL IV

ORIGINAL BASIS

O. NO.	1	2	3	4	5	6	7	8
ENERGY	-27.471	-21.562	-18.639	-15.666	-14.060	-14.006	-13.335	-11.875
OCCUPATION	2.000	2.000	2.000	2.000	2.000	2.000	2.000	2.000
(1)2S	-.366	.451	.189	.111	.005	-.036	.048	-.068
(1)2PZ	-.003	-.029	-.087	-.218	-.071	-.113	.192	-.379
(1)2PX	.014	-.011	-.034	-.054	.246	.115	-.098	-.038
(1)2PY	-.012	.010	.111	.118	.133	.316	.231	-.131
(2)2S	-.502	.000	-.379	-.071	-.017	.000	.146	.000
(2)2PZ	.000	-.153	-.005	-.000	-.000	-.124	-.000	.562
(2)2PX	.022	.000	-.058	-.107	.230	.000	-.412	-.000
(2)2PY	.003	.000	-.195	.142	-.256	.000	-.295	-.000
(3)2S	-.366	-.452	.189	.111	.005	.036	.048	.068
(3)2PZ	.003	-.029	.087	.218	.071	-.113	-.192	-.379
(3)2PX	.014	.011	-.034	-.054	.246	-.115	-.098	.038
(3)2PY	-.012	.010	-.111	.118	.133	-.316	.231	.131
(4)1S	-.116	.197	.242	-.075	-.288	-.581	-.178	.013
(5)1S	-.087	.224	.107	.347	.189	.258	-.049	.270
(6)1S	-.135	.000	-.403	.050	-.206	.000	-.341	-.000
(7)1S	-.116	-.197	.242	.075	-.288	.381	-.178	-.013
(8)1S	-.087	-.224	.107	.347	.188	-.258	-.049	-.270
(9)1S	-.154	-.000	.087	-.405	.358	.000	-.184	-.000
(10)4S	.142	.000	-.017	.037	-.069	.000	.034	.000
(10)4PZ	.000	.102	.000	-.000	.000	.022	.000	-.007
(10)4PX	.236	.000	-.022	.015	-.005	-.000	.013	.000
(10)4PY	-.029	-.000	.068	-.034	-.004	.000	.005	.000
(10)3DZ2	-.011	-.000	.011	.008	.016	-.000	-.023	.000
(10)3DXZ	.000	.000	-.000	-.000	.000	.026	-.000	.071
(10)3DX2Y2	.027	.000	-.009	.009	.000	-.000	.074	.000
(10)3DYZ	-.000	-.001	-.000	-.000	-.000	-.036	.000	.008
(10)3DXY	-.008	.000	-.008	-.004	.015	-.000	.028	-.000

O. NO.	9	10	11	12	13	14	15	16
ENERGY	-9.280	-7.738	-7.273	-7.169	-7.103	-6.236	-5.157	-4.488
OCCUPATION	2.000	2.000	2.000	2.000	2.000	.000	.000	.000
(1)2S	-.011	-.023	-.008	.010	.011	.004	-.036	.089
(1)2PZ	-.022	-.062	.028	-.010	-.054	.050	-.042	.162
(1)2PX	-.595	.201	-.064	.005	.045	-.332	.166	.213
(1)2PY	.223	-.120	-.026	-.065	-.112	.108	-.098	-.044
(2)2S	-.000	-.019	.023	-.000	-.021	.000	-.079	.113
(2)2PZ	.065	.000	.000	.032	-.000	-.123	.000	.000
(2)2PX	.000	-.406	.069	-.000	-.051	.000	-.790	-.075
(2)2PY	-.000	.173	.058	.000	.072	-.000	.326	-.133
(3)2S	.011	-.023	.008	-.010	.011	-.004	-.036	.089
(3)2PZ	-.022	.062	.028	-.010	.054	.050	.042	-.162
(3)2PX	.595	.201	-.064	-.005	.045	.332	.166	.213
(3)2PY	-.223	-.120	.026	.065	-.112	-.108	-.098	.044
(4)1S	-.031	.016	.021	.026	-.001	.067	-.098	.140
(5)1S	.008	.015	.043	-.015	.007	-.016	.015	.003
(6)1S	.000	-.023	.007	.000	.060	-.000	.035	.007
(7)1S	.031	.016	.021	-.026	-.001	-.067	.098	.140
(8)1S	-.008	.015	.043	.015	.007	.016	.015	.003
(9)1S	-.000	.152	.207	.000	.009	-.000	.494	-.676
(10)4S	-.000	-.045	.076	-.000	.247	-.000	.147	-.647
(10)4PZ	-.014	-.000	.000	-.012	.000	-.172	-.000	-.000
(10)4PX	.000	.025	.011	.000	.034	.000	.177	.393
(10)4PY	.000	-.025	.035	-.000	-.009	.000	.199	-.129
(10)3DZ2	.000	-.554	.735	.000	.043	.000	.274	.208
(10)3DXZ	-.353	.000	.000	-.194	.000	.916	-.000	-.000
(10)3DX2Y2	.000	-.038	.117	.000	.952	-.000	-.176	-.192
(10)3DYZ	.044	.000	.000	-.983	.000	-.191	.000	.000
(10)3DXY	-.000	.611	.623	.000	-.109	-.000	-.446	.084

TABLE 9 (Continued)
MODEL IV

ORIGINAL BASIS

MO. NO.	17	18	19	20	21	22	23	24
ENERGY	-3.962	-2.375	-.811	8.508	8.573	11.030	15.764	16.9
OCCUPATION	.000	.000	.000	.000	.000	.000	.000	.0
C (1)2S	-.049	-.058	-.079	-.122	.045	-.163	-.498	.1
C (1)2PZ	-.069	-.010	.187	-.149	-.837	.415	.616	.3
C (1)2PX	-.590	.142	.177	.244	-.051	-.006	.005	-.?
C (1)2PY	.168	.241	.006	.411	-.220	.385	-.275	.8
C (2)2S	-.057	.016	.000	-.006	.000	-.091	-.398	
C (2)2PZ	.000	-.000	.203	-.000	-.795	.000	.000	
C (2)2PX	.153	-.191	.005	.282	-.000	.118	.118	
C (2)2PY	-.008	.288	.005	.422	-.000	.592	.069	
C (3)2S	-.049	-.058	.079	-.122	-.045	-.163	-.498	
C (3)2PZ	.069	.010	.187	.148	-.837	-.415	-.616	
C (3)2PX	-.590	.142	.177	.244	.051	-.006	.005	
C (3)2PY	.168	.241	-.005	.411	.220	.385	-.275	
H (4)1S	-.070	.041	-.099	.314	-.228	.653	-.201	
H (5)1S	.016	-.023	-.047	-.418	-.672	.423	.712	
H (6)1S	.010	-.045	-.005	-.763	.000	-.322	.043	
H (7)1S	-.070	.041	.099	.314	.228	.653	-.201	
H (8)1S	.016	-.023	.047	-.418	.672	.423	.712	
H (9)1S	.521	.160	.005	-.083	.000	.243	-.119	
FE(10)4S	-.478	.048	.005	.560	-.000	-.521	.561	
FE(10)4PZ	.000	.000	-1.089	-.000	-.190	.000	-.000	
FE(10)4PX	.407	-.150	-.005	.803	-.000	-.693	.560	
FE(10)4PY	-.123	-.967	-.005	.180	-.000	.348	-.177	
FE(10)3DZ2	-.205	.055	.005	-.052	-.000	.060	-.008	
FE(10)3DX2	.000	.000	-.213	-.000	-.053	.000	-.000	
FE(10)3DX2Y2	-.081	-.009	-.005	.195	-.000	-.022	.116	
FE(10)3DY2	.000	-.000	.057	.000	.037	.000	-.000	
FE(10)3DXY	.081	-.198	.005	-.034	.000	.085	.019	

MO. NO.	25	26	27
ENERGY	21.653	31.906	53.957
OCCUPATION	.000	.000	.000
C (1)2S	.344	1.134	-1.013
C (1)2PZ	.245	.107	-.521
C (1)2PX	-.176	.080	-.088
C (1)2PY	-.513	.262	-.245
C (2)2S	.644	-.000	1.588
C (2)2PZ	.000	.970	.005
C (2)2PX	.347	.000	-.235
C (2)2PY	.964	.000	-.582
C (3)2S	.344	-1.134	-1.013
C (3)2PZ	-.245	.107	.521
C (3)2PX	-.176	-.080	-.088
C (3)2PY	-.513	-.262	-.245
H (4)1S	-.530	-.430	.296
H (5)1S	.133	-.541	.221
H (6)1S	-1.091	-.000	-.272
H (7)1S	-.530	.430	.296
H (8)1S	.133	.541	.221
H (9)1S	-.082	.000	.011
FE(10)4S	-.046	-.000	-.047
FE(10)4PZ	-.000	.021	.005
FE(10)4PX	-.013	.000	-.059
FE(10)4PY	-.084	.000	.043
FE(10)3DZ2	-.007	-.000	.079
FE(10)3DX2	-.000	.115	.005
FE(10)3DX2Y2	-.065	-.000	-.013
FE(10)3DY2	.000	-.007	.005
FE(10)3DXY	-.036	-.000	-.072

TABLE 9 (Continued)
MODEL V

ꞮGINAL BASIS

ꞮNO.Ɪ	1	2	3	4	5	6	7	8
ꞮERGY	-27,004	-21,401	-18,887	-16,110	-14,786	-13,915	-12,686	-12,331
ꞮCUPATION	2,000	2,000	2,000	2,000	2,000	2,000	2,000	2,000
(1)2S	-,302	,433	,225	-,074	,048	,020	-,024	,032
(1)2PZ	,007	-,015	-,079	,124	-,249	,045	-,084	,351
(1)2PX	,002	-,001	-,030	,148	,178	-,293	-,102	-,050
(1)2PY	-,004	-,012	-,078	-,047	,015	-,179	-,333	,166
(2)2S	-,522	,000	-,325	,035	-,124	-,000	-,153	,000
(2)2PZ	-,000	-,191	-,000	,000	-,000	,197	-,000	-,569
(2)2PX	,008	,000	-,099	,219	,272	-,000	,441	-,000
(2)2PY	,010	-,000	-,268	-,278	,094	,000	,222	-,000
(3)2S	-,302	-,433	,225	-,074	,048	-,020	-,024	-,032
(3)2PZ	-,007	-,015	,079	-,124	,249	,045	,084	,351
(3)2PX	,002	,001	-,030	,148	,178	,293	-,102	,050
(3)2PY	-,004	,012	-,078	-,047	,015	,179	-,333	-,166
(4)1S	-,039	,183	,180	-,206	-,149	,388	,274	,003
(5)1S	-,032	,192	,121	-,121	,359	-,233	-,141	-,295
(6)1S	-,116	,000	-,428	-,152	,060	,000	,287	-,000
(7)1S	-,039	-,183	,180	-,206	-,149	-,388	,274	-,003
(8)1S	-,032	-,192	,121	-,121	,359	,233	-,141	,295
(9)1S	-,119	,000	,084	,522	,044	-,000	,260	-,000

ꞮNO.Ɪ	9	10	11	12	13	14	15	16
ERGY	-8,567	-7,516	-3,806	8,489	9,431	13,027	15,308	21,338
CUPATION	2,000	,000	,000	,000	,000	,000	,000	,000
(1)2S	-,133	,018	-,139	-,046	,005	,113	-,402	-,464
(1)2PZ	-,224	-,028	-,243	,872	-,067	,173	,778	,002
(1)2PX	-,397	,188	-,335	,022	,739	-,785	-,085	,366
(1)2PY	,557	-,357	,424	,222	,282	-,485	-,010	,241
(2)2S	-,000	-,079	-,108	-,000	,096	,000	-,237	-,660
(2)2PZ	,085	-,000	,000	,808	,000	-,357	-,000	-,000
(2)2PX	-,000	-,838	-,092	,000	-,046	-,000	,241	-,430
(2)2PY	,000	,397	,211	-,000	,583	-,000	,684	-,849
(3)2S	,133	,018	-,139	,046	,005	-,113	-,402	-,464
(3)2PZ	-,224	,028	,872	,067	,173	-,778	-,002	
(3)2PX	,397	,188	-,335	-,022	,739	,785	-,085	,366
(3)2PY	-,557	-,357	,424	-,222	,282	,485	-,010	,241
(4)1S	,013	-,164	,019	,195	,665	-,827	,227	,526
(5)1S	,083	,057	-,064	,651	-,427	,428	,792	,111
(6)1S	,000	,116	-,116	,000	-,595	,000	-,434	1,114
(7)1S	-,013	-,164	,019	-,195	,665	,827	,227	,526
(8)1S	-,083	,057	-,064	-,651	-,427	-,428	,792	,111
(9)1S	,000	,377	,956	,000	-,030	-,000	,106	,085

TABLE 9 (Continued)
MODEL V

ORIGINAL BASIS

MO. NO.	17	18
ENERGY	33,374	54,223
OCCUPATION	,000	,000
C (1)2S	1,144	-1,012
C (1)2PZ	,059	-,478
C (1)2PX	,015	-,048
C (1)2PY	,344	-,346
C (2)2S	,000	1,572
C (2)2PZ	,968	-,000
C (2)2PX	-,000	-,219
C (2)2PY	-,000	-,536
C (3)2S	-1,144	-1,012
C (3)2PZ	,059	,478
C (3)2PX	-,015	-,048
C (3)2PY	-,344	-,346
H (4)1S	-,456	,247
H (5)1S	-,538	,234
H (6)1S	-,000	-,308
H (7)1S	,456	,247
H (8)1S	,538	,234
H (9)1S	,000	,010

TABLE 9 (Continued)
MODEL VI

RIGINAL BASIS

D. NO.	1	2	3	4	5	6	7	8
NERGY	-26.968	-21.371	-18.400	-15.375	-14.550	-14.081	-12.427	-11.783
CCUPATION	2.000	2.000	2.000	2.000	2.000	2.000	2.000	2.000
(1)2S	-.354	.440	.211	-.010	-.062	-.013	-.043	-.048
(1)2PZ	-.004	-.022	-.090	.102	.244	.074	-.089	-.342
(1)2PX	.005	-.009	-.047	.166	-.162	-.271	-.030	.016
(1)2PY	-.002	-.013	-.087	-.044	-.060	-.149	.291	-.189
(2)2S	-.520	.000	.358	.012	.102	.000	-.139	.000
(2)2PZ	-.000	-.152	-.000	-.000	.000	.148	.000	.562
(2)2PX	.019	.000	-.080	.154	-.145	-.000	.471	-.000
(2)2PY	.005	.000	.207	-.302	-.092	-.000	.205	-.000
(3)2S	-.354	-.440	.211	-.010	-.062	.013	-.043	.048
(3)2PZ	.004	-.022	.090	-.102	-.244	.074	.089	-.342
(3)2PX	.005	.009	-.047	.166	-.162	.271	-.030	-.016
(3)2PY	-.002	.013	-.087	-.044	-.060	.149	.291	.189
(4)1S	-.107	.227	.233	-.217	.167	.370	.204	.038
(5)1S	-.076	.204	.125	-.044	=.394	-.272	-.096	.257
(6)1S	-.140	.000	.410	-.223	=.052	.000	.297	-.000
(7)1S	-.107	-.227	.233	-.217	.167	-.370	.204	-.038
(8)1S	-.076	-.204	.125	.044	=.394	.272	-.096	-.257
(9)1S	-.138	-.000	.029	.501	.048	.000	.291	-.000
(10)4S	.122	.000	-.000	-.077	.054	.000	.032	-.000
(10)4PZ	-.000	.101	.000	.000	.000	-.018	-.000	-.016
(10)4PX	.205	.000	=.013	-.009	=.001	-.000	-.004	.000
(10)4PY	-.020	-.000	.044	.016	.014	-.000	-.017	.000
(10)3DZ2	-.001	=.000	.022	-.023	=.005	.000	.118	-.000
(10)3DXZ	.000	-.009	=.000	.000	.000	-.090	.000	.120
(10)3DX2Y2	.017	.000	=.012	.006	.015	.000	-.166	.000
(10)3DYZ	.000	-.001	.000	.000	.000	.114	.000	.048
(10)3DXY	-.008	.000	=.025	.005	-.018	.000	=.114	.000

D. NO.	9	10	11	12	13	14	15	16
NERGY	-8.782	-8.756	-8.489	-8.258	-8.085	-7.447	-6.101	-4.729
CCUPATION	2.000	2.000	2.000	2.000	2.000	.000	.000	.000
(1)2S	-.082	.019	=.020	-.023	.124	.001	-.023	.073
(1)2PZ	-.195	-.052	=.017	.074	.130	-.025	-.044	.045
(1)2PX	-.300	.048	.004	.017	.162	-.249	-.140	-.261
(1)2PY	.328	=.263	.016	.239	=.437	.159	.212	.042
(2)2S	.000	-.058	=.005	.065	.000	-.000	.041	.053
(2)2PZ	.155	.000	.000	-.000	=.029	-.108	.000	.000
(2)2PX	.000	-.312	-.065	.111	.000	-.000	.775	.120
(2)2PY	-.000	.189	.022	-.158	-.000	.000	-.268	-.138
(3)2S	.082	.019	=.020	-.023	-.124	-.001	=.023	.073
(3)2PZ	-.195	.052	.017	-.074	.130	-.025	.044	-.045
(3)2PX	.300	.048	.004	.017	-.162	.249	-.140	-.261
(3)2PY	-.328	=.263	.016	.239	.437	-.159	.212	.042
(4)1S	-.016	-.031	.021	.028	.002	.098	.138	.040
(5)1S	.072	.013	-.042	.007	-.081	-.049	-.039	-.024
(6)1S	-.000	.059	=.053	-.107	-.000	.000	-.068	.020
(7)1S	.016	-.031	.021	.028	-.002	-.098	.138	.040
(8)1S	-.072	.013	=.042	.007	.081	.049	-.039	-.024
(9)1S	-.000	.150	.167	-.134	.000	.000	-.218	-.213
(10)4S	-.000	.053	.097	.150	.000	-.000	.152	-.761
(10)4PZ	.027	.000	.000	-.000	-.062	-.167	.000	.000
(10)4PX	-.000	.036	=.002	-.043	=.000	.000	-.218	.578
(10)4PY	.000	-.026	-.041	.029	=.000	.000	-.164	-.223
(10)3DZ2	.000	=.729	.492	-.087	.000	.000	-.433	-.060
(10)3DXZ	-.439	=.000	.000	.000	-.068	.888	.000	-.000
(10)3DX2Y2	.000	=.192	-.171	-.899	-.000	-.000	.274	-.146
(10)3DYZ	-.534	=.000	-.000	.000	=.761	-.518	-.000	-.000
(10)3DXY	-.000	.353	.827	-.133	.000	-.000	.347	.134

TABLE 9 (Continued)
MODEL VI

ORIGINAL BASIS

MO. NO.	17	18	19	20	21	22	23	24
ENERGY	-3.507	-2.808	-1.130	8.730	8.843	14.808	14.831	16.4
OCCUPATION	.000	.000	.000	.000	.000	.000	.000	.0
C (1)2S	-.135	-.044	-.128	.008	.102	-.017	-.460	-.1
C (1)2PZ	-.248	.040	.171	.873	.158	-.071	.659	-.3
C (1)2PX	-.355	.195	.198	-.113	-.504	.756	-.271	-.1
C (1)2PY	.405	.143	.095	.151	-.161	.518	-.108	-.0
C (2)2S	-.113	.032	.000	.000	-.006	.000	-.312	-.2
C (2)2PZ	-.000	-.000	.223	.753	.000	.511	.000	-.0
C (2)2PX	-.081	-.227	.000	.000	-.147	-.000	.296	-.1
C (2)2PY	.204	.270	-.000	.000	-.310	-.000	.331	-1.0
C (3)2S	-.135	-.044	.128	-.008	.102	.017	-.460	-.1
C (3)2PZ	.248	-.040	.171	.873	-.158	-.071	-.659	-.3
C (3)2PX	-.355	.195	-.198	.113	-.504	-.756	-.272	-.3
C (3)2PY	.405	.143	-.095	-.151	-.161	-.517	-.108	-.0
H (4)1S	-.003	.006	-.050	.109	-.212	.864	-.153	-.1
H (5)1S	-.062	.006	-.043	.702	.477	-.357	.759	-.1
H (6)1S	-.110	-.004	.000	-.000	.641	.000	-.249	-.1
H (7)1S	-.003	.006	.050	-.109	-.212	-.864	-.154	-.1
H (8)1S	-.062	.006	.043	-.702	.477	.357	.759	-.1
H (9)1S	.949	.055	.000	-.000	.233	-.000	-.004	-.1
FE(10)4S	-.022	.046	.000	.000	-.604	-.000	.483	-.
FE(10)4PZ	.000	.000	-1.083	.173	.000	.245	.000	-.
FE(10)4PX	.071	-.160	-.000	.000	-.941	-.000	.430	-.
FE(10)4PY	.013	-.986	.000	.000	-.055	.000	-.069	-.
FE(10)3DZ2	-.137	.075	-.000	-.000	.087	-.000	.016	-.
FE(10)3DXZ	.000	-.000	-.232	.110	.000	.053	.000	-.
FE(10)3DX2Y2	.033	-.025	.000	.000	-.210	-.000	.086	.
FE(10)3DYZ	.000	-.000	.083	-.078	-.000	-.227	-.000	.
FE(10)3DXY	-.117	-.177	.000	-.000	.049	-.000	.035	-.

MO. NO.	25	26	27
ENERGY	22.501	33.651	52.968
OCCUPATION	.000	.000	.000
C (1)2S	-.415	1.145	-1.007
C (1)2PZ	.139	.053	-.461
C (1)2PX	.492	-.040	-.029
C (1)2PY	.308	.307	-.335
C (2)2S	-.658	-.000	1.583
C (2)2PZ	-.000	.925	.000
C (2)2PX	-.365	.000	-.252
C (2)2PY	-.488	.000	-.518
C (3)2S	-.415	-1.145	-1.007
C (3)2PZ	-.139	.053	.461
C (3)2PX	.492	.040	-.029
C (3)2PY	.308	-.307	-.335
H (4)1S	.843	-.546	.317
H (5)1S	.125	-.537	.245
H (6)1S	.824	.000	-.297
H (7)1S	.843	.546	.317
H (8)1S	.125	.537	.245
H (9)1S	.131	.000	.032
FE(10)4S	-.411	-.000	-.144
FE(10)4PZ	.000	-.114	-.000
FE(10)4PX	-.101	-.000	-.044
FE(10)4PY	.163	.000	.038
FE(10)3DZ2	-.118	-.000	.028
FE(10)3DXZ	-.000	.043	.000
FE(10)3DX2Y2	.094	.000	-.007
FE(10)3DYZ	-.000	.119	.000
FE(10)3DXY	.106	.000	-.056

REFERENCES

1. R. Hoffmann and R. B. Woodward, *Accounts Chem. Res.*, **1**, 17 (1968).
2. R. B. Woodward and R. Hoffmann, *J. Am. Chem. Soc.*, **87**, 395 (1965).
3. G. B. Gill, *Quart. Revs.*, **22**, 338 (1968).
4. D. R. Eaton, *J. Am. Chem. Soc.*, **90**, 4272 (1968).
5. T. H. Whitesides, *J. Am. Chem. Soc.*, **91**, 2395 (1969).
6. H. C. Longuet-Higgins and E. W. Abrahamson, *J. Am. Chem. Soc.*, **87**, 2045 (1965).
7. R. F. W. Bader, *Can. J. Chem.*, **40**, 1164 (1962).
8. R. G. Pearson, *J. Am. Chem. Soc.*, **91**, 4947 (1969).
9. F. D. Mango and J. H. Schachtschneider, *J. Am. Chem. Soc.*, **89**, 2484 (1967).
10. H. C. Volger and H. Hogeveen, *Rec. Trav. Chem.*, **86**, 830 (1967).
11. W. Merk and R. Pettit, *J. Am. Chem. Soc.*, **89**, 4788 (1967).
12. F. D. Mango and J. H. Schachtschneider, *J. Am. Chem. Soc.*, **91**, 1030 (1969).
13. F. D. Mango, *Tetrahedron Letters*, 4813 (1969).
14. F. D. Mango, *Advan. Catalysis*, **20**, 291 (1969).
15. R. Hoffmann and R. B. Woodward, *J. Am. Chem. Soc.*, **87**, 2046 (1965).
16. R. B. Woodward, *Symposium on Orbital Symmetry Correlations in Organic Chemistry*; Cambridge, (1969).
17. C. W. Bird, D. L. Collinere, R. C. Cookson, J. Hudec, and R. O. Williams, *Tetrahedron Letters*, 373 (1961); P. W. Jolly, F. G. H. Stone, and K. MacKenzie, *J. Chem. Soc.*, 7416 (1965).
18. C. W. Bird, R. G. Cookson, and Hudec, *J. Chem. and Ind.* (*London*), 20 (1960); L. G. Cannell, U.S. Patent 3,258,502 (1966).
19. G. N. Schrauzer and S. Eichler, *Chem. Res.*, **95**, 2764 (1962).
20. D. R. Arnold, D. J. Trecker and E. B. Whipple, *J. Am. Chem. Soc.*, **87**, 2596 (1965).
21. G. N. Schrauzer, *Advan. Catalysis*, **18**, 373 (1968).
22. G. N. Schrauzer and P. Glockner, *Chem. Res.*, **97**, 2451 (1964).
23. M. Green and D. C. Wood, *J. Chem. Soc.*, 1172 (1969).
24. R. L. Banks and G. C. Bailey, *Ind. Eng. Chem., Prod. Res. Develop.*, **3**, 170 (1964).
25. G. C. Bailey, *Catalysis Reviews*, **3**, 37 (1969).
26. C. P. C. Braddhaw, E. J. Howman, and L. Turner, *J. Catalysis*, **7**, 269 (1967),
27. N. Calderon, H. Y. Chen, and K. W. Scott, *Tetrahedron Letters*, 3327 (1967).
28. K. W. Scott, N. Calderon, E. A. Ofstead, W. A. Judy, and J. P. Ward, 155th National Meeting American Chemical Society, April 1968, 1b. L54.
29. G. C. Ray and D. L. Crain, Belgian Patent 694,420.
30. N. Calderon, E. A. Ofstead, J. P. Ward, W. A. Judy, and K. W. Scott, *J. Am. Chem. Soc.*, **90**, 4133 (1968).
31. British Petroleum Co. Ltd., British Patent 1,054,864.
32. E. A. Zuech, *Chem. Comm.*, 1182 (1968).
33. J. Wang and H. R. Menapace, *J. Org. Chem.*, **33**, 3704 (1968).
34. W. B. Hughes, *Chem. Comm.*, 431 (1969).
35. C. T. Adams and S. G. Brandenberger, *J. Catalysis*, **13**, 360 (1969).
36. H. Hogeveen and H. C. Volger, *J. Am. Chem. Soc.*, **89**, 2486 (1967).
37. G. S. Hammond, N. J. Turro, and A. Fischer, *J. Am. Chem. Soc.*, **83**, 4674 (1961).
38. H. Hogeveen and H. C. Volger, *Chem. Comm.*, 1133 (1967).

39. N. A. Bailey, R. D. Gillard, M. Keeton, R. Mason, and D. R. Russell, *Chem. Commun.*, 396 (1966); W. J. Irwin and G. J. McQuillin, *Tetrahedron Letters*, 1937 (1968).

40. H. C. Volger, H. Hogeveen, and M. M. P. Gaasbeck, *J. Am. Chem. Soc.*, **93**, 2137 (1969).

41. A. D. Wals, *Nature*, **159**, 165, 712 (1947); C. A. Coulson and W. E. Moffitt, *J. Chem. Phys.*, **15**, 151 (1947).

42. E. W. vonDoering and W. R. Roth, *Angew Chem. Intern. Ed. Engl.*, **2**, 115 (1963); G. Schröder, *Angew. Chem.*, **75**, 772 (1963).

43. H. C. Volger, H. Hogeveen, and M. M. P. Gaasbeck, *J. Am. Chem. Soc.*, **91**, 218 (1969).

44. H. E. Simmons quoted in H. Prinzbach, W. Eberbach, M. Klaus, G. V. Veh, and V. Scheidegger. *Tetrahedron Letters*, 1681 (1966).

45. P. K. Freeman, D. G. Kuper, and V. N. Mallikayana, *Tetrahedron Letters*, 3303 (1965).

46. L. Vaska and J. W. DiLuzio, *J. Am. Chem. Soc.*, **83**, 2784 (1961).

47. T. J. Katz and S. Cerefice, *J. Am. Chem. Soc.*, **91**, 2405 (1969).

48. W. J. Irwin and F. J. McQuilling, *Tetrahedron Letters*, 1937 (1968).

49. T. J. Katz and S. Cerefice, *J. Am. Chem. Soc.*, **91**, 6519 (1969).

50. T. J. Katz and S. Cerefice, *Tetrahedron Letters*, 2561 (1969).

51. D. M. Lemal, R. A. Lavald and Harrington, *Tetrahedron Letters*, 2770 (1965).

52. P. G. Gassman and D. S. Patton, *J. Am. Chem. Soc.*, **90**, 7276 (1968).

53. P. G. Gassman, D. H. Aue, and D. S. Patton, *J. Am. Chem. Soc.*, **90**, 7271 (1968).

54. J. Manassen *J. Catalysis*, **18**, 38 (1970).

55. J. E. Falk, *Porphyrins and Metalloporphyrins*, Elsevier, Amsterdam, London, New York, 1964.

56. M. Zerner and M. Gouterman, *Theor. Chem. Acta. (Berl.)*, **4**, 44 (1966).

57. R. B. Woodward and R. Hoffmann, *J. Am. Chem. Soc.*, **87**, 2511 (1965).

58. R. Hoffmann, *J. Chem. Phys.*, **39**, 1397 (1963).

59. R. S. Mulliken, *J. Chem. Phys.*, **23**, 1521 (1955).

60. G. V. Smith and J. R. Swoop, *J. Org. Chem.*, **31**, 3904 (1966); L. Roos and M. Orchin, *J. Am. Chem. Soc.*, **87**, 5502 (1965).

61. P. Abley and F. J. McQuillin, *Chem. Commun.*, 1503 (1969).

62. A. M. Khan, F. J. McQuillin, and I. Jardine, *J. Chem. Soc. (C)*, 136 (1967); S. Mitsui, Y. Kudo, and M. Kobayashi, *Tetrahedron*, **25**, 1921 (1969).

63. H. M. Gladney and A. Veillard, *Phys. Rev.*, **180**, 385 (1968); A. Veillard, *Chem. Commun.*, 1022, 1427 (1969); I. H. Hiller and V. R. Saunders, *Chem. Commun.*, 1275 (1969).

64. M. Wolfsberg and L. Helmholtz, *J. Chem. Phys.*, **20**, 837 (1952).

65. C. J. Ballhausen and H. B. Gray, *Inorg. Chem.*, **1**, 111 (1962); A. Viste and H. B. Gray, *Inorg. Chem.* **3**, 1113 (1964); H. Basch and H. B. Gray, *Inorg. Chem.* **4**, 639 (1967).

66. R. F. Fenske, K. G. Caulton, D. P. Radtke, and C. C. Sweeny, *Inorg. Chem.* **5**, 951 (1966); R. F. Fenske and C. C. Sweeny, *Inorg. Chem.* **3**, 1105 (1964); R. F. Fenske, *Inorg. Chem.* **4**, 33 (1965).

67. E. Clementi and D. L. Raimondi, *J. Chem. Phys.* **38**, 2686 (1963).

68. J. W. Richardson, W. C. Nieuwpoort, R. R. Powell, and W. F. Edgell, *J. Chem. Phys.* **35**, 1057 (1962); J. W. Richardson, R. R. Powell, and W. C. Nieuwpoort, *J. Chem. Phys.*, **38**, 796 (1963).

69. H. Basch and H. B. Gray, *Theoret. Chim. Acta* **4**, 367 (1966).
70. H. Basch, A. Viste, and H. B. Gray, *J. Chem. Phys.* **44**, 10 (1966).
71. P. O. Löwdin, *J. Chem. Phys.* **18**, 365 (1950).
72. J. Hinze and H. H. Jaffe, *J. Am. Chem. Soc.* **74**, 540 (1962).
73. L. C. Cusachs, *J. Chem. Phys.* **43**, 5157 (1965).
74. D. G. Carroll and S. P. McGlynn, *J. Chem. Phys.* **45**, 3827 (1966).

7

Electron-Transfer Catalysis

R. G. LINCK

Department of Chemistry
Revelle College
University of California, San Diego
La Jolla, California

I. Introduction

A. GENERAL SCOPE

The subject of this review is the homogeneous catalysis of reactions in which the active catalyst performs its function by undergoing a change in formal oxidation state. Such catalytic systems have been known for a long time. In 1894, for instance, Fenton (*1*) observed that in the presence of Fe^{2+} hydrogen peroxide exhibited much more powerful oxidizing power than in the absence of Fe^{2+}. Seven years later Marshall noted that oxidations by peroxydisulfate ion occurred much more rapidly in the presence of Ag^+ (*2*). These observations are but two early examples of a large number of catalytic systems involving metal ion complexes which undergo a change in formal oxidation state during reaction. Such catalytic systems are of considerable importance in modern technology. For instance, catalytic decompositions of peroxydisulfate ions are used to initiate polymerizations; and the conversion of ethylene to acetaldehyde by the Wacker process (*3,4*) utilizes two catalytic oxidation-reduction reactions

$$H_2O + H_2C{=}CH_2 + Pd(II) = Pd(0) + CH_3CHO + 2H^+ \qquad (1)$$

$$Pd(0) + 2Cu(II) = Pd(II) + 2Cu(I) \qquad (2)$$

$$\underline{2H^+ + 2Cu(I) + \tfrac{1}{2}O_2 = 2Cu(II) + H_2O} \qquad (3)$$

$$H_2C{=}CH_2 + \tfrac{1}{2}O_2 = CH_3CHO \qquad (4)$$

In addition to this economic importance, catalytic processes in general are of great interest in modern chemistry. Detailed understanding of the functioning of biological catalysts remains a desired goal. In this area catalysis of oxidation-reduction reactions by metal ions is an important subdivision (*5,6*). An attempt to review the field of catalysis by electron-transfer reactions thus appears warranted.

However, even this subdivision of homogeneous catalysis is a category much too broad to be adequately reviewed in the present context; accordingly, several restrictions on the subject matter are necessary. Perhaps the most drastic ones are the discussion of only those catalytic systems in which the active catalyst is a transition metal ion in aqueous solution and on which a reasonably detailed kinetic study has been made. While it is true that these restrictions narrow the broad field to that in which the author's primary interests lie—perhaps a sufficient reason in itself—it is also true that these restrictions make possible an approach that attempts to account for the individual steps in the catalytic mechanism. In this review the emphasis is on an understanding of the separate reactions that are components of a catalytic process. As such, there is no attempt at complete coverage of the literature. Rather, an extensive discussion of non-catalytic oxidation-reduction reactions is presented. This discussion compliments previous reviews; in order to achieve a coherent picture, however, some overlap is inevitable. Further, only a few selected samples of catalytic processes are considered in detail. These examples are chosen to illustrate various aspects of catalytic mechanisms. The success achieved in these attempts to understand the processes in detail serves as an internal check of speculations on less studied systems.

This review was completed in December, 1969.

B. Organization and Conventions

The understanding of catalytic oxidation-reduction processes should begin with the study of simple oxidation-reduction reactions of metal-ion complexes; Section II deals with this subject. That section accents the nature of the rate laws and mechanisms, the theory of redox reactions, and the empirical reactivity patterns that have evolved. In Section III a general introduction to catalytic systems is given, and Section IV treats several such systems.

Some conventions of notation have been adopted in this review. The indication of the formal valence state of an ion will be given by Roman numerals, such as Fe(III). This symbolism will be used when specification of the ligands surrounding the metal ion is unknown or when it is not desirable to make such a specification. Conversely, the symbol Fe^{3+} should be taken to mean $Fe(H_2O)_6^{3+}$; in cases such as Eu(II), where the coordination number is not known, Eu^{2+} should be taken to mean $Eu(H_2O)_n^{2+}$. Similarly, $FeCl^{2+}$ should be read $Fe(H_2O)_5Cl^{2+}$. The use of equal signs in chemical reactions will be used to signify stoichiometric equations or reactions in rapid equilibrium, whereas arrows will specify equations designed to carry mechanistic meaning. The symbol $\rightarrow \rightarrow$ will be used to imply several steps to products; such reactions will not, in general, be balanced. Throughout this review, K will designate an equilibrium constant and k a rate constant; these will be labeled by small integers or letters, but the repeated use of a given integer will be made when confusion will not result. Brackets will be used to denote concentrations. An asterisk to the upper left of a chemical symbol will indicate isotopic labeling of that element (or in the case of nuclear resonance experiments, spin labeling).

The following abbreviations for various ligands have been adopted: bip = 2,2′-bipyridyl;   en = ethylenediamine;   dien = diethylenetriamine; ox = oxalate; phen = 1,10-penanthroline; terpy = 2,2′,2″-tripyridyl; py = pyridine; EDTA = ethylenediamminetetraacetic acid, tetraanion; EDTAH = the trianion of EDTA; DMG = the monoanion of dimethylglyoxime; and I = ionic strength.

II. Elementary Oxidation-Reduction Systems

In order to understand the rate and mechanism of catalytic systems involving an oxidation-reduction reaction of a metal ion complex, it is necessary first to inquire about what is known concerning the rate and mechanism of oxidation-reduction reactions in noncatalytic systems. In general, questions about noncatalytic systems will more likely yield answers than those concerned with catalytic systems, if for no other reason than that in general one is dealing with fewer components in the problem and has greater freedom to vary parameters in the system. A thorough understanding of the rate of simple oxidation-reduction reactions is necessary to begin to appreciate why the reaction between one substrate and, e.g., the oxidized form of the catalyst, is more rapid than the reaction between the other substrate and the reduced form of the catalyst. This understanding is necessary to begin to understand why the catalytic system functions more efficiently than the direct reaction. This section on simple oxidation-reduction reactions is not meant to imply that everything is known about such reactions. It is, however, true that many simple systems have been investigated and some understanding has been obtained. Several earlier and more complete reviews of simple electron-transfer systems exist (*7–14*). Recently a compilation of much of the published data for homogeneous electron-transfer reactions has appeared (*15*).

A. General Observations

In most simple oxidation-reduction reactions, reactions in which the change in oxidation state of one component is mimicked by that on the other, e.g., reactions of 1 : 1 stoichiometry, the rate law is usually first order in each of the components.

$$A^{n+} + B^{m+} = A^{(n+1)+} + B^{(m-1)+} \tag{5}$$

$$\frac{-d[A^{n+}]}{dt} = k[A^{n+}][B^{m+}]. \tag{6}$$

There are, of course, numerous exceptions, but this rate law is usually found. This result makes considerable sense when one observes that in order to obtain a rate law independent of one component, the activation must occur at only one metal-ion center. Such a process could occur via the mechanism

$$A^{n+} \underset{k_2}{\overset{k_1}{\rightleftharpoons}} A^{(n+1)+} + e_{aq}^- \tag{7}$$

$$e_{aq}^- + B^{m+} \xrightarrow{k_3} B^{(m-1)+} \tag{8}$$

whereupon, with application of the steady-state assumption for $[e_{aq}^-]$, and under conditions where $k_3[B^{m+}] \gg k_2[A^{(n+1)+}]$,

$$\frac{-d[A^{n+}]}{dt} = k_1[A^{n+}]. \tag{9}$$

No reaction has been found to proceed by this mechanism. Indeed, even the powerful reductants Cr(II) and Eu(II) must overcome an activation free energy of about 50 kcal mole^{-1} in order to proceed by this mechanism. for the electrode potential of the reaction

$$\tfrac{1}{2}H_2(g) = H^+(aq) + e^-(aq) \tag{10}$$

is 2.8 V (16). (Note e^-(aq) refers to a real chemical species, not an arbitrary standard state.) Sufficient energy to overcome this barrier has been accumulated in a reactant only through photochemical excitation ($17,18$). Mechanisms somewhat related to this one, at least in the sense that only one reagent appears in the rate law and that some carrier is involved, have been observed. In these cases, however, the observation results from generation of an intermediate from a reagent that is not kinetically varied; and hence cannot be seen in the rate law even though it is present in the activated complex. (This type of ambiguity occurs, of course, in all kinetic schemes: one cannot know directly from the rate law the role played by the solvent medium in the system under investigation.) An example is the oxidation of Tl(I) by Ce(IV) in nitrate media in the absence of Ce(III) ($19,20$).

$$\frac{-dCe(IV)}{dt} = k_1[Ce(IV)] + k_2[Ce(IV)][Tl(I)] \tag{11}$$

where the first term involves either Eq. (12) or (13).

$$Ce(IV) + NO_3^- \longrightarrow Ce(III) + NO_3 \tag{12}$$

$$Ce(IV) + OH^- \longrightarrow Ce(III) + OH \tag{13}$$

The rate laws for reactions that do not involve reactants with $1:1$ stoichiometry, or in which a two or more electron change takes place through intermediate oxidation states, are sometimes more complicated. Complications arise from the competition between two reagents for a reactive intermediate; for example, consider the NpO_2^+ reduction of Cr(VI) (21)

$$7H^+ + 3NpO_2^+ + HCrO_4^- = 3NpO_2^{2+} + Cr^{3+} + 2H_2O \tag{14}$$

for which the rate law (at fixed $[H^+]$) is

$$\frac{-d[NpO_2^+]}{dt} = \frac{k[NpO_2^+][Cr(VI)]}{1 + k'[NpO_2^{2+}][NpO_2^+]^{-1}}. \tag{15}$$

This is consistent with the mechanism shown in Eqs. (16)–(18).

$$NpO_2^+ + Cr(VI) \underset{k_2}{\overset{k_1}{\rightleftharpoons}} NpO_2^{2+} + Cr(V) \tag{16}$$

$$NpO_2^+ + Cr(V) \overset{k_3}{\longrightarrow} NpO_2^{2+} + Cr(IV) \tag{17}$$

$$NpO_2^+ + Cr(IV) \overset{fast}{\longrightarrow} NpO_2^{2+} + Cr(III) \tag{18}$$

where k is to be associated with $3k_1$ and $k' = k_2 k_3^{-1}$. Although ambiguities in interpretation can arise from such rate laws, the added information in more complicated rate laws usually helps clarify the understanding of the mechanism.

Up to this time mention of rapid equilibria involving the metal ion reactants and other components of the reactive medium have been neglected. These equilibria play an important role in experimental rate laws. For instance, the isotopic exchange between Fe(II) and Fe(III) is catalyzed by Cl^-, presumably by the mechanism (22)

$$*Fe^{3+} + Cl^- \overset{K}{=} *FeCl^{2+} \tag{19}$$

$$*FeCl^{2+} + Fe^{2+} \overset{k}{\longrightarrow} *Fe(III) + Fe(II) + Cl^- \tag{20}$$

which yields, when $[Cl^-]$ is large compared to total ferric, $[Fe(III)]$,

$$\frac{d[*Fe(II)]}{dt} = \frac{kK[Fe(III)][Fe^{2+}][Cl^-]}{1 + K[Cl^-]}. \tag{21}$$

This is only one term in the experimental rate law in acidic chloride medium, the others involve transition states of composition (*22*) $\{FeFe^{5+}\}^{\ddagger}$ and $\{Fe(OH)Fe^{4+}\}^{\ddagger}$. It is transition states with gains or losses of protons that are most commonly found represented by additional terms in the rate law of reactions between metal-ion complexes. Such transition state compositions are often to be understood as reflections of overall stoichiometry: a reaction liberating several protons will usually have lost some of these in the transition state (*23*).

$$Np^{4+} + NpO_2^{2+} + 2H_2O = 2NpO_2^+ + 4H^+ \qquad (22)$$

$$\frac{-d[Np^{4+}]}{dt} = (k_1[H^+]^{-2} + k_2[H^+]^{-3})[Np^{4+}][NpO_2^{2+}] \qquad (23)$$

Newton and Baker (*24*) have summarized many examples of this behavior. The other situation wherein protons are gained or lost in the transition state is when a special stabilization of the transition state occurs via this mechanism, but such a proton change does not appear in the net stoichiometry. The loss of a proton from $Co(NH_3)_5OH_2^{3+}$ allows a transition state to develop in which the reductant, Cr(II), and the Co(III) center can more effectively share a ligand in the transition state (*25*)

$$\{(NH_3)_5Co-\overset{H}{\underset{}{O}}-Cr(H_2O)_5^{4+}\}^{\ddagger}$$

leading to a net stabilization of some 9 kcal mole^{-1} over the transition state

$$\{(NH_3)_5Co-\overset{H}{\underset{H}{O}}-Cr(H_2O)_5^{5+}\}^{\ddagger}$$

Such a difference manifests itself as a profound rate effect: in terms of catalytic phenomena, variations in pH can change the efficiency of a catalytic system simply by changing the dominant species in a hydrolytic equilibrium.

This survey of several features of rate laws that frequently manifest themselves in electron-transfer reactions should now be complimented by some discussion of stoichiometries. Of course, in a broad sense, an electron-transfer reaction can almost always be represented by formal valence state symbolism,

$$Fe(III) + Cr(II) \longrightarrow Fe(II) + Cr(III) \qquad (24)$$

but, as will be seen in the next section, this representation often overlooks a considerable amount of information about the process. It is useful to distinguish between three types of stoichiometries.

The first is well represented by Eq. (24). It is characterized by the transfer of an electron between the two reactants with no other apparent change in reactants or products in the stoichiometric equation

$$Fe(CN)_6^{3-} + Cr(bip)_3^{2+} = Fe(CN)_6^{4-} + Cr(bip)_3^{3+} \qquad (25)$$

A second type of stoichiometry is that in which a ligand, originally in the coordination sphere of one reagent is found after reaction in the coordination sphere of the other reagent, thus representing a transfer of ligand and electron

$$FeCl^{2+} + Cr^{2+} = CrCl^{2+} + Fe^{2+} \qquad (26)$$

Finally, the whole or a substantial part of one reagent may finish the reaction as part of the coordination sphere of the second reagent

$$2Cr^{2+} + Cl_2 = 2CrCl^{2+} \qquad (27)$$

$$V^{2+} + VO^{2+} = VOV^{4+} \qquad (28)$$

B. THE MECHANISM OF ELECTRON-TRANSFER REACTIONS

In certain cases, determination of the detailed stoichiometry allows one to gain a considerable amount of information about the mechanism of the electron-transfer process. However, it is clear that such stoichiometric information is not always available. If, for instance, the substitution of ligands into the coordination sphere of the products is rapid compared to the rate of the electron-transfer process, then the products will possess the equilibrium coordination sphere composition at any time of observation. In such a case the distinguishing features between [Eq. (25) and (26)] presence or absence of a ligand transfer process—would disappear. Hence the use of stoichiometry to help in elucidating information about the mechanism depends on substitution rates, the labilities of the various metal ion centers. This review is concerned with transion metal ions. Pursuant to the aforementioned importance of substitution rates, a brief discussion of the lability of such ions is warranted.

In 1950, Taube (26) pointed out that there were some striking correlations between electronic structure of the transition metal ion systems and their labilities. These effects are superimposed upon charge (Li^+ is more labile than Mg^{2+}) (27) and size (Cs^+ is more labile than Li^+) (27,28) effects, and dominate the chemistry of transition metal systems. Figure 1 illustrates the electronic effects for the $+2$ ions of the first-transition series. The values are for the rate of the process

$$M(H_2O)_5H_2^*O^{2+} + H_2O \longrightarrow M(H_2O)_6^{2+} + H_2^*O \qquad (29)$$

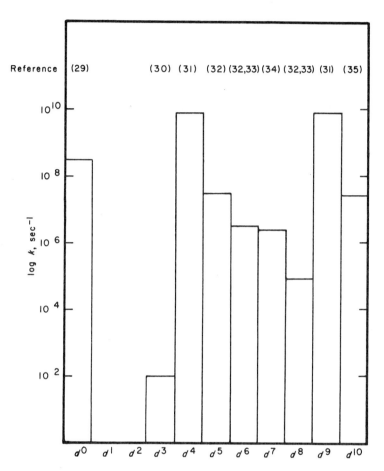

Fig. 1. The rate of water-exchange for the first row di-positive transition metal ions. The pertinent references are at the top of the figure.

and were obtained directly by nmr or indirectly by relaxation techniques. Table 1 illustrates the rates of reactions of some typical transition metal-ion complexes, as well as some data of complexes of the lanthanide and actinide series of metal ions. It should be noted particularly that d^3 and d^8, and low spin d^4, d^5, and d^6 complexes of pseudo-octahedral symmetry are generally slow to substitute (26), and that the lanthanides substitute quite rapidly. Satisfactory explanations of the gross features of the former trends can be obtained by use of crystal-field stabilization arguments after some assumption about the geometry of the transition state has been made

TABLE 1

The Rate of Substitution of Some Metal-Ion Complexes ($T = 25°C$)

Element	Electronic configuration[a]	Reaction	($M^{-1}\text{sec}^{-1}$)	$\Delta H^{\ddagger}$ (kcal mole^{-1})	$\Delta S^{\ddagger}$ (e.u.)	Ref.
Al(III)		$Al(H_2O)_5(H_2^*O)^{3+} + H_2O \longrightarrow$ exchange	1.6×10^{-1b}	27	28	(36)
Ga(III)		$Ga(H_2O)_5(H_2^*O)^{3+} + H_2O \longrightarrow$ exchange	1.8×10^{3b}	6.3	-22	(36)
Ti(III)	$d^1(t_{2g}^1)$	$Ti(H_2O)_5(H_2^*O)^{3+} + H_2O \longrightarrow$ exchange	10^{5b}	6.1	-15	(37)
V(III)	$d^2(t_{2g}^2)$	$V(H_2O)_6^{3+} + NCS^- \longrightarrow V(H_2O)_5NCS^{2+}$	1.1×10^2	7.6	-24	(38, 39)
Cr(III)	$d^3(t_{2g}^3)$	$Cr(H_2O)_5(H_2^*O)^{3+} + H_2O \longrightarrow$ exchange	5×10^{-6b}	27.6	10	(40)
		$Cr(NH_3)_5Cl^{2+} \longrightarrow Cr(NH_3)_5OH_2^{3+} + Cl^-$	9.3×10^{-6b}	21.8	-9.2	(41)
		$Cr(H_2O)_6^{3+} + NCS^- \longrightarrow Cr(H_2O)_5NCS^{2+}$	1.8×10^{-6}	25.1	-0.7	(42)
Fe(II)	$d^6(t_{2g}^6)$	$Fe(TPTZ)_2^{2+c} \longrightarrow Fe(TPTZ)(H_2O)_3^{2+} + TPTZ$	2.6×10^{-2b}			(43)
Fe(III)	$d^5(t_{2g}^3 e_g^2)$	$Fe(H_2O)_6^{3+} + NCS^- \longrightarrow Fe(H_2O)_5NCS^{2+}$	1.3×10^2	14.6	-10	(44)
		$Fe(H_2O)_5OH^{2+} + NCS^- \longrightarrow Fe(H_2O)_5NCS^{2+} + OH^-$	1.0×10^4	10	-7.6	(44)
Co(III)	$d^6(t_{2g}^6)$	$Co(NH_3)_5Cl^{2+} \longrightarrow Co(NH_3)_5H_2O^{3+} + Cl^-$	1.95×10^{-6b}	22	-9	(45)

V(IV)	d^1	VO(H$_2$O)$_4^{2+}$ + H$_2$O* $\longrightarrow$ exchange^d	5.0×10^{2b}	13.7	-0.6	(46, 47)
Cr(VI)	d^0	2HCrO$_4^-$ $\longrightarrow$ Cr$_2$O$_7^{2-}$ + H$_2$O	1.8			(48)
Ru(II)	$d^6(t_{2g}^6)$	[(NH$_3$)$_5$Ru-Cl-Cr(H$_2$O)$_6$]$^{4+}$ $\longrightarrow$ (NH$_3$)$_5$RuOH$_2^{2+}$ + Cr(H$_2$O)$_5$Cl^{2+}	4.5×10^{2b}			(49)
Ru(III)	$d^5(t_{2g}^5)$	Ru(NH$_3$)$_5$Cl^{2+} $\longrightarrow$ Ru(NH$_3$)$_5$OH$_2^{2+}$ + Cl$^-$	7.0×10^{-7b}	23	-11	(50)
Rh(III)	$d^6(t_{2g}^6)$	trans-Rh(en)$_2$Br$_2^+$ + F$^-$ $\longrightarrow$ trans-Rh(en)$_2$BrF$^+$ + Br$^-$	7.25×10^{-6e}	25	-4.3	(51)
Ir(III)	$d^6(t_{2g}^6)$	trans-Ir(H$_2$O)$_2$Cl$_4^-$ $\longrightarrow$ 1,2,6-Ir(H$_2$O)$_3$Cl$_3$ + Cl$^-$	8×10^{-9b}	29.9	4.9	(52)
Pd(II)	d^8	Pd(NH$_3$)$_4^{2+}$ + Cl$^-$ $\longrightarrow$ Pd(NH$_3$)$_3$Cl$^+$ + NH$_3$	7.1×10^{-3f}	18.5	-7	(53)
Pt(II)	d^8	Pt(dien)Br$^+$ + N$_3^-$ $\longrightarrow$ Pt(dien)N$_3^+$ + Br$^-$	7.7×10^{-3g}			(54)
Ce(III)	f^1	Ce^{3+} + Mux^h $\longrightarrow$ Ce(Mux)$^{2+}$	3×10^{9h}			(55)
Eu(III)	f^6	Eu^{3+} + An^{-i} $\longrightarrow$ EuAn^{2+}	1.05×10^{8i}	3.3	-12	(56)
UO$_2^{2+}$		UO$_2^{2+}$ + NCS$^-$ $\longrightarrow$ UO$_2$NCS^{+a}	2.9×10^{2j}			(57)

^{a} The octahedral notation is given in parentheses, even though the complexes are not truly octahedral in symmetry.
^{b} In units of sec^{-1}.
^{c} TPTZ is tris-(2-pyridyl)-s-triazine.
^{d} The exchange of the oxo ligand is slow.
^{e} $T = 50°$C, units of sec^{-1}.
^{f} There is also a term independent of [Cl$^-$], $k = 1.1 \times 10^{-3}$ sec^{-1}.
^{g} There is a term independent of [N$_3^-$], $k = 5 \times 10^{-4}$ sec^{-1}.
^{h} $T = 12°$, in NO$_3^-$ medium. Mux is murexide.
^{i} $T = 12.5°$, in ClO$_4^-$ medium. An$^-$ is o-aminobenzoate anion.
^{j} $T = 20°$.

(*12,58*). There is, however, still some doubt about several features of the lability of metal-ion complexes. Mechanistic conclusions generally favor dissociative activation for octahedral complexes, rather than associative activation, but some systems seem to use the latter mode of activation. Extensive reviews on the subject of substitution mechanisms are available (*11,12,59,60*).

It should be noted that the gross trends noted above require careful examination in individual cases. For example, whereas Co(III) complexes are generally inert, significant variation in lability occur as the ligand environment about the Co(III) center is varied. Consider Eqs. (30) and (31) in which no variation in charge type occurs.

$$trans\text{-}Co(en)_2NCSCl^+ \longrightarrow Co(en)_2NCSH_2O^{2+} + Cl^- \qquad (30)$$

$$trans\text{-}Co(en)_2NO_2Cl^+ \longrightarrow trans\text{-}Co(en)_2NO_2H_2O^{2+} + Cl \qquad (31)$$

The rate constants are 5×10^{-8} sec^1 (*61*) and 9.7×10^{-4} sec^{-1} (*62*) respectively at 25°; $trans\text{-}Co(en)_2OHCl^+$ reacts even faster (*63*), although $trans\text{-}Co(en)_2H_2OCl^{2+}$ doesn't substitute very rapidly. This latter datum is an example of a general phenomenon for those reactions in which it is thought that dissociative activation is important. In these cases there seems to be a general rate enhancement when an H_2O in the coordination sphere is replaced by an OH^-. This feature is shown by the previously-mentioned data and is illustrated for an Fe(III) substitution in Table 1. Similar considerations apparently apply to NH_2^- versus NH_3, an argument that has successfully accounted for the base catalyzed path in substitution reactions of Co(III)-amine complexes (*12,64*).

A knowledge of the lability characteristics of transition metal-ion complexes allows a meaningful application of the results of a study of stoichiometry to a determination of some properties of the transition state for a given oxidation-reduction reaction. The extensive studies of Cr(II) reductions of Co(III) complexes have been carried out to take advantage of the lability characteristics of the various reactant and product ions in determining mechanisms. Taube, Myers, and Rich (*65*) first established that the reduction of $Co(NH_3)_5Cl^{2+}$ by $Cr(H_2O)_6^{2+}$ took place via a transition state in which both ions were bound to the chloride ion. The argument establishing this point is made by noting that, relative to the rate of the reduction, the rate of loss of Cl^- from Co(III) is quite slow: during the course of the oxidation-reduction reaction no Cl^- is released to the solution from the coordination sphere of Co(III). On the other hand, reference to Fig. 1 shows that Co(II) will lose Cl^- quite rapidly once $CoCl^+$ is formed. Complimentary to this lability change at the cobalt center, is that at the chromium center; Cr(II) is quite labile, but once it becomes Cr(III), the ligands present in the coordination sphere of chromium are relatively frozen. Hence the only possible way for Cl^- to be transferred from the Co(III) reactant to the Cr(III) product is if both metal centers are bound to the Cl^- in the transition state:

$$Cr^{2+} + Co(NH_3)_5Cl^{2+} \longrightarrow [Co(NH_3)_5ClCr^{4+}]^\ddagger + H_2O \qquad (32)$$

$$[Co(NH_3)_5ClCr^{4+}]^\ddagger \longrightarrow \longrightarrow CrCl^{2+} + Co^{2+} \qquad (33)$$

This mechanism is called "inner-sphere" because the two metal ions have a ligand in their inner coordination spheres common to one another. The other type of mechanism is then defined: when two reactants do not share a common ligand in their first coordination spheres, the mechanism is called "outer-sphere." Figure 2 gives a schematic representation of the transition states for these two mechanisms.

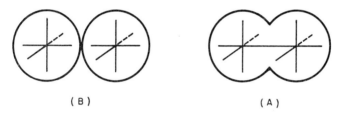

(B)   (A)

Fig. 2. Representation of inner-sphere (A) and outer-sphere (B) transition states.

Since the original demonstration of these two basic types of mechanisms, a great deal of effort has been expended on learning by which mechanism various reactions proceed. It is clear that different parameters can be important in determining the reactivity in the two schemes. First, there is the question of the timing of the substituion process needed to form the bond to the common ligand relative to the rate of the electron-transfer process. Figure 1 shows that reactions in which V(II) acts as a reductant on an inert oxidant should be particularly susceptible to studies of this feature of inner-sphere processes. This has proved to be so; in Table 2 are listed some data for a number of systems in which it has been established, by direct observation of the product V(III) complex, that V(II) reacts by an inner-sphere mechanism

$$Co(NH_3)_5ox^+ + V^{2+} \longrightarrow Co(II) + Vox^+ \qquad (34)$$

$$Vox^+ \longrightarrow V^{3+} + Hox^- \qquad (35)$$

TABLE 2

INNER-SPHERE REACTIONS OF VANADIUM(II)[a]

Oxidant	V(III) product	k (M^{-1} sec^{-1})	$\Delta H^{\ddagger}$ (kcal mole^{-1})	$\Delta S^{\ddagger}$ (e.u.)	Ref.
VO^{2+}	VOV^{4+}[b]	46[c]	12.3[d]	−16.5[d]	(66, 67)
Co(en)$_2$(N$_3$)$_2^+$	VN$_3^{2+}$	33			(68)
Co(NH$_3$)$_5$ox$^+$	Vox$^+$	45			(69)
Co(NH$_3$)$_5$N$_3^{2+}$	VN$_3^{2+}$[e]	13	11.7	−14	(70)
trans-Co(en)$_2$H$_2$ON$_3^{2+}$	VN$_3^{2+}$	17	13.7	−11.8	(71)

[a] $T = 25°$; I = 1.0 M.
[b] Partial product; some V(III) is also formed directly.
[c] This includes a minor term first-order in [H$^+$].
[d] At I = 2.0 M.
[e] Established in Ref. (71).

The peculiar feature of these data is the lack of change of the rate constant as the oxidant is varied. Such behavior has been suggested as being due to rate-limiting substration of the oxidant into the coordination-sphere of V(II)

$$V^{2+} + Co(NH_3)_5N_3^{2+} \underset{k_{-1}}{\overset{k_1}{\rightleftharpoons}} [(NH_3)_5Co(III)N_3V(II)(H_2O)_5^{4+}] + H_2O$$

$$\text{(36)}$$

$$[(NH_3)_5Co(III)N_3V(II)(H_2O)_5^{4+}] \overset{k_2}{\longrightarrow} [(NH_3)_5Co(II)N_3V(III)(H_2O)_5^{4+}] \quad \text{(37)}$$

$$[(NH_3)_5Co(II)N_3V(III)(H_2O)_5^{4+}] \longrightarrow 5NH_4^+ + Co^{2+} + VN_3^{2+} \quad \text{(38)}$$

$$H^+ + VN_3^{2+} \longrightarrow V^{3+} + HN_3 \quad \text{(39)}$$

that is, $k_2, k_{-1} > k_1$. Such a suggestion is in accord with data on the substitution on V^{2+} by NCS^- ($k = 28\ M^{-1}\ sec^{-1}$ at $25°$, $I = 0.84\ M$, $\Delta H^{\ddagger} = 13.5$ kcal mole^{-1}, $\Delta S^{\ddagger} = -7$ e.u.) (72). It has also recently been reported that saturation of an ion-pair precursor of the inner-sphere system (73) occurs because of slow substitution on V(II),

$$V^{2+} + Co(CN)_5N_3^{3-} \overset{K}{=} [V(H_2O)_6^{2+}, Co(CN)_5N_3^{3-}] \quad \text{(40)}$$

$$[V(H_2O)_6^{2+}, Co(CN)_5N_3^{3-}] \underset{k_{-1}}{\overset{k_1}{\rightleftharpoons}} [(H_2O)_5V(II)N_3Co(III)(CN)_5^-] + H_2O$$

$$\text{(41)}$$

$$[(H_2O)_5V(II)N_3Co(III)(CN)_5^-] \overset{k_2}{\longrightarrow} \text{Products} \quad \text{(42)}$$

where, as above, $k_2, k_{-1} > k_1$. There have been several suggestions that some rapid reactions of Cr(II) are limited by substitution on the Cr(II) center (74–76), but these suggestions are still the subject of some controversy.

A second parameter of some importance to inner-sphere reactions, but clearly of no importance to outer-sphere reactions, is the strength of the bond holding the two metal centers together, and the effect of that strength on the rate of the oxidation-reduction process. Coupled with this question is that of the actual electronic role played by the orbitals of the ligand in achieving electron-transfer (77). Answers to these questions are much less easy to find, and have been, in general, subordinate to the general problem of determining the mechanism of reactions in which the lability characteristics of the reactants and products are not suitable for direct determination.

The outer-sphere mechanism, on the other hand, offers no questions of these types. Indeed, the mechanism of these reactions is unique among chemical reactions in that no chemical bonds are broken or made in achieving the transition state. Such behavior should make these reactions ideally suitable for theoretical approaches. This subject has been explored extensively, with moderate success (as noted below). The need for cases in which a test of the theoretical predictions can be made has offered another impetus to the experimental determination of the mechanism of electron-transfer reactions.

As implied above, there are several reagents in which there is no ambiguity in the determination of the mechanism by which the electron-transfer process takes place. If both the oxidant and the reductant are inert to substitution (on the time scale appropriate for electron transfer to take place), the reaction must proceed by the outer-sphere mechanism.

$$\text{Fe(phen)}_3^{2+} + \text{IrCl}_6^{2-} \longrightarrow \text{Fe(phen)}_3^{3+} + \text{IrCl}_6^{3-} \tag{43}$$

The only assumption made in such a declaration is that an expansion of the coordination sphere (to seven ligands) does not occur in the transition state (9).

On the other hand, if substitution on either reagent is rapid compared to the rate of electron transfer, the mechanism may be either inner-sphere or outer-sphere. Within this large class, it is generally assumed that reactions with an inert reagent whose ligands lack a pair of electrons available for coordination to the other reagent will proceed by the outer-sphere mechanism (25).

$$6\text{H}^+ + \text{Cr}^{2+} + \text{Co(NH}_3)_6^{3+} = \text{Cr}^{3+} + \text{Co}^{2+} + 6\text{NH}_4^+ \tag{44}$$

A recent development indicates that some question regarding this assumption might be warranted. Williams and Hunt have studied the nitrogen and oxygen exchange of $\text{Co(NH}_3)_5\text{OH}^{2+}$ with solvent and the Co(II)–Co(III) exchange in this system, both in aqueous NH_3 (78). Consistent with an inner-sphere mechanism, the exchange of Co between Co(II) and $\text{Co(NH}_3)_5\text{OH}^{2+}$ does not lead to oxygen exchange with solvent. But interestingly, electron-exchange also does not lead to exchange of five ammonias as one would expect from the mechanism

$$*\text{Co}(*\text{NH}_3)_5\text{OH}^{2+} + \text{Co(NH}_3)_5^{2+} \longrightarrow *\text{Co}(*\text{NH}_3)_5^{2+} + \text{Co(NH}_3)_5\text{OH}^{2+} \tag{45}$$

$$*\text{Co}(*\text{NH}_3)_5^{2+} \longrightarrow *\text{Co}^{2+} + 5*\text{NH}_3 \tag{46}$$

Apparently, although the transition state contains ten NH_3 molecules, only 3–4 of those on the Co(III) center are made labile by electron-transfer. This implies bonding of NH_3 to both cobalt centers in the transition state.

When ligands on the inert reagent have available electron pairs, mechanistic declarations become difficult. There, is, however, a group of reagents in which determination of the reaction mechanism is simplified somewhat by their lability characteristics. The classic example of this type has already been discussed: the Cr(II) reduction of Co(III) complexes. What is required is that the electron-transfer process be rapid compared to substitution on one of the products and that some evidence be available to demonstrate that the ligand originated on the other metal ion system. This second requirement has usually been met by studying reduction of quite inert systems such as complexes of Cr(III) and Ru(III), and amine complexes of Co(III). But determinations have also been applied in systems more difficult to handle. It is found (79), for instance, that the reduction of $Co(H_2O)_6^{3+}$ by Fe^{2+} in the presence of chloride yields $FeCl^{2+}$ if the Cl^- is originally in the Co(III) solution, but only Fe^{3+} when the Cl^- is originally in the Fe^{2+} solution. This is taken to mean that the formation of $CoCl^{2+}$ is too slow to allow an inner-sphere chloride bridged transition state in the latter experiment. On the other hand, production of $FeCl^{2+}$ in both systems (i.e., satisfactory meeting of the first requirement) would not prove the presence of an inner-sphere mechanism, for the results could be interpreted by the path

$$Fe^{2+} + Cl^- = FeCl^+ \qquad (47)$$

$$FeCl^+ + Co(III) \longrightarrow FeCl^{2+} + Co(II) \qquad (48)$$

Besides Cr(II), one other ion that exhibits lability in the reduced state and sufficient inertness in the oxidized state to allow product identification to establish the nature of the transition state is $Co(CN)_5^{3-}$ (80). In addition, in several systems the actual primary product of the electron-transfer reaction is a binuclear intermediate, which then decays to the final products of the reaction $(24,81–84)$. Establishment of the presence of a binuclear intermediate strongly implies a transition state in which the two reactants were bound to a common ligand.

With the advent of flow systems, several other ions have been subjected to detailed product analysis by observing the disappearance of primary products. Typical examples are Eqs. (49) and (50) (85), and Eqs. (51) and (52) (68).

$$Fe^{2+} + Co(EDTAH)Cl^- \longrightarrow FeCl^{2+} + Co(II) \qquad (49)$$

$$FeCl^{2+} \longrightarrow Fe^{3+} + Cl^- \qquad (50)$$

$$V^{2+} + cis\text{-}Co(en)_2(N_3)_2^+ \longrightarrow VN_3^{2+} + Co(II) \qquad (51)$$

$$VN_3^{2+} + H^+ \longrightarrow V^{3+} + HN_3 \qquad (52)$$

But these methods fail if the concentration range cannot be made sufficiently high so as to generate a reasonable concentration of intermediate.

The inability to make a direct determination of the nature of the mechanism by which any given ion reacts has led to a wide variety of indirect methods of determining mechanisms. Some of these methods are better than others, but all should be viewed as offering only `an indication of mechanism, rather than a proof. In addition, most (but not all) suffer two rather general faults: (1) the usual variant in the relative rate ratio most methods use as an indicator of mechanism is the bridging ligand. This is the very ligand whose role in activation process is most poorly understood, and is probably most complicated. (2) The establishment of mechanism can at best apply to those systems under direct study by the relative rate measurement, but need have no relationship to other reactions of either of the reactants. Notwithstanding these general limitations on the indirect methods of determining mechanism, a discussion of these methods is of value: The results of attempts to determine mechanism indirectly have prompted experiments that continue to add to knowledge of the mechanism of electron-transfer reactions and have spurred activity in more sophisticated methods of mechanism determination.

1. Relative Rates as the Bridging Group Is Varied

The comparison of the relative rates of reduction of Co(III) complexes of the type $Co(NH_3)_5X^{n+}$ as X is varied was suggested by Candlin, Halpern, and Trimm (70). Two series of reactivities were established: (1) for reactions that were known to proceed by the inner-sphere mechanism— reactions of $Cr(H_2O)_6^{2+}$, and (2) for reactions that were known to proceed (or could be logically assumed to proceed) by the outer-sphere mechanism —reactions of $Cr(bip)_3^{2+}$ or the reduction of the complexes at the dropping mercury electrode. Then reactions with other reductants were compared with these two series. One outstanding feature of this comparison was that the rate of the outer-sphere reaction seemed to vary less with a change in the ligand X than did that of the inner-sphere reaction (70,86). Accordingly, the variation in rate occurring upon a change in X was suggested as an indication of the mechanism. Unfortunately, in support of the comment made above concerning the difficulty of understanding the variation in the bridging ligand, a feature of experiments such as these was that the relative rate of reduction by both $Cr(H_2O)_6^{2+}$ (87) and $Cr(bip)_3^{2+}$ (70) of the complexes with X a halide was observed to be $I^- > Br^- > Cl^- > F^-$. Similar orders are found for many other reducing agents, such as V^{2+} (70,88), $Co(CN)_5^{3-}$ (80), $Ru(NH_3)_6^{2+}$ (86), H atoms (89), and Cu^+ (90). In addition, the Cr^{2+} (91) and $Cr(EDTA)^{2-}$ (92) catalyzed substitutions of $Cr(NH_3)_5$-X^{2+}, the isotopic exchange reaction (93)

$$*Cr^{2+} + Cr(H_2O)_5 X^{2+} = *Cr(H_2O)_5 X^{2+} + Cr^{2+} \tag{53}$$

the $Pt(NH_3)_4^{2+}$ catalyzed substitution of $Pt(NH_3)_4 YX^{2+}$ (94–96) (where Y is not changed in a given series), and the Cr^{2+} reduction of $Fe(H_2O)_5 X^{2+}$ (97) all apparently occur with $I^- > Br^- > Cl^- > F^-$, although some halides are missing in various cases. This order has become known as the "normal" order (11). In contrast are the data on the Fe^{2+} (88,98), Eu^{2+} (70), and U^{3+} (99) reductions of $Co(NH_3)_5 X^{2+}$ complexes, where the relative rate order is $F^- > Cl^- > Br^- > I^-$. A similar ordering has been found in other reductions by Fe^{2+} (100–102) and Eu^{2+} (97). Attempts to rationalize this discrepancy have been made, including an exposition by Haim (103).

Haim has put correlations of relative reactivity as the X group in $Co(NH_3)_5 X^{2+}$ is varied from F^- to I^- on a quantitative basis. He considers the processes

$$M^{n+} + Co(NH_3)_5 F^{2+} = \{[(NH_3)_5 Co-F-M]^{(2+n)+}\}^{\ddagger} \qquad K_1^{\ddagger} \tag{54}$$

$$M^{n+} + Co(NH_3)_5 I^{2+} = \{[(NH_3)_5 Co-I-M]^{(2+n)+}\}^{\ddagger} \qquad K_2^{\ddagger} \tag{55}$$

$$Co(NH_3)_5 OH_2^{3+} + F^- = Co(NH_3)_5 F^{2+} \qquad K_3 \tag{56}$$

$$Co(NH_3)_5 OH_2^{3+} + I^- = Co(NH_3)_5 I^{2+} \qquad K_4 \tag{57}$$

Application of transition state theory yields

$$k_1 = \kappa \, \frac{\bar{k}T}{h} \, K_1^{\ddagger} \tag{58}$$

where $\bar{k}$ is Boltzman's constant, h is Planck's constant, and $\kappa = 1$ is the transmission coefficient. The value for the formal equilibrium constant for the process

$$\{[(NH_3)_5 Co-F-M]^{(2+n)+}\}^{\ddagger} + I^- = \{[(NH_3)_5 Co-I-M]^{(2+n)+}\}^{\ddagger} + F^- \tag{59}$$

can then be calculated as

$$K_{F^-, \, I^-} = \frac{K_2^{\ddagger} K_4}{K_1^{\ddagger} K_3} = \frac{k_2 K_4}{k_1 K_3} \tag{60}$$

Values for K for the various halides are available (104,105) and combined with the rate constants yield K_{X^-, Y^-} values for a variety of M's. Table 3 lists some of these values. Noteworthy is the observation that for Cr(II), V(II), Fe(II), and Eu(II), the K_{F^-, Y^-} values are less than unity, whereas for $Ru(NH_3)_6^{2+}$ and $Cr(bip)_3^{2+}$, K_{F^-, Y^-} values are greater than unity. Haim suggests this difference as an indirect means of determining mechanisms. The observation that $Co(CN)_5^{3-}$ has a K_{F^-, Y^-} value greater than

TABLE 3

VALUES FOR THE FORMAL EQUILIBRIUM CONSTANT, $K_{X-,Y-}$, FOR THE PROCESS

$$[[(NH_3)_5CoXM]^{(2+n)+}]^{\ddagger} + Y^- = [[(NH_3)_5Co\,YM]^{(2+n)+}]^{\ddagger} + X^{-a}$$

	$Y^- = F^-$	Cl^-	Br^-	I^-
$M = Cr^{2+}$ (87)				
$X = F^-$	1.0	1.2×10^{-1}	8.8×10^{-2}	5.7×10^{-2}
$X = Cl^-$		1.0	7.3×10^{-1}	4.8×10^{-1}
$X = Br^-$			1.0	6.5×10^{-1}
$M = V^{2+}$ (70, 88),				
$X^- = F^-$	1.0	9.6×10^{-2}	7.9×10^{-2}	2.0×10^{-1}
$M = Ru(NH_3)_6^{2+}$ (86, 106),				
$X^- = F^-$	1.0	14.5^b	27^b	32^b
$M = Co(CN)_5^{3-}$ (80),				
$X^- = F^-$	1.0	$\simeq 1.4 \times 10^3$	$< 1.7 \times 10^4$	
$M = Fe^{2+}$ (88)				
$X = F^-$	1.0	1.05×10^{-2}	1.9×10^{-3}	
$X = Cl^-$		1.0	1.8×10^{-1}	
$M = Eu^{2+}$ (70),				
$X^- = F^-$	1.0	9.8×10^{-4}	1.9×10^{-4}	2.8×10^{-5}
$M = Cr(bip)_3^{2+}$ (70),				
$X^- = F^-$	1.0	22^b	43^b	
$M = Cu^+$ (90),				
$X^- = F^-$	1.0	2.2×10^3	6.4×10^3	

^{a} Kinetic data from references listed after each metal ion. Equilibrium data used were, for $Co(NH_3)_5F^{2+}$, 25 M^{-1} (104, 107); for $Co(NH_3)_5Cl^{2+}$, 1.25 M^{-1} (105); for $Co(NH_3)_5Br^{2+}$, 0.39 M^{-1} (105); and for $Co(NH_3)_5I^{2+}$, 0.12 M^{-1} (103, 108).

^{b} There is, apparently, a mathematical error in Haim's table (103).

unity, although it is known by product analysis to proceed by an inner-sphere mechanism, is explained by the fact that $Co(CN)_5^{3-}$ is a "soft" metal ion center. A similar argument has been advanced concerning Cu(I) (90).

Although Haim's argument involves an interesting analysis of the long-troublesome problem of the "abnormal" order of reduction of $Co(NH_3)_5$-X^{2+} complexes by Fe^{2+} and Eu^{2+}, the validity of the $K_{F-,X-} > 1$ for outer-sphere reactions and $K_{F-,X-} < 1$ for inner-sphere reactions is not certain. Worth particular note is the fact that most $K_{Cl-,Br-}$ and $K_{Cl-,I-}$ values are between about 10^{-1} and 2 (with the exception of $Co(CN)_5^{3-}$, an exceptional case). Hence it is the reactivity of $Co(NH_3)_5F^{2+}$ compared to that of remaining complexes that shows significant variation. Further the reaction of V^{2+} with $Co(NH_3)_5I^{2+}$ is probably outer-sphere, for the reaction proceeds at too large a rate for water-substitution on V^{2+}. Yet the value of K_{F^-, I^-} for V^{2+} is only slightly greater than the value for Cr^{2+}. At the least this implies that the difference in energy between the inner-sphere and the outer-sphere transition states is slight. Finally, an independent means of analysis of the reaction of V^{2+} with $Co(NH_3)_5Cl^{2+}$ has led to the conclusion that this reaction is outer-sphere (71,109). Hence, there is some doubt about the use of K_{F^-, Y^-} values to determine mechanism.

2. Relative Rate of Reaction of Hydroxy-Complex Versus Aquocomplex

The suggestion has been made that the mechanism by which a given reductant reacts with a Co(III) complex can be determined by comparison of the acid independent and acid dependent paths for the reduction of $Co(NH_3)_5H_2O^{3+}$

$$5H^+ + Co(NH_3)_5H_2O^{3+} + Cr^{2+} = Cr^{3+} + Co^{2+} + 5NH_4^+ \qquad (61)$$

$$\frac{-d[Co(NH_3)_5H_2O^{3+}]}{dt} = \left[k_1 + \frac{k_2'}{[H^+]}\right][Co(III)][Cr(II)] \qquad (62)$$

If the term dependent on the $[H^+]$ is interpreted as due to the mechanism

$$Co(NH_3)_5H_2O^{3+} \overset{K}{=} Co(NH_3)_5OH^{2+} + H^+ \qquad (63)$$

$$Co(NH_3)_5OH^{2+} + Cr^{2+} \overset{k_2}{\longrightarrow} Co(II) + Cr(III) \qquad (64)$$

then under conditions where $[Co(NH_3)_5OH^{2+}]$ is small compared to the total concentration of Co(III),

$$\frac{-d[Co(III)]}{dt} = \left(k_1 + \frac{k_2 K}{[H^+]}\right)[Co)III)][Cr(II)] \qquad (65)$$

$$k_2 = \frac{k_2'}{K} \qquad (66)$$

On the basis of early experiments with reductants known to react by the inner-sphere or outer-sphere mechanisms and supported by more recent measurements on other systems (25,86), Taube and co-workers have suggested that large values of k_2/k_1 are to be associated with bridge formation, whereas small values of this ratio are to be associated with outer-sphere reactions. This argument has been used in a wide variety of cases. Some specific examples are listed in Table 4. Two reactions that have been established independently as reacting by the inner-sphere mechanism are

TABLE 4

VALUES FOR THE RELATIVE RATE OF REDUCTION OF $ML_5H_2O^{n+}$ COMPLEXES COMPARED TO $ML_5OH^{(n-1)+}$ COMPLEXES

ML_5H_2O	Reductant	k_{H_2O} [b]	ΔH^{*c}	$\Delta S^{\ddagger d}$	k_{OH} [e]	$\Delta H^{\ddagger c}$	Δ	$\dfrac{k_{OH}}{k_{H_2O}}$	Ref.
$Co(NH_3)_5H_2O^{3+}$	V^{2+}	5.3×10^{-1}	8.2	-32	$\leqslant 4$			$\leqslant 8$	(110)
	Cr^{2+}	$5 \times 10^{-1 f}$	2.9	-52	$1.7 \times 10^{6 f}$	4.6	-18	3.5×10^{6}	(111)
	Cu^+	1×10^{-3}			3.8×10^{2}			3.8×10^{5}	(90)
	Eu^{2+}	7.4×10^{-2}	9.3	-32	$< 2 \times 10^{3}$			$< 2.7 \times 10^{4}$	(112)
	$Cr(bip)_3^{3+}$	$2.1 \times 10^{3 g}$			$1 \times 10^{-3 g}$			4.7×10^{-1}	(113)
	$Ru(NH_3)_6^{2+}$	3			4×10^{-2}			1.3×10^{-2}	(86)
$Fe(H_2O)_6^{3+}$	V^{2+}	1.8×10^{4}			$< 4 \times 10^{5}$			$< 2.2 \times 10^{1}$	(114)
	Cr^{2+}	5.7×10^{2}	5.2	-28	4.4×10^{6}	4.6	-13	7.7×10^{3}	(97)
	Fe^{2+}	$8.7 \times 10^{-1 h}$	9.3	-25	$1.1 \times 10^{3 h}$	6.9	-18	1.3×10^{3}	(22)
		4^i	10.5	-21	$3.0 \times 10^{3 i}$	8.4	-14	7.5×10^{2}	(115)
	Cu^+	$\leqslant 2 \times 10^{4}$			9.8×10^{7}	2.2	-14.8	$\leqslant 5 \times 10^{3}$	(116)
	Eu^{2+}	3.4×10^{3}	3.5	-29	5.2×10^{6}	2.0	-20	1.5×10^{3}	(117)
	Np^{3+}	$3.2 \times 10^{11 j}$	4.7	-36	$4.0 \times 10^{5 j}$	4.2^j	-19^j	1.2×10^{4}	(118)
	$Ru(NH_3)_6^{2+}$	$3.1 \times 10^{5 k}$	3.2	-22	$5.0 \times 10^{4 k}$			1.6×10^{-1}	(120)
	$Ru(en)_3^{2+}$	$9.4 \times 10^{4 k}$	4.3	-22	$1.5 \times 10^{4 k}$			1.6×10^{-1}	(120)
	$Ru(NH_3)_5OH_2^{2+}$	$8.0 \times 10^{4 k}$			$4.8 \times 10^{5 k}$			6	(120)
$Cu(H_2O)_6^{2+}$	V^{3+}	$1.8 \times 10^{-2 l}$	21.3	-5	$1.7 \times 10^{2 l}$	9.4	-16	9.4×10^{3}	(121)
	Cr^{2+}	1.7×10^{-1}			5.9×10^{-1} $[H^+]^{-1 m}$				(122)

Cr(H$_2$O)$_6^{2+}$	Cr^{2+}	$\leqslant 2 \times 10^{-5}$			2.3	14.0	-10	$\geqq 1.1 \times 10^5$	(123, 124)
Ru(NH$_3$)$_5$OH$_2^{2+}$	Cr^{2+}	5×10^2			3.5×10^6			7×10^3	(125)
Np^{4+}	V^{2+}	2.6^n	9.7	-24	7^n			3	(126)
	Cr^{2+}	$<1 \times 10^{-2}$			8.6×10^{2o}			$>8.6 \times 10^4$	(127)
Co(terpy)$_2^{2+}$		7.4×10^4	3.4	-23	1.7×10^4	5.1	-20	2.3×10^{-1}	(129)
Co(H$_2$O)$_6^{2+}$		4	11.8	-17	$6\text{-}80^p$				(130)

^{a} $T = 25°$, $I = 1.0$ M.

^{b} Units are M^{-1} sec^{-1}.

^{c} Units are kcal mole^{-1}.

^{d} Units are e.u.

^{e} Units are M^{-1} sec^{-1}; note that these values depend upon the value of K in Eq. (6). The original literature should be consulted.

^{f} At 20°C.

^{g} At 40°C, $I = 0.05$ M.

^{h} At 0°, $I = 0.55$ M.

^{i} $I = 0.50$ M.

^{j} $I = 2.0$ M; data from Ref. (119) was used to calculate k_{OH}.

^{k} At 10°C, $I = 0.10$ M.

^{l} Note that the reductant loses the proton in this case. The [H$^+$] independent term may be only a medium effect: its value triples upon a change from LiClO$_4$ to NaClO$_4$. Values quoted are for LiClO$_4$.

^{m} The value of K [Eq. (65)] is not known well for either Cu^{2+} or Cr^{2+}, and it is not clear which reagent loses the proton. Units of this k are thus sec^{-1}; $T = 24.6°$C.

^{n} $I = 2.0$ M.

^{o} Calculated using $K = 5 \times 10^{-3}$ M (128).

^{p} There is considerable uncertainty in the value of K.

listed there: the Cr^{2+} reduction of $Co(NH_3)_5H_2O^{3+}$ and of $Ru(NH_3)_5$ H_2O^{3+}. Several systems listed must be outer-sphere because of low substitution rates. If these models are valid, then it would appear that most metal-ion reductions [at least of $Co(NH_3)_5OH_3^{2+}$ and $Fe(H_2O)_6^{3+}$] prefer an inner-sphere mechanism. Note that the V^{2+} reduction of $Co(NH_3)_5$ OH_2^{3+} may be inner-sphere with the rate limited water exchange, whereas $V^{2+} + Fe(H_2O)_6^{3+}$ is outer-sphere, as is $V^{2+} + Fe(H_2O)_5OH^{2+}$, unless substitution on the oxidant is effective (*114*).

3. Relative Rate as Complex is Changed from an Azide to a Thiocyanate

Ball and King (*93*) noted that for an inner-sphere reaction it was not unexpected that the $-NCS^-$ complex would react slower than the $-NNN^-$ one. Espenson (*98*) developed an indirect method of determining mechanism by suggesting that for outer-sphere systems, little discrimination should exist. Hence the relative rate ratio k_{N_3-}/k_{NCS-} should give an indication of the mechanism of a redox reaction. The justification for this argument rests with the preferential bonding to the N end of NCS^- by metal ions which are "hard," whereas the S end is preferred by those that are "soft" (*93,131*). Accordingly, if the reduction of $Co(NH_3)_5NCS^{2+}$ is to take place by an inner-sphere mechanism, the immediate product of the reaction [providing that attack takes place, as it appears to, at the remote site (*76*)] will, if the oxidized product is "hard," be an unstable isomer of that metal ion:

$$Co(NH_3)_5NCS^{2+} + Cr^{2+} \longrightarrow CrSCN^{2+} + Co(II) \tag{67}$$

$$CrSCN^{2+} \longrightarrow CrNCS^{2+} \tag{68}$$

On the other hand, when $Co(NH_3)_5N_3^{2+}$ is reduced, the symmetrical nature of the azide ligand makes it a perfectly suitable bridge with no adverse "hard-soft" interactions. A list of the relative rates available are given in Table 5. It should be stressed that these data do not reflect on the mechanism of M-NCS if k_{N_3-}/k_{NCS-} is large; then the only conclusion that can be drawn is that M-N_3 is inner-sphere, the M-NCS complex reacting slowly either because of the adverse "hard-soft" interactions or because it reacts by a slower outer-sphere mechanism.

4. The Effect of Added Ions

It is often found that the rate of an oxidation-reduction reaction between a labile reactant and an inert reactant is affected by other ions in the solution. Typical rate laws for both an inner-sphere and an outer-sphere reaction are (*135*)

TABLE 5

VALUES FOR THE RELATIVE RATE OF REDUCTION OF $ML_5N_3^{n+}$ COMPARED TO ML_5NCS^{n+} COMPLEXES[a]

$ML_5N_3^{n+}$	Reductant	k_{NCS^-}[b]	$\Delta H^{\ddagger c}$	$\Delta S^{\ddagger d}$	$k_{N_3^-}$[b]	$\Delta H^{\ddagger c}$	$\Delta S^{\ddagger d}$	$k_{N_3^-}/k_{NCS^-}$	Ref.
$Co(NH_3)_5N_3^{2+}$	V^{2+}	3×10^{-1}	6.9	−29	1.3×10^1	11.7	−14	4.3×10^1	(70)
	Cr^{2+}	1.9×10^1			$\cong 3 \times 10^5$			$\approx 1.6 \times 10^4$	(70)
	$Cr(bip)_3^{2+}$	1.0×10^{4e}			4.1×10^{4e}			4.1	(70)
	Fe^{2+}	$< 3 \times 10^{-6}$			8.8×10^{-3}			$> 3 \times 10^3$	(98)
	Cu^+	≈ 1			1.5×10^3	5.4	−26	$\approx 1.5 \times 10^3$	(90)
	Eu^{2+}	$\approx 7 \times 10^{-1}$			1.9×10^2	5.5	−30	$\approx 2.7 \times 10^2$	(70)
	$Co(CN)_5^{3-}$	$\cong 1.1 \times 10^{6f}$			1.6×10^{6f}			1.4	(80)
	U^{3+}	1.8×10^1			8.2×10^{5f}			4.5×10^4	(99)
$Fe(H_2O)_5N_3^{2+}$	V^{2+}	6.6×10^5			5.2×10^5			8×10^{-1}	(114)
	Cr^{2+}	2.8×10^7			2.9×10^7			1.0	(97)
	Fe^{2+}	$4.2 \to 12.2^g$	?	?	1.9×10^{3h}	13.3^h	7.0^h	$\cong 3 \times 10^2$	(101, 132, 133)
	Eu^{2+}	3.2×10^{5i}	4.4	−17	1.2×10^{7i}			3.7×10^1	(97)
$Cr(H_2O)_5N_3^{2+}$	Cr^{2+}	1.8×10^{-4j}			7.0^k	9.6	−23	5.3×10^4	(93, 134)

[a] $T = 25°$, $I = 1.0$ M.
[b] Units are $M^{-1} sec^{-1}$.
[c] Units are kcal mole^{-1}.
[d] Units are e.u.
[e] $I = 0.1$ M.
[f] $I = 0.2$ M.
[g] $T = 0°$, $I = 0.55$ M.
[h] At $0°$, $I = 0.55$ M. The activation parameters are said to be valid only below 15°C. Above this temperature, curvature of the log k versus $1/T$ plot occurs in such a manner as to lower $\Delta H^{\ddagger}$ and $\Delta S^{\ddagger}$.
[i] At 1.6°C.
[j] At 27°, $I = 1.0$ M.
[k] At 27°, $I = 0.5$ M.

$$Co(NH_3)_6^{3+} + Cr^{2+} = Cr(III) + Co(II) \qquad (69)$$

$$\frac{-d[Co(III)]}{dt} = (7.2 \times 10^{-3} \ M^{-1} \ sec^{-1} + 0.60 \ M^{-2} \ sec^{-1}[Cl^-])[Cr(II)]$$
$$[Co(III)] \quad (70)$$

$$Co(NH_3)_5(CH_3COO)^{2+} + Cr^{2+} = Cr(III) + Co(II) \qquad (71)$$

$$\frac{-d[Co(III)]}{dt} =$$

$$(3.4 \times 10^{-1} \ M^{-1} \ sec^{-1} + 2.0 \times 10^{-1} \ M^{-2} \ sec^{-1}[Cl^-])([Cr(II)]$$
$$[Co(III)] \quad (72)$$

It has been suggested that this situation may be a general one (9): outer-sphere reactions will be characterized by a value for the ratio of the rate constant associated with the transition state containing an ion from solution, relative to that transition state containing only those two metal ion centers, of about 5 to 100 M^{-1}, whereas smaller values of this ratio will be characteristic of inner-sphere reactions. As is true with several of the indirect methods of determining mechanism, this procedure is still open to test. Although a number of inner-sphere reactions have been found to have small values of his relative rate ratio, only a few outer-sphere systems have been studied. Pennington and Haim have surveyed most of the available data (136).

The question of the effect of added ligands is tied up with nonbridging ligand effects. These pertain because an explanation of the effect is that the added ligand bind to the labile reductant prior to the activated complex

$$Cr^{2+} + Cl^- = CrCl^+ \qquad (73)$$

$$CrCl^+ + Co(NH_3)_5OH_2^{3+} \longrightarrow \left[ClCr-\overset{\displaystyle H}{\underset{\displaystyle H}{O}}-Co(NH_3)_5^{4+} \right]^{\ddagger} \qquad (74)$$

$$\left[ClCr-\overset{\displaystyle H}{\underset{\displaystyle H}{O}}-Co(NH_3)_5^{4+} \right]^{\ddagger} \longrightarrow CrCl^{2+} + Co(II) \qquad (75)$$

The effect of this substitution of H_2O by Cl^- in the coordination sphere of Cr(II) probably has a significant effect on the rate. This is indicated by several examples in the literature (137–140). It is not clear at this time, however, that the difference between inner-sphere and outer-sphere mechanism will necessarily be reflected in this nonbridging ligand effect (106,109,141,142).

5. Volume of Activation

Because of the difference in the number of molecules of water in the transition state for an inner- and outer-sphere reaction,

$$Fe^{2+} + Co(NH_3)_5Cl^{2+} \xrightarrow{\text{i.s.}} [(H_2O)_5FeClCo(NH_3)_5^{4+}]^\ddagger + H_2O$$

$$\xrightarrow{\text{o.s.}} [(H_2O)_5FeH_2O,ClCo(NH_3)_5^{4+}]^\ddagger \tag{76}$$

it has been postulated that inner-sphere reactions should show a more positive volume of activation (143),

$$\Delta V^\ddagger = -RT \left(\frac{d \ln k}{dP} \right) \tag{77}$$

The results available are limited, partly because of the difficulty of studying rapid reactions under pressure. (Such difficulties may now be alleviated somewhat by Brower's (144) design for a high pressure "pressure jump" device in which the pressure dependence of the association of Fe(III) and NCS$^-$ has been measured; this device is useful only for reversible reactions however.) These limited results all show $\Delta V^\ddagger$ values which are positive for some reductions by Fe(II) that are suspected of being (143) or are now known to be (85) inner-sphere in mechanism. Unfortunately, no well-established outer-sphere reaction involving the Fe(II) reduction of a Co(III) complex was studied in order to calibrate the method. There has recently been a study of outer-sphere reactions (145); these do seem to show negative volumes of activation. The available data for the Fe^{2+}-Co(III) reactions show a variation in $\Delta V^\ddagger$ for various complexes which is somewhat mysterious (143): whereas $\Delta V^\ddagger$ for the Co(NH$_3$)$_5$N$_3^{2+}$ reduction by Fe^{2+} was 14 cm^3 mole^{-1}, that for trans-Co(NH$_3$)$_4$(N$_3$)$_2^+$, by either the acid independent or acid dependent path (146) was about only +3 cm^3 mole^{-1}.

6. Nonbridging Ligand Effects

The effect of a nonbridging ligand change on the rate of some electron-transfer reactions is very pronounced. For instance, the rate of reduction of trans-Co(en)$_2$NH$_3$Cl^{2+} by Fe^{2+} is 6.6×10^{-5} M^{-1} sec^{-1}, whereas the corresponding reduction of trans-Co(en)$_2$H$_2$OCl^{2+} is 0.24 M^{-1} sec^{-1} (147); the relative rate of reduction of trans-Co(NH$_3$)$_4$H$_2$OCl^{2+} compared to Co(NH$_3$)$_5$Cl^{2+} is about 7000 (98,142). Table 6 summarizes the effect on

TABLE 6

THE EFFECT OF NONBRIDGING LIGANDS ON THE RATE OF ELECTRON-TRANSFER REACTIONS[a]

Reaction:[b]	Fe^{2+} + Co(III)—Cl	V^{2+} + Co(III)—Cl	$Ru(NH_3)_6^{2+}$ + Co(III)—Cl	Cr^{2+} + Co(III)—N
Mechanism:	I.S.?	?	O.S.	I.S.
Conditions:	$T = 25°$	$T = 25°$,	$T = 20°$,	$T = 25°$,
	$\Sigma[ClO_4^-] = 1.0\ M$	$\Sigma[ClO_4^-] = 1.0\ M$	$I = 0.1\ M$	$I = 1.0\ M$
References:	(79, 98, 141, 142, 147, 148)	(109)	(86, 106)	(70, 149
Ligands				
cis-(en)$_2$NH$_3$	1.8×10^{-5}	1.9	1.2×10^1	3.1
trans-(en)$_2$NH$_3$	6.6×10^{-5}			3.8
cis-(en)$_2$NH$_2$CH$_2$C$_6$H$_5$	3.5×10^{-5}	2.8	3.7×10^1	
cis-(en)$_2$NCS	1.3×10^{-4}			3.0×10
trans-(en)$_2$NCS	1.7×10^{-4}			2.8×10^1
cis-(en)$_2$H$_2$O	4.6×10^{-4}	1.0×10^1	2.3×10^{2c}	4.5×10
cis-(en)$_2$Cl	1.6×10^{-3}	2.4×10^1	8.0×10^{2c}	
cis-(en)$_2$py	7.9×10^{-4}	3.2×10^1	6.6×10^2	
(NH$_3$)$_5$	1.4×10^{-3}	7.6	2.3×10^2	1.9×10
cis-(NH$_3$)$_4$Cl				
cis-(NH$_3$)$_4$H$_2$O	3.5×10^{-2}	5.8×10^1		
trans-(en)$_2$Cl	3.2×10^{-2}	1.3×10^2	8.0×10^{3c}	
trans-(en$_2$H$_2$O	2.4×10^{-1}	2.6×10^2	$>1.0 \times 10^5$	1.4×10
trans-(NH$_3$)$_4$Cl	2.21			
trans-(NH$_3$)$_4$H$_2$O	$\cong 1.0 \times 10^1$	$\cong 4.6 \times 10^2$		
(H$_2$O)$_5$	$>5 \times 10^3$			
trans-(H$_2$O)$_4$Cl				
cis-(H$_2$O)$_4$Cl				

[a] The K's are in units of $M^{-1}\ sec^{-1}$.
[b] The oxidant is represented by the metal ion and its valence and by the bridging ligand (inner-sphere reactions), or the ligand that is not varied (outer-sphere and reactions of unknown mechanism).

rate of a variation in the nonbridging ligands. The question of concern is whether the magnitude of this sensitivity to a change in nonbridging ligands is "relevant" to mechanism. Early suggestions (147) that the sensitivity was less for outer-sphere reactions than for inner do not seem correct, although a detailed test on complexes with "all other factors constant" has not been made (nor is it obvious which system is suitable for such a test). The only application of nonbridging ligand effects in determining mechanism that now appears fruitful is in rather limited areas. If a nonbridging ligand change is predicted (see Section II,D for a further discussion of these predictions) to increase the rate of a reaction past that rate which is the limit dictated by the substitution characteristics of one of the components, and such an increase takes place, then this latter reaction is surely, and the ones on which it was predicted are probably, outer-sphere in nature (71).

+ + III)—Cl	Cr²⁺ + Co(III)—CN	Cr²⁺ + Co(III)—OH	Fe²⁺ + Co(III)—N₃	V²⁺ + Ru(III)—Cl	Cu⁺ + Co(III)—Cl
I.S. $T=25°$, ·]=1.0 M 91, 93) 50–152)	I.S. $T=25°$ $I=0.15\,M$ (153)	I.S.? $T=25°$ $I=1.0\,M$ (111, 154)	I.S.? $T=25°$ $\Sigma[\text{ClO}_4^-]=0.5\,M$ (146, 155, 156)	O.S. $T=25°$ $I=0.10\,M$ (49, 125)	? $T=25°$ $I=0.2$ (90)
		2×10^5 2.2×10^5	$\cong 1 \times 10^{-4e}$		
		7.9×10^5	6.9×10^{-3e}		
5×10^{-2} $\times 10^{-2}$					2.13×10^4
$\cdot \times 10^{-2}$	3.6×10^1	1.7×10^6	8.7×10^{-3f}	3.05×10^3 9.8×10^3	4.88×10^4
$\cdot \times 10^{-1}$			3.55×10^{-1}		$>1.0 \times 10^7$
$\cdot$	1.12×10^3	2.6×10^6	8×10^{-1e}		
$\cdot \times 10^{-1}$	1.45×10^3		2.4×10^1	8.3×10^2 1.42×10^3	
$\cdot$ $\times 10^{14d}$ $\times 10^2$ $\times 10^2$					

^{c} These values are corrected from values measured at 25°; see Ref. (109).
^{d} Extrapolated to 25° from 0° using for $\Delta H^\ddagger$ the average value for the reaction of Cr^{2+} with cis-CrCl_2^+ and $trans$-CrCl_2^+.
^{e} $I=1.0\,M$.
^{f} $I=0.89\,M$.

C. THEORY OF ELECTRON-TRANSFER REACTIONS

Perhaps one of the most exciting areas of investigation in oxidation-reduction reactions of transition metal ions has been the generation and testing of theories of electron-transfer reactions. This interest arises because of the important feature of outer-sphere reactions—no chemical bonds are broken or formed—and the resultant simplification of theory. This subject has been well reviewed over the years (10,11,14,157–161) and the reader is referred to those treatments for more detail. It is the intent here to give only a general survey of the subject and to reiterate the general conclusions of the theories.

One important consideration is whether electron-transfer processes take place adiabatically or nonadiabatically. This is, do the energy surfaces defined by the electronic states of reactants and products as a function of nuclear displacements intersect with a small energy of interaction (perhaps several hundred calories (161)) or a larger energy of interaction? If the former case pertains, then the usual result of motion along the reaction coordinate will be a "jump" to the higher energy state, and the reaction rate will be small, not because of energy restrictions, but because of a low transmission coefficient. On the other hand, if the energy of interaction of the wave function describing reactants and products is large enough, all systems having the requisite energy will pass over the barrier from reactant to product—this is the adiabatic condition. The details of these two types of processes have been considered in depth by Reynolds and Lumry (161) and by Marcus (157). The experimental distinction is not easily made. Because of this difficulty, because all treatments can be made equivalent under certain assumptions, and because the theory of Marcus has been most widely tested, this discussion shall center on that theoretical approach.

One of the important early developments in the theoretical consideration of electron-transfer reactions was the recognition that nuclear motions would have to precede the electronic motion (162), and that the energy required to bring about reaction would be used for this purpose. Shortly thereafter, Marcus began (158,163) a series of papers based upon this concept, the assumption that the process was essentially adiabatic, but that interaction was weak (the potential energy surface was not split by enough to make calculation of the hypothetical intersection a poor approximation to the real activation free energy) and that the problem could be treated by statistical mechanical or continuum methods since bonds were not ruptured during the process. Marcus essentially calculated the energy necessary to create the nonequilibrium arrangement (of nucleii and electrons) of both the inner-coordination shell and the surrounding solvent. This nonequilibrium arrangement is not characteristic of either reactants or products, but of that hypothetical state in which the electron is migrating from one center to another: it is that arrangement in which the free energy of formation of reactant and product "activated complexes" are equal. Expressing the free energy change of forming the activated complex from reactant as ΔF^*, then

$$k = Z \exp[-\Delta F^* R^{-1} T^{-1}] \tag{78}$$

where Z is the number of collisions per unit time. The free energy change, ΔF^*, consists of three dominant terms: (1) The work necessary to bring the particles together in the medium of interest; (2) the energy necessary to rearrange the solvent electronic and nuclear positions to that of the non-equilibrium state demanded by the Franck-Condon governed activated complex; and (3) a similar term for the inner-coordination sphere of the ions involved. Expressions for each of these terms have been developed (*10,157,158,161*) and have been tested, with both rather successful (*10*) and rather unsuccessful results (*164*). It is difficult to test the theory because of the inability to guess necessary parameters.

By far the most dramatic testable result of Marcus's treatment is the so-called "cross reaction" equation. This equation relates the rate of a reaction between two reagents to that of the exchange reactions of each reagent and the net driving force of the reaction. It is applicable when Coulombic work terms cancel. Then

$$k_{12} = (k_{11}k_{22}K_{12}f)^{1/2} \tag{79}$$

where k_{12} is the rate of, for instance, the reaction

$$Fe^{2+} + Ce(IV) \longrightarrow Fe^{3+} + Ce(III) \tag{80}$$

k_{11} and k_{22} are the rates of

$$Fe^{2+} + {}^*Fe^{3+} \longrightarrow Fe^{3+} + {}^*Fe^{2+} \tag{81}$$

$${}^*Ce(IV) + Ce(III) \longrightarrow {}^*Ce(III) + Ce(IV) \tag{82}$$

respectively, and K_{12} is the equilibrium constant for reaction (80). The quantity f is related to the other parameters by equation (82)

$$\ln f = (\ln K_{12})^2/(4 \ln(k_{11}k_{22}/Z^2)) \tag{83}$$

Equation (79) follows directly from the fundamental equations of Marcus; or may be derived by the procedure introduced by Newton (*165*). This cross-exchange equation has been widely used; such use will be described in more detail below.

Finally, mention should be made of two recent articles on the theory of oxidation-reduction reactions. Ruff (*166*) has proposed a model for inner-sphere reactions that employs band theory. Although some success is claimed, there are several disturbing features of this development that require further study. Endicott (*167*) has also discussed inner-sphere reactions from the point of view of spin-exchange phenomena; data to test the results are meager.

D. Patterns in Metal-Ion Reactivity

The preceding discussion has been concerned with the rate laws that are typically found, substitution behavior of transition metal ions, the classification and determination of mechanism, and the theory of outer-sphere reactivity. How these results pertain to the prediction of reactivity patterns is of concern here. That the parameters influencing outer-sphere and inner-sphere reactions are somewhat different has been demonstrated; compare, for instance, the k_{OH}/k_{H_2O} argument for determining mechanism; accordingly, the discussion of trends in reactivity will be divided into these two topics. Before proceeding to that division, however, two more general topics require attention. First, an especially interesting correlation first pointed out by Rabideau and Newton (168) and recently summarized by Newton and Baker (24) will be presented; and secondly, a brief discussion of the mechanism by which various metal ion systems react will be developed.

It has been pointed out (24,168) that in a number of reactions involving actinide elements, the partial molar entropy of the transition state, $S^{\ddagger}$, is a function only of the charge of that transition state. $S^{\ddagger}$ is defined as

$$S^{\ddagger} = \Delta S^{\ddagger} - \sum_i S_i + \sum_r S_r, \tag{84}$$

where S_i is the partial molar entropy of species released in the activation process and S_r is the partial molar entropy of reactants. For charge types from 0 to $+6$ the values of $S^{\ddagger}$ fall in the ranges indicated in Table 7. Table

TABLE 7

Values for $S^{\ddagger}$ as Determined Empirically[a]

Charge type	Range of $S^{\ddagger}$ (eu)
0^b	$+ 20$ to $+ 40$
$+1$	$+ 10$ to $+ 40$
$+2$	$- 20$ to $+ 10$
$+3$	$- 60$ to $- 20$
$+4$	$- 90$ to $- 60$
$+5$	$- 70$ to -110
$+6$	-110 to -130

[a] Data from Ref. (24).
[b] Only one reaction is known, $S^{\ddagger} = 28$.

8 lists some values calculated for several reactions not previously correlated by Newton and Baker. The net activation process using dominant species has been used for all reactions involving aquo ions, but in order to obtain consistent results for other complex ions, Cobble's (181) "corrected" partial molar entropies have been used. This procedure defines the

TABLE 8

Values for the Partial Molar Entropy of Activated Complexes of
Oxidation Reduction Reactions of Transition Metal Ion Complexes[a,b]

Net activation process	$\Delta S^{\ddagger}$ (e.u.)	$S^{\ddagger}$ (e.u.)	Ref.
$V^{2+} + I_3^- = [VI_3^+]^{\ddagger}$	-14	$+20$	(170)
$V^{2+} + I_2 = [VI_2^{2+}]^{\ddagger}$	-21	-12	(170)
$V^{2+} + Br_2 = [VBr_2^{2+}]^{\ddagger}$	-26	-20	(170)
$V^{3+} + VO_2^+ + H_2O = [VO_3V^{2+}]^{\ddagger} + 2H^+$	8	-50^c	(171)
$V^{3+} + VO_2^+ + H_2O = [VO_2(OH)V^{3+}]^{\ddagger} + H^+$	7	-51^c	(171)
$Cu^+ + Co(NH_3)_5Cl^{2+} = [CuCo(NH_3)_5Cl^{3+}]^{\ddagger}$	-19	-29	(90)
$V^{2+} + VO^{2+} = [VOV^{4+}]^{\ddagger}$	-16	-72^c	(66)
$V^{2+} + CrNCS^{2+} = [VCrNCS^{4+}]^{\ddagger}$	-1	-71	(172)
$V^{2+} + Co(NH_3)_5Cl^{2+} = [VCo(NH_3)_5Cl^{4+}]^{\ddagger}$	-32	-58	(71)
$V^{3+} + VO^{2+} + H_2O = [VO(OH)V^{4+}]^{\ddagger} + H^+$	-2	-83	(173)
$Cr^{2+} + Fe^{3+} + H_2O = [CrFeOH^{4+}]^{\ddagger} + H^+$	-13	-79	(97)
$Cr^{2+} + V^{3+} + H_2O = [CrVOH^{4+}]^{\ddagger} + H^+$	0	-61	(174)
$Fe^{2+} + Fe^{3+} + H_2O = [FeFeOH^{4+}]^{\ddagger} + H^+$	11	-70	(22)
$Fe^{2+} + FeCl^{2+} = [FeFeCl^{4+}]^{\ddagger}$	-15	-92	(102)
$Fe^{2+} + Co(NH_3)_5Cl^{2+} = [FeCo(NH_3)_5Cl^{4+}]^{\ddagger}$	-30	-60	(98)
$Cu^+ + VO^{2+} + H^+ = [CuVOH^{4+}]^{\ddagger}$	-41	-79	(175)
$Cu^{2+} + V^{3+} + H_2O = [CuVOH^{4+}]^{\ddagger} + H^+$	4	-68	(121)
$Eu^{2+} + Fe^{3+} + H_2O = [EuFeOH^{4+}]^{\ddagger} + H^+$	1	-71	(117)
$Eu^{2+} + FeNCS^{2+} = [EuFeNCS^{4+}]^{\ddagger}$	-17	-78	(97)
$V^{2+} + VO^{2+} + H^+ = [V(OH)V^{+5}]^{\ddagger}$	-32	-87^c	(66)
$V^{2+} + Co(NH_3)_5H_2O^{3+} = [VCo(NH_3)_5H_2O^{5+}]^{\ddagger}$	-32	-65	(110)
$V^{3+} + VO_2^+ + H^+ = [VO(OH)V^{5+}]^{\ddagger}$	-9	-84^c	(171)
$Cr^{2+} + V^{3+} = [CrV_5^+]^{\ddagger}$	-40	-118	(174)
$Cr^{2+} + Fe^{3+} = [CrFe^{5+}]^{\ddagger}$	-28	-111^c	(97)
$Cr^{2+} + Co(NH_3)_5H_2O^{3+} = [CrCo(NH_3)_5H_2O^{5+}]^{\ddagger}$	-52	-85	(111)
$Cr^{2+} + Co(NH_3)_5H_2O^{3+} = [CrCo(NH_3)_5^{5+}]^{\ddagger} + H_2O$	-52	-102	(111)
$Eu^{2+} + Co(NH_3)_5H_2O^{3+} = [EuCo(NH_3)_5H_2O^{5+}]^{\ddagger}$	-32	-68	(112)
$V^{2+} + Np^{4+} = [VNp^{6+}]^{\ddagger}$	-24	-131	(126)

[a] Values of partial molar entropies of ions from Latimer (153) unless noted in footnote b.

[b] $\bar{S}^\circ$ values for ions not found in Latimer's tables are (ion, $\bar{S}^\circ$ in e.u., reference): V^{2+}, -23, (66); V^{3+}, -65, (168); VO^{2+}, -32, (176) (a value of -26 is widely used (177)); VO_2, -10, (176); Cr^{2+}, -13, estimated from the Powell-Latimer equation (178); $CrNCS^{2+}$, $\bar{S}^{\circ\prime} = -47$, see equation (10) and reference (179); $FeCl^{2+}$, $\bar{S}^{\circ\prime} = -50$, (102); $FeNCS^{2+}$, $\bar{S}^{\circ\prime} = -43$, (180); $Co(NH_3)_5H_2O^{3+}$, -10, (181); $Co(NH_3)_5Cl^{2+}$, -3, (181); Np^{4+}, -84, (168); Eu^{2+}, -18, (182).

[c] This value differs from that in the referenced literature because of a different choice of one of the partial molar ionic entropies. See footnote a.

process of formation of, for instance, $CrNCS^{2+}$, as Eq. (85) rather than Eq. (86).

$$Cr(H_2O)_6^{3+} + NCS^- = Cr(H_2O)_5NCS^{2+} + H_2O \tag{85}$$

$$Cr^{3+} + NCS^- = CrNCS^{2+} \tag{86}$$

Then the "corrected" entropy for $CrNCS^{2+}$ is related to the usual definition of partial molar entropy by

$$\bar{S}^{\circ\prime} = \bar{S}^\circ - S^\circ_{(H_2O)} . \tag{87}$$

It can be seen from Table 8 that a wide variety of electron-transfer reactions fit the correlation rather well. There are some exceptions. A correlation between charge and partial molar entropy of ionic species is observed in most empirical correlations (183); Cobble's work shows that some care is necessary in defining the process to calculate the entropy in the case of complex ions. It would seem possible, theoretically, once conventions regarding hydration are adopted, to use $S^{\ddagger}$ values to determine mechanism; but in general the spread of the values in the empirical correlation might be too large to be useful. The reaction of Cr^{2+} with $Co(NH_3)_5H_2O^{3+}$ is used to make such a comparison in Table 8. It is to be noted that both reactions fall into the proper range for a transition state of charge $+5$. It is interesting to note one case in which the inner-sphere and outer-sphere mechanisms are believed to occur in parallel. The appropriate $S^{\ddagger}$ values seem to differ by 17 e.u. (184). Further work on the question of the proper hydration numbers to use in calculating the $S^{\ddagger}$ values would appear appropriate.

The widths of the $S^{\ddagger}$ range for each charge type are sufficient to generate unacceptable uncertainties in the prediction of rate constant, even with an assumption of a $\Delta H^{\ddagger}$ value (20 e.u. at room temperature corresponds to a factor of about 2×10^4 in k). Nevertheless, the correlation does indicate that for a variety of simple oxidation-reduction reactions, the absolute entropy of the transition state is simply related to charge and that no "special" effects contribute to the entropy of activation: in particular if nonadiabatic processes are important, with the requisite effect on $\Delta S^{\ddagger}$ through a low value of κ, it must be equally important in all the reactions studied—an unlikely event.

Before discussing the reactivity patterns anticipated for various electron-transfer reactions, it is worthwhile to consider the mechanism by which various reactions take place. Because the majority of work concerning this difficult question has been performed on transition metal aquoions, these systems will define the limits of discussion.

The most striking feature in this area is the very strong preference for the inner-sphere activated complex shown by Cr^{2+}. In almost all reductions by Cr^{2+} (of oxidants with a suitable site for inner-sphere binding) that have been studied, the products of the reaction establish that an inner-sphere mechanism pertains. This result holds not only for reduction of complexes of Co(III), Cr(III), Fe(III), but also in Cr(II) reductive dehalogenation of organohalides (185,186). In contrast to this behavior is that of V^{2+} or $Co(CN)_5^{3-}$. The former ion appears to react by both inner-sphere and outer-sphere paths, sometimes with substantially equal rates—$Co(NH_3)_5$-Cl^{2+} (outer-sphere) compared to $Co(NH_3)_5N_3^{2+}$ (inner-sphere) (71)— and $Co(CN)_5^{3-}$ has been shown to use both inner-sphere and outer-sphere paths, sometimes in parallel (80). These results sound the warning: the choice of mechanism is probably a sensitive function of the reacting pair; nevertheless, a general summary does not seem inappropriate.

The reactions of Fe^{2+}, like those of Cr^{2+}, are generally thought to be inner-sphere. Several cases of inner-sphere mechanism have been established (85,187), and both the $k_{N_3^-}/k_{NCS^-}$ and k_{OH}/k_{H_2O} correlations, as well as the reactivity pattern correlations (70,188), support the conclusion that most Fe^{2+} reductions are inner-sphere. The data on Cu^+ are slight, but relative rate ratios and reactivity patterns favor inner-sphere reductions (90). In contrast to these systems, reductions with Eu^{2+} and V^{2+} are in a more confused state. In the latter case, the mechanism is tied to rate-limiting substitution reactions which dictate mechanism in certain cases: V^{2+} + $FeNCS^{2+}$ must have an outer-sphere activated complex (114). On the other hand, direct observation of product shows that some V^{2+} reactions are inner-sphere—Table 2. It would appear that factors other than rate-limiting water exchange play an important role in dictating the mechanism of reduction by V^{2+}. Neither k_{OH}/k_{H_2O} nor $k_{N_3^-}/k_{NCS^-}$ correlations are very meaningful because of the rate limit dictated by substitution. But oxygen fractionation experiments have supported an outer-sphere mechanism for the V^{2+} reduction of $Co(NH_3)_5H_2O^{3+}$ (189). A similar conclusion is valid for Eu^{2+}, but in this case $k_{N_3^-}/k_{NCS^-}$ data indicate an inner-sphere mechanism for the reduction of Fe(III) and Co(III) complexes. It is interesting to note that no highly effective rate term inverse is $[H^+]$ has been observed in the Eu^{2+}–$Co(NH_3)_5H_2O^{3+}$ reaction (see Table 4). Perhaps Eu^{2+} takes advantage of an inner-sphere path with N_3^-, but not with H_2O or OH^- as the ligand attached to the $Co(NH_3)_5^{3+}$ residue.

Certainly these data illustrate that the mode of mechanism chosen by metal ion reductants is not constant for a given reductant, but depends on both of the reaction partners. It seems likely that each system should be considered independently; this requires methods of determining mechanism that are not dependent on relative rate comparisons, but are based on a change in some parameter of the activated complex that depends critically on the mechanism. Such methods have not been exploited. Notwithstanding these difficulties, it is useful to discuss reactivity patterns for inner-sphere and outer-sphere mechanisms separately, for in each type of mechanism, if somewhat different criteria do not hold, certainly different accentuations seem advisable.

1. Outer-Sphere Reactivity Patterns

The dominant themes in the reactivity patterns of reactions that take place by the outer-sphere mechanism are the self-exchange rates and net driving force for reaction, parameters to be expected on the basis of Marcus's theory. This relationship has been tested in two ways several times. One manner of testing is to expand Eq. (79) to

$$\Delta F^\ddagger_{12} = \tfrac{1}{2}(\Delta F^\ddagger_{11} + \Delta F^\ddagger_{22} + \Delta F^\circ_{12}) - \frac{RT}{2}\ln f \qquad (88)$$

and to note the rearrangement

$$(\Delta F^\ddagger_{12} - \tfrac{1}{2}\Delta F^\ddagger_{11}) = \tfrac{1}{2}\Delta F^\ddagger_{22} + \tfrac{1}{2}\Delta F^\circ_{12} - \frac{RT}{2}\ln f \qquad (89)$$

yields a straight line plot of the left-hand side for variable complex "2," versus ΔF°_{12} if (1) $\Delta F^\ddagger_{22} \ll \Delta F^\circ_{12}$ or $\Delta F^\ddagger_{22}$ is a constant, and (2) $\ln f$ goes to zero compared to $\tfrac{1}{2}\Delta F^\circ_{12}$. Sutin and co-workers have demonstrated several such plots and have found slopes of about 0.5 ± 0.1 for reaction of $Fe(phen)_3^{2+}$ and ligand-substituted analogs with $Ce(IV)$ (190), $Mn(H_2O)_6^{3+}$ and $Mn(H_2P_2O_7)_3^{3-}$ (191), and $Co(III)$ (192). More recently measurements of the rate of variously substituted phenanthroline complexes of $Fe(II)$ and $Ru(II)$ with $Ce(IV)$ and $Tl(III)$ have been published (193–195). Approximate straight line plots of slope $\cong 0.5$ are again found. The intercepts of plots of this sort should give a measure of the "average" rate of self-exchange of the phen complexes. The values obtained are in striking disagreement with the lower limit established by ESR techniques for the $Fe(phen)_3^{2+}$-$Fe(phen)_3^{3+}$ exchange reaction (196).

A second manner of establishing the validity of equation (79) is by predicting values of k_{12} for known values of k_{11}, k_{22}, and K_{12}. Such applications were also made by Sutin and co-workers. Recently other data have been added to these. Table 9 summarizes some of the older data as well as

TABLE 9

<small>COMPARISON OF CALCULATED AND OBSERVED RATE CONSTANTS
FOR VARIOUS REACTIONS[a]</small>

Reaction	k_{12} (calcd)[b], $M^{-1} sec^{-1}$	k_{12} (obs), $M^{-1} sec^{-1}$	Ref.
$Fe(CN)_6^{4-} + Ce(IV)$	6.0×10^6	1.9×10^6	(192)
$Fe(CN)_6^{4-} + IrCl_6^{2-}$	5.7×10^5	3.8×10^5	(192)
$W(CN)_8^{4-} + Mo(CN)_8^{3-}$	1.7×10^7	5.0×10^6	(192)
$Fe^{2+} + Mn^{3+}$	3×10^4	1.5×10^4	(191)
$Fe(EDTA)^{2-} + Mn(CyDTA)^-$	6×10^6	4×10^5	(197)
$Co(EDTA)^{2-} + Mn(CyDTA)^-$	2.1	0.9	(197)
$Cr(EDTA)^{2-} + Co(EDTA)^-$	4×10^7	3×10^5	(197)
$Co(terpy)^{2+} + Co(bip)_3^{3+}$	3.2×10	6.4×10	(129)
$Co(terpy)^{2+} + Co(H_2O)_6^{3+}$	2×10^{10}	7.4×10^4	(129)
$Ru(NH_3)_6^{2+} + Fe(H_2O)_6^{3+}$	7.5×10^6	3.4×10^5	(120)
$Ru(en)_3^{2+} + Fe(H_2O)_6^{3+}$	4.2×10^5	8.4×10^4	(120)
$V(H_2O)_6^{2+} + Ru(NH_3)_6^{3+}$	4.2×10^3	8.2×10^1	(120)
$Fe(H_2O)_6^{2+} + Co(H_2O)_6^{3+}$	6×10^6	4.2×10^1	(190)

[a] It is assumed that all of these reactions and the corresponding self-exchange reactions are outer-sphere. This has not been demonstrated in all cases.

[b] See the original literature for the values used for the self-exchange rate constants for the K_{12} values.

some of the newer. In general it is seen that the agreement is satisfactory. If consideration is given to the approximations in the theory and the experimental errors in measuring the values, predictions within a factor of 5–10 are good. The exceptions to this fit are reactions of Co(III): it has been suggested quite often that spin changes necessary in the reduction of Co(III) (t_{2g}^6) to Co(II) $(t_{2g}^5 e_g^2)$ affect the rate significantly. Two recent studies are pertinent to this question. The reduction of *trans*-Co(py)$_4$Cl$_2^+$ by Co(bip)$_3^{3+}$ and Co(terpy)$_2^{2+}$ should test the spin multiplicity question, for the last complex is low spin (Co(II) $(t_{2g}^6 e_g^1)$ (198)), and there are no other obviously significant differences between the two reductants. The rates are $3.0 \times 10^5 M^{-1} sec^{-1}$ and $1.1 \times 10^4 M^{-1} sec^{-1}$ at $0°$, $I = 10^{-4}$. for Co(terpy)$_2^{2+}$ and Co(bip)$_3^{2+}$, respectively (129). These data would appear to argue against an important role for spin-multiplicity effects. Second, Waltz and Pearson have recently claimed to have generated Co(bip)$_3^{3+}$ in an excited state (199). Pulse radiolysis of Co(bip)$_3^{3+}$ in the presence of methanol to scavenge radicals other than e_{aq}^- yields a transient intermediate that absorbs strongly at 6000 Å, and which decays to $Co(bip)_3^{2+}$ with a half-life of about 15 μsec (depending on conditions). They believe this species is $[Co(bip)_2^{6+}]*$ $(t_{2g}^6 e_g^1)$.

$$Co(bip)_3^{3+} + e_{aq}^- \longrightarrow [Co(bip)_3^{2+}]* \tag{90}$$

One of the reagents that quenches this excited state is $Co(bip)_3^{3+}$. This suggests a test of the mechanism for exchange in the $Co(bip)_3^{2+,3+}$ system that was proposed earlier (200)

$$Co(bip)_3^{2+} \underset{}{\overset{K}{\rightleftharpoons}} [Co(bip)_3^{2+}]* \tag{91}$$

$$[Co(bip)_3^{2+}]* + Co(bip)_3^{3+} \overset{k}{\longrightarrow} Co(bip)_3^{3+} + Co(bip)_3^{3+} + \text{thermal energy} \tag{92}$$

where $k_{obs} = Kk$. [See Refs. (199) and (201) for remarks on microscopic reversibility.] Using their measured value of k ($8 \times 10^8 \ M^{-1} \ sec^{-1}$) and an estimation of K (a calculation based on a Boltzmann distribution) Waltz and Pearson calculate $k_{obs} \cong 160 \ M^{-1} \ sec^{-1}$ compared to the experimental value of about $7 \ M^{-1} \ sec^{-1}$. Although this agreement is satisfactory, it is clear that more experiments on the role of spin effects are needed (167).

Although the prediction of rate constants for cross-reactions using Eq. (79) has been shown to be, in general, correct, a fundamental problem remains in a priori predictions of reactivity patterns: one must know the self-exchange rate constants. What factors are important in determining these rates? Explicit answers can be obtained from the adiabatic theory. Given that most reactions are studied in systems of high ionic strength, the coulombic terms are small and can usually be neglected; similarly, the outer-sphere rearrangement terms, which depend, in relative rate comparisons, only on the radii of the reactants, can be subordinated to inner-sphere rearrangement energies. (An increase in radii of both reactants by a factor of 2 decreases the outer-sphere contribution to ΔF^* by a factor of about 2. This would, under typical conditions (10), increase the rate by about 200-fold.) Thus the pertinent factors are the necessary change in bond lengths and the force constants governing those changes. The outstanding reactivity pattern that is readily predictable from these considerations is that electronic systems, in pseudo-octahedral microsymmetry, in which the electron to be transferred is of a t_{2g} type—nonbonding—will react more rapidly than those in which the electron is antibonding—e_g. The available data can be used to argue that this hypothesis is correct. Table 10 lists the only set of data available with which to test this hypothesis. The basic concept seems to be substantiated. For those reactions in which an e_g electron is transferred in gaining the oxidized state of the element, the rate is quite

TABLE 10

RATES OF SELF-EXCHANGE REACTIONS, $M(H_2O)_6^{2+} + M(H_2O)_6^{3+}$

Reaction	Electronic configuration of reductant	k, $M^{-1}sec^{-1}$	Ref.
$V(H_2O)_6^{+2} + V(H_2O)_6^{3+}$	t_{2g}^3	1.0×10^{-2}	(202)
$Cr(H_2O)_6^{2+} + Cr(H_2O)_6^{3+}$	$t_{2g}^3 e_g^1$	$<2 \times 10^{-5}$	(123, 124)
$Mn(H_2O)_6^{2+} + Mn(H_2O)_6^{3+}$	$t_{2g}^3 e_g^2$	$\cong 3 \times 10^{-4}$ [a]	(191)
$Fe(H_2O)_6^{2+} + Fe(H_2O)_6^{3+}$	$t_{2g}^4 e_g^2$	3.7	(22)
$Eu(H_2O)_6^{2+} + Eu(H_2O)_6^{3+}$	f^7	$\cong 1 \times 10^{-5}$ [a]	(203, 204)

[a] $T = 25°$

[b] Estimated in the reference quoted.

low. There are, however, several disturbing features: (1) It has not been established that these reactions are outer-sphere; if the $Fe^{2+,3+}$ and $V^{2+,3+}$ exchanges took advantage of inner-sphere paths not open to the other reactions, it might well be that the rate of the outer-sphere paths of all the reactions would be nearly the same. (2) The rate of the $Eu^{2+,3+}$ exchange is strangely low on the basis of the argument advanced above. Because the removal of an f electron should not have any peculiar effects on the ligands, there is an expected similarity in rate of $V^{2+,3+}$ and $Fe^{2+,3+}$ compared to $Eu^{2+,3+}$; this is missing.

A second prediction can be obtained from consideration of first coordination sphere effects. This rationalization rests on consideration of the nature of bonding between the metal ion center and ligands. For those ligands with low-lying empty orbitals of π symmetry, back-bonding of t_{2g} electrons is an important contribution to the energy of ligand-metal interaction. Thus, increasing the number of t_{2g} electrons increases the strength of the π bonds, while, because of its influence on the effective nuclear charge of the metal ion, it decreases the strength of the σ bonds. Hence a compensation occurs which, in the context of rate of electron-transfer reactions, tends to make the two valence states more alike and, therefore, reduces the energy needed to achieve the transition state. An illustration of the effect on rate constants of these features is given by a comparison of the Fe(II)-Fe(III) exchange with H_2O as a ligand (no π-acceptor orbitals) compared to phen as a ligand. The rate of exchange of the former is $4 M^{-1}sec^{-1}$ (22), whereas that of the latter is $>3 \times 10^7 M^{-1}sec^{-1}$ (196), both at 25°.

This last effect, the increase in rate when the ligand is derived from an aromatic molecule, has often been interpreted in another fashion; the aromatic system gives rise to a low-lying path on which the electron can move from one center to another; it "conducts" the electron. Two recent reports have appeared which provide data to test this concept. First, the reaction of $Co(terpy)_2^{2+}$ with $Co(bip)_3^{3+}$ (π-bonding ligands) reacts with nearly the same rate as does $Co(H_2O)_6^{2+}$ with $Co(H_2O)_6^{3+}$. In this system, then, either the presence of π-bonding ligands has no effect or it is counter-balanced by some other parameter. Second, the blocking of approach to the aromatic system seems to cause no drastic change in rate. The rate of Cr^{2+} reduction of $Co(NH_3)_5py^{3+}$ is nearly the same as is that of $Co(NH_3)_5$ $(4\text{-}CH_3\text{-}py)^{3+}$ or $Co(NH_3)_5(3,5\text{-}(CH_3)_2\text{-}py)^{3+}$ (205). It would appear that these experiments support a role for aromatic ligands other than that of "electron-conductor."

Finally, some mention of a relatively new problem in outer-sphere reactivity patterns should be made. Increasing amounts of data on outer-sphere reactions of molecules with low symmetry are beginning to appear. Such systems will require careful consideration when relative rates are compared, for the "site" of attack becomes ambiguous. One reagent may prefer one orientation of the molecule, whereas a second may prefer another orientation. The size of the perturbation produced by such orientation effects has not been investigated.

2. Inner-Sphere Reactivity Patterns

The status of reactivity patterns in inner-sphere systems is not as well understood theoretically, although several attempts have been made to relate inner-sphere reactivity patterns with the Marcus cross-reaction equation. However, the considerable amount of data allows some empirical correlations to be made.

The foremost question to answer concerning reactivity patterns of inner-sphere reactions concerns the efficiency of a given bridging ligand. Examination of the data for reduction by Cr^{2+} and $Co(CN)_5^{3-}$ represent the only available data on compounds that definitely use the inner-sphere path. These data are most complete for the Cr^{2+} reduction of $Co(NH_3)_5X^{n+}$ complexes, and show the efficiency decreases along the series ($70,87,135$)

$$OH^-, F^-, Cl^-, Br^-, I^-, N_3^- > NCS^-, CN^- > H_2O, CH_3COO^-.$$

To illustrate the complexity of the role of the bridging ligand, the data for reduction of similar complexes with $Co(CN)_5^{3-}$ ($80,206$) can be examined; in this case, the efficiency for innersphere reaction decreases

$$Br^- > Cl^- > N_3^-, NCS^- > OH^- > F^- > CN^- > CH_3COO^-.$$

[It is likely that $Co(NH_3)_5I^{2+}$ reacts faster than $Co(NH_3)_5Br^{2+}$]. Some of the differences noted in these two series are explicable on the basis of consideration of a two-step process

$$M^{2+} + Co(NH_3)_5X^{2+} \overset{K}{=} [(NH_3)_5Co-X-M]^{4+} \tag{93}$$

$$[(NH_3)_5-Co-X-M]^{4+} \overset{k}{\longrightarrow} \text{Products} \tag{94}$$

Reaction (93) will be favored for $Co(CN)_5^{3-}$ when $X = I^-$, Br^-, because of the "soft" nature of the reductant and the bridging ligand. Accordingly' the observed rate constant, $k_{obs} = Kk$, will be greater; as noted above, this difference accounts for the $k_{N_3^-}/k_{NCS^-}$ ratio for $Co(CN)_5^{3-}$. What this comparison does imply is that the factors affecting the efficiency of bridging ligands are probably interwoven in a complicated pattern. If one looks at data on systems whose mechanism is thought to be inner-sphere, the same conclusion can be reached. In Table 11 are displayed some data for the

TABLE 11

RATES[a] OF REDUCTION OF SOME Fe(III) COMPLEXES

Reductant	Complex			Ref.
	$Fe(H_2O)_6^{3+}$	$Fe(H_2O)_5OH^{2+}$	$Fe(H_2O)_5Cl^{2+}$	
Cr^{2+}	2.3×10^3	3.3×10^6	2×10^7	(207)
Fe^{2+}	4^b	$3.0 \times 10^{3\,b}$	$(2.9-3.3) \times 10^c$	(22, 102, 208)
Eu^{2+}	$3.38 \times 10^{3\,d}$	$5.2 \times 10^{6\,d}$	$2 \times 10^{6\,d}$	(117)
Np^{3+}	$(1.6-2.9) \times 10^{1\,c}$	$1.8 \times 10^{5\,e}$	$3.3 \times 10^{3\,e}$	(118, 209)

^{a} k in units of $M^{-1}sec^{-1}$, $T = 25°$, $I = 1.0$ M.
^{b} $I = 0.55$ M.
^{c} $I = 3.0$ M; only about 50% of the observed rate (57.6 $M^{-1}sec^{-1}$) is via the chloride bridged inner-sphere path.
^{d} $T = 1.6°$.
^{e} $T = 0.9°$.

relative rate of reduction of various Fe(III) complexes. It is seen that for Cr^{2+} and Eu^{2+} as reductants, the hydroxide system as a bridge is about as equally effective as Cl^-; but for Np^{3+} and Fe^{2+}, OH^- is much more efficient than Cl^- is. No satisfactory explanation for these variations, nor those shown in Table 3, has been made.

Some success has been achieved in a correlation using the Marcus theory for cross reaction even though the mechanisms are inner-sphere (210,149, 118,116). The value used for the self-exchange reaction must bring in the bridging ligand. For instance, to predict the rate of reaction (95)

$$Fe^{2+} + Co(NH_3)_5Cl^{2+} \xrightarrow{k_{21}} Fe(III) + Co(II) \qquad (95)$$

under the assumption that the mechanism is inner-sphere, one uses reactions (96)–(99)

$$Cr^{2+} + Co(NH_3)_5Cl^{2+} \xrightarrow{k_{31}} Cr(III) + Co(II) \qquad (96)$$

$$*Cr^{2+} + CrCl^{2+} \xrightarrow{k_{33}} *CrCl^{2+} + Cr^{2+} \qquad (97)$$

$$*Fe^{2+} + FeCl^{2+} \xrightarrow{k_{22}} *FeCl^{2+} + Fe^{2+} \qquad (98)$$

$$Fe^{2+} + CrCl^{2+} \xrightleftharpoons{K_{23}} FeCl^{2+} + Cr^{2+} \qquad (99)$$

whereupon substitution into the cross-reaction equation, (79), assuming $f = 1$, and noting that $K_{21}K_{31}^{-1} = K_{23}$, one obtains

$$k_{21} = k_{31}\left(\frac{k_{22}}{k_{33}}\right)^{1/2}(K_{23})^{1/2}. \qquad (100)$$

Marcus has recently published an argument indicating the use of the cross-reaction equation may be valid for strong overlap cases (211).

If the center of one's attention shifts from the bridging ligand to the non-bridging ligands, the problem of predicting rate behavior is less severe. This is not unexpected; the perturbation is removed from the reaction center. Table 6 summarizes a good deal of the known information on "nonbridging ligand" effects. [For the purposes of this discussion, the nonbridging ligands in reactions that might proceed by the outer-sphere mechanism are pragmatically defined as those that are varied. The restrictions implicit in this definition have been discussed (142).] The ordering of nonbridging ligands in that table is that which leads to an increase in rate for various reductions of Co(III) and Cr(III) complexes. This ordering is of some significance, for it is seen that all reductants seem to follow the same pattern. This allows qualitative prediction of rate trends as nonbridging ligands are changed. More quantitative predictions can be had by application of linear-free energy arguments (109,142). The rationalization for nonbridging ligand effects has been based on an idea originally enunciated by Orgel (212). Benson and Haim (147) and others developed the concept using a combination of crystal field theory and qualitative measures of the ease of bond stretching, whereas Bifano and Linck (*141*) suggested a molecular orbital approach. In essence, both models are attempts to correlate reactivity with the energy of the lowest unoccupied orbital on the oxidant. The greater the energy of this orbital, the greater the barrier to reaction. It should be noted that both of these arguments are one-electron approximations to the complex problem. The similarity of reactivity of Co(III) and Cr(III) complexes is, thus, understood as a consequence of the similar character of the orbital to be populated. The order of increase in rate for a nonbridging ligand change is indicative of crystal field strength and ease of stretching in one case (*147*) and "σ-bonding strength" in the other (*141*). Both approaches must resort to an ordering of nonbridging ligand rate enhancing ability that is more empirical than theoretical, although there are exceptions. The fact that cis-Co(en)$_2$pyCl^{2+} reacts faster than cis-Co(en)$_2$NH$_3$Cl^{2+} was predicted and observed on the basis of "σ-bonding strength" (*141*).

In a sense complimentary to the arguments above, a change in nonbridging ligands on Cr(II) or Co(II) as reductants should have the opposite effect, for the perturbation by a strong σ donor ligand should raise the energy of the highest filled level, and hence lower the barrier to reaction. The data to test this concept are not as extensive, but a correlation between the rate of the reaction

$$\text{Co(DMG)}_2L + \left\langle\!\!\bigcirc\!\!\right\rangle\!\!-\!\text{CH}_2\text{Br} = \text{Co(DMG)}_2L\text{Br} + \text{Co(DMG)}L\left(\text{H}_2\text{C}\!\!-\!\!\left\langle\!\!\bigcirc\!\!\right\rangle\right) \quad (101)$$

and the pK$_b$ of the base, L, has been observed (*213*). If one applies the argument of Bifano and Linck that the pK$_b$ parallels the σ donor strength, these data fit the model. Guenther and Linck (*109*) observed that the sensitivity of a reductant to a given nonbridging ligand change varied and suggested that this sensitivity should increase as the net free energy change of reaction became more positive. The recently reported, although limited, data on Cu(I) reduction of various L_5Co(III)-Cl systems may require reexamination of this concept. Finally, it is to be noted that although only one case of a system with an electronic structure different from that of Co(III) and Cr(III) has been investigated, its behavior is different, a result that was to be anticipated. The reactivity pattern for the V^{2+} reduction of Ru(III)-Cl complexes presumably reflects "π-bonding strength" effects as the orbital to be populated has π-symmetry (*49*). Thus the change of nonbridging ligand, *trans*-NH$_3$ to *trans*-H$_2$O, a change that causes a significant rate of enhancement for reduction of Co(III) complexes, should decrease the rate of reduction of Ru(III) complexes since H_2O can donate π electrons whereas NH_3 cannot. This is the observed result in Table 6.

It thus appears that some general reactivity patterns may be applicable to inner-sphere systems. Detailed predictions of the effect of changing the bridging ligand do not yet appear unequivocal, but general trends can be seen in a large number of reactions. Detailed predictions of nonbridging effects seem on firmer ground, at least for Co(III) and Cr(III) complexes.

III. Classification and General Features of Catalytic Systems

The role that a catalyst plays in the acceleration (or inhibition) of a chemical reaction is of importance in attempts to understand those processes. Often this role is not easily determined. Nevertheless, classification of modes of catalytic activation is useful; this section outlines a classification of oxidation-reduction catalysts.

Upon inquiring into the role of a catalyst, several questions appear. How does the catalyst enter into the reaction? Is the catalyst the active component that attacks the second reagent? Does activation of the substrate occur by means of coordination to the metal ion or by actual electron-transfer? It would seem that the answer to this third question is of greatest importance. Those reactions in which two-well-defined valence states of the catalyst exist, both of which are independent of either reactant are designated *nonassociative activation*. For the stoichiometry

$$A + B = P_1 + P_2 \tag{102}$$

where P_i represent final products of reaction, nonassociative activation is

$$A + C \longrightarrow P_1 + C' \tag{103}$$

$$B + C' \longrightarrow P_2 + C \tag{104}$$

where C and C' are two valence states of the metal ion catalyst. Modifications of this simplest model involve the method by which the activation is *carried*. If A is converted by the catalyst to some intermediate, then either that species or C' can function as the carrier. The former is

$$A + C \longrightarrow I_1 + C' \tag{105}$$

$$I_1 + B \longrightarrow P + I_2 \tag{106}$$

$$C' + I_2 \longrightarrow P + C \tag{107}$$

The latter, catalyst carried, is

$$A + C \longrightarrow I_1 + C' \tag{108}$$

$$C' + B \longrightarrow I_2 + C \tag{109}$$

$$I_1 + I_2 \longrightarrow P \tag{110}$$

Such complexities are expected when the reactions are noncomplimentary; further complexities occur if chain reactions are considered.

A second type of catalytic action is that in which the catalyst and a reactant form a complex, and it is this species, or one derived from it, that causes further reaction. This mode of *associative activation* is represented by (111).

$$A + C \longrightarrow AC \tag{111}$$

There are several modifications of this type of activation. These depend on stoichiometric relationships as well as on the mechanism of reaction of AC with B. One way of carrying the reaction is by *redox-transfer*. In this pathway, the activated reagent (or a fragment of it) is transferred to B with regeneration of the free catalyst

$$AC + B \longrightarrow P + C \tag{112}$$

Another type of interaction of AC with B can occur if B coordinates with the metal complex, AC; reaction then takes place within the coordination sphere of the catalyst. This type of activation, *internal associative activation*, is presumed to occur in the Rh(I) catalyzed hydrogenation of olefins

$$\underset{A}{S} + H_2 + \underset{C}{((C_6H_5)_3P)_3RhCl} \longrightarrow \underset{AC}{(C_6H_5)_3P)_2RhSClH_2} + (C_6H_5)_3P \tag{113}$$

$$\underset{AC}{((C_6H_5)_3P)_2RhSClH_2} + \underset{B}{C_2H_4} \longrightarrow \underset{ACB}{((C_6H_5)_3P)_2RhH_2(C_2H_4)Cl} + S \tag{114}$$

$$(C_6H_5)_3P + ((C_6H_5)_3P)_2RhH_2C_2H_4Cl \longrightarrow ((C_6H_5)_3P)_3RhCl + C_2H_6 \tag{115}$$

where S is a solvent molecule. Note the important distinguishing feature of associative activation: there is no real existence of C' in the absence of A (or a fragment thereof).

This division of the modes of catalysis by electron-transfer reactions is in some cases arbitrary, as are all mechanistic divisions. But it is also of considerable practical importance. For the questions that one asks about catalysis by associative activation and nonassociative activation differ considerably. In the former case, knowledge of bond strengths and lability are of great importance, as is a knowledge of coordination-sphere geometries (*214–216*). If the bond between A and C is quite inert, it will not be easily broken when the demand by B is made. Further, in associative activation one is always faced with the question of the identity of the formal valence state of the catalyst. Within an AC complex the question of valence state is always arbitrary: consider $Co(CN)_5H^{3-}$

$$Co(III)-H^- \qquad Co(II)-H^0 \qquad Co(I)-H^+$$

For predictive purposes, however, it is often useful to have some knowledge of the "correct" valence assignment (*217*). It is also true that in associative activation one cannot easily study the catalyst's chemistry isolated from reactants, for at least one reactant is intimately involved with the catalyst.

In nonassociative activation, on the other hand, formal valence assignments are made with conventional rules, usually with success. In addition, these systems yield readily to the reactivity patterns discussed above because each part of the catalytic reaction is an "electron-transfer" reaction —reactions within the coordination sphere of metal ions are not of primary importance. Some *a priori* predictions concerning the reactivity requirements that a catalyst must exhibit to be of functional importance can be developed for the case of nonassociative activation. Consider the simplest case of nonassociative activation

$$A + C \xrightarrow{\quad k_1 \quad} P_1 + C' \tag{116}$$

$$C' + B \xrightarrow{\quad k_2 \quad} P_2 + C \tag{117}$$

$$A + B \xrightarrow{\quad k_3 \quad} P_1 + P_2 \tag{118}$$

where $k_1 < k_2$. Further, if the assumption that these reactions are outer-sphere is made, then from Eq. (79).

$$k_1 = (k_{AP_1} \, k_{CC'} \, K_{AC} \, f_{AC})^{1/2} \tag{119}$$

$$k_3 = (k_{AP_1} \, k_{BP_2} \, K_{AB} \, f_{AB})^{1/2} \tag{120}$$

and for catalysis to be effective

$$k_1 > k_3 \tag{121}$$

Substitution and rearrangement lead to the equation

$$\frac{k_{CC'}}{k_{BP_2}} > \frac{K_{AB}f_{AB}}{K_{AC}f_{AC}} \tag{122}$$

Predictions of effective catalytic systems can be made with this equation as a beginning approximation. Note that a high self-exchange rate constant is necessary; and the larger the net driving force for the reaction, the larger the ratio of the self-exchange constant for catalyst relative to that of reactant B must be for effective catalysis. An unfavorable $\Delta F°$ change for reaction (116) is not conducive to effective catalysis. Considerations of this type generate a reasonable start toward understanding catalytic phenomena. For instance, the catalytic role played by copper in many oxidation-reduction processes can be understood on the basis of a large self-exchange rate constant [the outer-sphere rate constant is calculated to be 7×10^5 $M^{-1}sec^{-1}$ from data on the reduction of $Ru(NH_3)_6^{3+}$ by Cu^+ ($86,120$)] and a potential that is neither powerfully reducing nor powerfully oxidizing. Such predictions of catalytic reactivity, however, must be applied with some care. The copper catalyzed reduction of Fe(III) by V(III) (218) is an illustrative case.

$$Cu^{2+} + V^{3+} \underset{k_2}{\overset{k_1}{\rightleftharpoons}} VO^{2+} + Cu^+ + 2H^+ \tag{123}$$

$$Cu^+ + Fe^{3+} \xrightarrow{k_3} Cu^{2+} + Fe^{2+} \tag{124}$$

In 1.0 M $HClO_4$, $k_3 > k_2 > k_1$ ($218,219$), so Eq. (122) should hold if the reactions are outer-sphere. The data are at variance with this prediction. The $V^{3+}-Cu^{2+}$ reaction is thermodynamically too unfavorable and the $V^{3+}-Fe^{3+}$ reaction too favorable for the catalysis by Cu^{2+} to be predicted on the basis of an outer-sphere mechanism. Nevertheless, such catalysis is quite effective, presumably by utilizing an inner-sphere mechanism. Such a mechanism appears common for Cu^+ reactions—see Section II,D. Considerations such as these emphasize the need for further work designed to establish reliable reactivity patterns.

Notwithstanding the difficulties in describing catalytic processes in terms of the individual steps in the system, several catalytic reactions have been examined quite carefully. In the next section three types of processes are discussed in detail. The emphasis is placed on obtaining information concerning the fundamental steps.

IV. Specific Catalytic Systems

A. HYDROGENATION by $Co(CN)_5^{3-}$ AND OTHER METAL IONS

The action of transition metal ions on hydrogen in aqueous solution was first conclusively demonstrated by Calvin ($220,221$). Since that time, the subject has developed considerably and many catalysts are now known.

Halpern, who has previously reviewed this topic (*214,222–224*) has pointed out that three modes of activation of hydrogen appear possible:

Homolytic

$$2Co(CN)_5^{3-} + H_2 = 2Co(CN)_5H^{3-} \qquad (125)$$

Heterolytic

$$Cu^{2+} H_2 = CuH^+ + H^+ \qquad (126)$$

Additive

$$Os(CO)_5 + H_2 = Os(CO)_4H_2 + CO \qquad (127)$$

The third mode seldom occurs in aqueous solution, has been discussed extensively (*225*), and is an example of internal association activation. Concern here is on reactions of the first two types, and principally the extensively studied catalysis by $Co(CN)_5^{3-}$.

The structure of $Co(CN)_5^{3-}$ in aqueous solution has been investigated by a number of workers. Recently Pratt and Williams (*226*), claimed, on the basis of similarities to some isonitrile-Co(II) complexes, infrared spectral measurements, and dehydration experiments, that the ion had a sixth ligand, a water molecule, occupying an octahedral coordination position. A number of other views, at various levels of disagreement, exist (*227–229*).

Iguchi appears to be the first to report the uptake of H_2 by alkali cyanide solutions of Co(II) (*230*). Later work (*231–232*) established that this solution contained, principally, $Co(CN)_5H^{3-}$; and that solutions of this substance could hydrogenate some alkenes (*233*). Hence $Co(CN)_5^{3-}$ is a catalyst for the reaction

$$H_2 + R_1R_2C{=}CR_3R_4 = R_1R_2CH{-}CHR_3R_4 \qquad (128)$$

A review of the scope of this reaction and related ones is available (*234*).

It now appears that the mechanism of the catalytic hydrogenation of olefins by $Co(CN)_5^{3-}$ may be of two types. In each the first step is associative activation of H_2 by homolytic bond breaking; but the two differ in the fate of $Co(CN)_5H^{3-}$. In the mechanism proposed for the hydrogenation of butadiene (*235*) the process is carried by insertion of $Co(CN)_5H^{3-}$ into the double bond of the olefin to form $Co(CN)_5C_4H_7^{3-}$ followed by subsequent displacement of $Co(CN)_5^{3-}$ by redox transfer of a hydrogen atom from a second $Co(CN)_5H^{3-}$.

$$2Co(CN)_5^{3+} + H_2 \underset{k_{-1}}{\overset{k_1}{\rightleftharpoons}} 2Co(CN)_5H^{3-} \qquad (129)$$

$$Co(CN)_5H^{3-} + C_4H_6 \underset{k_{-2}}{\overset{k_2}{\rightleftharpoons}} Co(CN)_5C_4H_7^{3-} \qquad (130)$$

$$Co(CN)_5H^{3-} + Co(CN)_5C_4H_7^{3-} \overset{k_3}{\longrightarrow} 2Co(CN)_5^{3-} + C_4H_8 \qquad (131)$$

In the mechanism proposed for the hydrogenation of cinnamate ion (236) or sorbate ion (237) the reaction is carried by a redox transfer mechanism only:

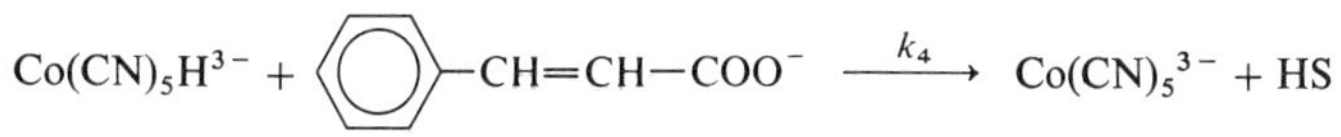

$$\text{Co(CN)}_5\text{H}^{3-} + \langle\bigcirc\rangle\!-\!\text{CH}\!=\!\text{CH}\!-\!\text{COO}^- \xrightarrow{\ k_4\ } \text{Co(CN)}_5{}^{3-} + \text{HS} \qquad (132)$$

$$\text{Co(CN)}_5\text{H}^{3-} + \text{HS}\cdot \xrightarrow{\ k_5\ } \text{Co(CN)}_5{}^{3-} + \langle\bigcirc\rangle\!-\!\text{CH}_2\!-\!\text{CH}_2\text{COO}^- \qquad (133)$$

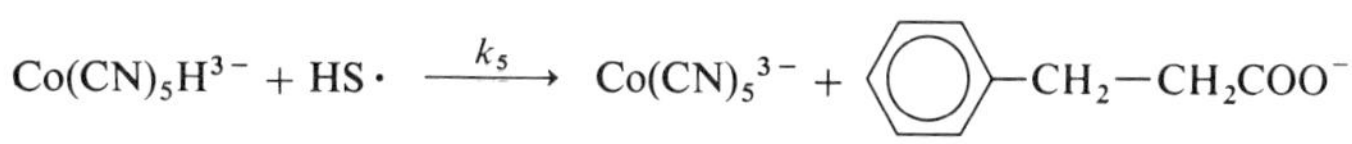

with the complication that a rapidly established equilibrium is set up between Co(CN)_5^{3-} and $\text{HS}\cdot$, the cinnamate radical

$$\text{Co(CN)}_5^{3-} + \text{HS}\cdot = \text{Co(CN)}_5(\text{HS})^{3-}. \qquad (134)$$

The outstanding point of difference between the two mechanisms concerns the question of whether $\text{Co(CN)}_5\text{H}^{3-}$ reduces a radical bound to another Co(CN)_5^{3-} residue or whether it reduces a radical free in solution. The steps in the proposed mechanisms are considered separately.

The first step Eq. (129), common to both schemes, is the reaction of molecular hydrogen with Co(CN)_5^{3-}. The rate law for this process is well established (231,238,239)

$$\frac{-d[\text{Co(CN}_5^{3-}]}{dt} = 2k_1[\text{Co(CN)}_5^{3-}]^2[\text{H}_2] - 2k_{-1}[\text{Co(CN)}_5\text{H}^{3-}]^2 \qquad (135)$$

where the value of k_1 is $4.2 \times 10^2 \ M^{-2}\text{sec}^{-1}$ at 20° and $I = 1.0 \ M$, but somewhat smaller, as expected, at $I = 0.4 \ M$. A mechanism consistent with this rate law is

$$\text{Co(CN)}_5^{3-} + \text{H}_2 \xrightleftharpoons{\text{fast}} \text{Co(CN)}_5\text{H}_2^{3-} \qquad (136)$$

$$\text{Co(CN)}_5^{3-} + \text{Co(CN)}_5\text{H}_2^{3-} \underset{k_{-1}}{\overset{k_1}{\rightleftharpoons}} 2\text{Co(CN)}_5\text{H}^{3-} \qquad (137)$$

Association of H_2 with the Co(CN)_5^{3-} apparently activates the H—H bond sufficiently to allow attack and rupture of that bond by a second Co(CN)_5^{3-}. The value of

$$K = \frac{[\text{Co(CN)}_5\text{H}^{3-}]^2}{[\text{Co(CN)}_5^{3-}]^2[\text{H}_2]} \qquad (138)$$

is $1.5 \times 10^5 \ M^{-1}$.

The reaction of $Co(CN)_5^{3-}$ with various other reagents is formally the same

$$2Co(CN)_5^{3-} + X - Y = Co(CN)_5 X^{3-} + Co(CN)_5 Y^{3-} \qquad (139)$$

However, in several of these cases, the rate law is different from that of the $Co(CN)_5^{3-}-H_2$ reaction (240–242),

$$\frac{-d[Co(CN)_5^{3-}]}{dt} = 2k_{X-Y}[Co(CN)_5^{3-}][XY] \qquad (140)$$

The mechanism of this reaction is proposed to be

$$Co(CN)_5^{3-} + XY \xrightarrow{\ k_{X-Y}\ } Co(CN)_5 X^{3-} + Y \qquad (141)$$

$$Co(CN)_5^{3-} + Y \cdot \xrightarrow{\ \text{fast}\ } Co(CN)_5 Y^{3-} \qquad (142)$$

In several cases radical trapping experiments (235,241) have given support to this suggestion. Table 12 lists several reactions that have been studied.

TABLE 12

THE REDUCTIVE CLEAVAGE OF SEVERAL MOLECULAR SPECIES BY $Co(CN)_5^{3-a}$

$X-Y$	$k^b (M^{-1}\text{sec}^{-1})$	$\Delta H^\ddagger$	$\Delta S^\ddagger$	E^c	Ref.
H_2^d	4.2×10^{2d}	0	-46	104	(231)
H_2O^e	1.6×10^{-2e}	4	-53	111	(231)
$HO-OH$	7.4×10^2	4.2	-31	51	(241)
$HO-NH_2$	5.3×10^{-3}	10.3	-35	$\cong 65$	(241)
$I-CN$	9.5	12.7	-1.2	82	(241)
$I-CH_3$	9.5×10^{-3f}				(242)
$I-CH_2C_6H_5$	3.8×10^{3f}				(242)
$Br-CH_2C_6H_5$	2.3^f				(242)
$Cl-CH_2C_6H_5$	4.9×10^{-4f}				(242)
$I-CH(CH_3)_2$	1.2^f				(242)
$I-C(CH_3)_3$	9.1^f				(242)

^{a} $T = 25°$.

^{b} Note the rate law is defined $\dfrac{-d[Co(II)]}{dt} = 2k[Co(II)][XY]$

^{c} The bond strength of the $X-Y$ bond, kcal mole^{-1}.
^{d} The rate law is that given in equation (31); the units of k are $M^{-2}\text{sec}^{-1}$.
^{e} The rate law is second order in $Co(CN)_5^{3-}$.
^{f} In 80% methanol, 20% H_2O.

For those reactions proceeding by rate law (140), it has been observed that the reactivity (as measured by $k(240)$ or $\Delta H^{\ddagger}$ (241)) parallels the bond strength of the $X-Y$ bond. The data in Table 12 illustrate this point. The pertinent question to ask is why the reactions of H_2 and H_2O with $Co(CN)_5^{3-}$ proceed by a rate law different from other XY systems. A possible explanation is that the higher bond strength found in these two molecules makes the mechanism second-order in $Co(CN)_5^{3-}$ somewhat lower in energy. A second feature of probable importance is the increased reactivity of $ICH_2C_6H_5$ compared to the corresponding benzyl bromide and chloride. This is consistent with reactivity patterns shown in Table 3.

The second step in the schemes for hydrogenation of olefins as catalyzed by $Co(CN)_5^{3-}$ involves the reaction of the hydridopentacyanocobalt(III) with the olefin, Eq. (130) and (132). These two reactions differ in their immediate products, but this difference is a point that is difficult to experi-mentally establish. Some evidence pertinent to the question of reaction (130) compared to (132) will be presented below; however, when reaction (134) is taken in combination with (132), the matter of most concern is the question of when the product from the insertion of $Co(CN)_5H^{3-}$ into the olefin is stable. This question clearly has relevance to the mechanism and rate of further reaction: equations (131) and (133).

Kwiatek and Seyler (243) have argued that organopentacyanocobalt(III) complexes are most stable if the bound carbon is primary or electron-withdrawing groups are bonded to this carbon. Thus, whereas the product of the reaction of $Co(CN)_5^{3-}$ with CH_3I or $C_6H_5CH_2I$ is the organocobalt complex [and $Co(CN)_5I^{3-}$ (240)], the reaction of isopropyliodide (in excess) with $Co(CN)_5^{3-}$ occurs dominantly by a path whose stoichiometry is $(242,243)$

$$2Co(CN)_5^{3-} + 2(CH_3)_2CHI = 2Co(CN)_5I^{3-} + CH_3CH=CH + (CH_3)_2CH_2 \quad (143)$$

The mechanism of this reaction is of some interest $(243,244)$. The olefin presumably arises from iodide abstraction to give the iso-propyl radical. which then reacts with a second $Co(CN)_5^{3-}$ by hydrogen atom abstraction. Reaction of this radical with $Co(CN)_5^{3-}$ to give an organocobalt complex is presumably unfavorable because of the instability of the secondary product. The $Co(CN)_5H^{3-}$ thus generated can react with further iso-propyl radicals as illustrated in Eq. (133). The reactivity pattern shown by the stoichiometry gives support to the concept of instability of unsubstituted secondary organocobalt systems. Further support for this concept is available in the data of Jackman, Hamilton, and Lawlor (245). They observed that reaction of $Co(CN)_5^{3-}$ with α,β-unsaturated carboxylate anions resulted in substantial adduct formation only if the α carbon was unsubstituted. Apparently, $-COO^-$ is not sufficiently electron-withdrawing to stabilize a tertiary residue—a result in agreement with earlier suggestions (*243*).

When insertion of $Co(CN)_5H^{3-}$ into an olefin occurs, it is in the direction expected for hydrogen atom transfer to form the most highly substituted radical. Further, it has been shown in at least two cases that the addition of $Co(CN)_5H^{3-}$ is stereospecific; the product results from *cis* addition. This means that the redox transfer step and addition of $Co(CN)_5^{3-}$ to the olefin are concerted or that the addition follows within the solvent cage before rotation about the $C-C$ bond can take place (*245*). This last result means that reaction (130) is favored over (132) in at least the cases studied.

The rate of reaction (130) or (132) depends on the nature of the attached organic groups. Studying systems in which adduct formation took place rapidly, and further reaction only slowly, if at all (the excess reagent was the alkene), Halpern and Wong have shown that the rate law is first order in each reagent (*246*)

$$\frac{-d[Co(CN)_5H^{3-}]}{dt} = k[Co(CN)_5H^{3-}][R_1R_2C=CR_3R_4] \quad (144)$$

Some of their data, and some related data from elsewhere, are summarized in Table 13. For systems with $R_1 = R_2 = H$, Halpern and Wong concluded that increased reactivity is observed when R_3 is CH_3 compared to $R_3 = H$; and when R_4 is less electron-withdrawing (*246*). These conclusions are understood on the basis of stability for the radical formed, and argue against a concerted process for steric reasons. Halpern and Wong favor a hydrogen atom transfer, without significant $Co-CR_3R_4$ interaction in the transition state; to accommodate the stereospecificity (*245*), if it exists in these systems, the collapse of the radical pair must be rapid. The low rate at which cinnamate ion reacts is compatible with the preliminary observations of Jackman, Hamilton, and Lawlor that substituents at R_1,R_2 slow the rate considerably. Sorbate anion, on the other hand, although it has a substituent ($R_1 = CH_3$, $R_2 = H$) has a weakly electron-withdrawing group as R_4 and hence reacts at a moderate rate. It should be noted that both of these rate data, and that for butadiene as reported by Burnett, Connolly, and Kemball, came from studies of the complete hydrogenation catalysis, and hence are obtained only by the use of several approximations (*236,237, 247*). These steady-state approximations, coupled as they are to equilibrium expressions, can be questioned (*248*).

It is this feature of the reported analysis of the rate data for hydrogenation of olefins that makes decisions about the mechanism of the final step difficult. The question as to whether this step is of the type represented by (131) or by (133), or if each mechanism is valid, depending on the system, will probably not be settled until some direct measure of this process can be made. Indeed the mechanism suggested for butadiene hydrogenation (*247*) is even more complex than that represented by Eqs. (*129*) to (*131*). In that mechanism, two sigma complexes,

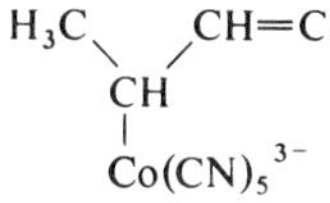

(1) 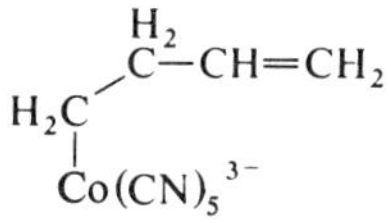   (2)

and a π complex,

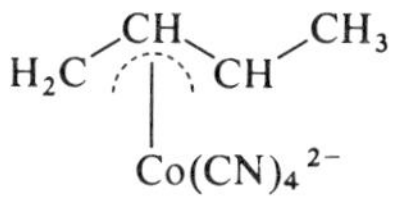

(3)

are said to exist, and each leads to distinctive products by reaction with $Co(CN)_5H^{3-}$. These authors suggest the attack of $Co(CN)_5H^{3-}$ on $Co(CN)_5C_4H_7^{3-}$ is related to the attack of $Co(CN)_5H^{3-}$ on $Co(CN)_5H^{3-}$ to give H_2 and the Co(II)-cyanide complex. Further experiments designed to establish the mechanism of the last step in the hydrogenation of olefins appear necessary.

As Halpern (*214*) has pointed out, several other metal ion complexes activate molecular hydrogen in solution. Cu(II) (*249–251*), Cu(I) (*252,253*), Ag(I) (*254*), and Ru(III) (*255*) and Rh(III) (*256*) in Cl^- media, and perhaps Pd(II) (*257*) activate H_2 by a heterolytic splitting of the bond, for example,

$$Cu^{2+} + H_2 \underset{k_2}{\overset{k_1}{\rightleftharpoons}} CuH^+ + H^+ \tag{145}$$

The fate of individual systems thereafter differs. Cu(II) presumably reacts via a second step,

$$CuH^+ + Cu^{2+} \overset{k_3}{\longrightarrow} 2Cu^+ + H^+ \tag{145}$$

with Cu(I) functioning as the active carrier. In the Ag(I) system, however, and probably in most of the other systems, the metal-hydride is the active

TABLE 13

THE RATE OF REDUCTION OF $R_1R_2C{=}CR_3R_4$ BY $Co(CN)_5H^{3-a}$

Alkene	$k(M^{-1}sec^{-1})$	Ref.
$CH_2{=}CHCOOH$	2.0×10^{-2}	(246)
$CH_2{=}CHCOO^-$	1.5×10^{-3}	(246)
$C_6H_5{-}CH{=}CHCOO^-$	$1.95 \times 10^{-4b,\,c,\,d}$	(236)
$CH_3{-}CH{=}CH{-}CH{=}CHCOO^-$	$3.0 \times 10^{-2b,\,d,\,e}$	(237)
$CH_2{=}CH{-}CH{=}CH_2$	1.6	(246)
	$7.6^{b,\,d,\,f}$	(247)
$CH_2{=}CH{-}C(CH_3){=}CH_2$	2.5	(246)
$CH_2{=}CHCN$	1.8×10^{-1}	(246)
$CH_2{=}C(CH_3)CN$	4.0	(246)
$CH_2{=}CHC_6H_5$	1.1	(246)

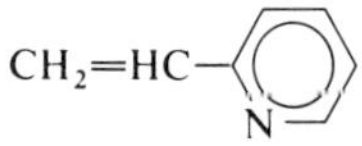

$CH_2{=}HC-$

1.0 (246)

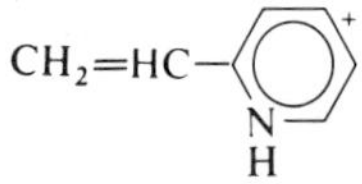

4.7 × 10² (246)

$CH_2{=}HC-$

^{a} $T = 25°$, 50 vol % CH_3OH; $I = 0.5\ M$.
^{b} In aqueous solution.
^{c} $T = 20°$, $I = 1.0\ M$, $\Delta H^{\ddagger} = 21.7$ kcal mole^{-1}, $\Delta S^{\ddagger} = -2$ e.u.
^{d} This number results only after certain assumptions are made concerning the overall mechanism of the $Co(CN)_5^{3-}$ catalyzed hydrogenation of this substrate; see text.
^{e} $T = 20°$, $I = 1.0\ M$, $\Delta H^{\ddagger} = 10.3$ kcal mole^{-1}, $\Delta S^{\ddagger} = -31$ e.u.
^{f} $T = 25°$, $I = 0.5\ M$.

carrier. The activation by Hg(I) and Hg(II) is believed to proceed without intervention of any hydride intermediate, Hg_{aq}° being the active carrier. Ag(I) also activates by the rate law (254)

$$\frac{-d[H_2]}{dt} = k[Ag^+]^2[H_2] \tag{147}$$

This is interpreted in terms of a homolytic splitting, compare the rate law Eq. (135). A striking feature of the reagents that serve to activate H_2 is their relative " softness." Readily polarizable systems that are substitution labile seem necessary. The activation by Ru(III) and Rh(III) occurs only in their higher chloride complexes: for instance, consider the reactivity order

$RhCl_6^{3-} > RhCl_5(H_2O)^{2-} > RhCl_4(H_2O)_2^- \gg RhCl_3(H_2O)_3$ (258). Saville (259) has offered a rationalization for these features in terms of four center transition states with proper matching of hard-hard and soft-soft interactions.

B. Oxidations with Peroxydisulfate Ion

In 1901 Marshall (2) noted that the presence of Ag^+ greatly enhanced the reactivity of $S_2O_8^{2-}$ with Mn^{2+}. Two comprehensive reviews (260,261) were published several years ago, but in the intervening time period a great deal has been learned. Of special importance is the surge of interest in radiation chemistry and flash photolysis: the chemistry of $S_2O_8^{2-}$ is, above all, the study of highly reactive species generated in low concentrations.

To understand the catalyzed reaction of $S_2O_8^{2-}$, it is useful first to inquire into the uncatalyzed decomposition of this ion

$$H_2O + S_2O_8^{2-} = 2SO_4^{2-} + \tfrac{1}{2}O_2 + 2H^+ \tag{148}$$

Even experimentally there are some questions open. An apparently definitive study of Kolthoff and Miller (262), over a pH range of 1 to 13, led to the rate law

$$\frac{-d[S_2O_8^{2-}]}{dt} = \{k_a + k_b[H^+]\}[S_2O_8^{2-}] \tag{149}$$

The first term leads to O_2 containing solvent oxygen whereas the second term leads to O_2 containing peroxide oxygen. In the intermediate pH ranges, these workers used buffers to control the pH. Recent work using a pH-stat to control pH has reported an unexplained maximum in the k versus pH curve (263). Table 14 illustrates the two sets of data. The first two columns of rate data serve to show the agreement at 50° between the two sets of investigations, whereas the third column illustrates the maximum in the $pH - k$ profile. It would be highly desirable to have further careful experiments to ascertain whether the maximum is real, and, if so, to determine its origin. [The emphasis on careful experiments must be even stronger in this field than in others. It has been an appalling experience to search the literature on the subject of peroxydisulfate kinetics: There are studies in the literature in which, even after the careful work of Kolthoff and co-workers—see the following—the authors have stated that there is no difference in the "nature of the reaction" between $S_2O_8^{2-}$ and As(III), catalyzed by Cu^{2+}, in the presence or absence of oxygen. There are studies on the thermal decomposition of $S_2O_8^{2-}$ in H_2O and D_2O, potentially important work, in which the data on the former system disagree with those in Table 14 by a factor of 10 without even an attempted rationalization of the difference.]

Assuming the rate law for decomposition of $S_2O_8^{2-}$ is as given in Eq. (149), there are still several mechanisms for the acid independent term. Bartlett and Cotman (264) and Kolthoff and Miller have suggested the scheme

TABLE 14

THE RATE OF THE UNCATALYZED THERMAL DECOMPOSITION OF PEROXYDISULFATE ION[a]

pH	$k \times 10^6$, sec^{-1b}	$k \times 10^6$, sec^{-1c}	$k \times 10^7$, sec^{-1d}
$[H^+] = 1.1 \times 10^{-2}$	1.83		
2			6.3
3	1.67		
3.3		1.80	
4			1.75
6			3.4
7	1.45	2.82	8.9
8			5.2
9			4.7
10	1.15	1.3	4.25
11		0.97	
13		0.96	1.75
$[OH^-] = 1.0 \times 10^{-1}$	1.04		

[a] $T = 50°$, $I = 0.4\ M$
[b] Data from Ref. (262).
[c] Data from Ref. (263)
[d] Data from Ref. (263), $T = 40°C$.

$$S_2O_8^{2-} \xrightarrow{k_2} 2SO_4^- \tag{150}$$

$$SO_4^- + H_2O \xrightarrow{k_3} HSO_4^- + OH \tag{151}$$

$$OH \longrightarrow \longrightarrow O_2 \tag{152}$$

where k_2 is to be associated with k_a of Eq. (149). A different initial step has been proposed by Froneaus and Östman on the basis of experiments with Ce(III) present (265,266)

$$H_2O + S_2O_8^{2-} \xrightarrow{k_5} HSO_4^- + SO_4^- + OH \tag{153}$$

$$Ce(III) + SO_4^- \xrightarrow{k_6} Ce(IV) + SO_4^{2-} \tag{154}$$

$$OH \longrightarrow \longrightarrow O_2 \tag{155}$$

where k_5 is to be associated with k_a. And House has suggested a more complicated mechanism, but one consistent with several other chain reactions of $S_2O_8^{2-}$ (261)

$$S_2O_8^{2-} \xrightarrow{k_2} 2SO_4^- \tag{156}$$

$$SO_4^- + H_2O \xrightarrow{k_3} HSO_4^- + OH \tag{157}$$

$$OH + S_2O_8^{2-} \xrightarrow{k_7} HSO_4^- + SO_4^- + \tfrac{1}{2}O_2 \tag{158}$$

$$SO_4^- + OH \xrightarrow{k_8} HSO_4^- + \tfrac{1}{2}O_2 \tag{159}$$

where $(k_2 k_3 k_7 k_8^{-1})^{1/2}$ is to be associated with k_a (this result is derived with the steady-state assumption for the radicals SO_4^- and OH, and the assumption that $k_3 k_7 k_8^{-1} \gg 1$).

Common to all of these mechanisms are the sulfate radical, SO_4^-, and the hydroxyl radical, OH. Fortunately, a great deal has been learned of these species. There is considerable evidence that the sulfate radical absorbs light at 4550 Å (267–272), although there is some question about the value of the extinction coefficient, $\varepsilon = 460\ M^{-1}\ cm^{-1}$ (269) or $\varepsilon = 1050$–$1100\ M^{-1}\ cm^{-1}$ (267,272). Observation of this absorption has allowed the study of the reactivity of this ion with a number of substrates; these values are compared with those of the hydroxyl radical (273) in Table 15. Two features are of importance: (1) in general, OH is slightly more reactive than SO_4^-. (2) The sulfate radical is generally much more reactive toward other substrates than toward water; Dogliotti and Hayon found the reaction of SO_4^- with itself accounted for the major part of the consumption of SO_4^- in the pH range 0.1 to 4.8 (267), although in more basic solutions its reactivity with the solvent (OH^-) is dominant (267,272). In addition to these features, it should be noted that the reaction of OH with $S_2O_8^{2-}$ has been reported to have a rate constant of less than $10^6\ M^{-1}\ sec^{-1}$ (272), and that OH does not apparently react with SO_4^{2-} at a measurable rate (277).

How do these data affect arguments about the mechanism of thermal decomposition of $S_2O_8^{2-}$? First, these experiments indicate that the rate of reaction of SO_4^- with H_2O in the equilibrium postulated by Wilmarth and Haim (260), Eq. (151) and its reverse, is established relatively slowly. This

TABLE 15

RATE CONSTANTS FOR REACTIONS OF SO_4^- AND OH

Substrate[a]	SO_4^-	References	OH	References
SO_4^-	$(3.7 \pm .5) \times 10^{8b,c}$	(269)		
	6×10^8	(270)		
HSO_4^-			8×10^{5b}	(267)
			$(1.18 \pm .33) \times 10^6$	(273)
OH^-	$\cong 5 \times 10^{7d}$	(269)		
	4.6×10^{7e}	(272)		
H_2O	$<2 \times 10^{3f}$	(269)		
Ce(III)	1.4×10^8	(270)	2.2×10^8	(275)
			3.8×10^8	(276)
Tl(I)	1.7×10^9	(270)	1×10^{10}	(277)
Fe(II)	9.9×10^8	(267)	5×10^9	(267)
			3×10^8	(276)
CH_3OH	2×10^7	(267)		
	2.5×10^7	(269)		
C_6H_6	8×10^8	(272)	4.3×10^9	(267)

[a] Temperature is room temperature.
[b] Rate constants are in units of $M^{-1}sec^{-1}$.
[c] Based on $\varepsilon(SO_4^-) = 460\ M^{-1}\ cm^{-1}$. The value becomes $8.9 \times 10^8\ M^{-1}$ sec^{-1} if $\varepsilon(SO_4^-) = 1100\ M^{-1}cm^{-1}$ (267,272).
[d] Calculated from Table 2 of Ref. (269) assuming the rate law is first order in each reagent.
[e] At pH = 11.
[f] Calculated from data in the indicated reference. See also other calculations in Ref. (274).

means that SO_4^- formed by reaction (150) and (153) probably exists for some time. Further, if an active reducing agent is present, reaction with SO_4^- will probably occur. This appears to be the case in the Fe^{2+} induced oxidation of As(III) by $S_2O_8^{2-}$ (278,280) (see the following). These experiments demonstrate that As(III) reacts more rapidly with SO_4^- than does Fe^{2+} by a factor of 21. This means, after consideration of the reactivity of Fe^{2+} with SO_4^-, that As(III) is a very efficient scavenger for this radical— nearly diffusion controlled. On the other hand, the uncatalyzed reaction of As(III) with $S_2O_8^{2-}$ is first order in $(S_2O_8^{2-})$, independent of [As(III)], and quite slow; after correction for a supposed Cu(II) impurity at $2 \times 10^{-8}\ M$. Woods, Kolthoff and Meehan find no acceleration at all (281). If this conclusion is valid, and the chain mechanism, reactions (150) (151), (158), and (159), pertained, one would expect a lower rate of reaction, because As(III) would react with SO_4^- much faster than does H_2O. Other radical scavengers give similar results; and similar conclusions have been drawn (264). The argument is not, however, conclusive since the evidence presented is, as Bartlett and Cotman point out, essentially negative.

The determination of the correctness of reactions (150) to (152) compared to reactions (153) to (155) is even less easily made. There is no proof of reaction (150) occurring thermochemically, although it is believed to occur photochemically (267). And the experimental data offered by Froneaus and Östman have no obvious invalidities. However, the interpretation requires that Ce(III) (in sulfate media) be highly reactive with SO_4^-, but comparatively unreactive toward OH. This assumption was questioned previously by Wilmarth and Haim (260), but the evidence from radiation chemists was less extensive then: The assumption is still questionable. Table 15 lists some recent determinations of the rate of reaction of Ce(III) with both SO_4^- and OH·; as is the usual case, OH· is slightly more reactive. Until this anomaly is explained there must remain some doubt concerning the interpretation of Froneaus and Östman (266). Recent work on this question has been reported by Crematy (340). He supports the mechanism of Bartlett and Cotman (264).

The acid dependent path, k_b, has been investigated less extensively. What is known about it is that the oxidation of As(III) is not inhibited by radical scavengers (281) and that the oxygen originally in the peroxo-linkage of the $S_2O_8^{2-}$ is not exchanged with solvent oxygen, but becomes O_2 in the thermal decomposition of the ion (262,282). This result has been suggested as being consistent with the mechanism (260).

$$H_2O + H^+ + O_3S^*O^*OSO_3^{2-} \longrightarrow HSO_4^- + H^+ + HO_3S^*O^*O^- \quad (160)$$

$$HO_3S^*O^*O^- \longrightarrow \longrightarrow {}^*O_2 \quad (161)$$

Consideration is now given to reactions in which metal-ion catalysis is important. In those catalytic systems in which chain reactions do not play a role, the reaction appears to go by the nonassociative activation path. The classic example of systems of this type are the Ag^+ catalyzed oxidations of Ce(III), Cr(III), Mn(II), and VO^{2+} (260,261). These oxidations proceed by the rate law

$$\frac{-d[S_2O_8^{2-}]}{dt} = k[S_2O_8^{2-}][Ag^+] \quad (162)$$

where k is independent of reductant after appropriate correction for ionic strength effects: $k = 3.6 \times 10^{-3} \, M^{-1} \sec^{-1}$ at $25°$ and $I = 1.0 \, M$ (260).

Several other systems have been studied since these previous summaries and yield similar data. The Ag^+ catalyzed oxidation of Tl(I) has $k = 3.6 \times 10^{-3} M^{-1} sec^{-1}$ at 25°, $I = 1.27 M$, with $\Delta H^{\ddagger} = 12$ and $\Delta S^{\ddagger} = -29$ (283). When no reductant is present, the catalyzed reaction leads to the oxidation of water: $k = 4.8 \times 10^{-3} M^{-1} sec^{-1}$ at 25°, $I = 0.68 M$, with $\Delta H^{\ddagger} = 13.8$ and $\Delta S^{\ddagger} = -22$ (284). The details of this oxidation of H_2O have recently been reported (285). There is a report (286) that the Ag^+ catalyzed oxidation of Se(IV) and Te(IV) by $S_2O_8^-$ proceeds by the rate law indicated in Eq. (162), but the data are limited and show changes in the second order rate constant as $[Ag^+]$ changes (in addition, the sign of $\Delta S^{\ddagger}$ reported in this paper is incorrect). The reported rate constant is somewhat smaller than those listed above.

The mechanism of the Ag^+ catalyzed oxidation by $S_2O_8^-$ could involve either Ag(II) or Ag(III) as the active oxidant, but there is some precedent for assuming the initial step to be

$$Ag^+ + S_2O_8^{2-} \longrightarrow Ag(II) + SO_4^- + SO_4^{2-} \tag{163}$$

This feature results from the studies of several other systems and their reduction of $S_2O_8^{2-}$. An important study by Pennington and Haim has made use of the substitution inertness of Cr(III) complexes to gain some insight into the detailed mechanism of the Cr(II) reduction of $S_2O_8^{2-}$ (274). These workers were able to show that the mechanism of the reaction,

$$2Cr(II) + S_2O_8^{2-} = 2Cr(III) + 2SO_4^{2-} \tag{164}$$

was best represented as

$$Cr^{2+} + S_2O_8^{2-} \longrightarrow CrSO_4^+ + SO_4^- \tag{165}$$

$$H^+ + SO_4^- + Cr^{2+} \longrightarrow Cr(H_2O)_6^{3+} + HSO_4^- \tag{166}$$

In the presence of Br^-, some $CrBr^{2+}$ is produced at the expense of Cr^{3+}; the yield of $CrSO_4^+$ remains at one mole per mole of $S_2O_8^{2-}$ destroyed. This effect is explained by competition between Cr^{2+} and Br^- for SO_4^-, Eq. (166) and

$$Br^- + SO_4^- \xrightarrow{k_a} Br^\circ + SO_4^{2-} \tag{167}$$

(k_a was estimated to be $4 \times 10^8 M^{-1} sec^{-1}$). The $CrBr^{2+}$ was then formed by reduction of Br° by Cr^{2+},

$$Cr^{2+} + Br^\circ \longrightarrow CrBr^{2+} \tag{168}$$

It is to be noted that although the reaction of Cr^{2+} with $S_2O_8^{2-}$ chooses an inner-sphere path, the oxidation of Cr^{2+} by SO_4^- is either outer-sphere or a hydrogen-atom-transfer reaction. Second, this reaction would seem to serve as a useful model for the initiation of a metal-ion catalyzed oxidation by $S_2O_8^{2-}$. Other metal ion systems react with $S_2O_8^{2-}$ with second-order rate laws, and at least in the case of Fe^{2+} and Cu^+, SO_4^- radicals are also generated (278). Table 16 lists some pertinent rate data for reduction of

TABLE 16

RATE OF REDUCTION OF PEROXYDISULFATE UNDER SECOND-ORDER CONDITIONS

Reductant	$E^{\circ b}$	Conditions	$k(M^{-1}\ \text{sec}^{-1})$	Ref.
Cr^{2+}	$+0.40$	$I = 1.0\ M,\ ClO_4^-;$ $[H]^+ = 0.1$	2.5×10^4	(274)
$Cu(I)$		$I = 1.0\ M,\ Cl^-$	1.3×10^3	(287)
$Fe(CN)_6^{4-}$	-0.36	$I = 10^{-3}\ M,\ LiOH$	4.6×10^{-3}	(288)
Fe^{2+}	-0.77	$I = 0.1\ M,\ ClO_4^-$	2.5×10^1	(281)
		$I = 0.5\ M,\ ClO_4^-$	3.0×10^1	(279)
$Fe(4,7\text{-}(CH_3)_2phen)_3^{2+}$	-0.86	$pH = 7$	1.9	(289)
$Os(bip)_3^{2+}$	-0.87	$I = 0$	4.9×10^1	(290)
$Fe(4,4'\text{-}(CH_3)_2bip)_3^{2+}$	-0.94	$I = 0$	6.7	(291)
$Fe(phen)_3^{2+}$	-1.14	$I = 0$	3.0×10^{-1}	(291)
	-1.07	$pH = 7$	1.1×10^{-1}	(289)
$Fe(5\text{-}Cl\ phen)_3^{2+}$	-1.11	$pH = 7$	4.2×10^{-2}	(289)
$Ru(bip)_3^{2+}$	-1.37	$I = 0$	9.0×10^{-3}	(291)
Ag^+	-1.9	$I = 1.0\ M,\ ClO_4^-$	3.6×10^{-3}	(260)

$^a\ T = 25°$.
^{b} These values are approximate; the conditions under which they were measured differ from those of the kinetic experiments. The values are for the half-cell reaction $Cr^{2+} = Cr^{3+} + e^-$.

$S_2O_8^{2-}$. It must be noted that the equation governing the rate constants in Table 16

$$\frac{-d[\text{Red}]}{dt} \quad \text{or} \quad \frac{-d[S_2O_8^{2-}]}{dt}$$

was, in general, not defined in the original literature; and hence comparisons of data that differ slightly have ambiguity. The earlier correlation of rate with net driving force of the reaction (291) does not hold over the entire range of reductants. There is, however, an increase in rate with a decrease in $\Delta F°$ for the reaction, although it would appear that charge effects [the reactions of $Fe(CN)_6^{4-}$ with $S_2O_8^{2-}$, like that of the $MnO_4^{-1,\ -2}$

exchange (292) is very sensitive to the nature and concentration of cation] and the possibility of inner-sphere paths $\{Fe^{2+}$ versus $Fe[4,7\text{-}(CH_3)_2\text{-}phen]_3^{2+}$—see also Ref. ($293$)} prevail over these thermodynamic barriers to reaction. Thus, although the initial step in the catalysis seems likely to be reaction (163), there is no conclusive proof that a two-electron process is absent

$$2H^+ + Ag^+ + S_2O_8^{2-} \longrightarrow Ag(III) + 2HSO_4^- \tag{169}$$

If the initial step in the Ag^+ catalyzed oxidation with $S_2O_8^{2-}$ is as represented in Eq. (163), then a number of possibilities for subsequent reactions exist. The sulfate radical could react with substrate in an intermediate-carried reaction, or with either Ag^+ or $Ag(II)$,

$$H^+ + Ag^+ + SO_4^- \longrightarrow Ag(II) + HSO_4^- \tag{170}$$

$$H^+ + Ag(II) + SO_4^- \longrightarrow Ag(III) + HSO_4^- \tag{171}$$

These latter options lead to catalyst-carried reactions; the two reaction products are coupled through the relatively rapid process (294)

$$Ag^+ + Ag(III) = 2Ag(II) \tag{172}$$

An earlier review has summarized the pertinent arguments concerning $Ag(II)$ and $Ag(III)$ as the carrier (260). The determination of the agent responsible for carrying the reaction has not yet been firmly established. Recent data on the rates of oxidation of a variety of metal ions by $Ag(II)$ (295) indicate that this reagent is certainly efficient enough to be the carrier. These data also show that $Ag(II)$ does not react via the pathway

$$Ag^{2+} + H_2O \xrightarrow{\text{slow}} Ag^+ + OH + H^+ \tag{173}$$

$$H^+ + OH + M^{n+} \xrightarrow{\text{fast}} M^{(n+1)+} + H_2O \tag{174}$$

at least at the acidity studied, ($4\,M$ $HClO_4$) (295). Further, there is no indication that the generation of $Ag(III)$ by the reverse of Eq. (172) offers a lower energy transition state. For a recent discussion of the role of Ag in persulphate oxidations, see Ref. (341).

A striking example of the difference in the path of reaction as well as the rate in the presence or absence of Ag^+ is afforded by some data reported by Thusius and Taube (296). They find that in the presence of Ag^+, $S_2O_8^{2-}$ "oxidizes" $Co(NH_3)_5H_2O^{3+}$ according to the stoichiometry

$$3H^+ + \tfrac{1}{2}S_2O_8^{2-} + Co(NH_3)_5H_2O^{3+} = Co^{2+} + SO_4^{2-} + 5NH_4^+ + \tfrac{1}{2}O_2 \tag{175}$$

The majority of the oxygen initially bound to $Co(III)$ is found in the O_2 gas. This reaction competes with oxidation of water by $S_2O_8^{2-}$ catalyzed by

Ag^+; but the overall rate constant for disappearance of $S_2O_8^{2-}$ is consistent with initiation by the Ag^+–$S_2O_8^{2-}$ reaction. In contrast to this observation, the stoichiometry that results when Ag^+ is absent {and $[Co(NH_3)_5OH_2^{3+}]$ is above about 2×10^{-3} M} is

$$H^+ + S_2O_8^{2-} + Co(NH_3)_5H_2O^{3+} = Co^{2+} + 4NH_4^+ + 2SO_4^{2-} + \tfrac{1}{2}N_2 + H_2O \quad (176)$$

These reactivities are explicable on the basis of Ag(II) (or (Ag(III)) and SO_4^- (or OH) as the reactive species. The Ag(II) path presumably involves oxidative coupling of two waters to give a peroxide,

$$(NH_3)_5CoOH_2^{3+} + H_2OAg(II) \longrightarrow Co^{2+} + Ag^+ + H_2O_2 + NH_4^+ \quad (177)$$

The peroxide is then rapidly oxidized to O_2. On the basis of isotope effects, Thusius and Taube believe the second path involves SO_4^- (or OH) abstraction of a hydrogen atom of the coordinated water to yield $(NH_3)_5CoOH^{3+}$. This Co(IV) complex then decomposes to give, presumably, the NH radical, which is then oxidized and dimerized (or vice versa) to N_2.

A very intriguing aspect of the chemistry of peroxydisulfate ion is the chain reactions that often are found. The uncatalyzed oxidation of formate ion (*297*) and oxalate ion (*298*) are recent examples; but catalyzed chain reactions are also well known. Such processes are made use of in the many polymerization initiation systems using $S_2O_8^{2-}$. Chain reactions are also commonly found in analytical techniques; they lead to titration errors in systems containing $S_2O_8^{2-}$ (*299*). A careful study of the Fe(II)-As(III)-$S_2O_8^{2-}$ systems has been made by Woods, Kolthoff, and Meehan.

These authors noted that the reduction of $S_2O_8^{2-}$ by Fe(II) induces the oxidation of As(III); and that the presence or absence of O_2 is a significant variable. In the absence of oxygen and in the presence of sufficient Fe(III), the induction factor, the equivalents of As(III) oxidized per equivalent of Fe(II) oxidized, approaches infinity. These facts can be explained (*278*) on the basis of reactions (178) to (180).

$$Fe^{2+} + S_2O_8^{2-} \xrightarrow{\ k_1\ } Fe(III) + SO_4^{2-} + SO_4^- \quad (178)$$

$$As(III) + SO_4^- \xrightarrow{\ k_2\ } As(IV) + SO_4^{2-} \quad (179)$$

$$As(IV) + Fe(III) \xrightarrow{\ k_3\ } As(V) + Fe^{2+}, \quad (180)$$

$$As(III) + S_2O_8^{2-} = As(V) + 2SO_4^{2-} \quad (181)$$

When the [As(III)] is lowered, there is a competition set up for SO_4^-, reactions (179) and (182)

$$Fe^{2+} + SO_4^- \xrightarrow{\ k_4\ } Fe(III) + SO_4^{2-} \quad (182)$$

The value of k_2/k_4 is 21 at $I = 0.5\,M$ and 25°. Further, at high [As(III)] and either low [Fe(III)] or high [Fe(II)], the induction factor goes down because of competition for As(IV): reactions (180) and (183)

$$\text{As(IV)} + \text{Fe}^{2+} \xrightarrow{k_5} \text{As(III)} + \text{Fe(III)} \tag{183}$$

Replacement of Cu^{2+} for Fe(III) leads to similar reactions. The reactions involving Fe(III) are generally acid dependent.

In the presence of oxygen, the Fe(II)-As(III)-$S_2O_8^{2-}$ system behaves in quite a different manner (279). Here a real chain reaction is set up; the result is that the stoichiometry is changed to the extent that $S_2O_8^{2-}$ is no longer consumed and merely acts as an initiator:

$$4\text{H}^+ + \text{O}_2 + 2\text{Fe}^{2+} + \text{As(III)} = 2\text{Fe(III)} + \text{As(V)} + 2\text{H}_2\text{O} \tag{184}$$

The proposed mechanism is the combinations of reactions (178) and (179) as initiation reactions, followed by the chain reactions

$$\text{As(IV)} + \text{O}_2 + \text{H}^+ \xrightarrow{k_6} \text{As(V)} + \text{HO}_2 \tag{185}$$

$$\text{HO}_2 + \text{Fe}^{2+} + \text{H}^+ \xrightarrow{k_7} \text{Fe(III)} + \text{H}_2\text{O}_2 \tag{186}$$

$$\text{Fe}^{2+} + \text{H}_2\text{O}_2 \xrightarrow{k_8} \text{Fe(III)} + \text{OH} + \text{OH}^- \tag{187}$$

$$\text{As(III)} + \text{OH} \xrightarrow{k_9} \text{As(IV)} + \text{OH}^- \tag{188}$$

The chain is terminated by the capture of the radical HO_2 by Fe(III).

$$\text{Fe(III)} + \text{HO}_2 \xrightarrow{k_{10}} \text{Fe(II)} + \text{O}_2 + \text{H}^+ \tag{189}$$

Again the Cu^{2+} system is similar.

Another chain reaction, this one involving the decomposition of $S_2O_8^{2-}$ in the chain, has been investigated by Kalb and Allen (300). These authors studied the Ag^+ catalyzed oxidation of $C_2O_4^{2-}$ by $S_2O_8^{2-}$ in the absence of O_2. They propose the mechanism

$$\text{Ag}^+ + \text{S}_2\text{O}_8^{2-} \longrightarrow \text{Ag}^{2+} + \text{SO}_4^- + \text{SO}_4^{2-} \tag{190}$$

$$\text{Ag}^+ + \text{SO}_4^- \longrightarrow \text{Ag}^{2+} + \text{SO}_4^{2-} \tag{191}$$

$$\text{Ag}^{2+} + \text{C}_2\text{O}_4^{2-} \longrightarrow \text{CO}_2 + \text{CO}_2^- + \text{Ag}^+ \tag{192}$$

$$\text{CO}_2^- + \text{S}_2\text{O}_8^{2-} \longrightarrow \text{CO}_2 + \text{SO}_4^- + \text{SO}_4^{2-} \tag{193}$$

where Eq. (190) is the initiation and Eqs. (191) to (193) are the chain steps. Termination occurs by trapping of the CO_2^- radical

$$2CO_2^- \longrightarrow C_2O_4^{2-} \tag{194}$$

$$CO_2^- + Ag^+ \longrightarrow CO_2 + Ag^\circ \tag{195}$$

$$Ag^\circ + Ag^{2+} \longrightarrow 2Ag^+ \tag{196}$$

Curiously, this reaction is inhibited by O_2, presumably through competition for the CO_2^- radical

$$CO_2^- + O_2 \longrightarrow CO_4^- \tag{197}$$

$$CO_4^- + CO_2^- \longrightarrow C_2O_4^{2-} + O_2 \tag{198}$$

The effects of coordination of the oxalate ion to Ag^+ have also been considered (300).

C. CATALYZED SUBSTITUTIONS

Reactions whose net stoichiometry is not oxidation-reduction can be catalyzed by oxidation-reduction processes. Generally such reactions are found in inert ligand-metal ion systems where variable oxidation states are available. Reference to Fig. 1 and Table 1 will illustrate the possibilities. If a metal-ion complex, say $M(II)L_5X$ is oxidized to $M(III)L_5X$, in which oxidation state lability is high, it may aquate before reduction to its original oxidation state.

The experiments dealing with this subject have been performed largely to gain an understanding of electron-transfer processes themselves. However, the results have been of much more general interest. One feature of special importance is their usefulness in synthesis. For instance, procedures taking advantage of oxidation-reduction catalysis have been devised for the synthesis of $Cr(en)_3^{3+}$ (301),

$$Cr^{3+} + 3en \xrightarrow{\ Cr^{2+}\ } Cr(en)_3^{3+} \tag{199}$$

and the malonaldehyde complex of Cr(III) (302). Catalysis of the formation of $Rh(py)_4Cl_2^+$ from $Rh(H_2O)Cl_5^{2-}$ and py by ethanol (303) have been explained by electron-transfer catalysis (304,305). Similarly, a number of important studies dealing with ligands with two possible coordination sites have used oxidation-reduction reactions for synthetic purposes—to generate unstable linkage isomers(131).

$$FeNCS^{2+} + Cr^{2+} = CrSCN^{2+} + Fe^{2+} \tag{200}$$

While reactions such as these are not catalytic, the isomerization reaction of the unstable product (306),

$$CrSCN^{2+} = CrNCS^{2+} \tag{201}$$

is catalyzed by an oxidation-reduction path, and hence is pertinent to this review.

Of the various substitution reactions catalyzed by oxidation-reduction reactions, three types are of principal importance. The catalyzed process can involve the substitution of ligands that are normally very inert

$$5H^+ + Cr(NH_3)_5Cl^{2+} = CrCl^{2+} + 5NH_4^+ \qquad (202)$$

or the ligand whose substitution is catalyzed could be the one that would normally substitute, but under catalytic conditions does so much more rapidly

$$Cr(H_2O)_5Cl^{2+} = Cr(H_2O)_6^{3+} + Cl^- \qquad (203)$$

(In this reaction, some or all of the water molecules substitute also; this feature is common to this type of reaction.) Finally, isomerization reactions, such as that illustrated in Eq. (201) can be catalyzed. The apparent preoccupation with Cr(III) chemistry here is no accident. Cr(III) is a metal-ion system in which many complexes are known, and these are usually inert. More importantly, however, Cr(III) can readily be reduced (or in some cases oxidized (307)) to an adjacent valence state that exhibits high lability. Hence the effort expended on studies of Cr(III) complexes has been great. There are, nevertheless, several other metal ions that exhibit catalyzed substitution reactions. Selective examples of metal-ion systems will be discussed separately under the three aforementioned general types of reactions.

1. Catalyzed Substitution of Normally Inert Ligands

a. Cr(III). The Cr(II) catalyzed substitution of the inert ligands of a Cr(III) complex is a well-known phenomenon in Cr(III) chemistry. It results from the known lability of Cr(II) and the use of an inner-sphere path for electron-transfer

$$5H^+ + Cr(NH_3)_5Cl^{2+} = CrCl^{2+} + 5NH_4^+ \qquad (204)$$

The mechanism of this reaction is

$$Cr^{2+} + Cr(NH_3)_5Cl^{2+} \longrightarrow Cr(NH_3)_5^{3+} + CrCl^{2+} \qquad (205)$$

$$Cr(NH_3)_5^{3+} \longrightarrow \longrightarrow Cr^{2+} + 5NH_4^+ \qquad (206)$$

Ogard and Taube (*91*) examined this reaction with a series of bridging ligands, F$^-$, Cl$^-$, Br$^-$, and I$^-$. More recently several other reactions have been studied kinetically, mostly with chloride ion as the bridging group (*150,152*). These data are listed in Table 6. A more complex type of substitution occurs when there is present in solution a ligand that complexes with Cr^{2+} before electron transfer. Then the substitution process becomes one of substitution of one ligand for another rather than aquation

$$5H^+ + Cr(EDTA)^{2-} + Cr(NH_3)_5Cl^{2+} = Cr(EDTA)^{1-} + Cr^{2+} + 5NH_4^+ + Cl^-$$
(207)

The kinetics of processes of this type have not been extensively investigated. Data for published studies are listed in Table 17. It is to be noted

TABLE 17

LIGAND SUBSTITUTION ON $Cr(NH_3)_5Cl^{2+}$ BY ELECTRON TRANSFER

Reductant	Product	$k\ (M^{-1}sec^{-1})$	Ref.
Cr^{2+}	$CrCl^{2+}$	5.1×10^{-2}	(91)
$Cr(CH_3COO)_3^-$	$Cr(CH_3COO)_3$	1.2	(308)
$Cr(EDTA)^{2-}$	$Cr(EDTA)^-$	2×10^3	(92)

$^a\ T = 25°, I = 1.0\ M.$

that in comparing the reactivity of Cr^{2+} and $Cr(EDTA)^{2-}$, the driving force for the reaction is larger for the latter (309). The increase in rate is compatible with this driving force change (K_{23} of Eq. 100 is much greater than one). It is clear that this type of reaction is of synthetic utility; the above-mentioned syntheses of $Cr(en)_3^{3+}$ and tris-(1,3-propanedialato)-chromium(III) are examples.

b. Co(III). Substitution of inert ligands of Co(III) complexes by electron-transfer catalysis was described in 1956. Adamson (310) observed that CN^- reacted with $Co(NH_3)_5Br^{2+}$ to form $Co(CN)_5Br^{3-}$. He ascribed this process to catalysis by Co(II), which, in the presence of CN^-, forms $Co(CN)_5^{3-}$; inner-sphere electron transfer yields the products

$$Co(CN)_5^{3-} + Co(NH_3)_5Br^{2+} \longrightarrow Co(CN)_5Br^{3-} + Co(II) \quad (208)$$

This catalysis of the replacement of NH_3 by CN^- is very efficient; hence only trace quantities of Co(II) need be present to cause substitution. This fact has led to literature reports that CN^- is a good nucleophile for Co(III) complexes (311). Careful purification of material shows, however, no attack by CN^- (312). Studies of the rate of electron-transfer processes such as reaction (208) have been reported (80). Depending on the initial Co(III) complex, $Co(NH_3)_5X^{n+}$, the inner-sphere path ($X = Cl^-, N_3^-, NCS^-$, OH^-, F^-) leads to exchange of five ammonias; the outer-sphere path

$(X = NH_3, SO_4^{2-}, CH_3COO^-)$ to replacement of all six ligands on the Co(III) center by CN^-.

Other systems leading to catalysis of substitution of Co(III) centers are less well studied. It is probable that future studies will show that the Co(II) catalysis of the process

$$4H^+ + Co(NH_3)_5Cl^{2+} + H_2EDTA^{2-} = Co(EDTAHCl)^- + 5NH_4^+ \qquad (209)$$

takes place (*313*). Recently it has been shown that the reaction

$$trans\text{-}Co(en)_2Cl_2^+ + 2NO_2^+ = trans\text{-}Co(en)_2(NO_2)_2^+ + 2Cl^- \qquad (210)$$

takes place more rapidly in the presence of Co(II) than in its absence (*314*). The details of the pathway for this catalysis are complex.

2. Catalyzed Substitution of the Normally Substitutable Ligand

a. Pt(IV). Substitution on Pt(IV) complexes has always been notoriously susceptible to catalysis by impurities and light. It is now well recognized that such catalysis often appears as a result of the presence of Pt(II) complexes and involves a mechanism that is electron-transfer in nature (*12*). The generally found rate law for a substitution on Pt(IV), for example (*315*),

$$trans\text{-}Pt(NH_3)_4Cl_2^{2+} + NH_3 = Pt(NH_3)_5Cl^{3+} + Cl^- \qquad (211)$$

is

$$\frac{d[Pt(NH_3)_5Cl^{3+}]}{dt} = k[Pt(NH_3)_4Cl_2^{2+}][Pt(NH_3)_4^+][NH_3] \qquad (212)$$

The mechanism usually associated with rate laws of this type is (*316*)

$$Pt(NH_3)_4^{2+} + NH_3 = Pt(NH_3)_5^{2+} \qquad (213)$$

where $K < 1$

$$Pt(NH_3)_5^{2+} + Pt(NH_3)_4Cl_2^{2+} \xrightarrow{k} [Cl-Pt(NH_3)_4-Cl-Pt(NH_3)_5^{4+}] \qquad (214)$$

$$[Cl-Pt(NH_3)_4-Cl-Pt(NH_3)_5^{4+}] \longrightarrow Cl^- + Pt(NH_3)_4^{2+} + Pt(NH_3)_5Cl^{2+} \qquad (215)$$

The generation of the transition state requires the formation of a bridge and the entry of an additional ligand to help convert the Pt(II) environment toward an environment compatible with Pt(IV). Several studies on systems of this type have been carried out. Table 18 lists some of the data.

Noteworthy is the trend in bridging ligand efficiency, $I^- > Br^- > Cl^-$ (compare in Table 18 reactions **4** and **5** and reaction **10** and **11**); surprisingly, NCS^- is more efficient than Cl^-—reactions **7** and **16**. Variation of

TABLE 18

The Rate of the Pt(II) Catalyzed Substitution of Pt(IV) Complexes[a]

Reaction	Pt(IV) complex[b]	Pt(II) complex	N[c]	I(M)	k(M^{-2} sec^{-1})[c]	Ref.
1	trans-NH$_3$Pt(NH$_3$)$_4$Cl^{2+}	Pt(NH$_3$)$_4^{2+}$	Cl$^-$	0.32	1.2×10^{-3}	(94)
2	cis-NH$_3$Pt(NH$_3$)$_3$ClCl^{2+}	Pt(NH$_3$)$_4^{2+}$	Cl$^-$	0.14	2.7×10^{-3}	(95)
3	trans-NH$_3$Pt(NH$_3$)$_4$I^{2+}	Pt(NH$_3$)$_4^{2+}$	Cl$^-$	0.35	9.9×10^2	(94)
4	trans-BrPt(NH$_3$)$_4$Cl^{2+}	Pt(NH$_3$)$_4^{2+}$	Cl$^-$	0.2	6.3	(96)
5	trans-BrPt(NH$_3$)$_4$Br^{2+}	Pt(NH$_3$)$_4^{2+}$	Cl$^-$	0.2	4.2×10^3	(96)
6	trans-ClPt(NH$_3$)$_4$Cl^{2+}	Pt(NH$_3$)$_4^{2+}$	Cl$^-$	0.18	6	(95)
7	trans-NCSPt(NH$_3$)$_4$Cl^{2+}	Pt(NH$_3$)$_4^{2+}$	Cl$^-$	1.1	2.6	(317)
8	trans-ClPt(NH$_3$)$_4$Cl^{2+}	Pt(NH$_3$)$_4^{2+}$	Br$^-$	0.2	1.1×10^2	(96)
9	trans-ClPt(NH$_3$)$_4$Br^{2+}	Pt(NH$_3$)$_4^{2+}$	Br$^-$	0.2	1.9×10^4	(96)
10	trans-NH$_3$Pt(NH$_3$) Br^{2+}	Pt(NH$_3$)$_4^{2+}$	Br$^-$	0.16	1.1×10	(94)
11	trans-NH$_3$Pt(NH$_3$)$_4$I^{2+}	Pt(NH$_3$)$_4^{2+}$	Br$^-$	0.016	1.2×10^4	(94)
12	trans-NH$_3$Pt(NH$_3$)$_4$I^{2+}	Pt(NH$_3$)$_4^{2+}$	I$^-$	0.2	9.7×10^2	(94)
13	trans-ClPt(NH$_3$)$_4$Cl^{2+}	Pt(NH$_3$)$_4^{2+}$	NH$_3$	0.19	6.3×10^{-1c}	(315)
14	trans-ClPt(en)$_2$Cl^{2+}	Pt(en)$_2^{2+}$	Cl$^-$	0.13	1.5×10	(95)
15	trans-ClPt(en)$_2$Cl^{2+}	Pt(en)$_2^{2+}$	NO$_2^-$	0.24	1.01^d	(318)
16	trans-NCSPt(NH$_3$)$_4$SCN^{2+}	Pt(NH$_3$)$_4^{2+}$	Cl$^-$	0.20	1.7×10^2	(317)
17	trans-BrPt(CN)$_4$Br^{2-}	Pt(CN)$_4^{2-}$	Cl$^-$	0.21	1.0×10	(319)
18	trans-BrPt(NO$_2$)$_4$Br^{2-}	Pt(NO$_2$)$_4^{2-}$	Cl$^-$	0.21	1.5×10^{-1}	(319)

[a] $T = 25°$.
[b] The complex is written to place the bridging ligand adjacent to the Pt(II) complex.
[c] The rate law is $\dfrac{-d[\text{Pt(IV)}]}{dt} = k[\text{Pt(IV)}][\text{Pt(II)}][\text{N}]$.
[d] $T = 22.5°$.
[e] $T = 50°$.

the *trans*-nonbridging ligand on the Pt(IV) center indicates $Br^- \cong Cl^- \cong$ $NCS^- > NH_3$ (reactions **1**, **4**, **6**, and **7**). The data also show an unusual behavior in rate as N is changed: both the comparison of reactions **3** and **11** and **6** and **8** show that $Br^- > Cl^-$ in rate enhancement. However, reactions **11** and **12** indicate $Br^- > I^-$! This unusual reversal in efficiency as one changes the halide ion may result from a competition between two roles played by the halide ion. A change in halide will affect the preequilibrium step, Eq. (213), and the electron-transfer step, Eq. (214). One would expect that K for reaction (213) would decrease in the series $I^- > Br^- > Cl^-$, for Pt(II) is generally regarded as " soft " (*320*); if $Cl^- >$ $Br^- > I^-$ in promoting the electron-transfer step then the maximum rate when N is Br^- may be explicable. Similar reasoning may be applicable to low reactivity of systems with $N = NO_2^-$—reactions **14** and **15**.

Several experiments demonstrating the effect of steric hindrance on the formation of the bridged activated complex have been reported (*95,318*). When en is substituted by 2,3-diamino-2,3-dimethylbutane, (TMen), the rate of the reaction

$$Pt(TMen)_2Cl_2^{2+} + *Cl^- \xrightarrow{\ Pt(TMen)_2^{2+}\ } \text{Exchange} \qquad (216)$$

is essentially completely quenched.

b. Cr(III). The use of electron-transfer reactions to catalyze the normal, although usually slow, substitution reactions of Cr(III) has been known for many years. Taube and Myers (*321*) noticed that Cr(II) catalyzed the aquation of $CrCl_2^+$; Taube (*137*), and later, Hunt and Earley (*138*) showed that Cr(III) complexes could be formed relatively rapidly by adding some Cr(II) to a mixture of Cr(III) and the ligand.

A quantitative experiment on the Cr(II) catalyzed exchange of chloride between Cl^- and $CrCl^{2+}$ was performed in 1954. Taube and King (*322*) suggested that this reaction, governed by the rate law

$$k[Cr^{2+}][CrCl^{2+}][Cl^-], \qquad (217)$$

was the reverse of the Cr(II) catalyzed aquation of $CrCl_2^+$. Subsequent work has lent support to this suggestion (*151,139*); the catalysis of the direct substitution of $X^- = Cl^-$, Br^- for I^- in CrI^{2+} has also been postulated as occurring by this mechanism (*136*).

$$\frac{-d[CrI^{2+}]}{dt} = k[CrI^{2+}][Cr^{2+}][X^-] \qquad (218)$$

$$X^- + Cr^{2+} + CrI^{2+} \rightleftharpoons [XCrICr^{3+}]^\ddagger \rightleftharpoons XCrI^+ + Cr^{2+} \qquad (219)$$

$$XCrI^+ + Cr^{2+} \longrightarrow CrX^{2+} + I^- + Cr^{2+} \qquad (220)$$

The values of $k([ClO_4^-] = 2.0\ M, 25°, H^+ = 1.0\ M)$ are for Cl^- and Br^- 275 and 62 $M^{-2}\ sec^{-1}$ respectively. The rate law for $X^- = F^-$ is different, but only, presumably, because of a difference in dominant species, HF being a weak acid. Correcting for this, one can calculate a value of $5.3 \times 10^4\ M^{-2}\ sec^{-1}$ for the F^- case. This order of effectiveness of the halides is as expected when consideration is given to the nonbridging ligand effects on the rate of reduction.

Early studies of catalyzed aquations dealt with disubstituted complexes (323):

$$H^+ + cis\text{-}CrF_2^+ + Cr^{2+} = CrF^{2+} + Cr^{2+} + HF \tag{221}$$

Under conditions of high [Cr(II)] and low $[H^+]$, net aquation of CrX^{n+} species are also catalyzed by Cr(II) (324,325)

$$CrI^{2+} + Cr^{2+} = Cr^{2+} + Cr^{3+} + I^- \tag{222}$$

The general rate law for reactions of this type is

$$\frac{-d[CrX^{n+}]}{dt} = \frac{k[CrX^{n+}][Cr^{2+}]}{[H^+]} + k'[CrX^{n+}][Cr^{2+}] \tag{223}$$

The mechanism for the first term presumably involves OH^- as a bridge between the two metal-ion centers

$$CrX^{n+} \overset{K}{=} HOCrX^{(n-1)+} + H^+ \tag{224}$$

$$Cr^{2+} + HOCrX^{(n-1)+} \xrightarrow{k_b} [CrOCrX^{(n+1)+}]^{\ddagger} \tag{225}$$

The available data are summarized in Table 19. If the mechanisms of these catalyzed aquations are as illustrated in Eqs. (224) and (225), then proper comparison of k_b values involves determining kK^{-1}, where K is the hydrolysis constant of CrX^{n+}. Unfortunately, these values are known for only a few of the complexes, and the geometry is unknown in all cases. If corrections are made for hydrolysis constants in the case of Cr^{3+} (124,312) (pK = 4.3) and $CrCl^{2+}$ (136,329) (pK = 5.5 or 6.1), the latter reacts from 20 to 200 times faster. This would seem to be consistent with a cis-$Cr(H_2O)_5OHCl^+$ configuration, as has been argued by Birk and Espenson (328) to account for the reactivity of $CrCN^{2+}$. But such arguments are of only slight value at present because of the uncertain value of the pK for $CrCl^{2+}$. More importantly, knowledge of the pK does not specify which proton is removed. For acidity purposes, a cis proton may be removed whereas removal of a $trans$ proton, although thermodynamically less favorable, may lead to a more stable transition state.

TABLE 19

THE RATE OF THE Cr(II) CATALYZED AQUATION OF SOME Cr(III) COMPLEXES[a]

Complex	k^b (sec^{-1})	k'^b (M^{-1} sec^{-1})	$\Delta H^\ddagger$ (kcal mole^{-1})	$\Delta S^\ddagger$ (e.u.)	$I(M)$	Ref.
A. The [H$^+$] path						
Cr(H$_2$O)$_5$NH$_3^{3+}$	5.9×10^{-5}		21.6	-5.6	2.0	(326)
Cr(H$_2$O)$_6^{3+}$	1.1×10^{-4}		22	-2	1.0	(123, 124)
Cr(H$_2$O)$_5$F^{2+}	1.6×10^{-6c}				2.0	(325)
Cr(H$_2$O)$_5$Cl^{2+}	3.2×10^{-4}		20	-7	2.0	(325, 327)
Cr(H$_2$O)$_5$Br^{2+}	1.7×10^{-3}		18.2	-10	2.0	(325)
Cr(H$_2$O)$_5$I^{2+}	2.1×10^{-2}		17.0	-9.4	2.0	(136)
	2.7×10^{-2}		16.2	-11	1.0	(325)
Cr(H$_2$O)$_5$OOCCH$_3^{2+}$	2.5×10^{-5}		22.1	-5.5	1.0	(124)
Cr(H$_2$O)$_5$CN^{2+}	$\cong 1 \times 10^{-4c}$				1.0	(328)
B. The hydrogen ion independent path						
Cr(H$_2$O)$_5$NH$_3^{3+}$		2.4×10^{-5}	13.9	-33	2.0	(326)
Cr(H$_2$O)$_6^{3+}$		$<2 \times 10^{-5}$			1.0	(123, 124)
Cr(H$_2$O)$_5$OOCCH$_3^{2+}$		1.6×10^{-5}	26.7	$+9$	1.0	(124)
Cr(H$_2$O)$_5$F^{2+}		4.4×10^{-5d}			2.0	(325)

[a] $T = 25°$.
[b] The rate law is as given in Eq. (223).
[c] Estimated from the value of 8.0×10^{-6} sec^{-1} at 40° using $\Delta H^\ddagger = 21$ kcal mole^{-1}.
[d] Estimated from the value of 2.8×10^{-3} sec^{-1} at 55° using $\Delta H^\ddagger = 21$ kcal mole^{-1}.
[e] At 40°.

Some further understanding of rate trends in these systems can be delineated. Pennington and Haim report that the hydrogen ion independent path for catalyzed aquation of CrI^{2+}, see Eq. (223), of less than $10^{-4} M^{-1} sec^{-1}$ *(324)*; k/k' has a value of $>270 M$; k/k' for $CrNH_3^{2+}$ is 2.5 *(326)*. Deutsch and Taube *(124)* report $k/k' = 1.6 M$ for the catalyzed aquation of $Cr(CH_3COO)^{2+}$. In addition, the activation parameters for this H^+ independent path are quite different from those of the H^+ dependent path and those reported by Carlyle and Espenson for $Cr(NH_3)^{3+}$. These data can be understood when comparison is made with the reduction of *trans*-$Co(NH_3)_4(N_3)_2^+$ by Fe^{2+} *(146)*:

$$\frac{-d[Co(III)]}{dt} = \{k_A + k_B[H^+]\}[Co(III)][Fe(II)] \qquad (226)$$

Similar rate behavior is seen in the Cr^{2+} reduction of *trans*-$Co(NH_3)_4$-$(CH_3COO)_2^+$ *(330)*. The hydrogen-ion catalyzed path presumably represents protonation of the *trans*-ligand, a feature which makes coordination sphere rearrangements easier. Protonation of bound acetate (measured directly by Deutsch and Taube) and bridging by hydroxide would be expected to give rise to a hydrogen-ion independent path of low energy, and would account for the low k/k' ratio. A similar argument will hold for the k/k' ratio of 0.18 in the reaction of Cr^{2+} with CrF^{2+} *(325)*.

The catalysis of substitution on Cr(III) can also be carried out with other reducing agents; V^{2+} and Eu^{2+} catalyze the equation of $CrCl^{2+}$ by oxidation-reduction mechanism. For the former *(331)*, the rate law is

$$\frac{-d[CrCl^{2+}]}{dt} = k_1[CrCl^{2+}][V^{2+}] + k_2 \frac{[CrCl^{2+}][V^{2+}]}{[H^+]} \qquad (227)$$

where $k_1 = 3.8 \times 10^{-2} M^{-1} sec^{-1}$ and $k_2 = 2.5 \times 10^{-3} sec^{-1}$ at $25°$, $I = 1.0 M$. These two terms are explained on the basis of outer-sphere or water-bridged, and hydroxide-bridged transition states, respectively. The reactions of Eu^{2+} with CrF^{2+}, $CrCl^{2+}$, $CrBr^{2+}$, and CrI^{2+} have also been reported, but only at $[H^+] = 1.0 M$ *(332)*. For comparison with the V^{2+} rate constant, the value of k in

$$\frac{-d[CrCl^{2+}]}{dt} = k[Eu^{2+}][CrCl^{2+}] \qquad (228)$$

is $1.4 \times 10^{-3} M^{-1} sec^{-1}$ at $25°$; the more powerful reductant is slower by a factor of $\cong 30$. A similar greater reactivity of V^2 in the catalyzed aquation of $Cr(CH_3COO)^{2+}$, compared to the isotopic exchange reaction

$$*Cr^{2+} + Cr(CH_3COO)^{2+} = *Cr(CH_3COO)^{2+} + Cr^{2+} \qquad (229)$$

is found. Deutsch and Taube (*124*) have argued that the symmetry of the electron to be lost from the reductant may be an important factor in the comparison of V^{2+} and Cr^{2+} in these reactions. V(II) also catalyzes the aquation of both $CrNCS^{2+}$ and $CrSCN^{2+}$ (*114,172*). By the path independent of $[H^+]$, the latter reacts considerably faster, a result consistent with both the greater negative free energy change and the availability of a better bridge in the latter complex. Proof of an inner-sphere reaction pertains in the $V^{2+} + CrSCN^{2+}$ reaction (*114*), whereas an OH^- bridged path is suggested as a part of the $V^{2+} + CrNCS^{2+}$ reaction (*172*).

  c Other Metal Ion Systems. Studies of the catalysis of substitution reactions of other metal-ion systems are less numerous. Sutin and co-workers have examined the Fe^{2+} catalyzed aquation of $FeCl^{2+}$ (*187,208*) and $FeNCS^{2+}$ (*333*). In the former case the rate law for the stoichiometry

$$FeCl^{2+} = Fe^{3+} + Cl^- \tag{230}$$

is, assuming complete reaction to the right,

$$\frac{-d[FeCl^{2+}]}{dt} = \left(k_1 + \frac{k_2}{[H^+]} + k_3[Fe^2] + k_4\frac{[Fe^{2+}]}{[H^+]} \right)[FeCl^{2+}] \tag{231}$$

The values for k_3 and k_4 are 6.2 M^{-1} sec^{-1} and 14.8 sec^{-1} at 25°, $I = 3.0\ M$ ClO_4^- (*208*). Since other similar dipositive ions do not show any catalysis of aquation (*187*), it is assumed that paths k_3 and k_4 operate through an electron-transfer mechanism

$$^*FeCl^{2+} + Fe^{2+} \;\rightleftharpoons\; ^*Fe^{2+} + Fe^{3+} + Cl^- \tag{232}$$

It is to be noted that reaction (232) leads to exchange of iron between the two valence states and is, therefore, at least part of the path for isotopic exchange with a transition state of composition $[FeFeCl^{4+}]^{\ddagger}$. Another part of this path with the same composition arises from the inner-sphere chloride-bridged mechanism

$$^*FeCl^{2+} + Fe^{2+} \;\xrightarrow{\;k_5\;}\; ^*Fe^{2+} + FeCl^{2+} \tag{233}$$

Indeed, it has been shown that both paths contribute, the values for k_5 being 33 M^{-1} sec^{-1} at 25°, I = 3.0 M ClO_4^- (*187*). What is striking is the low value for the ratio k_5/k_4. A similar comparison in the case of the Cr^{2+}-$CrCl^{2+}$ system gives a value of about $10^5\ M^{-1}$ (*93,325*). Similar effects are illustrated in Table 11.

$Co(CN)_5^{3-}$ catalyzes the reaction of $Co(CN)_5X^{3-}$ with CN^- to form $Co(CN)_6^{3-}$. This reaction presumably takes place by the mechanism (334)

$$Co(CN)_5^{3-} + Co(CN)_5X^{3-} \longrightarrow [Co(CN)_5-NC-Co(CN)_4X^{6-}]^{\ddagger} \quad (234)$$

$$Co(CN)_5-NC-Co(CN)_4X^{6-}]^{\ddagger} \longrightarrow Co(CN)_5NC^{3-} + Co(CN)_4X^{3-} \quad (235)$$

$$\underline{Co(CN)_5NC^{3-} \longrightarrow Co(CN)_6^{3-}} \qquad\qquad\qquad (236)$$

$$Co(CN)_4X^{3-} + CN^- \longrightarrow Co(CN)_5^{3-} + X^- \quad (237)$$

At $25°$, $I = 0.5\ M$, the rate constants are the order of $10^{-2}\ M^{-1}\ sec^{-1}$ for several X groups. It was suggested that charge effects are important in determining the rate constant.

Several other catalytic substitution reactions are known, but these have not been investigated in depth. V^{2+} catalyzes the loss of N_3^- from VN_3^{2+} (68), and acetate from $V(OOCCH_3)^{2+}$ (124). $Ru(NH_3)_5OH_2^{2+}$ is known to catalyze equations of $Ru(III)$ complexes ($335,125,49$) and it is presumably $Rh(I)$ that catalyzes the substitution process (336)

$$Rh(py)_4Cl_2^+ + Br^- = Rh(py)_4BrCl^+ + Cl^- \quad (238)$$

as well as the depolymerization of commercial "$RhCl_3 \cdot 3H_2O$" to $Rh(H_2O)_4Cl_2^+$ (337).

3. Catalyzed Isomerization Reactions

The use of inner-sphere oxidation reduction reactions to generate linkage isomers has been a productive area of research (131). In some cases, the resulting linkage isomer is unstable with respect to a catalyzed isomerization. Two examples are the isomerizations of $CrNC^{2+}$ and $CrSCN^{2+}$. The former ion, formed from reduction of $Co(NH_3)_5CN^{2+}$ by Cr^{2+}, isomerizes according to the rate law (153)

$$\frac{-d[CrNC^{2+}]}{dt} = (k_1 + k_2[Cr^{2+}])\left(\frac{K}{[H^+] + K}\right)[CrNC^{2+}] \quad (239)$$

where at $25°$, $I = 1.0\ M\ ClO_4^-$, $k_1 = 0.049\ sec^{-1}$, $k_2 = 1.60\ M^{-1}\ sec^{-1}$ and $K = 1.06\ M$. The value of k_2 should be compared to k_3 for the reaction

$$*Cr^{2+} + CrCN^{2+} \xrightarrow{k_3} *CrCN^{2+} + Cr^{2+} \quad (240)$$

where $k_3 = 7.7 \times 10^{-2}\ M^{-1}\ sec^{-1}$ at $25°$ and $I = 1.0\ M\ ClO_4^-$ (338). Consideration of the Cr^{2+}–$CrSCN^{2+}$ system yields $40\ M^{-1}\ sec^{-1}$ ($149,306$), and $1.4 \times 10^{-4}\ M^{-1}\ sec^{-1}$ (93), for k_2 and k_3 respectively. The ratios $k_2 k_3^{-1}$ reflect the driving force of the reaction that leads to isomerization, and, perhaps, the energetic advantage of a transition state in which a bond to the "right end" of the ligand is partially formed and one to the "wrong end" is partially broken. Haim and Sutin have presented an interesting argument concerning the efficiency of NCS⁻ as a "symmetrized" bridge compared to N_3^-. They conclude that the latter is superior (316).

A brief report on a catalyzed geometrical isomerization has appeared. The cis to trans isomerization of $Co(en)_2Cl_2^+$ in methanol has been reported to be catalyzed by Co(II) and en (339). No details are available.

ACKNOWLEDGMENTS

I wish to thank J. C. Sullivan for several helpful comments and T. W. Newton for enlightening correspondence; and I am grateful for several long and inspiring discussions with Jason.

REFERENCES

1. H. J. H. Fenton, J. Chem. Soc., 65, 899 (1894).
2. H. M. S. Marshall, Chem. News, 83, 76 (1901).
3. J. Smidt, W. Hafner, R. Jira, R. Sieber, J. Sedlmeier, and A. Sobel, Angew. Chemie Intl. Ed., 1, 80 (1962).
4. A. Aguilo, Advances in Organometallic Chemistry, 5, 321 (1967).
5. Oxidases and Related Redox Systems (T. E. King, H. S. Mason, and M. Morrison, eds.), Wiley, New York, Vol. 1 and 2, 1965.
6. The Biochemistry of Copper (J. Peisach, P. Aisen, and W. E. Blumberg, eds.), Academic, New York, 1966.
7. H. Taube, Advan. Inorg. Chem. and Radiochem., 1, 1 (1959).
8. J. Halpern, Quart. Rev. (London), 15, 207 (1961).
9. H. Taube, Proc. Robert. A. Welch Foundation Conf. on Chemical Research, VI, Houston, Texas, 1962, p. 7.
10. N. Sutin, Ann. Rev. Nucl. Sci., 12, 285 (1962).
11. N. Sutin, Ann. Rev. Phys. Chem., 17, 119 (1966).
12. F. Basolo and R. G. Pearson, Mechanisms of Inorganic Reactions, 2nd ed., Wiley, New York, 1967.
13. A. G. Sykes, Advan. Inorg. Chem. and Radiochem., 10, 153 (1967).
14. N. Sutin, Accounts of Chem. Res., 1, 225 (1968).
15. M. Matusek, F. Jaros, and A. Tockstein, Chem. Listy., 63, (a) 188, (b) 317, (c) 435, (1969).
16. J. H. Baxendale, Rad. Res. Supplement, 4, 114 (1964).
17. M. S. Matheson, W. A. Mulac, and J. Rabini, J. Phys. Chem., 67, 2613 (1963).
18. W. L. Waltz, A. W. Adamson, and Paul D. Fleischauer, J. Am. Chem. Soc., 89, 3923 (1967).

19. J. W. Gryder and M. K. Dorfman, *J. Am. Chem. Soc.*, **83**, 1254 (1961).
20. M. K. Dorfman and J. W. Gryder, *Inorg. Chem.*, **3**, 799 (1962).
21. J. C. Sullivan, *J. Am. Chem. Soc.*, **87**, 1495 (1965).
22. J. Silverman and R. W. Dodson, *J. Phys. Chem.*, **56**, 846 (1952).
23. J. C. Hindman, J. C. Sullivan, and D. Cohen, *J. Am. Chem. Soc.*, **81**, 2316 (1959) and references therein.
24. T. W. Newton and F. B. Baker, *Advances in Chemistry Series*, No. 71, 1967, p. 268 and references therein.
25. A. Zwickel and H. Taube, *J. Am. Chem. Soc.*, **73**, 793 (1961).
26. H. Taube, *Chem. Rev.*, **50**, 69 (1952).
27. M. Eigen, *Bunsenges physik. Chem.*, **67**, 753 (1963) and references therein.
28. M. Eigen and R. G. Wilkins, *Advances in Chemistry Series*, No. 49, 55 (1965).
29. M. Eigen and G. Maass, *Z. Physik. Chem.*, **49**, 163 (1966).
30. M. V. Olson, Y. Kanazawa, and H. Taube, *J. Chem. Phys.*, **51**, 289 (1969).
31. C. W. Merideth and R. E. Connick, *Abstracts, 149th Am. Chem. Soc. Meeting, Detroit, Mich., 1965*, Paper 106M.
32. T. J. Swift and R. E. Connick, *J. Chem. Phys.*, **37**, 307 (1962); *J. Chem. Phys.* **41**, 2553 (1964).
33. M. Eigen and K. Tamm, *Z. Elektrochem.*, **66**, 107 (1962).
34. A. M. Chmelnick and D. Fiat. *J. Chem. Phys.*, **47**, 3986 (1967).
35. D. B. Rorabacher, *Inorg. Chem.*, **5**, 1891 (1966).
36. D. Fiat and R. E. Connick, *J. Am. Chem. Soc.*, **90**, 608 (1968).
37. A. M. Chmelnick and D. Fiat, quoted in reference (*36*).
38. B. R. Baker, N. Sutin, and T. J. Welch, *Inorg. Chem.*, **6**, 1948 (1967).
39. W. Kruse and D. Thusius, *Inorg. Chem.*, **7**, 464 (1968).
40. J. P. Hunt and R. A. Plane, *J. Am. Chem. Soc.*, **76**, 5960 (1954).
41. M. A. Levine, T. P. Jones, W. E. Harris, and W. J. Wallace, *J. Am. Chem. Soc.*, **83**, 2453 (1961).
42. C. Postmus and E. L. King, *J. Phys. Chem.*, **59**, 1216 (1955).
43. G. K. Pagenkopf and D. W. Margerum, *Inorg. Chem.*, **7**, 2514 (1968).
44. J. F. Below, R. E. Connick, and C. P. Coppel, *J. Am. Chem. Soc.*, **80**, 2961 (1958).
45. A. W. Adamson and F. Basolo, *Acta. Chem. Scand.*, **9**, 1261 (1955).
46. J. Reuben and D. Fiat, *Inorg. Chem.*, **6**, 579 (1967).
47. K. Wüthrich and R. Connick, *Inorg. Chem.*, **6**, 583 (1967).
48. J. H. Swinehart and G. W. Castellan, *Inorg. Chem.*, **3**, 278 (1964).
49. W. G. Movius and R. G. Linck, *J. Am. Chem. Soc.*, **92**, 2677, (1970).
50. J. A. Broomhead, F. Basolo, and R. G. Pearson, *Inorg. Chem.*, **3**, 826 (1964).
51. H. L. Bott, E. J. Bounsall, and A. J. Pöe, *J. Chem. Soc.*, A, **1966**, 1275.
52. A. A. El-Awady, E. J. Bounsall, and C. S. Garner, *Inorg. Chem.*, **6**, 79 (1967).
53. R. A. Reinhardt and R. K. Sparkes, *Inorg. Chem.*, **6**, 2190 (1967).
54. H. B. Gray, *J. Am. Chem. Soc.*, **84**, 1548 (1962).
55. G. Geier, *Bunsenges Physik. Chem.*, **69**, 617 (1965) and references therein.
56. H. B. Silber, R. D. Farina, and J. H. Swinehart, *Inorg. Chem.*, **8**, 819 (1969).
57. P. Hurwitz and K. Kustin, *J. Phys. Chem.*, **71**, 324 (1967).
58. A. L. Companion, *J. Phys. Chem.*, **73**, 739 (1969) and references therein.
59. C. H. Langford and H. B. Gray, *Ligand Substitution Processes*, Benjamin, New York, 1965.
60. C. H. Langford and T. R. Stengle, *Ann. Rev. Phys. Chem.*, **19**, 193 (1968).

61. C. K. Ingold, R. S. Nyholm, and M. L. Tobe, *J. Chem. Soc.*, **1956**, 1691.

62. S. Asperger and C. K. Ingold, *J. Chem. Soc.*, **1956**, 2862.

63. M. E. Baldwin, S. C. Chan, and M. L. Tobe, *J. Chem. Soc.*, **1961**, 4637.

64. D. A. Buckingham, I. I. Olsen, and A. M. Sargeson, *J. Am. Chem. Soc.*, **88**, 5443 (1966).

65. H. Taube, H. Myers, and R. L. Rich, *J. Am. Chem. Soc.*, **75**, 4118 (1953).

66. T. W. Newton and F. B. Baker, *J. Phys. Chem.*, **68**, 228 (1964).

67. T. W. Newton and F. B. Baker, *Inorg. Chem.*, **3**, 569 (1964).

68. J. H. Espenson, *J. Am. Chem. Soc.*, **89**, 1276 (1967).

69. H. J. Price and H. Taube, *Inorg. Chem.*, **7**, 1 (1968).

70. J. P. Candlin, J. Halpern, and D. L. Trimm, *J. Am. Chem. Soc.*, **86**, 1019 (1964).

71. K. Hicks and R. G. Linck, unpublished observations.

72. J. M. Malin and J. H. Swinehart, *Inorg. Chem,*, **7**, 250 (1968). The $\Delta S^{\ddagger}$ is a result of a private communication from J. H. Swinehart.

73. K. M. Davies and J. H. Espenson, *J. Am. Chem. Soc.*, **91**, 3093 (1969).

74. J. Halpern, quoted in reference 75.

75. M. Green, K. Schug, and H. Taube, *Inorg. Chem.*, **4**, 1184 (1965).

76. M. Orhanović and N. Sutin, *J. Am. Chem. Soc.*, **90**, 4286 (1968)

77. J. Halpern and L. E. Orgel, *Disc. Faraday Soc.*, No. 29, 32 (1960).

78. T. J. Williams and J. P. Hunt, *J. Am. Chem. Soc.*, **90**, 7213 (1968).

79. T. J. Conocchioli, G. H. Nancollas, and N. Sutin, *J. Am. Chem. Soc.*, **86**, 1453 (1964).

80. J. P. Candlin, J. Halpern, and S. Nakamura, *J. Am. Chem. Soc.*, **85**, 2517 (1963).

81. D. H. Huchital and R. G. Wilkins, *Inorg. Chem.*, **6**, 1022 (1967) and references therein.

82. D. Seewald and N. Sutin, Paper M005, 155th ACS Meeting, San Francisco, April, 1968.

83. R. N. F. Thorneley and A. G. Sykes, *Chem. Commun.*, **1969**, 331.

84. W. G. Movius and R. G. Linck, *J. Am. Chem. Soc.*, **91**, 5394 (1969).

85. A. Haim and N. Sutin, *J. Am. Chem. Soc.*, **88**, 5343 (1966).

86. J. F. Endicott and H. Taube, *J. Am. Chem. Soc.*, **86**, 1686 (1964).

87. J. P. Candlin and J. Halpern, *Inorg. Chem.*, **4**, 766 (1965).

88. H. Diebler and H. Taube, *Inorg. Chem.*, **4**, 1029 (1965).

89. J. Halpern and J. Rabani, *J. Am. Chem. Soc.*, **88**, 699 (1966).

90. O. J. Parker and J. H. Espenson, *J. Am. Chem. Soc.*, **91**, 1968 (1968).

91. A. E. Ogard and H. Taube, *J. Am. Chem. Soc.*, **80**, 1084 (1958).

92. H. Ogino and N. Tanaka, *Bull. Chem. Soc. Japan*, **41**, 2411 (1968).

93. D. L. Ball and E. L. King, *J. Am. Chem. Soc.*, **80**, 1091 (1958).

94. W. R. Mason, III, and R. C. Johnson, *Inorg. Chem.*, **4**, 1258 (1965).

95. F. Basolo, M. L. Morris, and R. G. Pearson, *Disc. Faraday Soc.*, No. 29, 80 (1960).

96. R. R. Rettew and R. C. Johnson, *Inorg. Chem.*, **4**, 1565 (1965).

97. D. W. Carlyle and J. H. Espenson, *J. Am. Chem. Soc.*, **91**, 599 (1969).

98. J. H. Espenson, *Inorg. Chem.*, **4**, 121 (1965).

99. J. H. Espenson and R. T. Wang, private communication.

100. J. Hudis and A. C. Wahl, *J. Am. Chem. Soc.*, **75**, 4153 (1953).

101. R. A. Horne, Ph.D. Thesis, Columbia University, New York, 1955.

102. N. Sutin, J. K. Rowley, and R. W. Dodson, *J. Phys. Chem.*, **65**, 1248 (1961).

103. A. Haim, *Inorg. Chem.*, **7**, 1475 (1968).

104. C. H. Langford, *Inorg. Chem.*, **4**, 265 (1965) and references therein.

105. D. A. Buckingham, I. I. Olsen, A. M. Sargeson, and H. Satrapa, *Inorg. Chem.*, **6**, 1027 (1967).

106, R. C. Patel and J. F. Endicott, *J. Am. Chem. Soc.*, **90**, 6364 (1968).

107. S. C. Chan, *J. Chem. Soc.*, **1964**, 2375.

108. R. G. Yalman, *Inorg. Chem.*, **1**, 16 (1962); there is an apparent typographical error in Table III of this reference.

109. P. R. Guenther and R. G. Linck, *J. Am. Chem. Soc.*, **91**, 3769 (1969).

110. P. H. Dodel and H. Taube, *Zeit. f. Phys. Chemie.*, **44**, 92 (1965).

111. A. Zwickel and H. Taube, *J. Am. Chem. Soc.*, **81**, 1288 (1959).

112. J. Doyle and A. G. Sykes, *J. Chem. Soc.*, *A*, **1968**, 2836.

113. A. Zwickel and H. Taube, *Disc. Faraday Soc.*, N7. 29, 42 (1960).

114. B. R. Baker, M. Orhanović, and N. Sutin, *J. Am. Chem. Soc.*, **89**, 722 (1967).

115. S. Fukushima and W. L. Reynolds, *Talanta*, **11**, 283 (1964).

116. O. J. Parker and J. H. Espenson, *Inorg. Chem.*, **8**, 1523 (1969).

117. D. W. Carlyle and J. H. Espenson, *J. Am. Chem. Soc.*, **90**, 2272 (1968).

118. T. W. Newton and N. A. Daugherty, *J. Phys. Chem.*, **71**, 3768 (1967).

119. R. M. Milburn, *J. Am. Chem. Soc.*, **79**, 537 (1957).

120. T. J. Meyer and H. Taube, *Inorg. Chem.*, **7**, 2369 (1968).

121. O. J. Parker and J. H. Espenson, *J. Am. Chem. Soc.*, **91**, 1313 (1969).

122. K. Shaw and J. H. Espenson, *Inorg. Chem.*, **7**, 1619 (1968).

123. A. Anderson and N. A. Bonner, *J. Am. Chem. Soc.*, **76**, 3826 (1954).

124. E. Deutsch and H. Taube, *Inorg. Chem.*, **7**, 1532 (1968).

125. J. A. Stritar and H. Taube, *Inorg. Chem.*, **8**, 2281 (1969).

126. M. J. Burkhart and T. W. Newton, *J. Phys. Chem.*, **73**, 1741 (1969).

127. R. C. Thompson and J. C. Sullivan, *J. Am. Chem. Soc.*, **89**, 1096 (1967).

128. J. C. Sullivan and J. C. Hindman, *J. Phys. Chem.*, **63**, 1332 (1959).

129. R. Farina and R. G. Wilkins, *Inorg. Chem.*, **7**, 514 (1968).

130. H. S. Habib and J. P. Hunt, *J. Am. Chem. Soc.*, **88**, 1668 (1966).

131. J. L. Burmeister, *Coord. Chem. Rev.*, **3**, 225 (1968) and references therein.

132. G. S. Laurence, *Trans. Faraday Soc.*, **53**, 1326 (1957).

133. D. Bunn, F. S. Dainton, and S. Duckworth, *Trans. Faraday Soc.*, **57**, 1131 (1961).

134. R. Snellgrove and E. L. King, *Inorg. Chem.*, **3**, 288 (1964).

135. P. V. Manning and R. C. Jarnagin, *J. Phys. Chem.*, **67**, 2884 (1963).

136. D. E. Pennington and A. Haim, *Inorg. Chem.*, **6**, 2138 (1967).

137. H. Taube, *J. Am. Chem. Soc.*, **77**, 4481 (1955).

138. J. B. Hunt and J. E. Earley, *J. Am. Chem. Soc.*, **82**, 5312 (1960).

139. D. E. Pennington and A. Haim, *Inorg. Chem.*, **7**, 1659 (1968).

140. R. D. Cannon and J. E. Earley, *J. Chem. Soc.*, *A*, **1968**, 1102.

141. C. Bifano and R. G. Linck, *J. Am. Chem. Soc.*, **89**, 3945 (1967).

142. R. G. Linck, *Inorg. Chem.*, **7**, 2394 (1968).

143. J. P. Candlin and J. Halpern, *Inorg. Chem.*, **4**, 1086 (1965).

144. K. R. Brower, *J. Am. Chem. Soc.*, **90**, 5401 (1968).

145. W. H. Jolley and D. R. Stranks, private communication.

146. A. Haim, *J. Am. Chem. Soc.*, **85**, 1016 (1963).

147. P. Benson and A. Haim, *J. Am. Chem. Soc.*, **87**, 3826 (1965).

148. C. Bifano and R. G. Linck, unpublished observations.

149. A. Haim and N. Sutin, *J. Am. Chem. Soc.*, **88**, 434 (1966).

150 D. E. Pennington and A. Haim, *Inorg. Chem.*, **5**, 1887 (1966).

151. J. H. Espenson and S. G. Slocum, *Inorg. Chem.*, **6**, 906 (1967).

152. M. J. DeChant and J. B. Hunt, *J. Am. Chem. Soc.*, **89**, 5988 (1967); *J. Am. Chem. Soc.*, **90**, 3695 (1968).

153. J. P. Birk and J. H, Espenson, *J. Am. Chem. Soc.*, **90**, 1153 (1968).

154. R. D. Cannon and J. E. Earley, *J. Am. Chem. Soc.*, **87**, 5264 (1965); *J. Am. Chem. Soc.*, **88**, 1872 (1966).

155. A. Haim, *J. Am. Chem. Soc.*, **86**, 2352 (1964).

156. R. G. Linck, unpublished observations.

157. R. A. Marcus, *Ann. Rev. Phys. Chem.*, **15**, 155 (1964) and references therein

158. R. A. Marcus, *J. Chem. Phys.*, **43**, 679 (1965) and references therein.

159. N. S. Hush, *Trans. Faraday Soc.*, **59**, 396 (1963).

160. N. S. Hush, *Progress in Inorganic Chemistry*, Vol. 8 (F. A. Cotton, ed.) Interscience, New York, 1967, p. 391.

161. W. L. Reynolds and R. W. Lumry, *Mechanisms of Electron Transfer*, Ronald Press, New York, 1966.

162. W. F. Libby, *J. Phys. Chem.*, **56**, 863 (1952).

163. R. A. Marcus, *J. Chem. Phys.*, **24**, 966 (1956).

164. P. G. Rasmussen and C. H. Brubaker, Jr., *Inorg. Chem.*, **3**, 977 (1964).

165. T. W. Newton, *J. Chem. Ed.*, **45**, 571 (1968).

166. I. Ruff, *Quart. Rev.*, **22**, 199 (1968) and references therein.

167. J. F. Endicott, *J. Phys. Chem.*, **73**, 2594 (1969).

168. T. W. Newton and S. W. Rabideau, *J. Phys. Chem.*, **63**, 365 (1959).

169. W. M. Latimer, *The Oxidation States of the Elements and Their Potentials in Aqueous Solution*, Prentice-Hall, Englewood Cliffs, N.J., 1952.

170. J. M. Malin and J. H. Swinehart, *Inorg. Chem.*, **8**, 1407 (1969).

171. N. A. Daugherty and T. W. Newton, *J. Phys. Chem.*, **68**, 612 (1964).

172. M. Orhanovic, H. N. Po, and N. Sutin, *J. Am. Chem. Soc.*, **90**, 7224 (1968).

173. S. C. Furman and C. S. Garner, *J. Am. Chem. Soc.*, **74**, 2333 (1952).

174. J. H. Espenson. *Inorg. Chem.*, **4**, 1025 (1965).

175. K. Shaw and J. H. Espenson, *J. Am. Chem. Soc.*, **90**, 6622 (1968).

176. G. L. Bertrand, G. W. Stapleton, C. A. Wulff, and L. G. Hepler, *Inorg. Chem.*, **5**, 1283 (1966).

177. M. J. LaSalle and J. W. Cobble, *J. Phys. Chem.*, **48**, 519 (1955).

178. R. Powell and W. M. Latimer, *J. Chem. Phys.*, **19**, 1139 (1951).

179. C. Postmus and E. L. King, *J. Phys. Chem.*, **59**, 1208 (1955).

180. R. H. Betts and F. S. Dainton, *J. Am. Chem. Soc.*, **75**, 5721 (1953).

181. J. W. Cobble, *J. Chem. Phys.*, **21**, 1446 (1953).

182. R. L. Montgomery, U.S. Bureau of Mines Report of Investigations, 5468, 1959.

183. G. N. Lewis and M. Randall, revised by K. S. Pitzer and L. Brewer, *Thermodynamics*, 2nd ed., McGraw-Hill, New York, 1961, pp. 522–524.

184. T. W. Newton and F. B. Baker, *J. Phys. Chem.*, **67**, 1425 (1963).

185. F. A. L. Anet and C. Leblanc, *J. Am. Chem. Soc.*, **79**, 2649 (1957).

186. J. K. Kochi and D. M. Singleton, *J. Am. Chem. Soc.*, **90**, 1582 (1968) and references therein.

187. R. J. Campion, T. J. Conocchioli, and N. Sutin, *J. Am. Chem. Soc.*, **86**, 4591 (1964).

188. J. Halpern, *Advances in Chem. Series*, **49**, 120 (1965).

189. H. Diebler, P. H. Dodel, and H. Taube, *Inorg. Chem.*, **5**, 1688 (1966).
190. G. Dulz and N. Sutin, *Inorg. Chem.*, **2**, 917 (1963).
191. H. Diebler and N. Sutin, *J. Phys. Chem.*, **68**, 174 (1964).
192. R. J. Campion, N. Purdie, and N. Sutin, *Inorg. Chem.*, **3**, 1091 (1964).
193. J. D. Miller and R. H. Price, *J. Chem. Soc.*, *A*, **1966**, 1048.
194. J. D. Miller and R. H. Price, *J. Chem. Soc.*, *A*, **1966**, 1370.
195. J. Burgess, *J. Chem. Soc.*, *A*, **1968**, 3123.
196. D. W. Larson and A. C. Wahl, *J. Chem. Phys.*, **43**, 3765 (1965).
197. R. G. Wilkins and R. E. Yelin, *Inorg. Chem.*, **7**, 2667 (1968).
198. See footnote 27 of reference 129.
199. W. L. Waltz and R. G. Pearson, *J. Phys. Chem.*, **73**, 1941 (1969).
200. B. R. Baker, F. Basolo, and H. M. Neumann, *J. Phys. Chem.*, **63**, 371 (1959).
201. R. L. Burwell, Jr., and R. G. Pearson, *J. Phys. Chem.*, **70**, 300 (1966).
202. K. V. Krishnamurty and A. C. Wahl, *J. Am. Chem. Soc.*, **80**, 5921 (1958).
203. D. J. Meier and C. S. Garner, *J. Phys. Chem.*, **46**, 853 (1952).
204. A. Adin and A. G. Sykes, *J. Chem. Soc.*, *A*, **1966**, 1230.
205. Y. Wang and E. S. Gould, *J. Am. Chem. Soc.*, **91**, 4998 (1969).
206. J. Halpern and S. Nakamura, *J. Am. Chem. Soc.*, **87**, 3002 (1965).
207. G. Dulz and N. Sutin, *J. Am. Chem. Soc.*, **86**, 829 (1964).
208. E. G. Moorhead and N. Sutin, *Inorg. Chem.*, **6**, 428 (1967).
209. T. W. Newton, G. E. McCrary, and W. G. Clark, *J. Phys. Chem.*, **72**, 4333 (1968).
210. N. Sutin, *Exchange Reactions*, International Atomic Energy Agency, Vienna, 1965, p. 7.
211. R. A. Marcus, *J. Phys. Chem.*, **72**, 891 (1968).
212. L. E. Orgel, *Report of the Tenth Solvay Conference, Brussels, 1965*, p. 289.
213. P. W. Schneider, P. F. Phelan, and J. Halpern, *J. Am. Chem. Soc.*, **91**, 77 (1969).
214. J. Halpern, *Ann. Rev. Phys. Chem.*, **16**, 103 (1965).
215. J. P. Collman, *Accounts of Chem. Res.*, **1**, 136 (1968), and references therein.
216. R. Cramer, *Accounts of Chem. Res.*, **1**, 186 (1968).
217. L. Vaska, *Accounts of Chem. Res.*, **1**, 335 (1968).
218. W. C. E. Higginson and A. G. Sykes, *J. Chem. Soc.*, **1962**, 2841.
219. J. H. Espenson, K. Shaw, and O. J. Parker, *J. Am. Chem. Soc.*, **89**, 5730 (1967).
220. M. Calvin, *Trans. Faraday Soc.*, **34**, 1181 (1938).
221. M. Calvin, *J. Am. Chem. Soc.*, **61**, 2230 (1939).
222. J. Halpern, *Quart. Rev.*, **10**, 463 (1956).
223. J. Halpern, *Advances in Catalysis*, **11**, 301 (1959).
224. J. Halpern, *Advances in Chemistry Series*, **70**, 1 (1968).
225. J. P. Collman and W. R. Roper, *Adv. in Organometallic Chem.*, **7**, 53 (1968).
226. J. M. Pratt and R. J. P. Williams, *J. Chem. Soc.*, *A*, **1968**, 1291.
227. J. J. Alexander and H. B. Gray, *J. Am. Chem. Soc.*, **89**, 3356 (1967).
228. K. G. Caulton, *Inorg. Chem.*, **7**, 392 (1968).
229. J. P. Maher, *J. Chem. Soc.*, *A*, **1968**, 2918.
230. M. Iguchi, *J. Chem. Soc. Japan*, **63**, 634 (1942); *Chem. Abstracts*, **41**, 2975 (1947).
231. B. deVries, *J. Catalysis*, **1**, 489 (1962).
232. R. G. S. Banks and J. M. Pratt, *J. Chem. Soc.*, *A*, **1968**, 854 and references therein.
233. J. Kwiatek, I. L. Mador, and J. K. Seyler, *J. Am. Chem. Soc.*, **84**, 304 (1962).
234. J. Kwiatek, *Catalysis Rev.*, **1**, 37 (1967).
235. J. Kwiatek and J. K. Seyler, *J. Organometalic Chem.*, **3**, 421 (1965).

236. L. Sinándi and F. Nagy, *Acta Chim. Acad. Sci. Hung.*, **46**, 137 (1965).
237. L. I. Sinándi, F. Nagy, and E. Budo, *Acta Chim. Acad. Sci. Hung.*, **58**, 39 (1968)
238. L. Sinándi and F. Nagy, *Acta. Chim. Acad. Sci. Hung.*, **46**, 101 (1965).
239. M. G. Burnett, P. J. Connolly, and C. Kemball, *J. Chem. Soc.*, *A*, **1967**, 800.
240. J. Halpern and J. P. Maher, *J. Am. Chem. Soc.*, (a) **86**, 2131 (1964); (b) *J. Am. Chem. Soc.*, **87**, 5361 (1965).
241. P. B. Chock, R. B. K. Dewar, J. Halpern, and L-Y. Wong, *J. Am. Chem. Soc.*, **91**, 82 (1969).
242. P. B. Chock and J. Halpern, *J. Am. Chem. Soc.*, **91**, 582 (1969).
243. J. Kwiatek and J. K. Seyler, *Advances in Chem. Ser.*, **70**, 207 (1968).
244. J. Halpern and N. P. Johnson, as quoted in reference (*242*).
245. L. M. Jackman, J. A. Hamilton, and J. M. Lawlor, *J. Am. Chem. Soc.*, **90**, 1914 (1968).
246. J. Halpern and L-Y. Wong, *J. Am. Chem. Soc.*, **90**, 6665 (1968).
247. M. G. Burnett, P. J. Connolly, and C. Kemball, *J. Chem. Soc.*, *A*, **1968**, 991.
248. S. W. Benson, *The Foundations of Chemical Kinetics*, McGraw-Hill, 1960, p. 654.
249. R. G. Dakers and J. Halpern, *Can. J. Chem.*, **32**, 969 (1954).
250. E. Peters and J. Halpern, *J. Phys. Chem.*, **59**, 793 (1955).
251. J. Halpern, F. R. MacGregor, and E. Peters, *J. Phys. Chem.*, **60**, 1455 (1956).
252. W. J. Dunning and P. E. Potter, *Proc. Chem. Soc.*, **1960**, 244.
253. E. A. von Hahn and E. Peters, *J. Phys. Chem.*, **69**, 547 (1965).
254. A. H. Webster and J. Halpern, *J. Phys. Chem.*, **61**, 1239 (1957).
255. J. F. Harrod, S. Ciccone, and J. Halpern, *Can. J. Chem.*, **39**, 1372 (1961).
256. J. F. Harrod and J. Halpern, *Can. J. Chem.*, **37**, 1933 (1959).
257. J. Halpern, J. F. Harrod, and P. E. Potter, *Can. J. Chem.*, **37**, 1446 (1959).
258. B. R. James and G. L. Rempel, *Can. J. Chem.*, **44**, 233 (1966).
259. B. Saville, *Angew. Chem. Intl. Ed.*, **6**, 928 (1967).
260. W. K. Wilmarth and A. Haim, in *Peroxide Reaction Mechanisms* (J. O. Edwards, ed.), Interscience, New York, 1962, p. 175.
261. D. A. House, *Chem. Rev.*, **62**, 185 (1962).
262. I. M. Kolthoff and I. K. Miller, *J. Am. Chem. Soc.*, **73**, 3055 (1951).
263. M. M. Breuer and A. D. Jenkins, *Trans. Faraday Soc.*, **59**, 1310 (1963).
264. P. D. Bartlett and J. D. Cotman, Jr., *J. Am. Chem. Soc.*, **71**, 1419 (1949).
265. S. Froneaus and C. O. Östman, *Acta Chem. Scand.*, **9**, 902 (1955).
266. S. Froneaus and C. O. Östman, *Acta Chem. Scand.*, **22**, 2827 (1968).
267. E. Heckel, A. Henglein, and G. Beck, *Ber. Bunsenges physik. Chem.*, **70**, 149 (1966).
268. E. Hayon and J. J. McGarvey, *J. Phys. Chem.*, **71**, 1472 (1967).
269. L. Dogliotti and E. Hayon, *J. Phys. Chem.*, **71**, 2511 (1967).
270. L. Dogliotti and E. Hayon, *J. Phys. Chem.*, **71**, 3802 (1967).
271. R. Devonshire and J. J. Weiss, *J. Phys. Chem.*, **72**, 3815 (1968).
272. W. Roebke, M. Renz, and A. Henglein, *Int. J. Radiat. Phys. Chem.*, **1**, 39 (1969)
273. D. G. Marketos, *Zeit. f. Physik. Chemie*, **65**, 306 (1969).
274. D. E. Pennington and A. Haim, *J. Am, Chem. Soc.*, **90**, 3700 (1968).
275. T. J. Sworski, *J. Phys. Chem.*, **57**, 2858 (1963) and references therein.
276. Quoted in M. Anbar and P. Neta, *Int. J. Appl. Rad. and Isotopes*, **18**, 493 (1967).
277. E. Hayon, *Trans. Faraday Soc.*, **61**, 723 (1965).
278. R. Woods, I. M. Kolthoff, and E. J. Meehan, *J. Am. Chem. Soc.*, **85**, 2385 (1963).
279. R. Woods, I. M. Kolthoff, and E. J. Meehan, *J. Am. Chem. Soc.*, **85**, 3334 (1963).

280. R. Woods, I. M. Kolthoff, and E. J. Meehan, *J. Am. Chem. Soc.*, **86**, 1698 (1964).
281. R. Woods, I. M. Kolthoff, and E. J. Meehan, *Inorg. Chem.*, **4**, 697 (1965).
282. V. A. Lunenok-Burmakina and A. P. Potemskaya, *Ukr. Khim. Zh.*, **30**, 1262 (1964). *Chem. Abstracts*, **62**, 8655h (1965).
283. H. G. S. Sengar and Y. K. Gupta, *J. Indian Chem. Soc.*, **43**, 223 (1966).
284. H. G. S. Sengar and Y. K. Gupta, *J. Indian Chem. Soc.*, **44**, 769 (1967).
285. H. N. Po, J. H. Swinehart, and T. L. Allen, *Inorg. Chem.*, **7**, 244 (1968) and references therein.
286. M. C. Agrawal, R. K. Shingal, and S. P. Mushran, *Z. Physik. Chemie*, **62**, 112 (1968).
287. I. M. Kolthoff and R. Woods, *J. Am. Chem. Soc.*, **88**, 1371 (1966).
288. R. W. Chlëbek and M. W. Lister, *Can. J. Chem.*, **45**, 2411 (1967).
289. J. Burgess and R. H. Prince, *J. Chem. Soc.*, *A*, **1966**, 1772.
290. D. H. Irvine, *J. Chem. Soc.*, **1958**, 2166
291. D. H. Irvine, *J. Chem. Soc.*, **1959**, 2977.
292. L. Gjertsen and A. C. Wahl, *J. Am. Chem. Soc.*, **81**, 1572 (1959); J. C. Shepard and A. C. Wahl, *J. Am. Chem. Soc.*, **79**, 1020 (1957).
293. A. A. Green, J. O. Edwards, and P. Jones, *Inorg. Chem.*, **5**, 1858 (1966).
294. B. M. Gordon and A. C. Wahl, *J. Am. Chem. Soc.*, **80**, 273 (1958).
295. D. H. Huchital, N. Sutin, and B. Warnqvist, *Inorg. Chem.*, **6**, 838 (1967).
296. D. D. Thusius and H. Taube, *J. Phys. Chem.*, **71**, 3845 (1967).
297. T. J. O'Flynn and D. A. House, *New Zealand J. of Science*, **12**, 276 (1969).
298. H. N. Po and T. L. Allen, *J. Am. Chem. Soc.*, **90**, 1127 (1968).
299. See, for example, L. J. Csányi, J. Batyai, and F. Solymosi, *Zeit. f. Anal. Chemie*, **195**, 9 (1963).
300. A. J. Kalb and T. L. Allen, *J. Am. Chem. Soc.*, **86**, 5107 (1964).
301. R. D. Gilland and P. R. Mitchell, *J. Chem. Soc.*, *A*, **1968**, 2129.
302. J. P. Collman, E. T. Kittleman, W. S. Hurt, and N. A. Moore, *Inorg. Syn.*, **8**, 141 (1966).
303. M. Delepine, *Bull. Soc. Chim. France*, **45**, 235 (1929).
304. J. V. Rund, F. Basolo, and R. G. Pearson, *Inorg. Chem.*, **3**, 658 (1964).
305. J. V. Rund, *Inorg. Chem.*, **7**, 24 (1968).
306. A. Haim and N. Sutin, *J. Am. Chem. Soc.*, **87**, 4210 (1965).
307. A. E. Ogard and H. Taube, *J. Phys. Chem.*, **62**, 357 (1958).
308. R. D. Cannon, *J. Chem. Soc.*, *A*, **1968**, 1098.
309. R. L. Pecsok, L. D. Shields, and W. P. Schaefer, *Inorg. Chem.*, **3**, 114 (1964).
310. A. W. Adamson, *J. Am. Chem. Soc.*, **78**, 4260 (1956).
311. H. S. Nagarajaiah, A. G., Sharpe, and D. B. Wakefield, *Proc. Chem. Soc.*, **1959**, 385.
312. S. C. Chan and M. L. Tobe, *J. Chem. Soc.*, **1963**, 966.
313. See, for instance, R. G. Wilkins and R. Yelin, *J. Am. Chem. Soc.*, **89**, 5496 (1967).
314. A. R. Norris and M. L. Tobe, *Inorg. Chem. Acta.*, **1**, 41 (1967).
315. R. C. Johnson and E. R. Berger, *Inorg. Chem.*, **4**, 1262 (1965).
316. F. Basolo, P. H. Wilks, R. G. Pearson, and R. G. Wilkins, *J. Inorg. Nucl. Chem.*, **6**, 161 (1958).
317. W. R. Mason, III, E. R. Berger, and R. C. Johnson, *Inorg. Chem.*, **6**, 248 (1967).
318. H. R. Ellison, F. Basolo, and R. G. Pearson, *J. Am. Chem. Soc.*, **83**, 3943 (1961).
319. W. R. Mason, *Inorg. Chem.*, **8**, 1756 (1969).

320. U. Belluco, L. Cattalini, F. Basolo, R. G. Pearson, and A. Turco, *J. Am. Chem. Soc.*, **87**, 241 (1965).
321. H. Taube and H. Myers, *J. Am. Chem. Soc.*, **76**, 2103 (1954).
322. H. Taube and E. L. King, *J. Am. Chem. Soc.*, **76**, 4053 (1954).
323. Y.-T. Chia and E. L. King, *Disc. Faraday Soc.*, No, **29**, 109 (1960).
324. D. E. Pennington and A. Haim, *J. Am. Chem. Soc.*, **88**, 3450 (1966).
325. A. Adin, J. Doyle, and A. G. Sykes, *J. Chem. Soc.*, *A*, **1967**, 1504.
326. J. H. Espenson and D. W. Carlyle, *Inorg. Chem.*, **5**, 586 (1966).
327. A. Adin and A. G. Sykes, *J. Chem. Soc.*, *A*, **1966**, 1518.
328. J. P. Birk and J. H. Espenson, *Inorg. Chem.*, **7**, 991 (1968).
329. G. Schwarzenbach and H. Wenger, in *Progress in Coordination Chemistry* (M. Cais, ed.), Elsevier, Amsterdam, 1968, p. 108.
330. K. D. Kopple and R. R. Miller, *Proc. Chem. Soc.*, **1962**, 306.
331. O. J. Parker and J. H. Espenson, *J. Am. Chem. Soc.*, **90**, 3689 (1968).
332. A. Adin and A. G. Sykes, *J. Chem. Soc.*, *A*, **1968**, 354.
333. T. J. Conocchioli and N. Sutin, *J. Am. Chem. Soc.*, **89**, 282 (1967).
334. J. P. Birk and J. Halpern, *J. Am. Chem. Soc.*, **90**, 305 (1968).
335. J. F. Endicott and H. Taube, *J. Am. Chem. Soc.*, **84**, 4985
336. R. D. Gillard, B. T. Heaton, and D. H. Vaughan, *Chem. Comm.*, **1969**, 974.
337. B. R. James, M. Kastner, and G. L. Rempel, *Canad. J. Chem.*, **47**, 349 (1969).
338. J. P. Birk and J. H. Espenson, *J. Am. Chem. Soc.*, **90**, 2266 (1968).
339. S. Kawaguchi and H. Fujioka, *Inorg. and Nuclear Chem. Letters*, **2**, 243 (1966).
340. E. P. Crematy, *Experimentia*, **26**, 124 (1970).
341. J. M. Anderson and J. K. Kochi, *J. Amer. Chem. Soc.*, **92**, 1651 (1970).

Author Index

Numbers in parentheses are reference numbers and indicate that an author's work is referred to although his name is not cited in the text. Numbers in italics show the page on which the complete reference is listed.

A

Abley, P., 37(117), 39(130), 40(130), *55*, 269(61), *294*
Abrahamson, E. W., 224(6), *293*
Adams, C. T., 243(35), *293*
Adams, R. H., 40(131), *55*
Adamson, A. W., 301(18), 306(45), 363 (310), *372, 373, 379*
Adin, A., 335(204), 367(325, 328), 368 (325, 327), 369(325, 332), 370(325), *377, 380*
Adkins, H., 26(32), *52*, 158(51), *218*
Agrawal, M. C., 356(286), *379*
Aguiló, A., 96(9), 103(9), *142*, 298(4), *372*
Albanesi, G., 192(161), *221*
Alderson, T., 69(80), 73(80), *87*
Aldridge, C. L., 48(176), *56*, 165(78), *219*
Alexander, J. J., 344(227), *377*
Allegra, G., 69(78), 71(91), *87*
Allen, T. L., 356(285), 359(298), 360(300), 361(300), *379*
Anbar, M., 254(276), *378*
Anderson, A., 319(123), 335(123), 358 (341), 368(123), *375, 377–380*
Anderson, C. B., 98(32), *142*
Andreeta, A., 27(39), 40(39), *53, 55*, 60(9), *85*, 161(62), *218*
Anet, F. A. L., 331(185), *376*
Angoletta, M., 139(184), *146*
Aoki, D., 73(113), *88*
Aresta, M., 22(6), 49(6), *52*
Arnold, D. R., 238(20), *293*
Aryoshi, J., 50(200), *57*, 62(34), 81(185) *90*
Asano, R., 112(91), *144*

Asinger, F., 161(66), *218*
Asperger, S., 308(62), *374*
Aue, D. H., 259(53), *294*
Augustine, R. L., 33(95b), 34(99a), 51 (205), *54*

B

Baardman, F., 173(103b), *219*
Babos, B., 15(1), *51*
Bacha, J. D., 133(150), *145*
Baddley, W. H., 38(121), *55*
Bader, R. F. W., 224(7), *293*
Bagga, M. M., 49(186), *57*
Baikie, P. E., 49(187), *57*
Bailar, J. C., 40(131, 132), 48(172), *55, 56*, 80 (153), *89*
Bailey, G. C., 239(24, 25), 240(25), 241 (25), 243(25), *293*
Baird, M. C., 42(141), *56*, 216(204), *222*
Baird, W. C., Jr., 98(34), 99(34), *142*
Baker, B. R., 306(38), 318(114), 331(114), 334(200), 370(114), *373, 375, 377*
Baker, F. B., 303(24), 310(66, 67), 313 (24), 329(66), 330(184), *373, 374*
Baldwin, M. E., 308(63), *374*
Ball, D. L., 314(93), 321(93), 370(93), *374, 375*
Ballard, D. G. H., 84(190, 191), *91*
Ballhausen, C. J., 270(65), *294*
Banks, R. G., 28(50, 51), *53*
Banks, R. G. S., 344(232), *377*
Banks, R. L., 239(24), *293*
Bargmann, E. D., 201(182), *222*
Barney, A. L., 73(110), *88*

381

398

Author Index

Subject Index

A

753